Structural Analysis in Molecular Biology

Structural Analysis in Molecular Biology

Editor: Gildroy Swan

R CALLISTO
REFERENCE

www.callistoreference.com

Callisto Reference,
118-35 Queens Blvd., Suite 400,
Forest Hills, NY 11375, USA

Visit us on the World Wide Web at:
www.callistoreference.com

ISBN: 978-1-63239-813-0 (Hardback)

The publisher's policy is to use permanent paper from mills that operate a sustainable forestry policy. Furthermore, the publisher ensures that the text paper and cover boards used have met acceptable environmental accreditation standards.

Cataloging-in-publication Data

Structural analysis in molecular biology / edited by Gildroy Swan.
    p. cm.
Includes bibliographical references and index.
ISBN 978-1-63239-813-0
1. Molecular structure. 2. Molecular biology. 3. Structural bioinformatics. 4. Proteins--Structure. I. Swan, Gildroy.
QH506 .S77 2017
572.8--dc23

Table of Contents

Preface..VII

Chapter 1   **IMPACT_S: Integrated Multiprogram Platform to Analyze and Combine Tests
            of Selection** 1
            Emanuel Maldonado, Kartik Sunagar, Daniela Almeida, Vitor Vasconcelos,
            Agostinho Antunes

Chapter 2   **Computational Design of Protein-Based Inhibitors of *Plasmodium vivax*
            Subtilisin-Like 1 Protease**..10
            Giacomo Bastianelli, Anthony Bouillon, Christophe Nguyen, Dung Le-Nguyen,
            Michael Nilges, Jean-Christophe Barale

Chapter 3   **Computational Insights into the Inhibitory Mechanism of Human AKT1 by an
            Orally Active Inhibitor, MK-2206** 24
            Mohd Rehan, Mohd A. Beg, Shadma Parveen, Ghazi A. Damanhouri, Galila F. Zaher

Chapter 4   **An Integrated Model of Transcription Factor Diffusion Shows the Importance of
            Intersegmental Transfer and Quaternary Protein Structure for Target Site Finding** 36
            Hugo G. Schmidt, Sven Sewitz, Steven S. Andrews, Karen Lipkow

Chapter 5   **A Polyketide Synthase Acyltransferase Domain Structure Suggests a
            Recognition Mechanism for its Hydroxymalonyl-Acyl Carrier Protein Substrate** 47
            Hyunjun Park, Brian M. Kevany, David H. Dyer, Michael G. Thomas,
            Katrina T. Forest

Chapter 6   **Understanding the Mechanism of Atovaquone Drug Resistance in *Plasmodium
            falciparum* Cytochrome b Mutation Y268S using Computational Methods** 57
            Bashir A. Akhoon, Krishna P. Singh, Megha Varshney, Shishir K. Gupta,
            Yogeshwar Shukla, Shailendra K. Gupta

Chapter 7   **Exploring the Genes of Yerba Mate (Ilex paraguariensis A. St.-Hil.) by NGS
            and *De Novo* Transcriptome Assembly**..69
            Humberto J. Debat, Mauro Grabiele, Patricia M. Aguilera, Rosana E. Bubillo,
            Mónica B. Otegui, Daniel A. Ducasse, Pedro D. Zapata, Dardo A. Marti

Chapter 8   **Bayesian Model of Protein Primary Sequence for Secondary Structure Prediction** 85
            Qiwei Li, David B. Dahl, Marina Vannucci, Hyun Joo, Jerry W. Tsai

Chapter 9   ***In Vitro, In Silico* and *In Vivo* Studies of Ursolic Acid as an Anti-Filarial Agent** 97
            Komal Kalani, Vikas Kushwaha, Pooja Sharma, Richa Verma, Mukesh Srivastava,
            Feroz Khan, P. K. Murthy, Santosh Kumar Srivastava

Chapter 10 **How Structure Defines Affinity in Protein-Protein Interactions** 110
            Ariel Erijman, Eran Rosenthal, Julia M. Shifman

Chapter 11   **On the Importance of the Distance Measures used to Train and Test Knowledge-Based Potentials for Proteins** 120
Martin Carlsen, Patrice Koehl, Peter Røgen

Chapter 12   **How Does Domain Replacement Affect Fibril Formation of the Rabbit/Human Prion Proteins** 138
Xu Yan, Jun-Jie Huang, Zheng Zhou, Jie Chen, Yi Liang

Chapter 13   **Structural and Biophysical Characterization of *Bacillus thuringiensis* Insecticidal Proteins Cry34Ab1 and Cry35Ab1** 151
Matthew S. Kelker, Colin Berry, Steven L. Evans, Reetal Pai, David G. McCaskill, Nick X. Wang, Joshua C. Russell, Matthew D. Baker, Cheng Yang, J. W. Pflugrath, Matthew Wade, Tim J. Wess, Kenneth E. Narva

Chapter 14   **Comparative Analysis of Human γD-Crystallin Aggregation under Physiological and Low pH Conditions** 166
Josephine W. Wu, Mei-Er Chen, Wen-Sing Wen, Wei-An Chen, Chien-Ting Li, Chih-Kai Chang, Chun-Hsien Lo, Hwai-Shen Liu, Steven S.-S. Wang

Chapter 15   **Frequency and Fitness Consequences of Bacteriophage Φ6 Host Range Mutations** 183
Brian E. Ford, Bruce Sun, James Carpino, Elizabeth S. Chapler, Jane Ching, Yoon Choi, Kevin Jhun, Jung D. Kim, Gregory G. Lallos, Rachelle Morgenstern, Shalini Singh, Sai Theja, John J. Dennehy

Chapter 16   **Combining Physicochemical and Evolutionary Information for Protein Contact Prediction** 196
Michael Schneider, Oliver Brock

Chapter 17   **TALEs from a Spring – Superelasticity of Tal Effector Protein Structures** 211
Holger Flechsig

Chapter 18   **Solution NMR of MPS-1 Reveals a Random Coil Cytosolic Domain Structure** 218
Pan Li, Pan Shi, Chaohua Lai, Juan Li, Yuanyuan Zheng, Ying Xiong, Longhua Zhang, Changlin Tian

Permissions

List of Contributors

Index

Preface

Molecular biology provides the basis for understanding biological functions that occur at a molecular level. It covers a wide range of topics related to molecular and cell biology such as transcriptomics, bioinformatics, biomedicine, etc. This book paves way for a thorough understanding of molecular structures and their functions. Recent research in this field has accelerated with the aid of technological advances that provide accurate data for the analysis of newer and complex structures. Different approaches, evaluations, methodologies and advanced studies on molecular biology have been included in this book. It includes contributions of experts and scientists which will provide innovative insights into this discipline.

This book unites the global concepts and researches in an organized manner for a comprehensive understanding of the subject. It is a ripe text for all researchers, students, scientists or anyone else who is interested in acquiring a better knowledge of this dynamic field.

I extend my sincere thanks to the contributors for such eloquent research chapters. Finally, I thank my family for being a source of support and help.

Editor

IMPACT_S: Integrated Multiprogram Platform to Analyze and Combine Tests of Selection

Emanuel Maldonado[1⊙], Kartik Sunagar[1,2⊙], Daniela Almeida[1,2⊙], Vitor Vasconcelos[1,2], Agostinho Antunes[1,2]*

1 CIIMAR/CIMAR – Interdisciplinary Centre of Marine and Environmental Research, University of Porto, Porto, Portugal, 2 Department of Biology, Faculty of Sciences, University of Porto, Porto, Portugal

Abstract

Among the major goals of research in evolutionary biology are the identification of genes targeted by natural selection and understanding how various regimes of evolution affect the fitness of an organism. In particular, adaptive evolution enables organisms to adapt to changing ecological factors such as diet, temperature, habitat, predatory pressures and prey abundance. An integrative approach is crucial for the identification of non-synonymous mutations that introduce radical changes in protein biochemistry and thus in turn influence the structure and function of proteins. Performing such analyses manually is often a time-consuming process, due to the large number of statistical files generated from multiple approaches, especially when assessing numerous taxa and/or large datasets. We present IMPACT_S, an easy-to-use Graphical User Interface (GUI) software, which rapidly and effectively integrates, filters and combines results from three widely used programs for assessing the influence of selection: Codeml (PAML package), Datamonkey and TreeSAAP. It enables the identification and tabulation of sites detected by these programs as evolving under the influence of positive, neutral and/or negative selection in protein-coding genes. IMPACT_S further facilitates the automatic mapping of these sites onto the three-dimensional structures of proteins. Other useful tools incorporated in IMPACT_S include Jmol, Archaeopteryx, Gnuplot, PhyML, a built-in Swiss-Model interface and a PDB downloader. The relevance and functionality of IMPACT_S is shown through a case study on the toxicoferan-reptilian Cysteine-rich Secretory Proteins (CRiSPs). IMPACT_S is a platform-independent software released under GPLv3 license, freely available online from http://impact-s.sourceforge.net.

Editor: Konrad Scheffler, University of California, San Diego, United States of America

Funding: This research was supported in part by the European Regional Development Fund (ERDF) through the COMPETE - Operational Competitiveness Program and national funds through FCT under the project PEst-C/MAR/LA0015/2013, by the FCT projects PTDC/AAC-AMB/104983/2008 (FCOMP-01-0124-FEDER-008610), PTDC/AAC-CLI/116122/2009 (FCOMP-01-0124-FEDER-014029) and PTDC/AAC-AMB/121301/2010 (FCOMP-01-0124-FEDER-019490), and by the University of Porto and Santander Totta project PP-IJUP2011-315 to AA. The funders had no role in study design, data collection and analysis, decision to publish, or preparation of the manuscript.

Competing Interests: The authors have declared that no competing interests exist.

* Email: aantunes@ciimar.up.pt

⊙ These authors contributed equally to this work.

Introduction

The nature and strength of evolutionary selection pressures can be estimated at the molecular level, as a non-synonymous to synonymous substitution rate ratio omega ($\omega = dN/dS$), where ω greater than, equal to and less than 1 is indicative of positive, neutral and negative selection, respectively [1]. This approach often fails to detect subtle adaptations that only affect certain regions of the protein and/or take place over a very short period of evolutionary time [2]. Moreover, the evaluation of selective pressures solely at the nucleotide level and the assumption that all mutations affect the fitness of the organism equally could be misleading. Although non-synonymous substitutions introduce variations in coding regions, a novel amino acid could have identical or similar biochemical and/or structural properties to that of the ancestral residue. Such substitutions are unlikely to influence the structure or the function of the protein and hence are least likely to affect the fitness of an organism. Thus, it is important to discern the nature of mutations to precisely understand the evolution of a protein. By employing mapping strategies of mutational sites onto the three-dimensional (3D) structure of the protein, it is possible to gain further insights into how its structure and/or function are affected by the changes in certain residues or regions [3]. For example, using a similar integrative approach, we and others have demonstrated that most predatory venom-components in a diversity of animal lineages adopt Rapid Accumulation of Variations in Exposed Residues of Toxins [4] and accumulate mutations on the molecular surface under the influence of positive Darwinian selection, while preserving the key functional and structural residues that stabilize the overall structure of the protein [4–13]. The mutation of the molecular surface may increase the toxin's ability to target novel molecular receptors and aid in evading immune response upon injection into prey animals [8]. Phenomena like this can be easily detected by mapping sites under positive and negative selection on 3D structures of proteins. Mapping of sites onto multiple sequence alignments of protein-coding genes is also beneficial, as it enables the identification of differential evolution of domains by revealing mutational hotspots [5,10,11]. Hence, to efficiently assess the influence of natural selection on protein-coding genes and to accurately identify regions under various regimes of natural selection, it is essential to adopt a complementary approach [14]. Although several software applications have been proposed to independently evaluate selection pressures at the codon-level [15–18], an integrative approach, which additionally evaluates the strength and radicalness of non-synonymous substitutions at the amino acid-level, is still missing. In addition, the manual integration of results from various approaches is time-consuming.

To address these shortcomings and to facilitate the integration of results from various selection assessments, we propose IMPACT_S, a free platform-independent user-friendly GUI software that integrates results of nucleotide and amino acid-level assessments by employing three widely used softwares: Codeml from Phylogenetic Analysis by Maximum Likelihood (PAML) package [17], Datamonkey, a web-server of the HyPhy [20] package - www.datamonkey.org [18,19], and Selection on Amino Acid Properties using phylogenetic trees (TreeSAAP) [15]. Sites detected as positively and negatively selected can be automatically mapped onto the 3D structure of the protein. If experimentally determined protein structures are unavailable, IMPACT_S facilitates the prediction of models, through homology, using the built-in Swiss-Model [21–23] interface. Homology modeling is based on evolutionary relationships between target and template sequences. The template sequences result from homology searches of experimentally determined protein structures, through successive BLAST-p [34] searches. This method involves the construction of an atomic resolution model for the protein under study using an experimentally determined 3D-structure of a similar homologous protein [23]. The following criteria are automatically considered by Swiss-Model in order to obtain a single high quality template: (i) identification of related proteins with experimentally solved structures, (ii) mapping of corresponding residues of target and template structures, (iii) building of the 3D model on the basis of the alignment and finally, (iv) evaluation of the quality of the resulting model [23].

In the following sections, we introduce the foundations, the structure and the development of IMPACT_S and demonstrate the relevance and functionality of IMPACT_S using a case study on Cysteine-rich Secretory Proteins (CRiSPs), a class of reptilian and mammalian glycoproteins with extremely diversified functions. CRiSPs have been theorized to play important functions in the mammalian reproductive pathways, while in toxicoferan reptiles (venomous snakes and lizards) they have been hypothesized to participate in prey envenoming and capture [11].

Design and Implementation

IMPACT_S Foundation

IMPACT_S integrates three broadly used bioinformatic softwares for assessing selection pressures shaping the evolution of protein-coding genes: (i) PAML and (ii) Datamonkey for assessing selection at the codon-level, and (iii) TreeSAAP to detect selection at the protein-level.

Codeml from the PAML package [17], implements powerful site-specific [24,25], branch-specific [26,27] and branch-site specific models [25,28] that detect the influence of natural selection, easily and reliably. Site-specific models, which are the current focus of this work, comprise the alternative models – model 2a (M2a) and model 8 (M8) and the null models – model 1a (M1a) and model 7 (M7). The alternative models include the Bayes Empirical Bayes (BEB) [25] approach for identifying positively selected sites. The computed ω value and the sites detected by the alternative models as positively selected are only considered by the user if the likelihood ratio test (LRT) is significant [24,25]. The

LRT is conducted by comparing the null models with the alternative models (M1a vs. M2a and/or M7 vs. M8). Both M2a and M8 have one additional class or category when compared to their null counterparts M1a and M7, respectively [24]. In case the LRT is significant in both tests, with the alternative models as fitting the data better, then it is possible to assume stronger evidence for the presence of sites under positive selection [25]. However, PAML is incapable of identifying negatively selected sites, in contrast to certain models implemented in the Datamonkey web-server [18,19]: Single-Likelihood Ancestor Counting (SLAC) [16], Fixed Effects Likelihood (FEL) [16], Random Effects Likelihood (REL) [16] and Fast Unconstrained Bayesian ApproXimation (FUBAR) [29] methods. The recently proposed Mixed Effects Model of Evolution (MEME) [2] is a state-of-the-art method for detecting sites that evolve under episodic selection pressures, which are often difficult to identify using traditional site-specific methods. This method allows the distribution of ω to vary not only across sites, but also from branch to branch at a site [2]. SLAC infers the number of non-synonymous and synonymous substitutions that have occurred at each site using Maximum Likelihood (ML) reconstructions of ancestral sequences, while FEL estimates the ratio of non-synonymous to synonymous substitutions, not assuming a priori distribution of rates across sites substitution, on a site-by-site basis. REL involves fitting a distribution of substitution rates across sites and then inferring the rate at which individual sites evolve based on ML estimates. FUBAR on the other hand, which supersedes SLAC, FEL and REL, infers the rate at which individual sites evolve based on an approximate hierarchical Bayesian approach using a Markov Chain Monte Carlo sampler.

The choice of the method not only depends on the question being addressed, but also on the size of the dataset. When the positive selection on the evolution of protein-coding genes is sufficiently strong, the site-specific models of Codeml [17,25,30] can efficiently identify positively selected sites. However, due to the episodic or transient nature of natural selection on protein-coding genes, the precise identification of the regions that have undergone adaptive evolution is often difficult [2]. Moreover, a strong influence of purifying selection on a majority of lineages can mask the signal of positive selection on others [2]. MEME was proposed to address these shortcomings and to reliably identify sites that are influenced by both episodic and pervasive influence of positive selection at the level of an individual site [2]. In order to identify sites influenced by positive and negative selection, the following methods can be chosen, considering the size of datasets to avoid biased results [16]: SLAC for large datasets (over 40 sequences), REL for datasets of intermediate size (20–40 sequences), FEL for intermediate to large datasets (over 50 sequences) and FUBAR for analyzing very large datasets (the authors tested this model on a dataset of 3142 sequences, which is much faster when compared with the other methods [29]). While Datamonkey and Codeml assess the influence of natural selection at the codon-level, TreeSAAP measures the selective influences on 31 biochemical and structural amino acid properties – such as hydrophobicity, polarity, solvent accessible reduction ratio and buriedness – during cladogenesis and performs goodness-of-fit and categorical statistical tests [15]. It enables the estimation of radicalness of a mutation at a particular site by revealing the influence of the novel amino acid, introduced by non-synonymous nucleotide substitution, on the structural and biochemical properties of the protein. The result files generated by TreeSAAP are organized in two main subdirectories: Evpthwy and Substs. The former includes the sliding window analyses, while the latter contains results for particular sites [31,32]. Both identify amino acid properties under

A. PAML tab

IMPACT_S

File Edit Tools Run Help

TreeSAAP | Datamonkey | PAML | Results & 3D

Alignment: package/IMPACT_S/IDataSets/CaseStudy/IPAMLTabData/Caenophidian_CRiSPs_Sequence.fas

Tree File: 2/Documents/java/IMPACT_S_package/IMPACT_S/IDataSets/CaseStudy/IPAMLTabData/tree.txt

Options: ● Options (M7,M8) ○ Options (M1,M2) ○ Options (M0,M3) ○ File ○ Other

Output File: ts/java/IMPACT_S_package/IMPACT_S/IDataSets/CaseStudy/IPAMLTabData/CRiSPsResults.txt

Run Codeml | View Details

Options: ○ Options (M7,M8) ○ Options (M1,M2) ○ Options (M0,M3) ● All (M0-M8)

Output File: nts/java/IMPACT_S_package/IMPACT_S/IDataSets/CaseStudy/IPAMLTabData/CRiSPsResults.txt

Lnl Table: ava/IMPACT_S_package/IMPACT_S/IDataSets/CaseStudy/IPAMLTabData/LnLTable_GENID22.csv

Ratios Table: /IMPACT_S_package/IMPACT_S/IDataSets/CaseStudy/IPAMLTabData

BEB Results: nents/java/IMPACT_S_package/IMPACT_S/IDataSets/CaseStudy/

Select: P>95% (*) ▼ | Extract BEB | View | Filter Align

Likelihood Values:
-6718.705160 -6598.579842

Chi²: 5.9915

LRT: 230.250636000000067

B. BEB Results table

File View Columns

View File: [File1:]/home/labpc2/Documents... ▼

M8Site	AA	Pr(w>1)	PostMean	+-	SE For w
30	Q	0.999**	3.496	+-	0.095
73	N	0.968*	3.417	+-	0.453
76	L	0.969*	3.421	+-	0.444
81	D	1.000**	3.500	+-	0.011
82	Y	1.000**	3.500	+-	0.016
83	S	0.986*	3.465	+-	0.297
87	E	1.000**	3.499	+-	0.053
100	N	0.999**	3.497	+-	0.092
102	R	1.000**	3.500	+-	0.031
103	A	0.992**	3.480	+-	0.226
106	E	0.998**	3.494	+-	0.121
110	L	1.000**	3.500	+-	0.005
115	Y	0.999**	3.497	+-	0.081
119	V	0.991**	3.478	+-	0.237
		1.000**	3.500	+-	0.011
				+-	0.124
				+-	0.005
				+-	0.025
				+-	0.004
				+-	0.057
				+-	0.475
				+-	0.429
				+-	0.004
				+-	0.058
				+-	0.004
				+-	0.308
				+-	0.322
				+-	0.050
				+-	0.535
				+-	0.360
				+-	0.005
				+-	0.035
				+-	0.278
				+-	0.148
				+-	0.190
				+-	0.259
				+-	0.445
				+-	0.019

Close

E. InL table

File View Columns

View File: [File]

Model	np	InL
0	91	-6964.393838
1	92	-6711.896261
2	94	-6595.953989
3	95	-6593.227401
7	92	-6713.705160
8	94	-6598.579842

C. Results & 3D

Residue Range: 24 - 299

☐ Minimum Size

Last Color:

Select Background:

Reset   Close

D. Ratios table

CSV Viewer - IMPACT_S Lnl & Ratios Table <2>

File View Columns

View File: [File2:]/home/labpc2/Documents/java/IMPACT_S/IDat

Model	p/w	K=1	K=2	K=3	K=4	K=5	K=6	K=7
0	p:							
0	w:	1.14056						
1	p:	0.48460	0.51540					
1	w:	0.07157	1.00000					
2	p:	0.39687	0.36260	0.24054				
2	w:	0.05714	1.00000	3.58421				
3	p:	0.43544	0.38890	0.17566				
3	w:	0.08774	1.42279	4.48642				
7	p:	0.10000	0.10000	0.10000	0.10000	0.10000	0.10000	0.10000
7	w:	0.00000	0.00079	0.01639	0.11335	0.39120	0.74533	0.94095
8	p:	0.07450	0.07450	0.07450	0.07450	0.07450	0.07450	0.07450
8	w:	0.00000	0.00033	0.00763	0.05835	0.23724	0.56619	0.85466

665 x 952   21.3/61.4 Mb   17/-1 .as   jmol

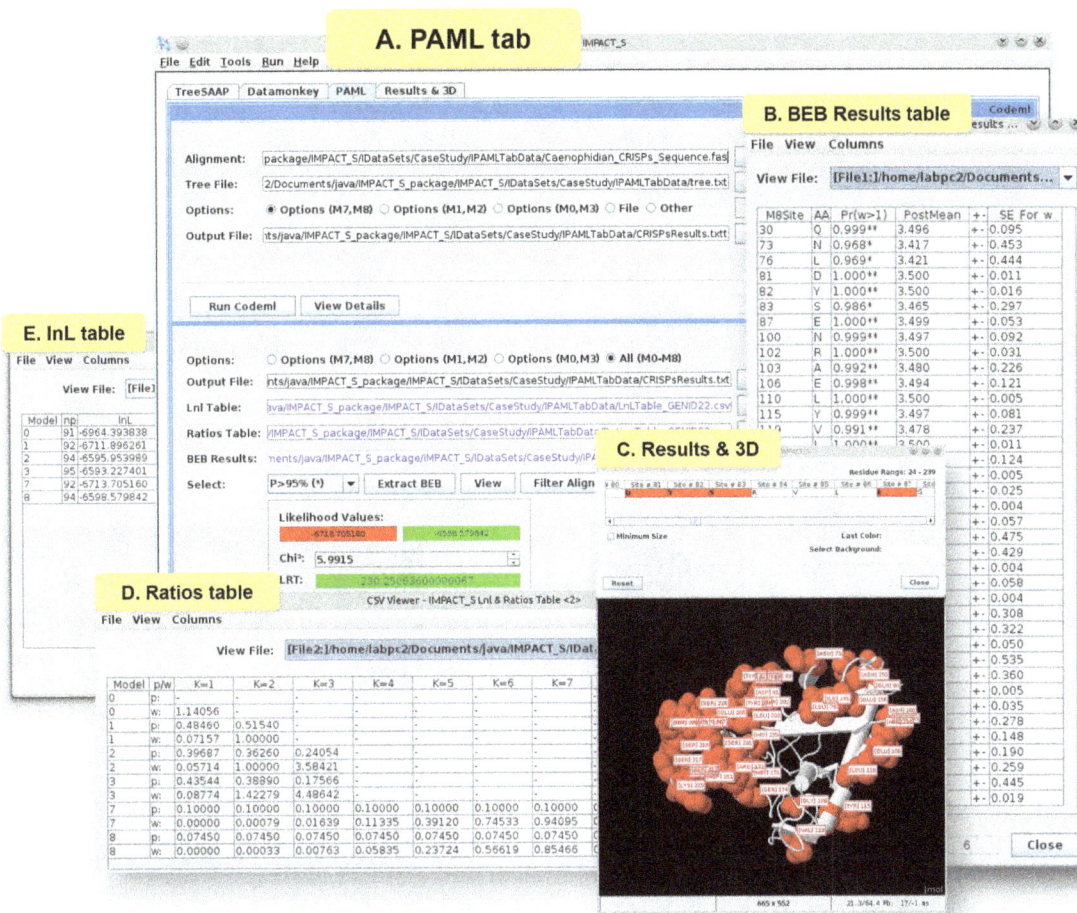

Figure 1. PAML tab and related results. (A) The 'PAML' tab from IMPACT_S showing the results of the LRT test (M7 *vs.* M8), (B) Automatically extracted and tabulated 'BEB results' table, (C) Mapping of the positively selected sites onto the 3D protein-structure of CRiSPs (2GIZ-A), (D) Tabulated results of ω estimation under the site-specific models (M0, M1, M3, M7 and M8) and (E) The 'InL' table, showing all the log-likelihood values for these models.

selection, scored in every category, typically from 1 to 8. The assessment of biochemical and structural properties of the novel amino acids, introduced by non-synonymous substitutions, results in generation of numerous statistical files, especially when a large number of amino acid properties are evaluated. In order to detect sites that recurrently fall under the influence of natural selection or to detect sites with a defined number of selected properties, the user has to assess an array of statistical files. Thus, the interpretation of the result files generated by TreeSAAP for downstream analyses can be an extremely time-consuming process.

Manually noting down log-likelihood (lnL) values, various maximum-likelihood parameter estimates and ω values from the result files of Codeml for every model analyzed is also tedious. These values are essential for conducting likelihood-ratio tests and the identification of the appropriate regime of selection (for additional information regarding models and LRT tests, see [24]). This process is required to be repeated for each dataset being analyzed. Similarly, tables generated by the Datamonkey web-server, which contain various site statistics, are required to be manually downloaded and processed to retrieve sites detected as positively and negatively selected at the desired significance level.

It has been suggested that the application of all the major selection assessment methods in HyPhy [20] (SLAC, FEL, REL,

FUBAR and MEME), followed by the consensual identification of sites detected as positively selected among the methods, can minimize false positives, especially when analyzing smaller datasets [16]. This feature is indeed available on the Datamonkey web-server. However, the results of these analyses can only be integrated manually with the results of Codeml and TreeSAAP selection assessments and the compilation of the sites detected in common by these approaches becomes a difficult task. Moreover, to our knowledge, there is no software to export these results for the downstream analyses, such as the mapping of the detected mutational sites onto the 3D structures of proteins.

Therefore, IMPACT_S has been designed to be a single platform, with which the user can employ various state-of-the-art selection assessments and identify the regime of natural selection on protein-coding genes. Furthermore, these sites could be automatically mapped onto the alignment and 3D structure (either in a homology model or a known crystallographic structure of the protein), making the entire process extremely rapid and efficient, even when a large number of datasets are analyzed.

IMPACT_S Structure

IMPACT_S implements four main tabs: (i) 'PAML', (ii) 'Datamonkey', (iii) 'TreeSAAP' and (iv) 'Results & 3D'. The first three tabs enable the user to import the corresponding result files

A. Datamonkey tab

C. Results & 3D

B. Common Sites table

Codon	SLAC (P-value)	SLAC (dN-dS)	FEL (P-value)	FEL (dN-dS)	REL (Bayes Factor)	REL (dN-dS)	MEME (P-value)	MEME (dN
75	0.027731	2.57123	0.0163848	2.32669	942.128	3.29446	0.0155013	Infinity
77	0.05241	2.58373	0.0413775	2.0434	347.028	3.12622	0.0292505	Infinity
99	0.0789237	2.31647	0.0739573	1.72272	234.348	2.50343	0.00168166	Infinity
103	0.0272323	2.65808	0.00743416	2.00692	9094.79	3.25329	0.013153	Infinity
110	0.0155536	3.78931	0.00273875	4.79271	314.636	3.30674	0.000107048	Infinity
122	0.0593173	2.21349	0.0400477	2.05997	494.823	2.98397	0.000474173	Infinity
128	0.0952326	1.75964	0.0867963	1.20013	683.892	1.0043	0.0261716	Infinity
145	0.00206772	4.39339	0.00165779	4.00304	1153.12	3.41042	0.000872301	Infinity
150	0.0309194	2.39439	0.00366912	1.7997	186542	2.97153	0.00148156	Infinity
168	0.0384652	2.17742	0.00946813	1.66196	30910	2.44413	0.0056715	Infinity
171	0.0577045	2.74405	0.00538175	2.40207	4344.99	3.45375	0.003677	Infinity
172	0.00563617	5.23958	0.00142015	5.5318	320.902	3.3178	0.00384099	Infinity
174	0.0169056	3.16082	0.00501529	3.1118	1662.24	3.41955	0.000253356	Infinity
184	0.0585277	1.88637	0.0545633	1.29235	1677.34	1.29108	0.0367663	Infinity
202	0.0438415	2.39658	0.0124635	1.79806	7200.23	2.81761	0.918752	Infinity
226				3857	11561.3	3.44669	0.00117421	Infinity
231				7354	1109.93	2.8988	0.0122011	Infinity

Figure 2. Datamonkey tab and related results. (A) 'Datamonkey' tab showing all the selected options, (B) 'Common Sites' table generated from the positive-selection analyses under various methods (SLAC, FEL, REL, MEME and FUBAR) – The sites 110, 172, 202 and 231 (highlighted in pink), were simultaneously detected by Datamonkey, PAML (Codeml) and TreeSAAP as positively-selected (Table S1) [1], (C) 'Results & 3D' tab showing the mapping of sites found in the 'Common Sites' table.

and automatically extract the sites detected as positively and negatively selected. IMPACT_S has been designed to organize these results in Comma Separated Values (CSV) format, which can be viewed and exported with the built-in CSV Viewer tool. Sites detected as positively and negatively selected by various selection assessment methods can be mapped individually for each method, or mapped in combination, as common sites, onto the 3D structure of proteins under the 'Results and 3D' tab. To map sites from two or more methods (considering the type of selection) in conjunction, as common sites, these must be firstly merged into a final Merged Results (MR) table. IMPACT_S provides several mapping schemes (please refer to Table 4 in the tutorial included in the package, for a list of all the available schemes and descriptions), which are applied to the results and presented using the molecular visualization software, Jmol [33] (http://jmol. sourceforge.net/). The 3D structure can be obtained through homology modeling using the built-in Protein Modeler tool that acts as an interface to the Swiss-Model web-server [21–23] (http://swissmodel.expasy.org). Alternatively, these sites can be mapped onto the experimentally deduced crystal structures by importing PDB files from RCSB Protein Data Bank (RCSB PDB) [35,36] (http://www.pdb.org) or the Swiss-Model Template Library (SMTL or ExPDB) [21–23] into IMPACT_S. Sites can also be marked onto the two-dimensional (2D) codon or protein sequence alignment using the built-in Alignment Filter tool. This useful feature enables the assessment of the influence of domain-specific accumulation of mutations and helps to generate publication quality figures (Figure 10 in [3]). This tool is available in every tab of IMPACT_S, except in the TreeSAAP Evpthwy, where the user can find the built-in Gnuplot (http://www.gnuplot. info/) Options tool, which is specifically designed to plot graphs for the sliding window analysis using any amino acid property file. IMPACT_S also incorporates phylogenetic resources such as PhyML [37] and Archaeopteryx (http://www.phylosoft.org/ archaeopteryx), the successor of ATV [38], and other built-in tools that deal with Multiple Sequence Alignment (MSA) and CSV files [39]. Throughout the process of dataset analyses the user may have different requirements and hence employ these tools to perform any alignment and/or phylogenetic tree alterations, as well as the generation of new phylogenetic trees, which are required for PAML and TreeSAAP and optional in Datamonkey analyses.

IMPACT_S Development

The IMPACT_S program was implemented in Java and has been tested on Linux, Windows and MacOS workstations. Java SwingWorker threads (from Java version 1.6.0) are used to run the entire integrated software, handle outputs and to allow GUI updates. Even though it is possible to process and filter PAML results using BioPerl libraries, it is a difficult task for users with no

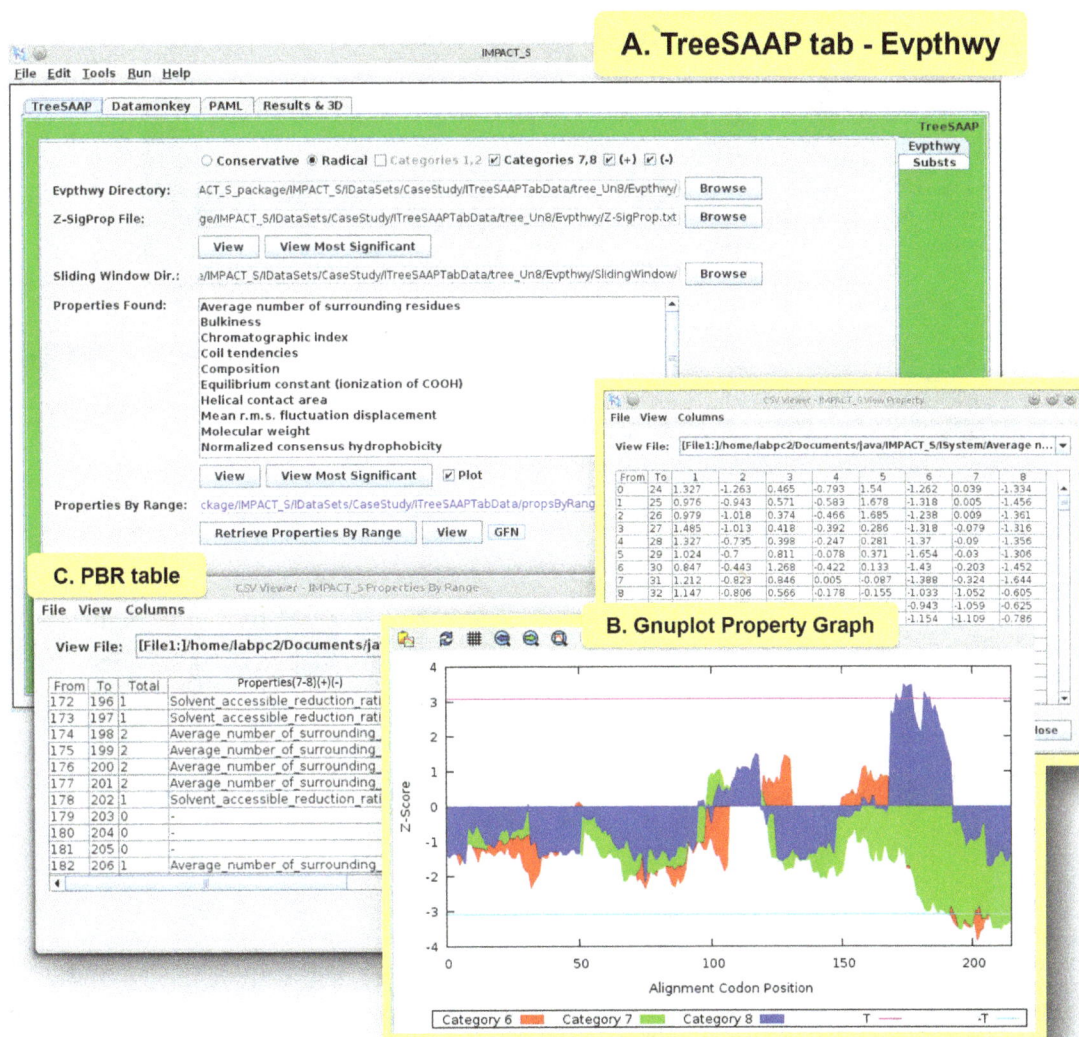

Figure 3. Evolutionary Pathway (Evpthwy) tab under TreeSAAP tab and related results. (A) 'TreeSAAP' tab showing the 'Evpthwy' tab with results of the CRiSP case study, (B) Gnuplot graph for the property "Average number of surrounding residues" with Z-scores in y-axis and Codon positions in x-axis, (C) 'PBR' table showing all ranges ('From' - 'To') retrieved from the sliding window analysis with the 'Total' count of properties for each range and their names [Properties (7–8) (+) (−) – names of properties under significant positive (Z-score ≥3.09) and negative (Z-score ≤3.09) selection, found in the categories 7 and 8].

programming skills. Several tools found in IMPACT_S were improved from our previously developed software IMPACT [39], which include the MSA Editor/Viewer, the MSA Format Converter and the CSV Viewer. IMPACT_S uses the Internet to connect to the Swiss-Model [21–23] and RCSB PDB [35,36] (http://www.pdb.org/) web-servers. The functionalities related to these web-servers are found in two distinct built-in tools, the Protein Modeler tool which implements Swiss Model functionalities and the PDB Downloader tool which includes the RSCB PDB and the SMTL.

IMPACT_S Availability and Requirements

IMPACT_S software was developed in Java with the platform-independent context in mind, providing the use of the software by a wider audience of users. The source code can be used by anyone with skills to manipulate and extend the program capabilities, or even used in other projects. It is an open source software distributed under GNU General Public License version 3.0 (GPLv3) and is freely available with an example dataset, a detailed tutorial and manual, to allow its user-friendly application, at http://impact-s.sourceforge.net. IMPACT_S is an integrated platform, enabling users to access results and knowledge from any stage of the user controlled workflow [40]. Being an integrated platform, users are not required to install independent programs, since they are already provided in the download package, but are allowed to replace (or include) the existing binaries. IMPACT_S requires the following:

Operating System: Linux/UNIX, Windows or MacOS;

Programming Language: Java;

Optional Programs Included: Codeml (PAML), TreeSAAP, PhyML, Archaeopteryx, Jmol and Gnuplot;

Other Requirements: Java version (minimum) 1.6.0.

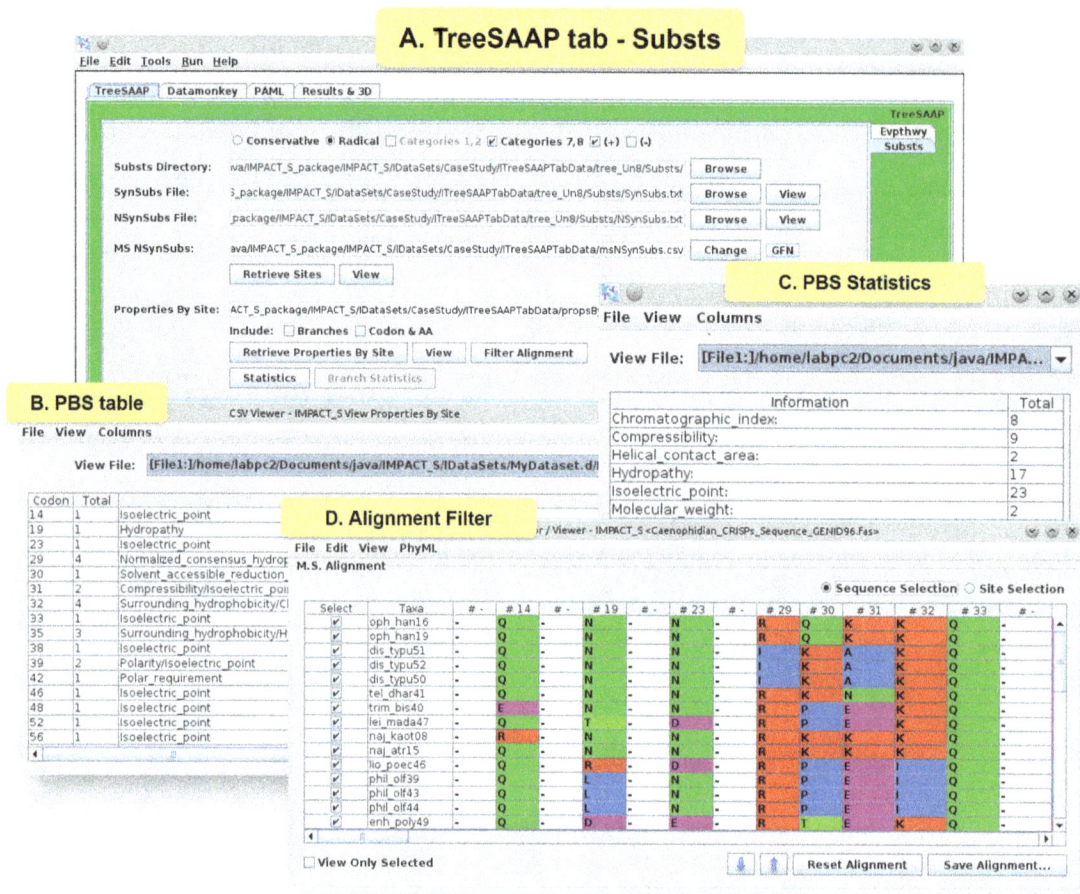

Figure 4. Substitutions (Substs) tab under TreeSAAP tab and related results. (A) 'TreeSAAP' tab showing the 'Substs' tab and the results of the CRiSPs case study, (B) 'PBS' table showing all the significant codons ('Codon' column) and the total count ('Total' column) of properties for each codon and their respective names ['Properties (7–8) (+)' – names of properties under significant (Z-score ≥3.09) positive selection], (C) 'PBS statistics', with respect to the PBS table, showing the number of times ('Total' column) that the same property ('Information' column) is selected across the data set provided and (D) Alignment Filter option with the final view of the alignment containing only the codons from the PBS table with their original positions.

Results

CRiSPs belong to a class of glycoproteins exclusive to vertebrates and have been implicated in a broad range of functions. Previously, we have demonstrated that toxicoferan-reptilian CRiSPs are significantly influenced by positive selection, and in snakes more than Anguimorpha and helodermatid lizards, while their mammalian homologues exhibit extreme coding sequence conservation (Table S1) [11]. By using a toxicoferan-reptilian CRiSPs dataset we demonstrate the applications of IMPACT_S and its useful designs and features. It should be noted that we have used the same dataset only to present all the tabs and associated methods. Nonetheless, in a real evolutionary study the researcher must choose the method that best fits the dataset.

Site-specific analyses were executed under the 'PAML' tab (Figure 1A) of IMPACT_S, and the value of ω was estimated using the following pairs of models: i) M0 and M3; ii) M1a and M2a and iii) M7 and M8. IMPACT_S was then used to automatically extract all the relevant information from the generated result files, such as a table of ω, the ML parameter estimates associated with each model (Figure 1D) and lnL table for every model containing the corresponding number of parameters (np) (Figure 1E). The LRT was conducted in IMPACT_S to compare alternate models that allow $\omega>1$ (M3, M2a and M8)

with their null models that do not (M0, M1a and M7). The user can select from three significance thresholds, namely: $p = 0.05$ (by default selected), $p = 0.01$ and $p = 0.001$. The degrees of freedom (df) are calculated and may differ according to the pair of models selected due to the different np. For example, if the pairs (M1a $vs.$ M2a) or (M7 $vs.$ M8) are selected the df is 2. The (M0 $vs.$ M3) pair is an exception, for which the df is 4. The sites detected as evolving under positive selection according to the BEB [25] approach (Figure 1B) were then extracted using IMPACT_S and automatically mapped onto the 3D structure of CRiSPs (PDB: 2GIZ-A; Figure 1C).

All the models implemented in the Datamonkey web-server (SLAC, FEL, REL, MEME and FUBAR) were executed and the corresponding results were processed under the 'Datamonkey' tab of IMPACT_S (Figure 2A). Using the default significance cut-off, the results of positive and negative selection analyses were tabulated automatically for each method. The default significance cut-offs used in IMPACT_S are the same as on the Datamonkey web-server: $p = 0.1$ for SLAC, FEL and MEME; posterior probability = 0.9 for FUBAR and Bayes Factor = 50 for REL. The significance given by default can be changed by selecting from the available options or by typing the desired value. The positively selected sites detected using the default significance under these methods were tabulated, combined (Figure 2B: 'Common Sites'

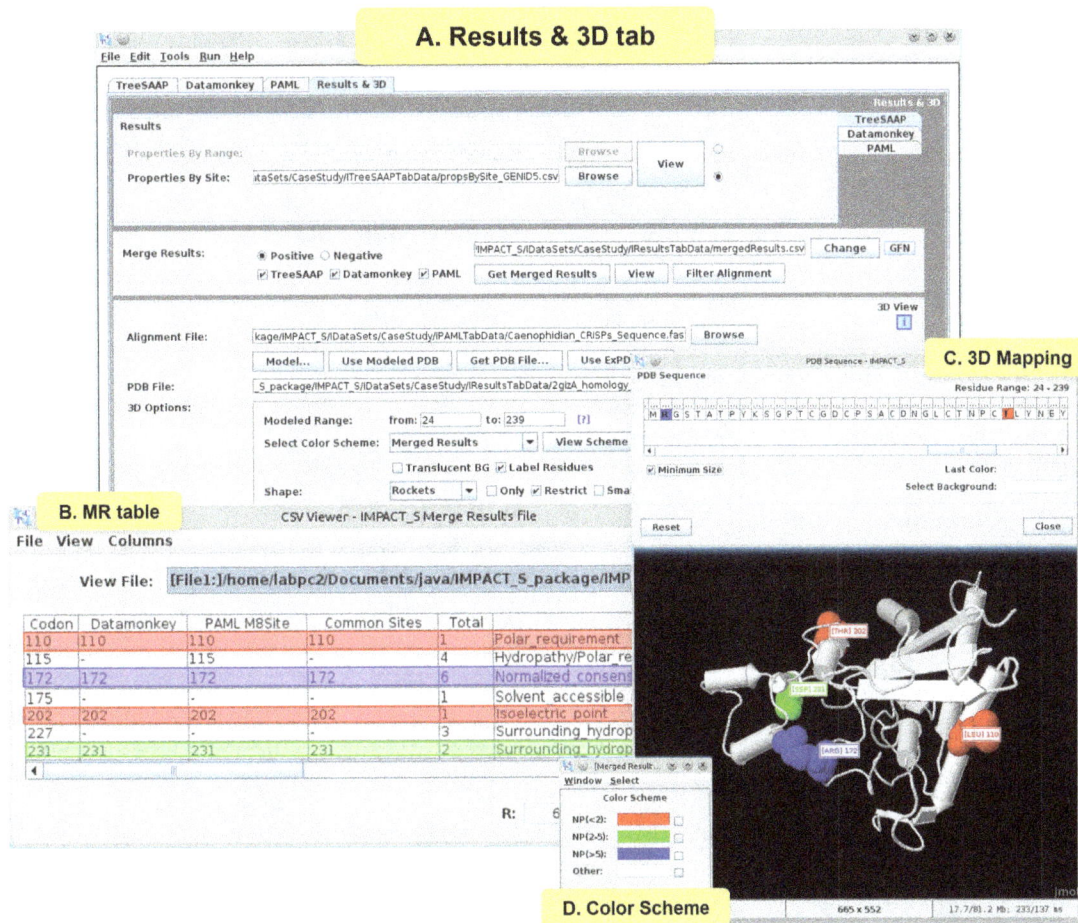

Figure 5. 3D mapping and related functionalities under Results & 3D tab. (A) 'Results & 3D' tab with the results of the CRiSPs case study, (B) 'MR table' (Table S3) presenting six columns: 'Codon' containing the site found as positive selected (TreeSAAP – Z-score ≥3.09); 'Datamonkey' with the positive selected common sites positions from SLAC, FEL, REL, MEME and FUBAR; 'PAML M8Site' with the information from BEB results M8; 'Common Sites' – reporting the site position common to all the previous mentioned columns; 'Total' - count of the number of properties per site; 'Properties (7–8) (+)' with the names of properties found under categories 7 and 8 for Z-score ≥3.09, (only common sites are shown and highlighted) (C) 3D structure and 'PDB Sequence', mapping the sites positions from the 'Common Sites' column following the (D) Merged Results 'Color Scheme' according to a pre-defined number of properties: red – less than 2 (NP<2), green - between 2 and 5 (NP(2–5)) and blue - more than 5 (NP>5).

table) and then mapped onto the 3D structure of the protein (Figure 2C) in IMPACT_S.

Furthermore, we conducted a protein-level analyses under the 'TreeSAAP' tab to measure the influence of selection on 31 biochemical properties. Here, the data was processed from the 'Evpthwy' tab for the sliding window analyses and from the 'Substs' tab for mapping the substitutions on the 3D structure of the protein. In the Evpthwy directory under 'TreeSAAP' tab (Figure 3A), the 'Properties By Range (PBR)' table was generated (Figure 3B), which consists of the most significant properties under categories 7 and 8 (most radical changes). A graph was then plotted using the Gnuplot Options dedicated tool for each significant property (Figure 3C). The 'Substs' tab (Figure 4A) was used to generate the 'Properties By Site (PBS)' table (Figure 4B) that consists of all the significant codons under categories 7 and 8. This tab provides the statistical files (Figure 4C) for counting properties. Using the 'Alignment Filter' tool, the sites reported in the PBS table as positively selected were highlighted in a new alignment view that keeps only these sites and their original positions in the provided alignment (Figure 4D). The information regarding the radical physicochemical amino acid changes varying

across the reptilian-CRiSPs phylogenetic tree can also be obtained under this tab (Table S2).

In order to identify sites that were detected in common as positively selected by the aforementioned analyses (PAML, Datamonkey and TreeSAAP), their resultant files were combined in IMPACT_S under the 'Results & 3D' tab (Figure 5A) as the MR table (Figure 5B; Table S3). The evolution of such sites are most likely to be influenced by natural selection, since they are detected in common by a diversity of selection assessment methodologies. Such sites are then automatically mapped onto the 3D structure (2GIZ-A; Figure 5C) using the 'Merged Results' Color Scheme option (Figure 5D).

Conclusion

A diversity of selection assessment methodologies are employed to identify and corroborate sites that are influenced by various regimes of natural selection. Analyzing and integrating the result files from these methods is a tedious and time-consuming process. In this context, IMPACT_S offers a user-friendly interface to rapidly manage, filter, organize and interpret the results in different ways. Moreover, IMPACT_S provides the user the freedom to choose different available methods, which should follow the respective authors' recommendations from programs integrated in this GUI. If properly used, this software will automatically map the positively and negatively selected sites onto the sequence alignment as well as the 3D protein structure. This enables the user to infer which sites are important, according to their specific location and significance for all types of selection. IMPACT_S is a new integrated multiprogram platform, which allows the user to combine the results from different bioinformatic programs. Merging PAML or Datamonkey with TreeSAAP may provide additional information about amino acid properties, such as hydrophobicity, at the sites identified by the aforementioned codon methods. Overall, amino acid properties can provide crucial information that can be extremely relevant in protein-protein interactions [41] and thus a change in this biochemical property can compromise the folding of the protein interfering in its stability and function.

Supporting Information

Table S1 Amino-acid sites under positive selection in toxicofera-reptilian CRiSPs.

References

1. Yang Z, Nielsen R, Goldman N, Pedersen AM (2000) Codon-substitution models for heterogeneous selection pressure at amino acid sites. Genetics 155(1): 431–449.
2. Murrell B, Wertheim JO, Moola S, Weighill T, Scheffler K, et al. (2012) Detecting individual sites subject to episodic diversifying selection. PLoS Genet 8(7): e1002764. doi:10.1371/journal.pgen.1002764.
3. da Fonseca RR, Johnson WE, O'Brien SJ, Ramos MJ, Antunes A (2008) The adaptive evolution of the mammalian mitochondrial genome. BMC Genomics 9: 119. doi:10.1186/1471-2164-9-119.
4. Sunagar K, Jackson TN, Undheim EA, Ali SA, Antunes A, et al. (2013) Three-Fingered RAVERs: Rapid Accumulation of Variations in Exposed Residues of Snake Venom Toxins. Toxins 5(11): 2172–2208. doi:10.3390/toxins5112172.
5. Brust A, Sunagar K, Undheim E, Vetter I, Yang D, et al (2013) Differential evolution and neofunctionalization of snake venom metalloprotease domains. Mol Cell Proteomics 12(3): 651–663. doi:10.1074/mcp.A112.023135.
6. Kini RM, Chan YM (1999) Accelerated evolution and molecular surface of venom phospholipase A2 enzymes. J Mol Evol 48(2): 125–132.
7. Kozminsky-Atias A, Zilberberg N (2012) Molding the business end of neurotoxins by diversifying evolution. FASEB J 26(2): 576–586. doi:10.1096/fj.11-187179.
8. Low DH, Sunagar K, Undheim EA, Ali SA, Alagon AC, et al. (2013) Dracula's children: molecular evolution of vampire bat venom. J Proteomics 89: 95–111. doi:10.1016/j.jprot.2013.05.034.
9. Ruder T, Sunagar K, Undheim EA, Ali SA, Wai TC, et al. (2013) Molecular phylogeny and evolution of the proteins encoded by coleoid (cuttlefish, octopus, and squid) posterior venom glands. J Mol Evol 76(4): 192–204. doi:10.1007/s00239-013-9552-5.
10. Sunagar K, Fry BG, Jackson T, Casewell NR, Undheim EAB, et al. (2013) Molecular evolution of vertebrate neurotrophins: Co-option of the highly conserved nerve growth factor gene into the advanced snake venom arsenal. PloS ONE 8(11): e81827. doi:10.1371/journal.pone.0081827.
11. Sunagar K, Johnson WE, O'Brien SJ, Vasconcelos V, Antunes A (2012) Evolution of CRISPs associated with toxiceroran-reptilian venom and mammalian reproduction. Mol Biol Evol 29(7): 1807–1822. doi:10.1093/molbev/mss058.
12. Tian C, Yuan Y, Zhu S (2008) Positively selected sites of scorpion depressant toxins: possible roles in toxin functional divergence. Toxicon 51(4): 555–562. doi:10.1016/j.toxicon.2007.11.010.
13. Zhu S, Bosmans F, Tytgat J (2004) Adaptive evolution of scorpion sodium channel toxins. J Mol Evol 58(2): 145–153. doi:10.1007/s00239-003-2534-2.
14. Antunes A, Ramos MJ (2007) Gathering computational genomics and proteomics to unravel adaptive evolution. Evol Bioinform Online 3: 207–209.
15. Woolley S, Johnson J, Smith MJ, Crandall KA, McClellan DA (2003) TreeSAAP: selection on amino acid properties using phylogenetic trees. Bioinform 19(5): 671–672. doi:10.1093/bioinformatics/btg043.
16. Pond SL, Frost SD (2005) Not so different after all: a comparison of methods for detecting amino acid sites under selection. Mol Biol Evol 22(5): 1208–1222. doi:10.1093/molbev/msi105.
17. Yang Z (2007) PAML 4: Phylogenetic analysis by maximum likelihood. Mol Biol Evol 24(8): 1586–1591. doi:10.1093/molbev/msm088.
18. Delport W, Poon AF, Frost SD, Pond SL (2010) Datamonkey 2010: a suite of phylogenetic analysis tools for evolutionary biology. Bioinform 26(19): 2455–2457. doi:10.1093/bioinformatics/btq429.
19. Pond SL, Frost SD (2005) Datamonkey: rapid detection of selective pressure on individual sites of codon alignments. Bioinform 21(10): 2531–2533. doi:10.1093/bioinformatics/bti320.
20. Pond SL, Frost SD, Muse SV (2005) HyPhy: hypothesis testing using phylogenies. Bioinform 21(5): 676–679. doi:10.1093/bioinformatics/bti079.
21. Schwede T, Kopp J, Guex N, Peitsch MC (2003) SWISS-MODEL: an automated protein homology-modeling server. Nucl Acids Res 31(13): 3381–3385. doi:10.1093/nar/gkg520.
22. Arnold K, Bordoli L, Kopp J, Schwede T (2006) The SWISS-MODEL workspace: a web-based environment for protein structure homology modelling. Bioinform 22(2): 195–201. doi:10.1093/bioinformatics/bti770.
23. Bordoli L, Kiefer F, Arnold K, Benkert P, Battey J, et al. (2009) Protein structure homology modeling using SWISS-MODEL workspace. Nat Prot 4, 1–13. doi:10.1038/nprot.2008.197.
24. Wong WS, Yang Z, Goldman N, Nielsen R (2004) Accuracy and power of statistical methods for detecting adaptive evolution in protein coding sequences and for identifying positively selected sites. Genetics 168(2): 1041–1051. doi:10.1534/genetics.104.031153.
25. Yang Z, Wong WS, Nielsen R (2005) Bayes empirical bayes inference of amino acid sites under positive selection. Mol Biol Evol 22(4): 1107–1118. doi:10.1093/molbev/msi097.
26. Yang Z (1998) Likelihood ratio tests for detecting positive selection and application to primate lysozyme evolution. Mol Biol Evol 15(5): 568–573.
27. Yang Z, Nielsen R (1998) Synonymous and nonsynonymous rate variation in nuclear genes of mammals. J Mol Evol 46(4): 409–418.
28. Zhang J, Nielsen R, Yang Z (2005) Evaluation of an improved branch-site likelihood method for detecting positive selection at the molecular level. Mol Biol Evol 22(12): 2472–2479. doi:10.1093/molbev/msi237.
29. Murrell B, Moola S, Mabona A, Weighill T, Sheward D, et al. (2013) FUBAR: a fast, unconstrained bayesian approximation for inferring selection. Mol Biol Evol 30(5): 1196–1205. doi:10.1093/molbev/mst030.
30. Yang Z (1997) PAML: a program package for phylogenetic analysis by maximum likelihood. Computer applications in the biosciences: CABIOS 13(5): 555–556.
31. McClellan DA, McCracken KG (2001) Estimating the influence of selection on the variable amino acid sites for the cytochrome b protein functional domains. Mol Biol Evol 18(6): 917–925.
32. McClellan DA, Palfreyman EJ, Smith MJ, Moss JL, Christensen RG, et al. (2005) Physicochemical evolution and molecular adaptation of the cetacean and artiodactyl cytochrome b proteins. Mol Biol Evol 22(3): 437–455. doi:10.1093/molbev/msi028.
33. Herráez A (2006) Biomolecules in the computer: Jmol to the rescue. Biochem Mol Biol Educ 4(4): 255–261. doi:10.1002/bmb.2006.494034042644.
34. Altschul SF, Madden TL, Schaffer AA, Zhang J, Zhang Z, et al. (1997) Gapped BLAST and PSI-BLAST: a new generation of protein database search programs. Nucl Acids Res 25(17): 3389–3402.

Table S2 IMPACT_S count of the positively-selected properties varying across the toxicofera-reptilian CRiSPs phylogenetic tree.

Table S3 Merged results from the IMPACT_S integrative approach for the analyses of toxicofera-reptilian CRiSPs.

Acknowledgments

We would like to thank Rui Borges, Cidália Gomes and Imran Khan for their helpful discussions and support. We also thank João Paulo Machado, Joana Pereira, Anoop Alex, Siby Philip and Susana Pereira for their useful suggestions. We are indebted to Dr. Sergei L. Kosakovsky Pond from the University of California, San Diego for his assistance with the use of Datamonkey. We are thankful to the Academic Editor Konrad Scheffler and to the two anonymous reviewers for their valuable comments and suggestions, which led to the improvement of the manuscript.

Author Contributions

Conceived and designed the experiments: EM KS DA AA. Performed the experiments: EM. Analyzed the data: EM KS DA AA. Contributed reagents/materials/analysis tools: VV AA. Wrote the paper: EM KS DA AA.

35. Rose PW, Beran B, Bi C, Bluhm WF, Dimitropoulos D, et al. (2011) The RCSB Protein Data Bank: redesigned web site and web services. Nucl Acids Res 39 (suppl 1): D392-D401. doi:10.1093/nar/gkq1021.

36. Rose PW, Bluhm WF, Beran B, Bi C, Dimitropoulos D, et al. (2011) The RCSB Protein Data Bank: site functionality and bioinformatics use cases. NCI-Nat Path Interac Datab. doi:10.1038/pid.2011.1.

37. Guindon S, Gascuel O (2003) A simple, fast, and accurate algorithm to estimate large phylogenies by maximum likelihood. Syst Biol 52(5): 696–704. doi:10.1080/10635150390235520.

38. Zmaseck CM, Eddy SR (2001) ATV: display and manipulation of annotated phylogenetic trees. Bioinform 17(4): 383–384. doi:10.1093/bioinformatics/17.4.383.

39. Maldonado E, Dutheil JY, da Fonseca RR, Vasconcelos V, Antunes A (2011) IMPACT: integrated multiprogram platform for analyses in ConTest. J Hered 102(3): 366–369. doi:10.1093/jhered/esr003.

40. Gosh S, Matsuoka Y, Asai Y, Hsin K, Kitano H (2012) Software for systems biology: from tools to integrated platforms. Nat Rev Genet 12(12): 821–832. doi:10.1038/nrg3096.

41. Nabholz B, Ellegren H, Wolf JB (2013) High levels of gene expression explain the strong evolutionary constraint of mitochondrial protein-coding genes. Mol Evol Biol 30(2): 272–284. doi:10.1093/molbev/mss238.

Computational Design of Protein-Based Inhibitors of *Plasmodium vivax* Subtilisin-Like 1 Protease

Giacomo Bastianelli[1,2]**, Anthony Bouillon**[3,4¤]**, Christophe Nguyen**[5]**, Dung Le-Nguyen**[5]**,
Michael Nilges[1,2*ჟ]**, Jean-Christophe Barale**[3,4*ჟ¤]

1 Institut Pasteur, Unité de Bioinformatique Structurale, Département de Biologie Structurale et Chimie, Paris, France, 2 CNRS UMR 3528, Paris, France, 3 Institut Pasteur, Unité d'Immunologie Moléculaire des Parasites, Département de Parasitologie et de Mycologie & CNRS URA 2581, Paris, France, 4 CNRS, URA2581, Paris, France, 5 SYSDIAG, CNRS UMR3145 CNRS-BioRad, Montpellier, France

Abstract

Background: Malaria remains a major global health concern. The development of novel therapeutic strategies is critical to overcome the selection of multiresistant parasites. The subtilisin-like protease (SUB1) involved in the egress of daughter *Plasmodium* parasites from infected erythrocytes and in their subsequent invasion into fresh erythrocytes has emerged as an interesting new drug target.

Findings: Using a computational approach based on homology modeling, protein–protein docking and mutation scoring, we designed protein–based inhibitors of *Plasmodium vivax* SUB1 (PvSUB1) and experimentally evaluated their inhibitory activity. The small peptidic trypsin inhibitor EETI-II was used as scaffold. We mutated residues at specific positions (P4 and P1) and calculated the change in free-energy of binding with PvSUB1. In agreement with our predictions, we identified a mutant of EETI-II (EETI-II-P4LP1W) with a *Ki* in the medium micromolar range.

Conclusions: Despite the challenges related to the lack of an experimental structure of PvSUB1, the computational protocol we developed in this study led to the design of protein-based inhibitors of PvSUB1. The approach we describe in this paper, together with other examples, demonstrates the capabilities of computational procedures to accelerate and guide the design of novel proteins with interesting therapeutic applications.

Editor: Laurent Rénia, Agency for Science, Technology and Research - Singapore Immunology Network, Singapore

Funding: This work was partly supported by MEST-CT-05-020311, a Marie Curie Early Stage Research Training Fellowship (EIMID) of the Framework Program 6 by the European Commission. AB is a fellow of the "Direction Generale pour l'Armement" from the French Ministry of Defense. This work was partly supported by the "Fond dédié: combattre les maladies parasitaires" granted by Sanofi-Aventis and the French Ministry of Research and the Institut Carnot-Pasteur Maladies Infectieuses. The funders had no role in study design, data collection and analysis, decision to publish, or preparation of the manuscript. SYSDIAG, CNRS UMR3145 CNRS-BioRad provided support in the form of salaries and operating costs for authors CN and DLN, but did not have any additional role in the study design, data collection and analysis, decision to publish, or preparation of the manuscript. The specific roles of these authors are articulated in the "author contributions" section.

Competing Interests: The authors have the following interests. This work was partly supported by the "Fond dédié: combattre les maladies parasitaires" granted by the French Ministry and Sanofi-Aventis. Christophe Nguyen and Dung Le-Nguyen are employed by SYSDIAG, CNRS UMR3145 CNRS-BioRad. There are no patents, products in development or marketed products to declare.

* Email: nilges@pasteur.fr (MN); jcb@pasteur.fr (JCB)

ჟ These authors participated equally in the supervision of this work.

¤ Current address: Institut Pasteur, Unité de Biologie et Génétique du Paludisme, Team Malaria Targets and Drug Development, Département de Parasitologie et de Mycologie, Paris, France

Introduction

With more than 400 millions infections worldwide, malaria remains a major public health issue, principally in sub-Saharan Africa. An effective vaccine would help reduce disease burden, but the best candidates are still in development or evaluation phase [1,2]. The rapid development of multidrug-resistant *Plasmodium* [3] parasites necessitates accelerating the discovery of novel anti-malarial compounds to meet the needs of the agenda for malaria control and eradication [4].

In humans, *Plasmodium sp.* development comprises different stages, with the asexual intra–erythrocytic forms being responsible for the symptoms of the disease, such as fever, anemia, and cerebral malaria that can lead to death [5]. The erythrocyte invasion by *Plasmodium* merozoites critically depends on protease activities involved in both the daughter parasites egress from erythrocytes, and invasion into another erythrocyte. The parasite subtilisin-like protein 1 (SUB1) plays a critical role during both the hepatic and erythrocytic phases of *Plasmodium* biological cycle and is hence considered an interesting multi-stage target for developing a new class of anti–malarials [6] [7].

Most of the ancient therapies against *Plasmodium* are based on small molecules such as chloroquine, quinolones, antifolate, artemisinin derivatives, or atovaquone. The development of new classes of active molecules such as protein–based drugs or peptidomimetics [8,9] is an active and promising field of research. Among protein–based drugs, dermaseptin S4 (DS4) was shown to irreversibly inhibit the *in vitro* parasite growth through a cytotoxic hemolytic activity. Dermaseptin S3 acts in a similar manner as DS4 but did not present hemolytic activity through a cytotoxic hemolytic activity [10].

STRUCTURES MODELING

SUB1 (Homology Modeling)

EETI-II (Substrate Mutant)

DOCKING

Ensemble Docking

Conformational Sampling (MDs)

Docking Refinement

SCORING

Mutate

Conformational Sampling (MDs)

Free Energy Calculations (GBSA)

EXPERIMENTAL TESTING

Protein Production

Enzymatic testing

Figure 1. Computational Protein Design Strategy. Step 1: Prediction of the structure of the enzyme (PvSUB1) by comparative modelling and of the scaffold for mutational analysis (EETI-II-sub) by replacing one of the loops with a substrate sequence. Step 2: docking of EETI-II-sub to the target protein by ensemble docking procedure with several conformations from molecular dynamics simulations for each protein partner, and refinement of the best solutions. Step 3: mutation of the scaffold, conformational sampling and scoring of the mutants. Step 4: experimental testing by an enzymatic inhibitory assay on the recombinant enzyme of PvSUB1.

In the design of protein–based drugs, most approaches use combinatorial libraries based on different screening methods such as phage [11], ribosome [12] or mRNA display [13]. Their use is wide–spread, in particular for selecting high-affinity protein binders, despite their limitations due to the library size and the large quantities of the target protein needed to perform screening. Moreover, when the selection is not based on binding but on inhibiting a crucial enzyme of the biological cycle, a rather complex selection system has to be employed. Computational protein design can be used to reduce the sequence/structure space that needs to be explored and thus accelerate the process of screening and selection of target inhibitors.

Here, we present a strategy for the computational design of protein-based inhibitors targeting the subtilisin–like 1 protease of the human parasite *Plasmodium vivax* (PvSUB1). PvSUB1 can be expressed as a recombinant active enzyme [14] [15], and a specific enzymatic assay allows one to evaluate specific inhibitors. To search for potential inhibitors of PvSUB1, we used a computational design strategy, employing as scaffold the small protein EETI-II (*Ecballium elaterium* trypsin inhibitor II) [16], a trypsin inhibitor extracted from *Ecballium elaterium*. The family of cystein–knot proteins, to which EETI-II belongs, and in particular the cyclotides [17], possesses interesting biochemical properties [18]. EETI-II is composed of 28 amino-acids and its three-dimensional structure is tightly constrained by 3 disulphide bridges that contribute to its rigidity and biological stability [19]. We opted for this scaffold because several studies showed the possibility to engineer this protein to obtain specific mutants [20], *via* the extension of the EETI bioactive loop [21] or by changing its sequence to change its specificity towards the targeted enzyme [22] [23] [24] [25].

Compared to studies using an iterative computational design procedure focused on electrostatic binding contributions and single mutants [26], or on re–designing a scaffold protein to bind to a specified region on a target protein [27], we here faced the additional challenge that the 3D structure of the target itself or a close sequence homologue was not known. Nonetheless, the use of state–of–the–art structure prediction, docking and scoring methods allowed us to successfully identify mutants of the scaffold EETI-II that inhibited the target PvSUB1 enzyme.

Results and Discussion

The computational protein design approach involved four steps (see Figure 1). The first step was the modeling of the structure of the enzyme (PvSUB1) and the scaffold (EETI-II). Because of the lack of an experimental PvSUB1 structure, we built structures based on sequence homology. We also generated the model of a mutant of EETI-II containing the substrate sequence of PvSUB1, which we called EETI-II- sub. The second step was the docking of EETI-II-sub to the target protein. We employed an ensemble docking procedure with several conformations obtained from molecular dynamics (MD) simulations for each protein partner to implicitly include flexibility in the docking, and refined the best docking solutions by molecular dynamics to obtain high-quality structures of the complex. The third step aimed at identifying mutants of EETI-II-sub that had higher binding affinity towards PvSUB1. In this step, we mutated residues in EETI-II-sub at the protein–protein interface of the complex, ran conformational sampling of the mutant with molecular dynamics, and calculated the free energy of binding via implicit solvent models based on the Generalized Born approximation (GBSA). The last step consisted in the experimental testing of the inhibitor by an enzymatic inhibitory assay specific for the PvSUB1 recombinant enzyme.

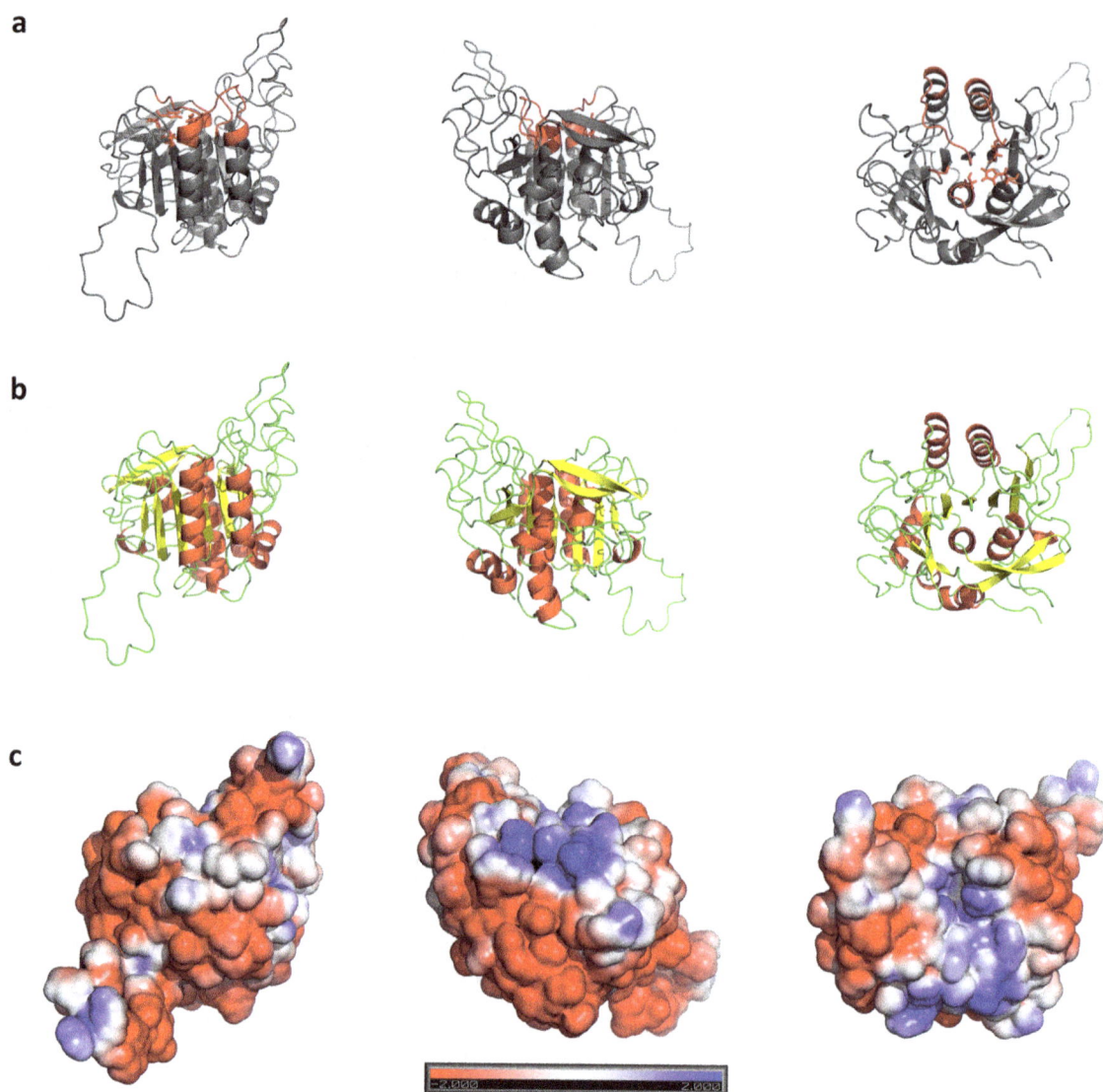

Figure 2. 3D model of PvSUB1 catalytic domain. A: Highlighted in red is the region forming the substrate binding pocket and red sticks correspond to the residues that form the catalytic triad; **B:** Cartoon representation of secondary structures; **C:** APBS surface electrostatic representation.

Table 1. Catalytic site distances along MD simulations.

Distance	1R0R	1T02	PvSUB1
HIS@CA-SER@CA	8.52±0.34	9.38±0.36	9.89±0.24
HIS@CB-SER@CB	6.56±0.43	7.55±0.41	7.91±0.31
HIS@CA-ASP@CA	7.78±0.52	7.91±0.29	8.33±0.18
HIS@CB-ASP@CB	6.59±0.66	6.73±0.41	7.46±0.24
SER@CA-ASP@CA	10.24±0.35	9.18±0.32	10.35±0.25
SER@CB-ASP@CB	7.42±0.42	8.83±0.49	8.72±0.34
SER@OG-HIS@NE2*	7.54±1.11	4.69±1.19	5.11±0.71
ASP@CG-HIS@ND1	8.52±0.34	9.38±0.36	6.74±0.54
ASN@CG-SER@CB	6.70±0.71	7.5±0.91	6.52±0.41

Values are expressed in Å. The distance SER@OG-HIS@NE2 shows the largest fluctuation for both subtilisins with known structure and for our models, consistent with variations of this distance in subtilisin crystal structures (Table 2). Catalytic triad: Asp316, His372, Ser549; Asp137, His168, Ser325 and Asp139, His171, Ser328 for PvSUB1, 1R0R and 1T02 respectively.

Table 2. Subtilisin catalytic site geometries.

Distance	lower-range	upper-range
HIS@CA-SER@CA	8.3	8.72
HIS@CB-SER@CB	6.44	6.89
HIS@CA-ASP@CA	7.22	7.46
HIS@CB-ASP@CB	5.83	6.61
SER@CA-ASP@CA	9.87	10.11
SER@CB-ASP@CB	8.15	8.53
SER@OG-HIS@NE2	2.57	3.36
ASP@CG-HIS@ND1	3.15	3.35
ASN@CG-SER@CB	6.34	6.74
SER@OH-surface	1.58	1.60

Smallest (lower-range) and longest (upper-range) measures from experimentally determined structures used as templates in the modeling of PvSUB1. The asparagine (Asn) is the residue forming the oxyanion hole. Catalytic triads of PvSUB1, bacterial subtilisins Carlsberg (1R0R) and BPN (1TO2): Asp316, His372, Ser549; Asp137, His168, Ser325 and Asp139, His171, Ser328, respectively.

Modeling and molecular dynamics simulations of PvSUB1

In order to generate a reliable 3D-model of PvSUB1, we used the procedure described in our previous publication where we modeled the structure of PfSUB1, a close homologous of PvSUB1 [14]. A similar homology modeling strategy generated 3D-models of PfSUB1 used to identify small-molecule inhibitors of PfSUB1 with an *in silico* screening approach [15]. The particular challenge of obtaining a high quality 3D model of PvSUB1 was the low sequence identity with the available templates (only ~30%, just above the "twilight zone" for homology modeling). Using state of the art modeling methods, it is possible to generate homology models with a Cα RMSD<1.0 Å when the sequence identity with the template is >50%. With sequence identity below 25%, larger divergences from the target structure can appear, making the model less precise [28]. However, our previous analysis had shown that major divergences between the PvSUB1 sequence and the structural templates were localized in regions distant from the catalytic groove that binds the substrate, and that the sequence identity in the substrate binding area is >30% [14,15]. In addition, to evaluate the stability and the overall quality of the PvSUB1 model (Figure 2), we performed multiple molecular dynamics simulations of our model PvSUB1 and of two of the templates used in the modeling, subtilisins BPN (1TO2) and Carlsberg (1R0R).

Tables 1 and 2 show distance ranges among the residues of the catalytic triad, in the MD simulations of PvSUB1 and in the crystal structures of the templates. All distances fell within the ranges observed in the experimental X–ray crystal structures apart from the distances HIS372@CB-ASP316@CB and ASP316@CG-HIS372@ND1, which are slightly outside (less than 1 Å). This is consistent with the fact that this distance shows the highest variation in subtilisin X–ray crystal structures (Table 2). Stability can be also measured by analyzing the RMSD along the MD trajectories from the starting structure. In Figure 3.A, we plotted the average RMSDs obtained for the 5 MD trajectories of 10 ns each.

The trajectories of the PvSUB1 model diverge more than those of BPN (1TO2) and Carlsberg (1R0R). This is primarily due to the regions in the model for which there is no structural information in the templates. In the model, these regions are unstructured, solvent exposed and distant from the binding pocket. When we removed them from the analysis, the RMSD reduced to values similar to those observed in the MD trajectories of subtilisin BPN (1TO2). The trend of the RMSD from the average structure shows a stabilization of the model along the MD trajectories (Figure 3.B). The per–residue fluctuation (RMSF) in Figure 3.C shows some very flexible regions, for example the region 50–80, where structural information in the templates is absent.

Most of residues forming the binding region (orange rectangles) were much less flexible than the rest, similar to what we observed for the other two subtilisins (data not shown). The fact that the model of the binding region showed similar stability in MD simulations as the X–ray crystal structures is an indication that there are no major errors in the model of this region. Obviously, despite the care we took in generating and validating our model, there may be structural errors with an effect on the success of our computational design procedure. For this reason, we included an additional step to refine the model of the structure of PvSUB1.

Refinement of PvSUB1 model

To obtain a refined structure of the PvSUB1 model, we performed MD simulations of the complex of PvSUB1 and its substrate hexapeptide (P4-VGADDV-P2'). The hexapeptide was docked according to the X-ray structure of Subtilisin E with its pro-peptide (1SCJ), and refined with MD, where we restrained a few distances between the protein and the peptide (see [14] for details). Even though subtilisins do not undergo major conformational changes upon binding [29], small rearrangements might take place at the interface at the level of side–chains for example. The refinement allowed us to obtain bound–like conformations of PvSUB1, which facilitated the subsequent docking step (Figure 4). Figure 5 shows the catalytic triad Ser-His-Asp of the PvSUB1 model.

Docking of EETI-II-sub

The wild type EETI-II did not inhibit PvSUB1 (Table 3). We then replaced the EETI-II residues involved in the binding to the protease catalytic groove with the sequence of the PvSUB1 natural hexapeptide substrate. This EETI-II-sub mutant inhibited PvSUB1 activity with a Ki >0.75 mM (Table 3). We derived snapshots (50 for each protein) for the ensemble docking of EETI-II-sub onto PvSUB1 by a cluster analysis (based on the residues at the interface) of the 5×2 ns molecular dynamics trajectories to obtain the best representative structures from the simulations.

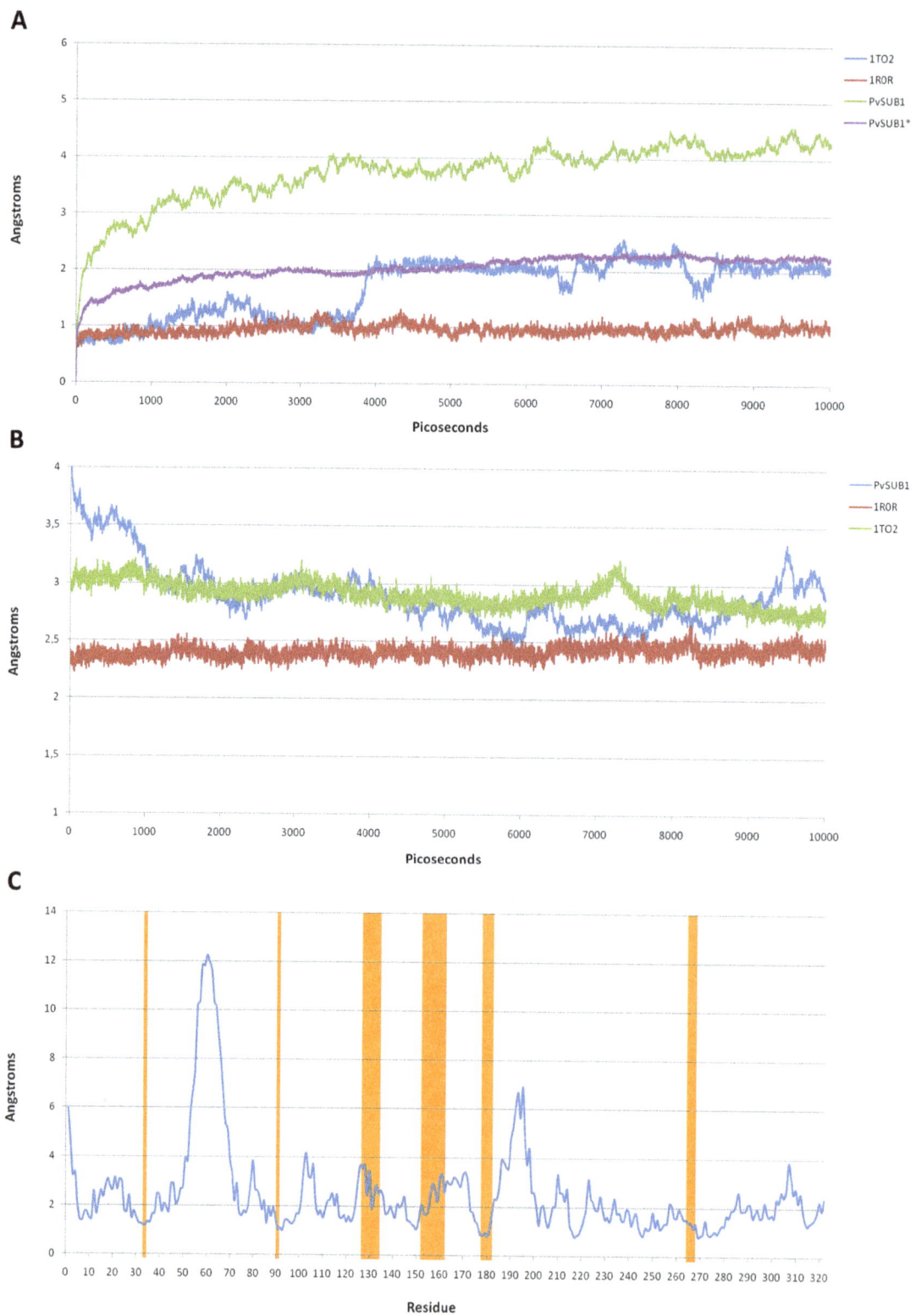

Figure 3. PvSUB1 molecular dynamics simulations. A: Average RMSD values for PvSUB1 and the 3D structure of two homologous bacterial subtilisins (1TO2, 1ROR). PvSUB1* shows the RMSD calculated without the regions missing template structural information; **B:** Fluctuation of the RMSD from the average structure. **C:** Root mean square fluctuation (RMSF) on a per-residue basis. In orange are highlighted PvSUB1 residues involved in the substrate-binding region.

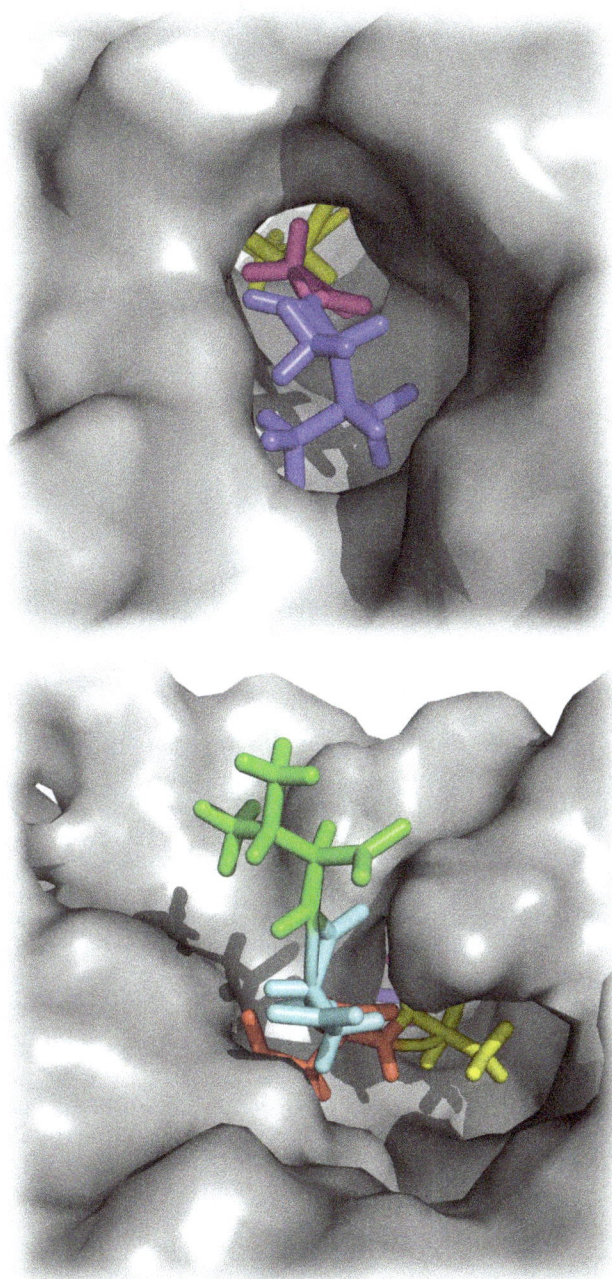

Figure 4. Docking of PvSUB1 hexapeptide substrate into PvSUB1 catalytic groove. Blue: P4, Violet: P3, Yellow: P2, Red: P1, Cyan: P1′, Green: P2′.

Figure 6 shows the score distribution among the solutions, with the 5 solutions with the highest geometric score (see Methods) highlighted with red circles. The relatively low geometric score indicated that most docking solutions were sub-optimal, although we had used an elaborate ensemble docking procedure. To optimize and refine the docking results we took the best 5 docking poses (red circles in 6) and refined them by a restrained molecular dynamics procedure similar to what we used for the refinement of the PvSUB1 model. This procedure helped re-establish native contacts when compared to regular unrestrained MD simulations. Subsequently we selected the solution that fulfilled all distances. Figure 7 shows the docked complex.

Scoring mutants

A preliminary free energy calculation was performed with snapshots from multiple MD simulations of EETI-II-sub docked onto PvSUB1 to obtain more consistent MM/GBSA results [30]. A free-energy decomposition (Figure 8) shows the contribution of each single residue to the total free energy of binding.

The biggest contribution to the free energy of binding came from the main-chain contacts of residues P4, P3, P2 and P1. This is in agreement with previous observations of important interactions between a protein-inhibitor and a serine protease active site, where important contacts are made by main-chain atoms [31]. For the case of EETI-II-sub the highest contribution originates from the cysteine in P3 and its main-chain, accounting for −4.34 kcal/mol.

We then tried to identify the most favorable mutations that could improve the binding affinity of EETI- II-sub to PvSUB1. The cysteine in P3 cannot be mutated because its side-chain is involved in a disulphide bridge that has an important function in stabilizing the EETI-II scaffold and maintaining the loop rigid, whereas the alanine in P2 already contributes with −4.17 kcal/mol to the total binding energy. We also looked at the parasite sequences that are natural substrates of PvSUB1 or PfSUB1 to suggest positions to introduce mutations in the EETI-II scaffold. Table 4 lists these sequences (experimentally derived or deducted from sequence alignments) for PfSUB1 and PvSUB1. While the sequences of several PfSUB1 substrates were experimentally determined [32,33], few were identified for PvSUB1 [15]. Considering the evolutionary proximity of *P. vivax* and *P. falciparum*, with active sites displaying >60% sequence identity [15], these predicted sequences can be considered reliable. Comparing the cleavage sites, we observed that only alanine and glycine appeared in P2, suggesting that only small residues are tolerated in this position. Position P1′ has a negative contribution to the energy and therefore is an interesting position to mutate. However, considering the lack of a specific pocket for this residue (Figure 7) we can consider this position almost a secondary contact residue [31] and we decided to keep the wild–type residue. Finally, SUB1 cleavage site sequences have a fairly high similarity at the P1 and P4 positions (Table 4) and we therefore focused on these positions to mutate the EETI-II inhibitory loop. It is worth mentioning that the contribution of these P1 and P4 positions within the substrate-PfSUB1 interaction has recently been experimentally established [34].

We performed 10×100 ps MD simulations and MM/GBSA free energy calculations for all possible residues in position P4 and P1 independently, assuming that the effect of the two mutations would be additive. The free energy calculations for the mutants in P4 showed that hydrophobic and bulky residues were preferred for this position (Figure 9.A). This result fits with the fact that pocket S4 is composed of six hydrophobic residues (L131, M134, F153, I161, F162, P205) and seems to have enough space to accommodate larger hydrophobic side-chains than valine (Figure 10). Position P1 instead presents as favorable mutations aromatic residues with polar groups (Tyr, Trp), glutamate and positively charged residues (Lys, Arg). Surprisingly we found as favorable mutations some positively charged residues, whereas most of sequences recognized by the homologous PfSUB1 present negatively charged (Asp, Glu) or neutral polar (Gln, Asn) side-chains at P1. This might be explained by either the low substrate specificity common to some subtilisins or imprecisions in the structure of the complex.

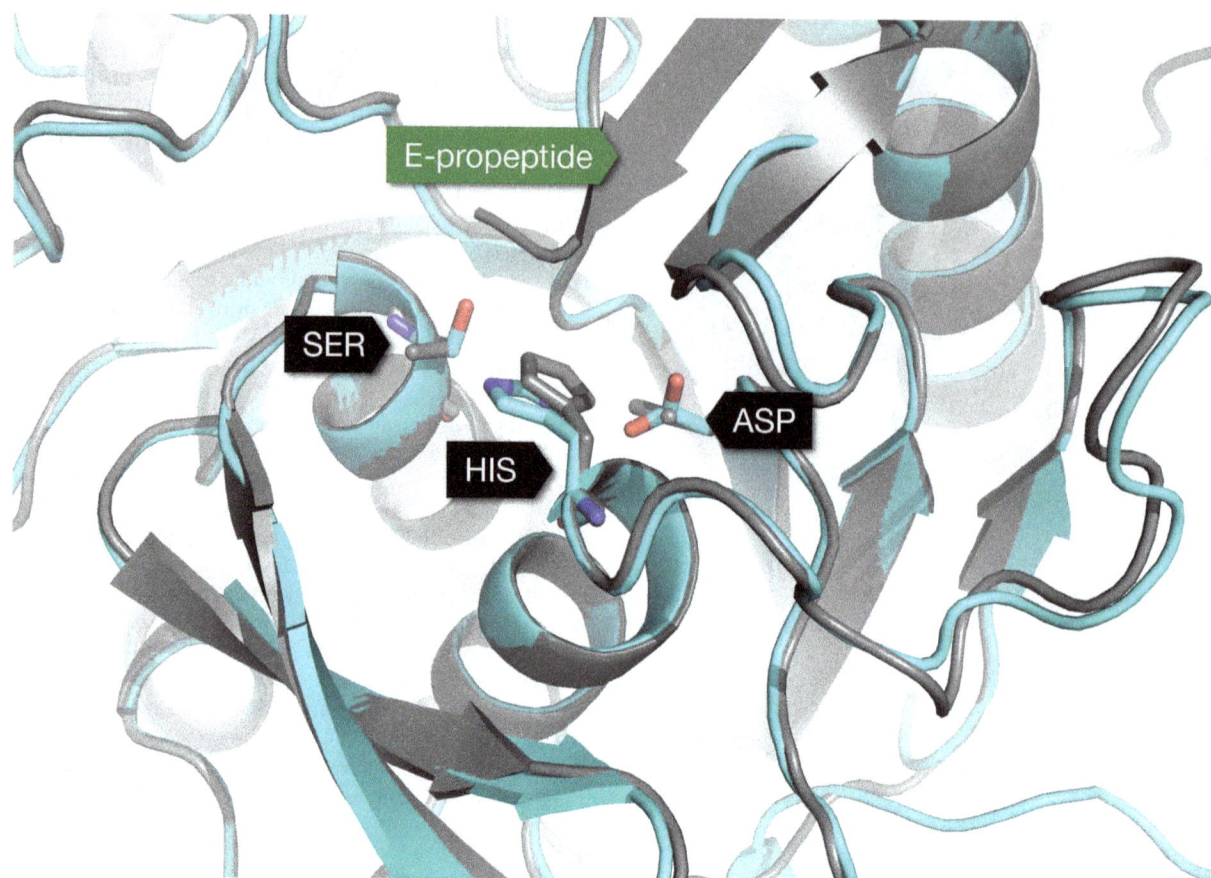

Figure 5. Structural alignment of the obtained PvSUB1 model (cyan) with the 3D-structure of Subtilisin E (gray, PDB 1SCJ) that was used as a template in the homology modelling. The catalytic triads in both proteins are highlighted with a stick representation. PvSUB1 catalytic triad: Asp 316, His 372 and Ser 549. Subtilisin E catalytic triad: Asp 32, His 64, Ser 221.

Evaluation of EETI-II mutants on PvSUB1 enzymatic activity

All mutants of EETI-II were produced by chemical synthesis, folded and purified by reverse–phase HPLC. In Table 3 we present the results of PvSUB1 inhibition of the different synthesized EETI-II mutants. We initially tested mutants in position P4 according to our scoring results and found that the EETI-II with a leucine in P4 (EETI-II-P4L) inhibits PvSUB1 with a Ki of 147 µM, i.e., about one order of magnitude higher that the

Table 3. Sequence and inhibitory activity of EETI-II mutants on PvSUB1.

Name of tested EETI-II	EETI-II active site sequences P_4 P_3 P_2 P_1 P'_1 P'_2	Ki on PvSUB1 (mM)
EETI-II-WT	G C P R I L	NI
PvS1-WT	V C A D D V	>0.75
PvS1-P_{4W}	W C A D D V	>0.75
PvS1-P_{4P}	P C A D D V	>0.75
PvS1-P_{4M}	M C A D D V	>0.75
PvS1-P_{4L}	L C A D D V	0.15±0.03
PvS1-P_{4I}	I C A D D V	0.6±0.01
PvS1-P_{4L} P_{1E}	L C A E D V	0.34±0.05
PvS1-P_{4L} P_{1K}	L C A K D V	0.39±0.14
PvS1-P_{4L} P_{1R}	L C A R D V	0.75±0.035
PvS1-P_{4L} P_{1Y}	L C A Y D V	0.24±0.02
PvS1-P_{4L} P_{1W}	L C A W D V	0.08±0.01

Active site sequences of the tested EETI-II mutants and their Ki for PvSUB1. NI: No Inhibition.

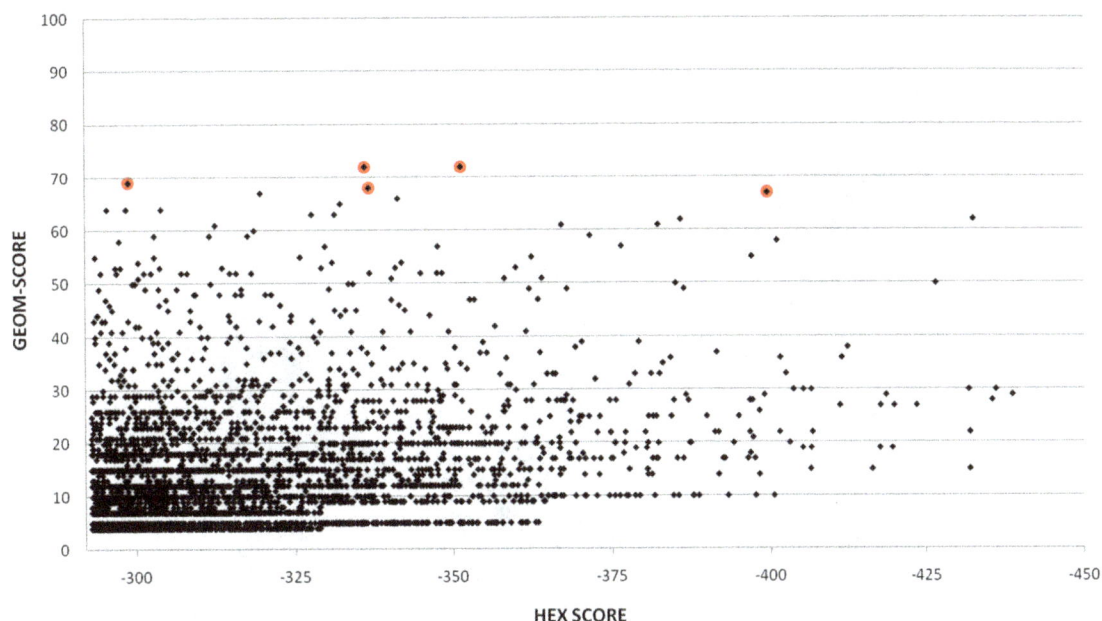

Figure 6. Docking results. The red circles indicate the docking poses that have been selected for refinement.

valine (Table 3). This could be explained by the higher flexibility of the leucine and its bulkier side-chain, which could fit more tightly into the hydrophobic S4 pocket (Figure 10). The fact that the isoleucine mutant shows a Ki of 591 μM, which is three times less than that of leucine, could be the effect of its beta–branched side–chain. We decided to keep the leucine mutant in P4 and test mutations in P1. Among all the EETI-II-P4L mutants for the P1 position, only one showed a better Ki of 86 μM, the EETI-II-P4L-P1W, which contained a tryptophane in P1. All other mutants had higher Ki suggesting the importance of keeping the aspartate in this position. We found the comparison of the active mutants with the list of substrate sequences of PvSUB1 in Table 4 particularly intriguing. In position P4 of the substrate there is a preference for valine and threonine, while leucine is only present in 2 out of 18 sequences; the mutant with a leucine in P4 is more than 5x more active on PvSUB1 than the one with a valine (PvS1-WT). This might be explained by the presence of important structural constraints (disulphide bonds) that are present in EETI-II compared to a more flexible conformation of the substrate sequence [29]. According to the cleavage sites predicted by sequence alignment of the substrates, we observe a prevalence of negatively charged residues such as glutamate and aspartate while there was a preference for aromatic residues in the designed inhibitors, with the tryptophan having the best Ki. Even in this case, the explanation might lie in the structural constraints that are present in the inhibitors and/or a set of new interactions at position P1 (Figure 10).

Conclusions

Subtilisin-like proteins of *Plasmodium* are promising biological targets for developing novel therapeutics. One of these proteins, SUB1 plays an essential role in both the hepatic and erythrocytic stages of *Plasmodium*, making this enzyme a particularly interesting drug target. With the aim to develop an inhibitor of *Plasmodium vivax* SUB1 (PvSUB1), we redesigned a protein scaffold, the natural trypsin inhibitor EETI-II. Despite challenges in this project (the 3D structure of the target protein had to be

modeled from homologues with only 30% sequence identity), the computational procedure allowed us to predict mutants that proved to be inhibitors of PvSUB1 in experimental tests.

However, some predicted good inhibitors did not show any improvements in binding to PvSUB1 when experimentally tested. This might be caused by the flexibility in the binding as proteases and their S# pockets are inherently flexible and known to alter their shape to accommodate various substrates. Imprecisions in the homology model of PvSUB1 can also have obvious consequences for the precision of the docking and of the energetic analysis. The free energy analysis based on implicit solvent methods is in itself approximate and neglects important factors influencing binding. It is encouraging that, nonetheless, we have obtained protein–based inhibitors of PvSUB1 and this opens new ways to further improve our best mutants by computational or experimental protein engineering protocols. The protein design approach described in this work demonstrates the capabilities of computational proce-dures to accelerate and guide the design of novel proteins with potential future therapeutic applications.

Methods

Modeling PvSUB1

Models of PvSUB1 were generated and validated with the same protocol as in a previous study [14]. Binding pocket residues of PvSUB1 were defined according to the interaction of canonical inhibitors with subtilisins: 152–162, 127–134, 265–268, 179–182, 90 (His), 34 (Asp), 267 (Ser). Regions were structural information from template structures is lacking are 307–324, 1–9, 51–73, 190–199, 227–231. The models were further evaluated by performing molecular dynamic simulations and comparing their dynamic behavior to that of two subtilisins 3D structures used in the homology modeling (Carlsberg: 1R0R, BPN: 1TO2). The validated model of PvSUB1 was further refined in order to obtain bound-like conformations for the ensemble docking. We docked the hexapeptide of the sequence recognized by PvSUB1 to its structure, using 1SCJ (subtilisin E + pro-peptide) as template. We performed restrained molecular dynamic simulation to refine the

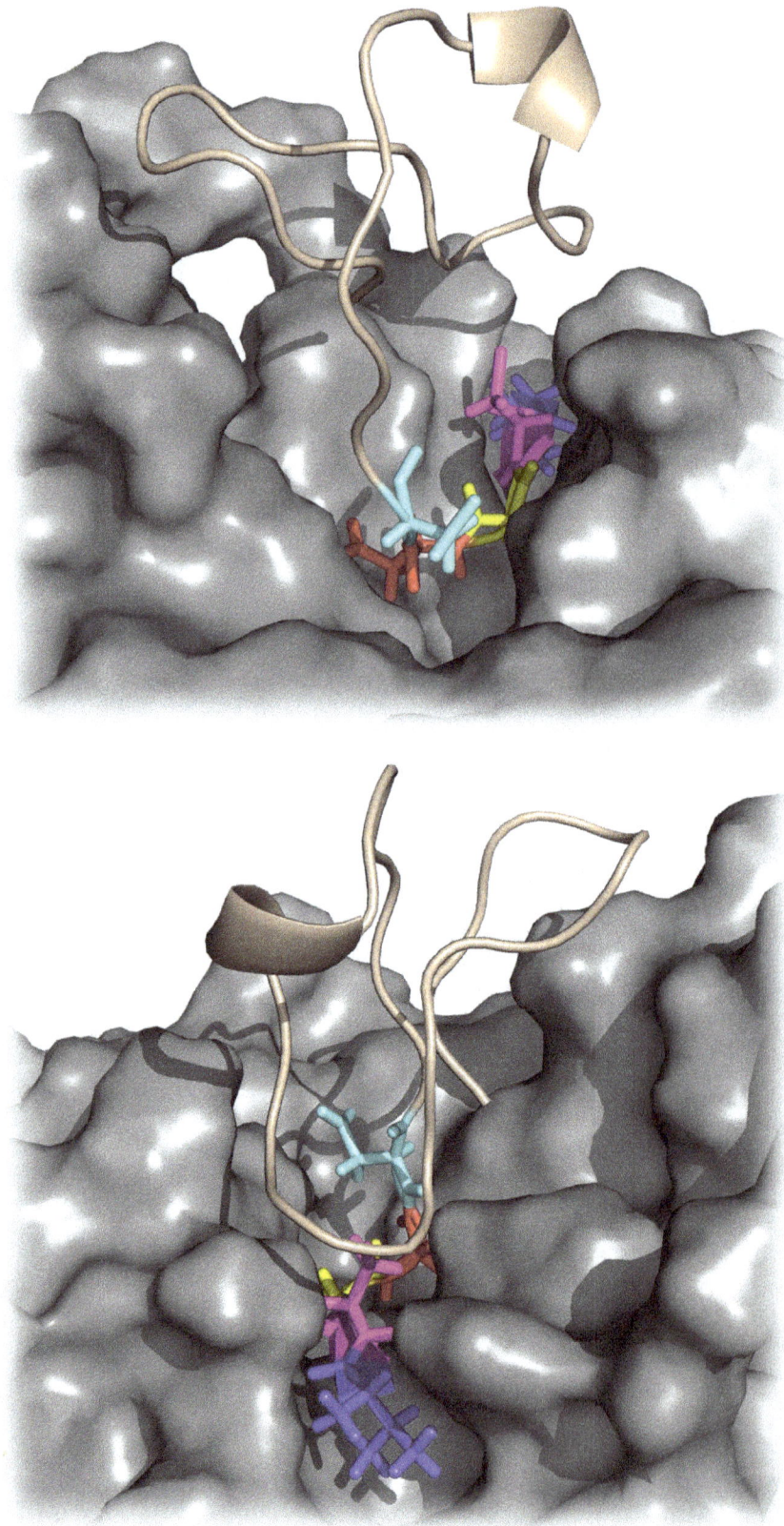

Figure 7. EETI-II-sub docked to PvSUB1. Blue: P4, Violet: P3, Yellow: P2, Red: P1, Cyan: P1′.

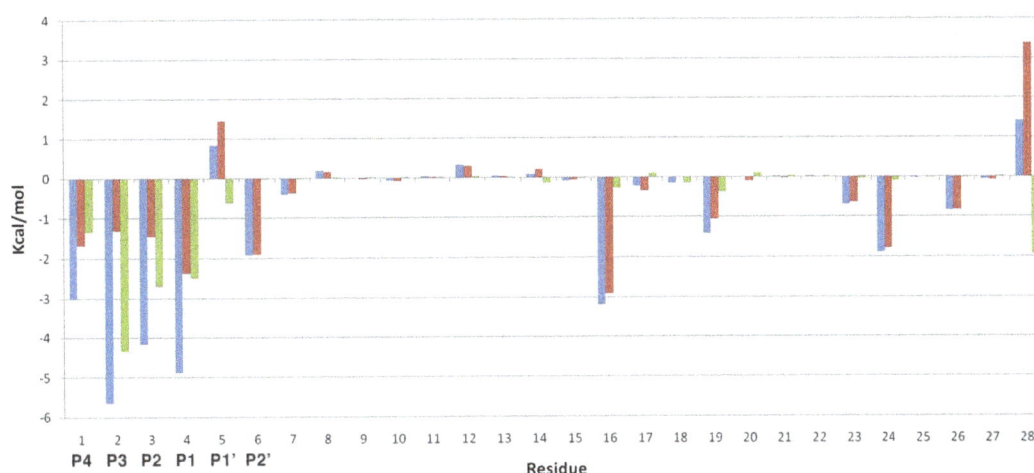

Figure 8. Free energy decomposition. Blue: All atoms, Red: Side chain atoms, Green: Backbone atoms. The largest contribution to the free energy of binding comes from the main-chain contacts of residues P4, P3, P2 and P1. The highest contribution comes from the cysteine in P3 and its main-chain, accounting for -4.34 kcal/mol.

complex, using the protocol described in the Section "Docking Refinement". Surface electrostatic distributions on models were calculated with the APBS [35] module implemented in Pymol.

EETI-II. The structure of EETI-II (2IT7) was retrieved from the PDB (http://www.rcsb.org) [36] and mutations at position P1 and P4 of the inhibitory loop were generated with the toolkit MMTSB [37]. We built the EETI-II substrate-like mutant (EETI-II-sub) by replacing its inhibitory loop (GCPRIL) with the sequence recognized by PvSUB1 (VGADDV). 3D images of the protein complexes were rendered with the molecular modeling software Pymol.

Molecular Dynamics (MD) Simulations

All molecular dynamics simulations were performed with the SANDER module from the AMBER9 [38] package and the force–field ff99SB [39]. After minimization *in vacuo* the complexes were hydrated with TIP3P [40] water molecules and neutralized by adding an appropriate number of monovalent counterions. The MD unit cell was a truncated octahedral box with a minimum distance of 10 Å between the solute and the cell boundary. To minimize the water molecules we ran a two stages minimization protocol in which we first applied positional restraints with an energy constant of 10 kcal/(mol Å^2) to the solute followed by a stage with 1 kcal/(mol Å^2) energy constant. Both minimization stages consisted of 10 steepest-descent and 490 conjugate-gradient steps. During the equilibration/heating and production dynamics all covalent bonds to hydrogen atoms were constrained with the SHAKE [41] algorithm, and we used a time step of 2 fs. We used periodic boundary conditions with a distance cutoff of 8.0 Å for the direct part of the non-bonded interaction and PME [42] (Particle Mesh Ewald) to account for long-range electrostatic interactions. The minimized system was then thermalized and equilibrated by heating from 0 to 300°K over 20 ps under constant-volume conditions followed by 10 ps at constant-pressure. The production MD phase was launched from the final configuration after equilibration with a relaxation time of 2.0 ps for heat bath coupling and a pressure relaxation time of 2.0 ps.

Docking

The pool of conformers used for the ensemble docking was obtained from multiple (5×2 ns) MD trajectories of PvSUB1 (receptor) and EETI-II (ligand). We extracted snapshots every 10 ps of MD for a total of 1000 snapshots of each protein and clustered these structures with a single-linkage algorithm implemented in GROMACS [43] (g-cluster tool), where the RMSD was calculated only for the binding interface residues. We selected for cross-docking the centroids of each of the 50 clusters identified. The ensemble docking was performed with the rigid-body docking software HEX [44] (version 4.5) where the search was restricted to the binding pocket by positioning EETI-II structures around the interface and limiting the search to 30° for twist range, receptor and ligand range. We used a shape + electrostatic correlation type while the other parameters of HEX were left as default. The best 5000 docking solutions according to the HEX score were selected and re-ranked by a mixed score based on the geometry of the interaction between a canonical loop inhibitor and a subtilisin. This score (geometric score) was composed for 40% of the HEX score (based on surface complementarity) and for 60% of an empirical score defined by the conserved distances between atoms in the inhibitory loop of a canonical inhibitor and a subtilisin. A mixed score permits to take into consideration the shape complementary and the conserved structural feature of the interaction. The best 5 solutions were selected for refinement.

Docking Refinement

The refinement is based on a protocol that uses restrained molecular dynamics simulations. The refinement procedure consisted of a total of 360.000 steps based on specific distance restraints at the ligand/receptor interface and used the NMR refinement tools in AMBER9. The chosen conserved distance restraints were 267SER@OG-P1@C (lower bounds = 2.5 Å, lower-intermediate = 3.0 Å, intermediate- upper = 4.0 Å, upper bounds = 5.5 Å), 154SER@HN-P3@O, 154SER@O-P3@HN, 129GLY@HN-P4@O, 129GLY@O-P4@HN, 127LYS@O-P2@ HN (lower bounds = 1.5 Å, lower-intermediate = 2.0 Å, intermediate- upper = 3.0 Å, upper bounds = 3.5 Å). In all cases, an energy constant of 20 kcal/mol Å^2 was employed. The refinement consisted of three phases. In the first phase (120.000 steps), the

Table 4. SUB1 natural substrates.

Plasmodium protein containing a SUB1 processing site	P4P3P2P1 ↓ P'1P'2	References
Pro-region first maturation site of PfSUB1	**VSAD ↓ NI**	[48]
Pro-region second maturation site of PfSUB1	**VSAD ↓ NI**	[48]
PfSERA1 site 1	IKAE ↓ AE	[6]
PfSERA2 site 1	TKGE ↓ DD	[6]
PfSERA3 site 1	VKAA ↓ SV	[6]
PfSERA4 site 1	ITAQ ↓ DD	[6]
PfSERA5 site 1	**IKAE ↓ TE**	[6] [49]
PfSERA6 site 1	VKAQ ↓ DD	[6]
PfSERA7 site 1	FKGE ↓ DE	[6]
PfSERA9 site 1	VKGS ↓ TE	[6]
PfSERA1 site 2	IYSQ ↓ ED	[6]
PfSERA2 site 2	IWGQ ↓ ET	[6]
PfSERA3 site 2	LYGQ ↓ EE	[6]
PfSERA4 site 2	VYGQ ↓ DT	[6]
PfSERA5 site 2	**IFGQ ↓ DT**	[6] [49]
PfSERA6 site 2	VHGQ ↓ SN	[6]
PfSERA7 site 2	ISAQ ↓ DE	[6]
PfSERA9 site 2	VHGQ ↓ SG	[6]
PvSERA1 site 1	TKGE ↓ DE	
PvSERA2 site 1	MKAQ ↓ DE	
PvSERA3 site 1	AKGE ↓ DE	
PvSERA4 site 1	RKAQ ↓ QQ	
PvSERA5 site 1 TKGE ↓ DE		
PfSERA5 site 3	VRGD ↓ TE	[6]
PfMSP1 clone 3D7 junction MSP1₈₃ -MSP1₃₀	**LVAA ↓ SE**	[50]
PfMSP1 clone 3D7 junction MSP1₃₀ -MSP1₃₈	**ITGT ↓ SS**	[50]
PfMSP1 clone 3D7 junction MSP1₃₈ -MSP1₄₂	**VTGE ↓ AI**	[50]
PfMSP1 clone FCB1 junction MSP1₃₀ -MSP1₃₈	**VSAN ↓ DD**	[51]
PfMSP1 clone FCB1 junction MSP1₃₈ -MSP1₄₂	**VTGE ↓ AV**	[51]
PcMSP1 junction MSP1₈₃ -MSP1₃₀	ATGE ↓ SE	[52]
PcMSP1 junction MSP1₃₀ -MSP1₃₈	VSAE ↓ SE	[52]
PcMSP1 junction MSP1₃₈ -MSP1₄₂	ANAQ ↓ ST	[52]
PvMSP1 clones Sal1 and (Belem) junction MSP1₈₃ -MSP1₃₀	LRGA(S) ↓ SA	
PvMSP1 clones Sal1 and Belem junction MSP1₃₀ -MSP1₃₈	VGGN ↓ SE	
PvMSP1 clones Sal1 and Belem junction MSP1₃₈ -MSP1₄₂ TTGE ↓ AE		
PfMSP6 clone 3D7	**VQAN ↓ SE**	[31]
PfMSP7 clone 3D7 site 1	**VKAQ ↓ SE**	[31]
PfMSP7 clone 3D7 site 2	**TQGQ ↓ EV**	[31]
PfRAP-1 clone 3D7	**IVGA ↓ DE**	[32]
PfMSRP2 clone 3D7	**LKGE ↓ SE**	[32]
Pro-region first maturation site of PvSUB1	**VGAD ↓ NI**	[15]
Pro-region second maturation site of PvSUB1	**SHAA ↓ SS**	[15]
Pro-region third maturation site of PvSUB1	**HLAG ↓ SK**	[15]
Pro-region first maturation site of PbSUB1	**VGAD ↓ SI**	[15]
Pro-region first maturation site of PySUB1	VGAD ↓ SI	[15]

The table shows the sequences of the cleavage sites recognized by SUB1 in in *Plasmodium falciparum* and *Plasmodium vivax*. Cleavage site sequences in bold characters have been experimentally determined, while the ones in normal characters are deducted from sequence alignments. The arrow indicates the site of cleavage between the P1 and the P1'. Cleavage site sequences have a fairly high similarity in particular at the P1, P2 and P4 positions.

A

B

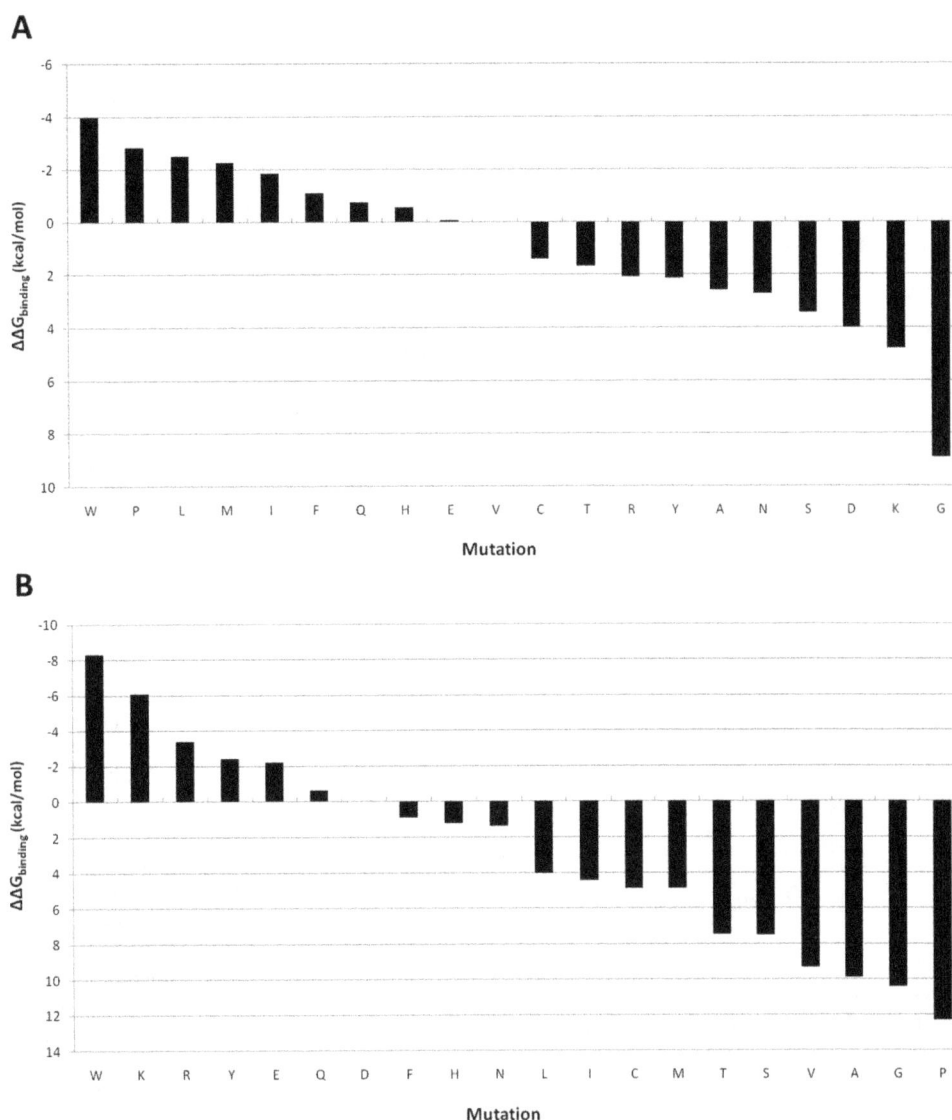

Figure 9. Scoring mutations on P4 and P1. A: mutants in position P4. The mutational profile of P4 shows that hydrophobic and bulky residues are preferred for this position. **B:** mutants in position P1. Position P1 instead prefers aromatic residues with polar groups (Tyr, Trp), glutamate and positively charged residues (Lys, Arg).

receptor (PvSUB1) was kept rigid by applying a Cartesian restraint with an energy constant of 10 kcal/(mol Å^2) and the distance restraints were switched on/off every 15.000 steps. In the second phase a Cartesian restraint with an energy constant of 10 kcal/(mol Å^2) was applied only to heavy main-chain atoms keeping side-chains fully flexible, and in the third phase the energy constant of this Cartesian restraint was reduced to 0.01 kcal/(mol Å^2). During these phases, the ligand (EETI-II or hexapeptide) was kept completely flexible. An additional 1 ns of regular MD simulation was performed to allow the system to relax into its final configuration. In the validation, this resulted in a high accuracy complex structure, with no distortions at the interface.

Scoring

The MM/GBSA protocol [45] in AMBER9 was used to calculate the relative free energy of each mutant. The default GBSA model used in the calculations was that of Tsui and Case [46] with an external dielectric of 80 and internal dielectric of 1.0.

For calculating the nonpolar contribution, the surface tension coefficient was set to 0.0072 and the surface offset to 0.0. The solvent accessible surface area was calculated with the ICOSA method. We calculated the relative free energy of binding from snapshots extracted each 10 ps from 10×100 ps trajectories.

Protein Production

EETI-II and mutants. All peptides (desalted, 35%–60% pure as assessed by HPLC) were obtained from GenScript Corporation, Piscatway, NJ, USA. In a typical procedure, peptide (50 mg) was dissolved in 75 mL of KH$_2$PO$_4$ buffer (0.2 M, pH 8.2) and allowed to air-oxidize at room temperature under gentle stirring. Monitoring was achieved with Ellman's test [47] and analytical HPLC (column ACE C18, 5 μm×4.6 mm, eluent A: 0.1% TFA/H$_2$O, eluent B: 60% CH$_3$CN/H$_2$O/0.1% TFA) with a 30 min linear gradient of 25% to 55% B at 1 mL flow rate (monitoring at 210 nm). When Ellman's tests are negative and the HPLC monitoring shows no more trace of starting materials (after

A

B

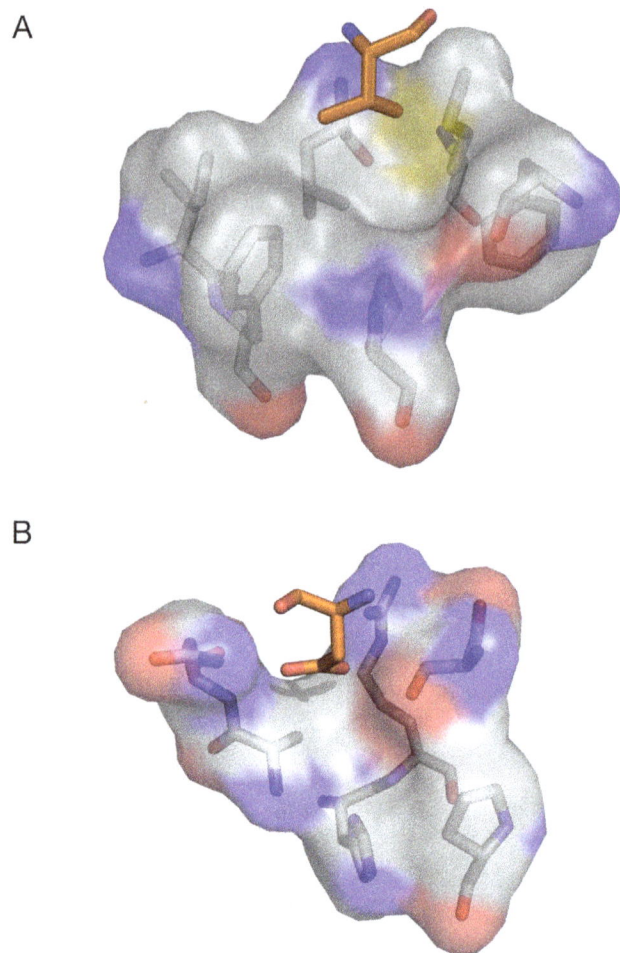

Figure 10. Residues forming the S1 and S4 pockets. The residue P4 (A) and P1 (B) of EETI-II are shown with an orange stick representation.

3 to 5 days), the reaction mixture was centrifuged and the supernatant loaded onto a preparative HPLC column (Merck Lichrospher C_{18}, 10 μm, 250×25 mm). Elution was achieved with a 90 min linear gradient of 25% to 55% B at 10 mL flow rate (monitoring at 220 nm). The fractions containing the oxidized peptide were combined and lyophilized to yield 15 to 30% of the desired peptide. Successful oxidation was confirmed by mass spectrometry (MALDI-Tof, Bruker Biflex III).

Production and purification of the PvSUB1 recombinant enzyme. The production and purification of the PvSUB1 (Genbank accession number FJ536585) recombinant enzyme was performed essentially as previously described [14,15]. Briefly for large-scale protein production, *Spodoptera frugiperda* Sf9 insect cells (1L at 3×106 cells/mL, Invitrogen) were infected for 72 h with recombinant PvSUB1-recombinant baculovirus at a Multiplicity Of Infection (MOI) of 10 in Insect XPRESS medium (Lonza) supplemented with 50 μg/mL gentamycin and 0.5 μg/

mL tunicamycin (Sigma-Aldrich). Culture supernatant containing the secreted PvSUB1 recombinant protein was harvested, centrifuged 30 min at 2150 g to remove cells and cellular debris and concentrated/diafiltrated against D-PBS 0.5 M NaCl, 5 mM Imidazole (loading buffer). The proteins were purified on an AKTA purifier system (GE Healthcare). The sample was loaded onto a 3 mL TALON Metal affinity resin (Clontech Laboratories) equilibrated in loading buffer. After extensive washes with loading buffer, the bound protein was eluted with a linear gradient of 5–200 mM imidazole in D-PBS, 0.5 M NaCl. Fractions containing PvSUB1 were pooled and concentrated by using Amicon Ultra 15 (10000 MWCO) and size-fractionated onto a HiLoad 16/60 Superdex 75 column equilibrated with 20 mM Tris pH 7.5, 100 mM NaCl. Fractions were monitored by absorbance (280 nm) and analyzed by Coomassie blue staining of SDS-PAGE gels and enzyme activity assay. The fractions containing the recombinant enzyme activity were pooled and the protein concentration was determined using the BCA Protein Assay following manufacturers recommendations (Bio Basic). Purified PvSUB1 recombinant protein was stored at −20°C following the addition of 30% v/v of pure Glycerol.

Enzymatic Test

For the kinetic assays we used the purified recombinant PvSUB1 enzyme and its specific peptide sub- strate whose sequence is deduced from PvSUB1 auto-maturation site: KLVGADDVSLA, with cleavage occurs between the two aspartates for PvSUB1. The KLVGADDVSLA sequence was coupled to the fluorophore/quencher dyes Dabsyl/Edans (Exc/Em 360/500 nm) at each edge. The enzymatic assays were performed in 20 mM Tris pH 7.5 and 25 mM $CaCl_2$ at 37°C as previously described [15]. For the determination of the Ki, the compounds, previously resuspended in ultra-pure distilled water at 10 mM, were tested at ten different concentrations ranging from 1 mM to 2 μM following sequential 1:2 dilutions. The final mixture was distributed in duplicate into a 384-well black microtiter plate (Thermo Scientific) and the fluorescence was monitored every 3 minutes for 90 min at 37°C in a Labsystems Fluoroskan Ascent spectro-fluorometer. The slope of the linear part of the kinetic was determined in an Excel (Microsoft) spreadsheet. Every steps of the enzymatic assay were done on ice to make sure that the protein was not active before the measure of the fluorescence. The Ki and IC50 values were determined (N = 3) using GraphPad Prism software.

Acknowledgments

We are grateful to Odile Puijalon for constant support and critical reading of the manuscript. This work is dedicated to the memory of our colleague Dung Le-Nguyen.

Author Contributions

Conceived and designed the experiments: GB DLN MN JCB. Performed the experiments: GB AB CN DLN. Analyzed the data: GB AB CN DLN MN JCB. Contributed reagents/materials/analysis tools: GB AB CN DLN MN JCB. Contributed to the writing of the manuscript: GB DLN MN JCB.

References

1. Greenwood BM, Targett GA (2011) Malaria vaccines and the new malaria agenda. Clin Microbiol Infect 17: 1600–1607.

2. Agnandji ST, Lell B, Soulanoudjingar SS, Fernandes JF, Abossolo BP, et al. (2011) First results of phase 3 trial of RTS,S/AS01 malaria vaccine in African children. N Engl J Med 365: 1863–1875.

3. Noedl H, Socheat D, Satimai W (2009) Artemisinin-resistant malaria in Asia. N Engl J Med 361: 540–541.

4. Alonso PL, Brown G, Arevalo-Herrera M, Binka F, Chitnis C, et al. (2011) A research agenda to underpin malaria eradication. PLoS Med 8: e1000406.

5. Mishra SK, Newton CR (2009) Diagnosis and management of the neurological complications of *falciparum* malaria. Nat Rev Neurol 5: 189–198.

6. Yeoh S, O'Donnell RA, Koussis K, Dluzewski AR, Ansell KH, et al. (2007) Subcellular discharge of a serine protease mediates release of invasive malaria parasites from host erythrocytes. Cell 131: 1072–1083.

7. Tawk L, Lacroix C, Gueirard P, Kent R, Gorgette O, et al. (2013) A Key Role for *Plasmodium* Subtilisin-like SUB1 Protease in Egress of Malaria Parasites from Host Hepatocytes. J Biol Chem 288: 33336–33346.

8. Keizer DW, Miles LA, Li F, Nair M, Anders RF, et al. (2003) Structures of phage-display peptides that bind to the malarial surface protein, apical membrane antigen 1, and block erythrocyte invasion. Biochemistry 42: 9915–9923.

9. Zhu S, Hudson TH, Kyle DE, Lin AJ (2002) Synthesis and in vitro studies of novel pyrimidinyl peptidomimetics as potential antimalarial therapeutic agents. J Med Chem 45: 3491–3496.

10. Dagan A, Efron L, Gaidukov L, Mor A, Ginsburg H (2002) In vitro antiplasmodium effects of dermaseptin S4 derivatives. Antimicrob Agents Chemother 46: 1059–1066.

11. Smith GP, Petrenko VA (1997) Phage Display. Chem Rev 97: 391–410.

12. He M, Taussig MJ (2002) Ribosome display: cell-free protein display technology. Brief Funct Genomic Proteomic 1: 204–212.

13. Roberts RW, Szostak JW (1997) RNA-peptide fusions for the in vitro selection of peptides and proteins. Proc Natl Acad Sci U S A 94: 12297–12302.

14. Bastianelli G, Bouillon A, Nguyen C, Crublet E, Petres S, et al. (2011) Computational reverse-engineering of a spider-venom derived peptide active against *Plasmodium falciparum* SUB1. PLoS One 6: e21812.

15. Bouillon A, Giganti D, Benedet C, Gorgette O, Petres S, et al. (2013) In Silico Screening on the Three-dimensional Model of the *Plasmodium vivax* SUB1 Protease Leads to the Validation of a Novel Anti-parasite Compound. J Biol Chem 288: 18561–18573.

16. Heitz A, Chiche L, Le-Nguyen D, Castro B (1989) 1 H 2D NMR and distance geometry study of the folding of *Ecballium elaterium* trypsin inhibitor, a member of the squash inhibitors family. Biochemistry 28: 2392–2398.

17. Craik DJ, Daly NL, Waine C (2001) The cystine knot motif in toxins and implications for drug design. Toxicon 39: 43–60.

18. Craik DJ, Clark RJ, Daly NL (2007) Potential therapeutic applications of the cyclotides and related cystine knot mini-proteins. Expert Opin Investig Drugs 16: 595–604.

19. Kolmar H (2009) Biological diversity and therapeutic potential of natural and engineered cystine knot miniproteins. Curr Opin Pharmacol 9: 608–614.

20. Le-Nguyen D, Mattras H, Coletti-Previero MA, Castro B (1989) Design and chemical synthesis of a 32 residues chimeric microprotein inhibiting both trypsin and carboxypeptidase A. Biochem Biophys Res Commun 162: 1425–1430.

21. Christmann A, Walter K, Wentzel A, Kratzner R, Kolmar H (1999) The cystine knot of a squash-type protease inhibitor as a structural scaffold for *Escherichia coli* cell surface display of conformationally constrained peptides. Protein Eng 12: 797–806.

22. Hilpert K, Wessner H, Schneider-Mergener J, Welfle K, Misselwitz R, et al. (2003) Design and characterization of a hybrid miniprotein that specifically inhibits porcine pancreatic elastase. J Biol Chem 278: 24986–24993.

23. Reiss S, Sieber M, Oberle V, Wentzel A, Spangenberg P, et al. (2006) Inhibition of platelet aggregation by grafting RGD and KGD sequences on the structural scaffold of small disulfide-rich proteins. Platelets 17: 153–157.

24. Souriau C, Chiche L, Irving R, Hudson P (2005) New binding specificities derived from Min-23, a small cystine-stabilized peptidic scaffold. Biochemistry 44: 7143–7155.

25. Wentzel A, Christmann A, Kratzner R, Kolmar H (1999) Sequence requirements of the GPNG beta-turn of the *Ecballium elaterium* trypsin inhibitor II explored by combinatorial library screening. J Biol Chem 274: 21037–21043.

26. Lippow SM, Wittrup KD, Tidor B (2007) Computational design of antibody-affinity improvement beyond in vivo maturation. Nat Biotechnol 25: 1171–1176.

27. Jha RK, Leaver-Fay A, Yin S, Wu Y, Butterfoss GL, et al. (2010) Computational design of a PAK1 binding protein. J Mol Biol 400: 257–270.

28. Chung SY, Subbiah S (1996) A structural explanation for the twilight zone of protein sequence homology. Structure 4: 1123–1127.

29. Otlewski J, Jelen F, Zakrzewska M, Oleksy A (2005) The many faces of protease-protein inhibitor interaction. EMBO J 24: 1303–1310.

30. Genheden S, Ryde U (2010) How to obtain statistically converged MM/GBSA results. J Comput Chem 31: 837–846.

31. Komiyama T, VanderLugt B, Fugere M, Day R, Kaufman RJ, et al. (2003) Optimization of protease-inhibitor interactions by randomizing adventitious contacts. Proc Natl Acad Sci U S A 100: 8205–8210.

32. Koussis K, Withers-Martinez C, Yeoh S, Child M, Hackett F, et al. (2009) A multifunctional serine protease primes the malaria parasite for red blood cell invasion. Embo J 28: 725–735.

33. Silmon de Monerri NC, Flynn HR, Campos MG, Hackett F, Koussis K, et al. (2011) Global identification of multiple substrates for *Plasmodium falciparum* SUB1, an essential malarial processing protease. Infect Immun 79: 1086–1097.

34. Fulle S, Withers-Martinez C, Blackman MJ, Morris GM, Finn PW (2013) Molecular Determinants of Binding to the *Plasmodium* Subtilisin-like Protease 1. J Chem Inf Model.

35. Baker NA, Sept D, Joseph S, Holst MJ, McCammon JA (2001) Electrostatics of nanosystems: application to microtubules and the ribosome. Proc Natl Acad Sci U S A 98: 10037–10041.

36. Berman HM, Westbrook J, Feng Z, Gilliland G, Bhat TN, et al. (2000) The Protein Data Bank. Nucleic Acids Res 28: 235–242.

37. Feig M, Karanicolas J, Brooks CL, 3rd (2004) MMTSB Tool Set: enhanced sampling and multiscale modeling methods for applications in structural biology. J Mol Graph Model 22: 377–395.

38. Case DA, Darden TA, Cheatham TEI, Simmerling CL, Wang J, et al. (2006) AMBER9. University of California, San Francisco.

39. Hornak V, Abel R, Okur A, Strockbine B, Roitberg A, et al. (2006) Comparison of multiple Amber force fields and development of improved protein backbone parameters. Proteins 65: 712–725.

40. Jorgensen WL, Chandrasekhar J, Madura JD (1983) Comparison of simple potential functions for simulating liquid water. J Chem Phys 79: 926.

41. Ryckaert J, Ciccotti G, Berendsen H (1977) Numerical integration of the cartesian equations of motion of a system with constraints: molecular dynamics of n-alkanes. J Comput Phys 23: 327–341.

42. Darden T, York D, Pederson L (1993) Particle mesh ewald-an nlog(n) method for ewald sums in large systems. J Chem Phys 98: 10089–10092.

43. Van Der Spoel D, Lindahl E, Hess B, Groenhof G, Mark AE, et al. (2005) GROMACS: fast, flexible, and free. J Comput Chem 26: 1701–1718.

44. Ritchie DW, Kemp GJ (2000) Protein docking using spherical polar Fourier correlations. Proteins 39: 178–194.

45. Gohlke H, Kiel C, Case DA (2003) Insights into protein-protein binding by binding free energy calculation and free energy decomposition for the Ras-Raf and Ras-RalGDS complexes. J Mol Biol 330: 891–913.

46. Tsui V, Case DA (2000) Theory and applications of the generalized Born solvation model in macromolecular simulations. Biopolymers 56: 275–291.

47. Ellman GL (1959) Tissue sulfhydryl groups. Arch Biochem Biophys 82: 70–77.

48. Sajid M, Withers-Martinez C, Blackman MJ (2000) Maturation and specificity of *Plasmodium falciparum* subtilisin-like protease-1, a malaria merozoite subtilisin-like serine protease. J Biol Chem 275: 631–641.

49. Debrabant A, Maes P, Delplace P, Dubremetz JF, Tartar A, et al. (1992) Intramolecular mapping of *Plasmodium falciparum* P126 proteolytic fragments by N-terminal amino acid sequencing. Molecular and Biochemical Parasitology 53: 89–96.

50. Stafford WH, Blackman MJ, Harris A, Shai S, Grainger M, et al. (1994) N-terminal amino acid sequence of the *Plasmodium falciparum* merozoite surface protein-1 polypeptides. Mol Biochem Parasitol 66: 157–160.

51. Heidrich HG, Miettinin-Bauman A, Eckerskorn C, Lottspeich F (1989) The N-terminal amino acid sequences of the *Plasmodium falciparum* (FCBI) merozoite surface antigen of 42 and 36 kilodaltons, both derived from the 185–195 kilodalton precursor Molecular and Biochemical Parasitology 34: 147–154.

52. O'Dea KP, McKean PG, Harris A, Brown KN (1995) Processing of the *Plasmodium chabaudi chabaudi* AS merozoite surface protein 1 in vivo and in vitro. Molecular and Biochemical Parasitology 72: 111–119.

Computational Insights into the Inhibitory Mechanism of Human AKT1 by an Orally Active Inhibitor, MK-2206

Mohd Rehan[1]*, Mohd A. Beg[1], Shadma Parveen[2], Ghazi A. Damanhouri[1], Galila F. Zaher[3]

1 King Fahd Medical Research Center, King Abdulaziz University, Jeddah, Kingdom of Saudi Arabia, **2** Bareilly College, M. J. P. Rohilkhand University, Bareilly, Uttar Pradesh, India, **3** Department of Haematology, Faculty of Medicine, King Abdulaziz University, Jeddah, Kingdom of Saudi Arabia

Abstract

The AKT signaling pathway has been identified as an important target for cancer therapy. Among small-molecule inhibitors of AKT that have shown tremendous potential in inhibiting cancer, MK-2206 is a highly potent, selective and orally active allosteric inhibitor. Promising preclinical anticancer results have led to entry of MK-2206 into Phase I/II clinical trials. Despite such importance, the exact binding mechanism and the molecular interactions of MK-2206 with human AKT are not available. The current study investigated the exact binding mode and the molecular interactions of MK-2206 with human AKT isoforms using molecular docking and (un)binding simulation analyses. The study also involved the docking analyses of the structural analogs of MK-2206 to AKT1 and proposed one as better inhibitor. The Dock was used for docking simulations of MK-2206 into the allosteric site of AKT isoforms. The Ligplot+ was used for analyses of polar and hydrophobic interactions between AKT isoforms and the ligands. The MoMa-LigPath web server was used to simulate the ligand (un)binding from the binding site to the surface of the protein. In the docking and (un)binding simulation analyses of MK-2206 with human AKT1, the Trp-80 was the key residue and showed highest decrease in the solvent accessibility, highest number of hydrophobic interactions, and the most consistent involvement in all (un)binding simulation phases. The number of molecular interactions identified and calculated binding energies and dissociation constants from the co-complex structures of these isoforms, clearly explained the varying affinity of MK-2206 towards these isoforms. The (un)binding simulation analyses identified various additional residues which despite being away from the binding site, play important role in initial binding of the ligand. Thus, the docking and (un)binding simulation analyses of MK-2206 with AKT isoforms and its structure analogs will provide a suitable model for studying drug-protein interaction and will help in designing better drugs.

Editor: Mohammad Saleem, Hormel Institute, University of Minnesota, United States of America

Funding: The authors have no funding or support to report.

Competing Interests: The authors have declared that no competing interests exist.

* Email: mrehan786@gmail.com

Introduction

The PI3K/AKT/mTOR signaling pathway is an important pathway for normal cellular functions in the human body and is the most commonly dysregulated pathway in cancer [1,2]. The AKT is one of the key proteins of this pathway belonging to the serine/threonine AGC protein kinase family and is also known as Protein Kinase B (PKB). The human AKT is found in three isoforms AKT1, 2, and 3, also known as PKB-α, -β and -γ and these isoforms are highly homologous multi-domain proteins possessing both common and distinct cellular functions [3,4]. The AKT is involved in several functions in the body such as metabolism, growth, proliferation, differentiation, and survival of the cells [5,6]. Conversely in regards to cancer, the constant activation and/or over-expression of AKT frequently contributes to the resistance to cancer chemotherapy or radiotherapy [7,8]. Recently, in vitro and in vivo studies with small molecule inhibitors of the AKT have been successful in attenuating chemotherapeutic resistance when combined with the standard chemotherapy [9,10]. Therefore, specific inhibition of AKT activity may be a good alternative approach to treat cancer and increase the efficacy of chemotherapy. In this regard, significant

efforts have been made to generate chemical compounds designed specifically to target AKT or other targets in the AKT signaling pathway and some of these compounds are in clinical trials for cancer treatment [2]. The majority of known AKT inhibitors are ATP competitive and have poor specificity against other closely related kinases. The increasing attention for AKT specific inhibitors or even AKT-isoform specific inhibitors led to the discovery of allosteric AKT inhibitors [11–13]. One such compound, MK-2206 (IUPAC name: 8-[4-(1-aminocyclobutyl)-phenyl]-9-phenyl-2H-[1,2,4]triazolo[3,4-f][1,6]naphthyridin-3-one), is a highly potent, selective, and orally active allosteric inhibitor of AKT which has been recently identified [14–16] and is effective at nanomolar concentration against purified recombinant human AKT1, 2, and 3 [17]. The compound is almost equally potent for human AKT1 and human AKT2 (IC_{50}, 5 nmol/L and 12 nmol/L, respectively) and is about five-fold less potent against human AKT3 (IC_{50}, 65 nmol/L) [17]. Various preclinical studies have demonstrated that MK-2206 effectively inhibited AKT and promoted cancer cell death when used alone or augmented the efficacy of several anti-cancer agents when used in combination [14–15,18–22]. The MK-2206 is orally active and has been shown to be safe in humans [23–24]. The preclinical results with this compound were highly successful and it is now in phase I/II clinical trials for treatments of solid tumors and acute myelogenous leukemia (http://clinicaltrials.gov/ct2/results?term=MK2206).

There is conclusive evidence that MK-2206 is a promising compound for inclusion in the standard cancer therapy protocols. Several studies have shown the MK-2206 mediated inhibition of the key cancer-regulatory-protein AKT, however, the exact binding mechanisms and the molecular interactions of MK-2206 with AKT have not been studied. Recently [25], a molecular dynamic simulation study of MK-2206 was performed to calculate the binding free energy at the binding site of AKT1. The modeling involved sketching of MK-2206 at the AKT1 binding site by modifying the structure of bound Inhibitor VIII in the co-complex structure. However, this model may not correctly predict the binding mode of MK-2206 as it is not a derivative of Inhibitor VIII. In this regard, the present study was proposed to investigate the structural and molecular details of MK-2206 binding against AKT1 using molecular docking and ligand (un)binding simulation approach. In the current study we used a molecular docking program which takes into account the shape complementarity, and van der Walls and Coulombic electrostatic energies at the binding site to identify the correct pose of MK-2206. We also validated the MK-2206 binding mode with the previously existing knowledge of key interacting residue in allosteric inhibition of AKT1. Therefore, we believe that our modeling provides better representation of the binding mode and gives more accurate prediction of molecular interaction of MK-2206 with AKT1. Further, the homology modeling and docking analyses with the AKT2 and AKT3 were also carried out to explain the varying affinity of MK-2206 towards these isoforms. Finally, we also performed docking study of structure analogs of MK-2206 to AKT1 and proposed an structural analog as a better inhibitor. This study will shed light on inhibitory mechanism of human AKT isoforms by MK-2206 and its structure analogs, and will help experimental biologist in testing and designing better inhibitors.

Materials and Methods

Data retrieval

The molecular structure of MK-2206 was retrieved from PubChem compound database (CID, 24964624). The 3-D structure of human AKT1 was obtained from Protein Data Bank (PDB, http://www.rcsb.org/; PDB ID: 3O96). This structure is a co-complex structure containing an allosteric inhibitor, Inhibitor VIII. This AKT1 structure (PDB ID: 3O96) was chosen as our study involved the docking of the allosteric inhibitor, MK-2206 to AKT1 and required a clue for allosteric site from the bound allosteric inhibitor. Further, the information of the bound inhibitor can be used to compare the binding of the docked inhibitor. The amino acid sequences for all three isoforms AKT1, AKT2, and AKT3 were obtained from UniProtKB/Swiss-Prot (http://www.uniprot.org/; IDs: P31749, P31751, Q9Y243, respectively).

Structural analogs of MK-2206

To retrieve structural analogs of MK-2206, a search was performed using option "Similar Compounds" in the PubChem. The "Similar Compound" search involves calculation of the Tanimoto coefficient which requires PubChem dictionary-based binary fingerprint (https://pubchem.ncbi.nlm.nih.gov/search/help_search.html). The fingerprint consists of series of chemical substructure "keys" and each key (in binary form) indicates the presence or absence of a particular substructure in a compound. Thus, the binary keys together form a "fingerprint" of a particular

chemical compound. The fingerprints do not take into consideration the variation in stereochemical or isotopic information. The Tanimoto coefficient measures the degree of similarity and a threshold value is set to retrieve the compounds similar to a query structure. A threshold of "100%" refers to "exact match" (ignoring stereo or isotopic information), whereas a threshold of "0%" would return all compounds present in PubChem database. The threshold utilized for "Similar Compounds" search of MK-2206 was the pre-programmed default (80%), which retrieved 45 similar compounds. These compounds further on filtering using Lipinski's Rule-of-five (for evaluating druglikeness of compounds) shortlisted to 33 compounds. On visual analyses, irrelevant and redundant structures were removed and, finally 30 of 33 structures were selected for further study.

Homology modeling of human AKT2 and AKT3

The 3-D structures of human AKT2 and AKT3 which were available in PDB were truncated versions. In order to model the full proteins, Modeller9v11 package [26] was used. The templates were identified using 'blastp' against PDB database. A close homologous structure of human AKT1 with PDB Id: 3O96 was identified for AKT2 (83% identity, 93% similarity) and AKT3 (83% identity, 90% similarity) both. This structure of human AKT1 (PDB Id: 3O96) is same which we selected for docking and (un)binding simulation analyses. In addition to the common template, the truncated AKT2 structure (partial kinase domain) covering 143–481 residues (PDB Id: 1MRY) and the truncated AKT3 structure (Ph domain) covering 1–118 residues (PDB Id: 2×18) were also considered while modeling AKT2 and AKT3 proteins respectively. A total of 100 three-dimensional models were generated and best 5 models were picked in each case. The selection of best 5 structure models out of 100 generated models was performed on the basis of lower value of the Modeller objective function or the DOPE assessment score and with the higher value of GA341 assessment score. To evaluate and select the single best model, steriochemical properties of the five best models were assessed using PROCHECK [27].

Molecular docking

Dock v.6.5 (University of California, San Francisco) was used for docking simulations of MK-2206 into the allosteric site of AKT1 [28]. The best docked conformation search strategy used was Random Conformation Search which utilizes the grid-based scoring functions of Coulombic and Lennard-Jones forces. Chimera v.1.6.2 [29] was used in the structure preparation of the protein and the ligand initially required by Dock and also in visualizing the structures at various stages of docking process.

Analyses of docked protein-ligand complex

To generate an illustration and analyze the whole protein-ligand complex, PyMOL v.1.3 was used [30]. For the polar and hydrophobic interactions between AKT1 and the ligand MK-2206, illustrations were generated and the analyses were performed by Ligplot+ v.1.4.3 program [31–32]. For further confirmation and calculating the extent of involvement of interacting residues obtained from Ligplot+, loss in ASA (Accessible Surface Area) was evaluated after the MK-2206 binding to AKT1. It is known that for a residue to be involved in interaction, it should lose more than 10 Å^2 ASA in the direction from unbound to the bound state [33]. The ASA calculations of unbound protein and the protein-ligand complex were performed by Naccess v.2.1.1 [34]. The loss in ASA, ΔASA of the ith residue in the direction from unbound to bound state was calculated using the expression:

Table 1. The human AKT1 residues interacting with MK-2206 are listed with the number of non-bonding contacts and the loss in Accessible Surface Area (ASA).

Interacting residues	No. of hydrophobic contacts	ΔASA (Å^2)
Asn-53	3	48.65[2]
Gln-59	1	16.44[7]
Leu-78	1	5.61[8]
Trp-80*	13	77.85[1]
Val-201	3	38.37[3]
Leu-264	1	19.16[5]
Val-270	2	35.8[4]
Tyr-272	1	16.61[6]

The ranking of residues on the basis of loss in solvent accessibility is indicated by superscripts with the value of ΔASA.
*The most common residue in all phases of (un)binding simulation.

$$\Delta\text{ASA}_i = \text{ASA}_i^{\text{Protein}} - \text{ASA}_i^{\text{Protein}-\text{ligand}}$$

In addition to the Dock score (Grid score) obtained from Dock v.6.5 [28], the binding energy and dissociation constants were also calculated using X-Score v.1.2.11 [35–36].

Protein-ligand (un)binding simulation

To simulate the ligand (un)binding from the binding site to the surface of the protein, a Molecular Motion Algorithms (MoMA) based web server, MoMa-LigPath (http://moma.laas.fr), was used [37–38]. The MoMa-LigPath takes into consideration the flexibility for the protein side-chains and the ligand and involves geometric constraints only. The program simulates how the ligand is driven to the binding site from the surface of the protein or from the binding site to the surface. The program also provides snapshots of molecular interactions bringing the ligand from the surface of the protein to the binding site. During the process of (un)binding simulation, the program also identifies the important residues of the target protein which despite being away from the binding site, still help in driving the ligand to the binding site of the protein.

Protein sequence alignment and analyses

The amino acid sequences of three AKT isoforms were aligned using Muscle v.3.8.31 [39], and further analyses and illustration were prepared by Jalview v.2.8 [40–41].

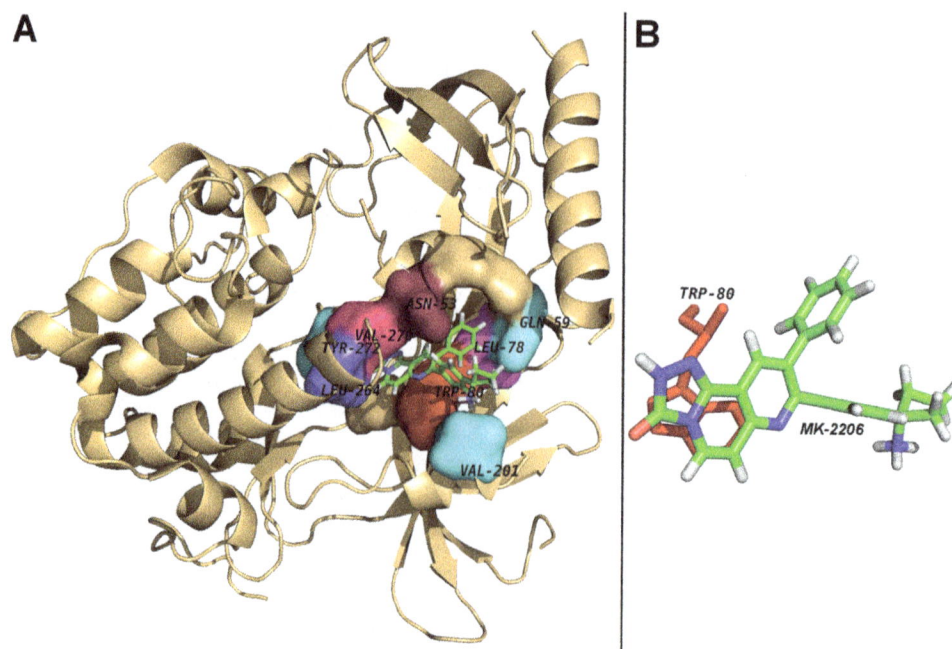

Figure 1. Molecular docking analyses of MK-2206 to the allosteric site of human AKT1. Panel A: Human AKT1 is illustrated in cartoon representation and MK-2206 is in stick representation. The interacting residues are labeled and are shown as surface in different colors. Panel B: The possible aromatic stacking interaction of the amino acid residue, Trp-80 through its indole group with naphthyridin moiety of MK-2206 is shown.

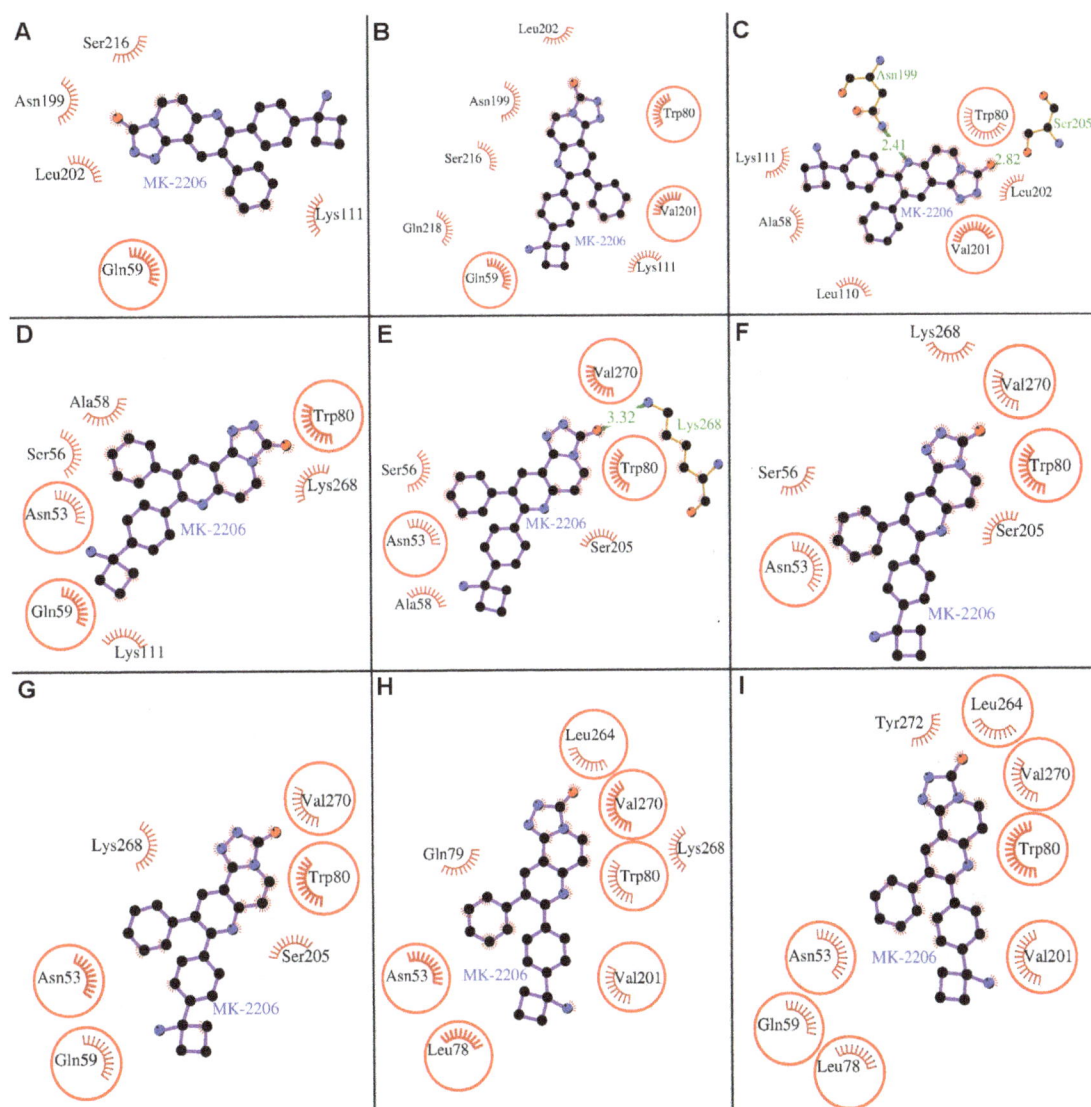

Figure 2. (Un)binding simulation analyses of MK-2206 binding to the allosteric site of human AKT1. Panels A–I: The (un)binding simulation phases of MK-2206; 'A' denotes farthest phase from the binding site, 'H' - the closest to the binding site, and 'I' - the binding site phase. The hydrogen bonds are shown as green-dashed lines with indicated bond length and the residues involved in hydrophobic interactions are shown as red arcs. The residues which are common to the last phase (F) are encircled.

Table 2. Procheck analyses for quality of structure models of AKT2 and AKT3 with the common template.

Protein	Ramachandran Plot Analyses				Labeled residues	
	Most favorable	Additional Allowed	Generously Allowed	Disallowed	All Ramachandrans	Chi1-Chi2
Template	87.0%	11.7%	0.6%	0.6%	12 (out of 355)	9 (out of 252)
AKT2	89.0%	9.0%	1.4%	0.6%	15 (out of 389)	8 (out of 263)
AKT3	88.8%	9.5%	1.4%	0.3%	18 (out of 391)	3 (out of 266)

Ramachandran plot analyses showing the percentage of residues lying in each of the four different regions. In disallowed region, the number of residues is also given in parantheses with percentage. The number of labeled residues in all Ramachandrans and Chi1–Chi2 are also given in the table.

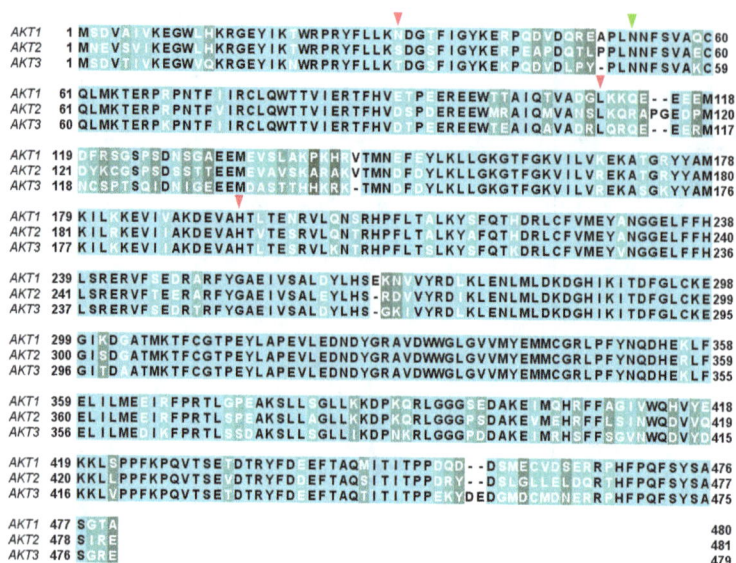

Figure 3. Multiple Sequence Alignment of the three AKT isoforms, AKT1, AKT2, and AKT3. The conserved positions are shown in light green and the corresponding amino acids in black font, whereas the less-conserved positions are shown in gray color with the corresponding amino acids in white font. The initial and final position of each isoform in all the rows of the alignment is also provided. The position-equivalent-residues (residues of different isoforms falling at same column position in the isoform alignment) overlapping among the interacting residues of MK-2206 are marked by triangles; the green triangle (Asn-53) indicates the residue overlapping between AKT1 and AKT2 binding, while the red triangles indicate the position-equivalent-residues overlapping among the interacting residues of AKT2 and AKT3 which are Ser-31, Leu-110, and His-196 of AKT2 (corresponding to Thr-31, Leu-109, and His-192 of AKT3 respectively).

Results and Discussion

Molecular docking analyses of MK-2206

The docking analyses of MK-2206 revealed that the compound packed against the residues Asn-53, Gln-59, Leu-78, Trp-80, Val-201, Leu-264, Val-270, and Tyr-272 of AKT1 and was stabilized by the hydrophobic interactions (Fig. 1A). The Dock score was negative with high absolute value and number of hydrophobic interactions that kept MK-2206 bound in the cavity was also reasonably high (25 interactions from 8 different residues, Table 1). All identified MK-2206 interacting residues of AKT1 with the loss in solvent accessibility and the total number of hydrophobic interactions are listed in Table 1. The higher the loss in solvent accessibility for a residue in the direction from unbound to the bound state, the more involved is the residue in the ligand binding [33]. The importance of AKT1 residues for MK-2206 binding were also ranked on the basis of loss in solvent accessibility (Table 1). The Trp-80 was identified as the key residue of AKT1 and was involved in the majority of hydrophobic interactions and showed highest decrease in its solvent accessibility after MK-2206 binding (approx. 77.85 Å^2) as shown (Table 1). The Trp-80 was also the most common residue through all phases of MK-2206 (un)binding simulation (Fig. 2), demonstrating its importance in initial binding and finally bringing the drug into the active site of AKT1. Furthermore, Trp-80 also seemed to make aromatic stacking interactions between the indole group and naphthyridin moiety of MK-2206 (Fig. 1B). All the findings of Trp-80 as key interacting residue validates MK-2206 binding mode as these findings were consistent with a previous study [42] in which it is shown that the inhibition of AKT1 by Akti (an allosteric inhibitor of AKT) is critically dependent upon a solvent-exposed tryptophan residue (Trp-80) present in all three AKT isoforms and whose mutation to alanine yields an Akti-resistant kinase.

Human AKT isoforms and comparison of interacting residues for MK-2206

All the three human AKT isoforms showed a high degree of amino acid sequence homology (Fig. 3). The calculated percentage identity of AKT1 with AKT2 and AKT3 was 81.12% and 82.37% respectively. The 3-D structure models of AKT2 and AKT3 were generated as described in Materials and Methods section. On evaluation by Ramachandran plot, the resultant models (Fig. 4) revealed that there are few residues in generously allowed and disallowed regions (Table 2). All the criteria including the high percentage of residues in allowed regions of the Ramachandran plot, DOPE energy profile comparison with the template, and less numbers of labeled residues (unfavorable conformation) deduced from Ramachandran (Fig. 4, Table 2) and Chi1-Chi2 plots provide confidence to the models. In essence, we found that the models generated were of good quality and can effectively be used for further studies.

When MK-2206 was docked to the structure models of AKT2 and AKT3, we found its binding mode was different in different isoforms. This may be observed because of change in conformation induced by variation in amino acids for these isoforms as shown in the isoform alignment (Fig. 3). However, there were some position-equivalent-residues (residues of different isoforms falling at same column position in the isoform alignment) among the interacting residues of these isoforms which were overlapping (Fig. 3). As shown in Fig. 5, the interacting residues of the isoforms AKT1 and AKT2 shared only one residue Asn-53 as common. Whereas, in case of AKT2 and AKT3, three position-equivalent-residues (residues of the isoforms falling at same column position in the isoform alignment, Fig. 3) were common viz. Ser-31, Leu-110, and His-196 of AKT2 (corresponding to Thr-31, Leu-109, and His-192 of AKT3 respectively) as displayed in Fig. 5. This showed AKT1 shared binding site with AKT2 but not with AKT3 within the allosteric site and AKT3 shared binding site with AKT2 but

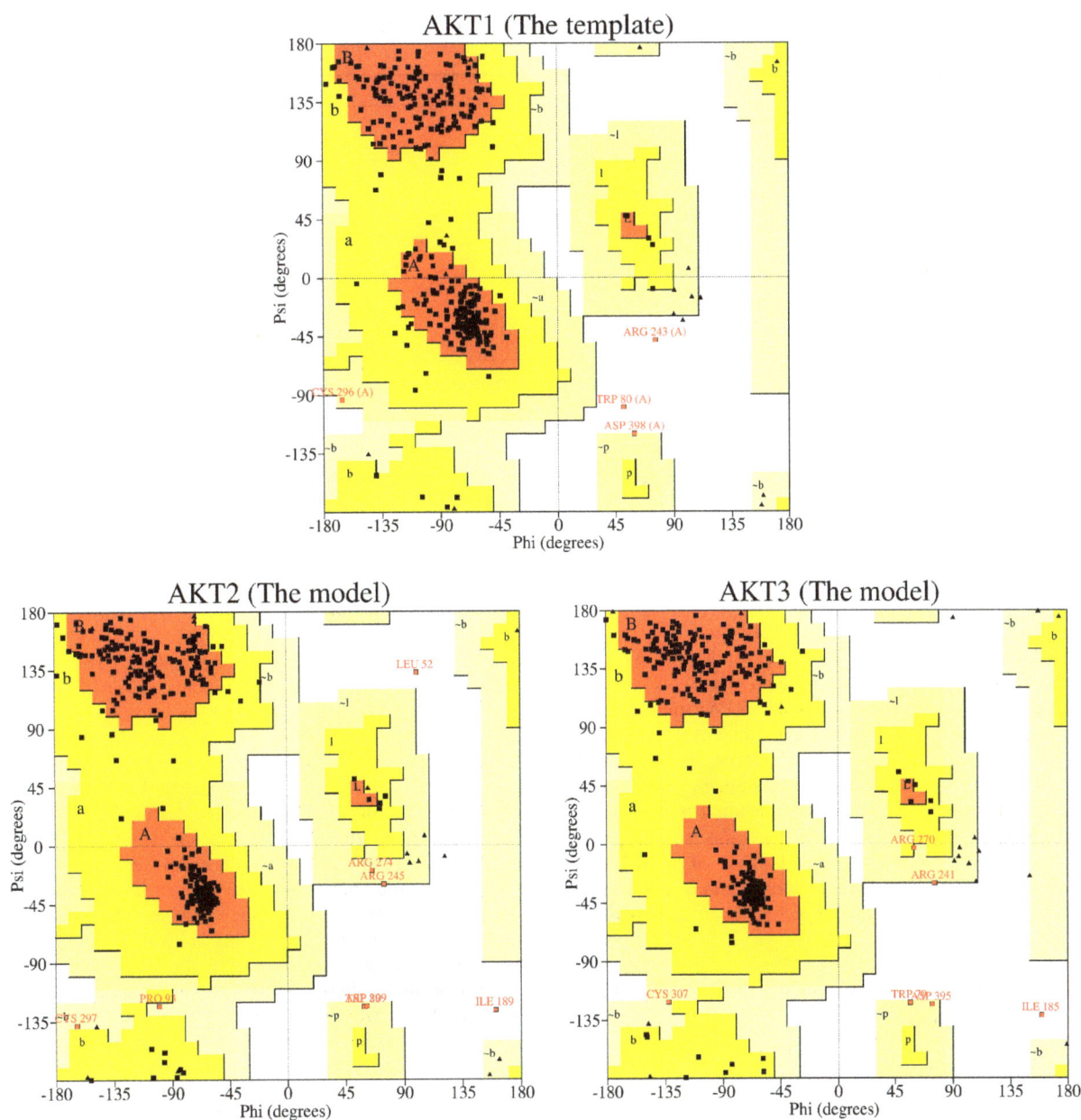

Figure 4. Ramachandran plot showing the residues as square dots lying in the four different regions, most favorable, additional allowed, generously allowed, and disallowed regions.

Table 3. The human AKT2 residues interacting with MK-2206 are listed with the number of non-bonding contacts and the loss in Accessible Surface Area (ASA).

Interacting residues	No. of hydrophobic contacts	ΔASA (Å^2)
Ser-31	2	25.89[5]
Asp-32	12	57.88[2]
Asn-53	2	61.70[1]
Ser-56	2	17.52[6]
Leu-110	4	39.73[4]
His-196	2	47.72[3]

The ranking of residues on the basis of loss in solvent accessibility is indicated by superscripts with the value of ΔASA.

Table 4. The human AKT3 residues interacting with MK-2206 are listed with the number of non-bonding contacts and the loss in Accessible Surface Area (ASA).

Interacting residues	No. of hydrophobic contacts	ΔASA (Å^2)
Met-1	1	23.72[3]
Thr-31	1	16.91[6]
Leu-109	1	11.77[7]
His-192	2	23.41[5]
Thr-193	2	23.54[4]
Leu-194	2	29.20[2]
Arg-198 (H-bond)	2	33.25[1]

The ranking of residues on the basis of loss in solvent accessibility is indicated by superscripts with the value of ΔASA. The residue forming hydrogen-bond is indicated with the residue name in parentheses.

not with AKT1. To sum up, the isoform AKT2 is sharing binding site with AKT1 and AKT2 both but it is doing so through different overlapping position-equivalent-residue pairs. When we looked at the molecular-interactions and interacting residues of these isoforms, we found that the number of hydrophobic interactions of AKT1 and AKT2 was similar (25 interactions, AKT1; 24 interactions, AKT2) but it decreased to a higher degree for AKT3 (11 interactions) as shown in Table 3–4. We also calculated binding energies and dissociation constants for the co-complex structures of these isoforms (Table 5). We found that the binding energy order was AKT1 at highest, then AKT2, followed by AKT3 with greater difference. Furthermore, the dissociation constant of AKT1 was also slightly more than that of AKT2 but it was ten times of AKT3. These findings of the number of molecular interactions, the binding energy, and the dissociation constant were corroborating with one another and in agreement with what is reported in literature [17] that the binding affinity of MK-2206 is less for AKT2 with respect to that of AKT1 but decreased to a higher degree for AKT3.

Comparison between binding mode of MK-2206 and AKT inhibitor, inhibitor VIII

The human AKT1 structure chosen for docking analyses (PDB ID: 3O96) is available in PDB as co-complex structure crystallized with an allosteric inhibitor, inhibitor VIII [13]. In order to determine the difference between the binding mode of MK-2206 to AKT1 from that of inhibitor VIII, a comparative analyses was performed. It was found that the common interacting residues for both the ligands were Trp-80, Tyr-272 and Leu-264 (Fig. 6A–B). With respect to these common three interacting residues, inhibitor VIII binding site is towards the region encompassing residues Ile-84, Glu-85, Val-183, Thr-211, Arg-273, Asp-274, Asp-292, and Cys-296 whereas MK-2206 binding site is in opposite direction, the region involving the residues Asn-53, Glu-59, Leu-78, Val-201,

and Val-270 (Fig. 6A–B). Alternatively, with respect to the terminal three-ring moiety of the ligands localized at the same small region in the allosteric site, their orientations are in opposite direction to each other (Fig. 6C–D).

(Un)binding simulation analyses of MK-2206

The docked complex of AKT1 with MK-2206 was subjected to (un)binding simulation using MoMA-LigPath. The (un)binding simulation analyses of MK-2206 binding provided snapshots of varying molecular interactions with respect to decreasing distance from the binding site (Fig. 2A–I). While describing the (un)binding simulation analyses, we introduced two terminologies, 'Common residues' and 'Additional residues' to describe the two kinds of residues playing role in different phases of (un)binding simulation. The Common residues are the residues which are overlapping with the identified interacting residues of AKT1 (Table 1) whereas, the Additional residues are the residues which play role in binding at a certain phase of (un)binding simulation but they are not part of listed interacting residues in Table 1. The (un)binding simulation analyses of MK-2206 is briefly summarized as follows: In the first phase, Phase A (Fig. 2A), in addition to a Common residue, Gln-59, which is one of the identified listed interacting residues, there were Additional residues viz Ser-216, Asn-199, Leu-202, and Lys-111 that played role in initial binding of the ligand to the surface of the protein. In Phase B (Fig. 2B), another residue, Gln-218, was also involved as an Additional residue besides the Common residues, Trp-80 and Val-201, that interacted with the ligand. In Phase C (Fig. 2C), the ligand was bound by both the hydrogen bonds and hydrophobic interactions. The hydrogen bonds were formed by two Additional residues Asn-199 and Ser-205, whereas, the hydrophobic interactions were exerted by both Common residues, Trp-80 and Val-201, and Additional residues, Lys-111, Leu-202, Ala-58, and Leu-110. In Phase D (Fig. 2D), the Additional residues Leu-110, Leu-202 and, those forming hydrogen bonds (Asn-199, Ser-205) and one

Table 5. The binding strength of MK-2206 to the three AKT isoforms given by various scores are listed in the table.

AKT isoform	Binding energy	pK$_d$ or −log(K$_d$)	Dock score
AKT1	−8.83	6.47	−26.55
AKT2	−8.29	6.07	−37.05
AKT3	−7.48	5.48	−25.00

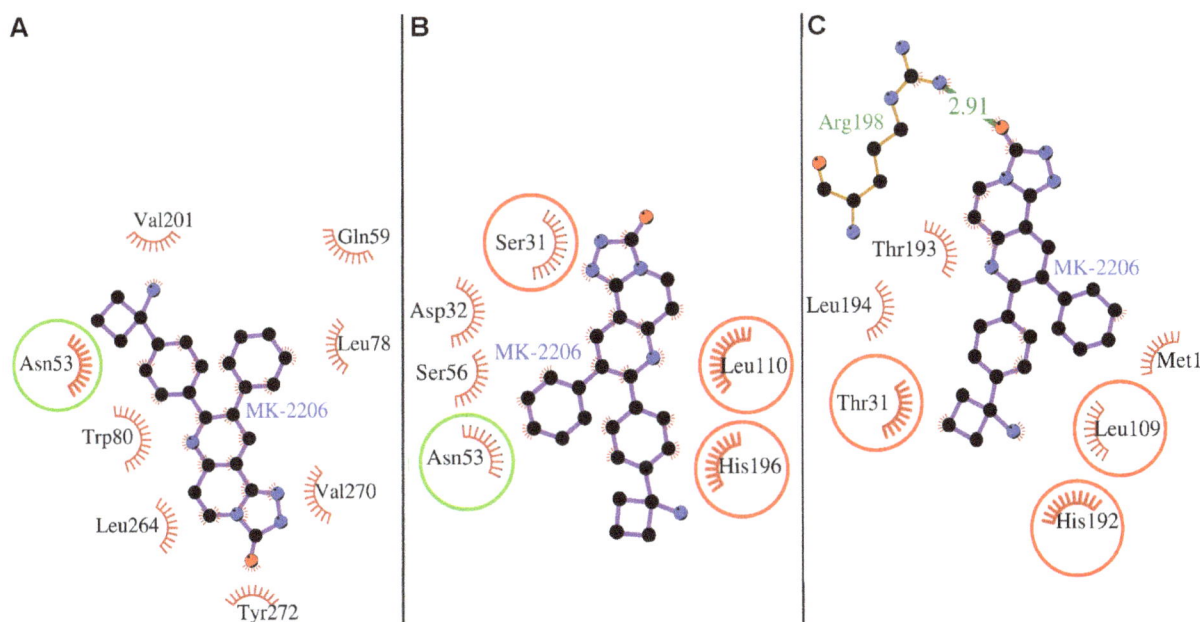

Figure 5. Comparison of MK-2206 binding and interacting residues of all three isoforms of AKT. Only one residue Asn-53 was shared among interacting residues of AKT1 and AKT2, and shown as encircled in green. Three position-equivalent-residues (residues of different isoforms falling at same column position in the isoform alignment) were shared among interacting residues of AKT2 and AKT3, and shown as encircled in red.

Common residue Val-201 disappeared. Whereas two Additional residue (Ser-56, Lys-268) and two Common residues (Asn-53, Glu-59) appeared forming interactions. It can be observed as the ligand is approaching towards the binding site, the number of Common residues increases with the decrease in the number of Additional residues. In the Phase E (Fig. 2E), an Additional residue, Lys-268, which was present in Phase D also, formed a hydrogen-bond with the ligand. The Common residue Gln-59 is replaced by Val-270 and the Additional residue Lys-111 is replaced by Ser-205. However, the number of Additional and Common residues remained the same at this phase similar to Phase D. In Phase F (Fig. 2F), the hydrogen bond disappeared but the Additional residue forming this hydrogen bond, Lys-268 was still there forming hydrophobic interactions in this phase. Other residues remained same except the Additional residue Ala-58 which disappeared at this phase. In Phase G (Fig. 2G), one Common residue Glu-59 appeared with the disappearance of one Additional residue Ser-56. In Phase H (Fig. 2H), the Common residues appeared were Leu-78, Val-201, and, Leu-264 with the disappearance of one Common residue Glu-59. In Additional residues, Gln-79 appeared with the disappearance of Ser-205. The number of Additional residues remained same but the number of Common residues increased. Finally, in the binding site phase, Phase I (Fig. 2I), all the Additional residues disappeared. In the Common residues, Gln-59 which was present at Phase G reappeared and another Common residue Tyr-272 appeared completing the quorum of the Common residues.

Molecular docking study of structural analogs of MK-2206 to AKT1

The careful analyzes of all the 30 structural analogs for common scaffold led us to devise rules to classify the compounds in three groups. We found that the compounds are derivatives of common scaffold with varying R_1 and R_2 group as shown in Fig. 7. These compounds were, therefore, classified in three groups as R_1-, R_2- and R_1R_2- structural analogs of MK-2206 depending on the

substitutions made at R_1- or R_2- or both on the common scaffold. The docking of all these structural analogs to the allosteric site of AKT1 was carried out. The ligand-interaction plots of all the 30 structures grouped according to R_1-, R_2- and R_1R_2- classes are provided as Files S1–S3. Except for one belonging to R_1-class which has no overlapping interacting residues with MK-2206 i.e. binding to different site, all the structure analogs have Trp-80 as common interacting residues. This may gives an idea about the binding of the common scaffold and it also underscores the importance of Trp-80 in the scaffold binding. In case of R_1-class, despite Trp-80, the other residue Asn-53 was also common. Whereas, in R_2-class, in addition to Trp-80 and Asn-53 which were common in R_1-class, another residue Val-270 was also found common. Finally, in R_1R_2-class, Trp-80 was the only common residue whereas Asn-53 and Val-270 including Tyr-272 were frequently appearing as common residues among many structural analogs. The dock score is directly obtained from Dock v.6.5 [28], whereas binding energy and pK_d (for dissociation constant) were calculated using X-Score v.1.2.11 [35–36]. All these scores for all the structural analogs are tabulated in Table S1. With few exceptions, the dock scores are not varying much with respect to that of MK-2206. In order to identify the better inhibitor than the MK-2206, we found a compound with CID 67256123 as the best binder with highest binding energy and pK_d among the structural analogs of MK-2206. We believe this R_2-structural analog can prove to be better inhibitor than MK-2206 provided it qualifies the low toxicity and better oral-availability criteria in vivo.

Comparison between binding mode of MK-2206 and a proposed better inhibitor, an MK-2206 analog

This R_2-structural analog of MK-2206 is having cyclohexa-2,4-dien-1-one group on R_2 instead of mere ketonic oxygen in MK-2206 (Fig. 8A–B). The common interacting residues between the two ligands were Asn-53, Trp-80, Leu-264, Val-270, and Tyr-272 (Fig. 8C–D). The two ligands bind in the same location within the allosteric site, however, the orientations of the analogs were

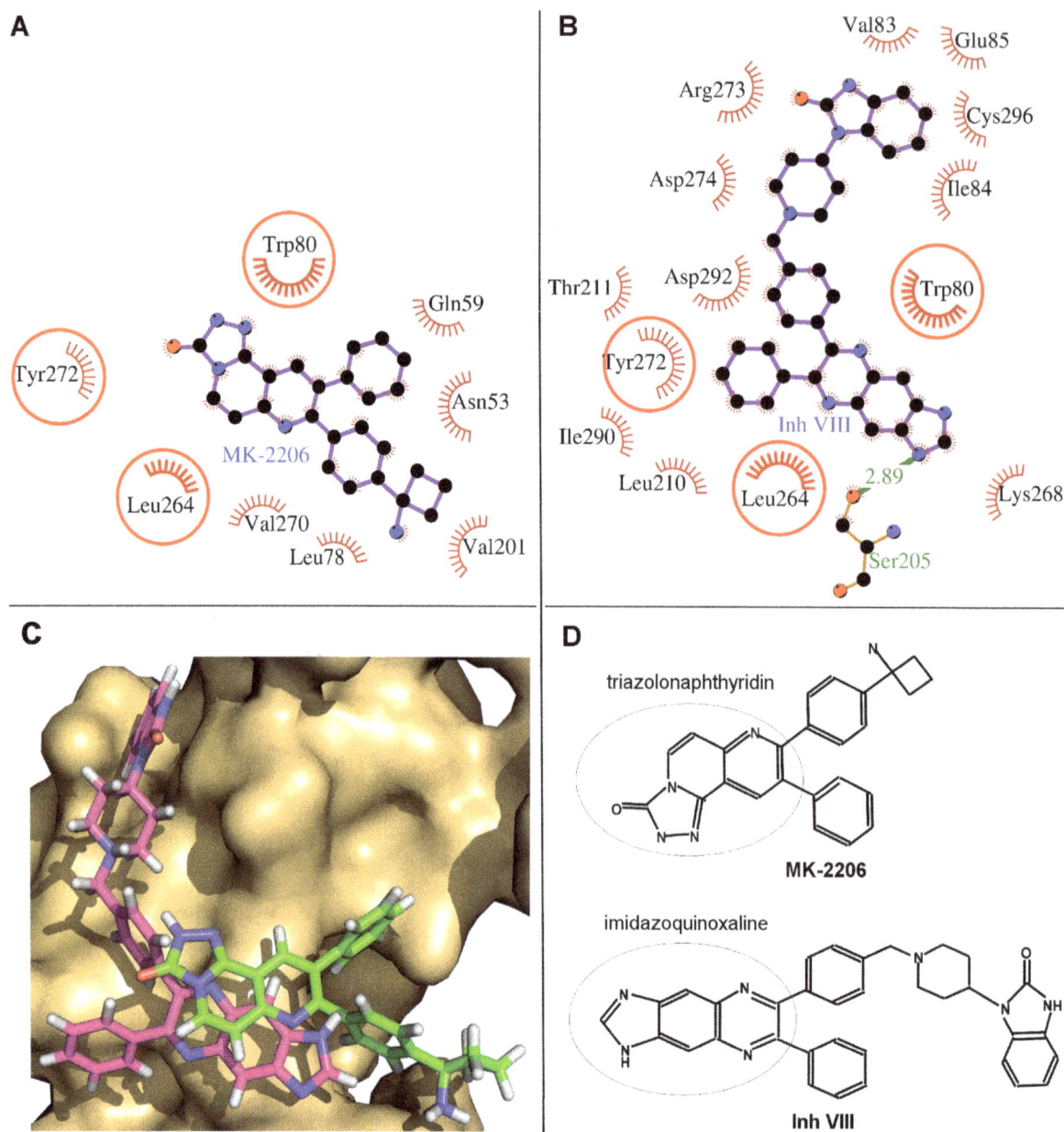

Figure 6. Comparative binding analyses of MK-2206 and inhibitor VIII. Panels A–B: The binding of MK-2206 and inhibitor VIII are displayed. The hydrogen bond is shown as green-dashed line with indicated bond length and the residues involved in hydrophobic interactions are shown as red arcs. The interacting residues which are common for both the ligands are encircled. Panel C: The exact orientation of binding for both the ligands in the binding site of the protein is shown. Panel D: Schematic structure of MK-2206 and inhibitor VIII are shown. The three ring moieties of both the molecules are encircled.

inverted. In case of MK-2206, the two phenyl groups protruding from three ring-structure triazolonaphthyridin moiety face towards the surface of the protein within the allosteric site and not involved in molecular interactions with many residues. Whereas, in case of the MK-2206 analog, the two protruding phenyl groups from triazolonaphthyridin moiety face towards deep inside the cavity and involved in multiple interactions and thus provide better fit than the original drug MK-2206. This is also evident from the total number of molecular interactions in MK-2206 analog (40 interactions out of 10 residues, Table 6) which drastically increased from that of MK-2206 (25 interactions out of 8 residues, Table 1). Although the dock scores of both the drugs were similar

but the binding energy of MK-2206 analog was higher than that of MK-2206 and the dissociation constant of analog was approximately 10 times of the drug MK-2206. All these findings suggest that the MK-2206 analog is proposed to be a better inhibitor than the original drug MK-2206.

Conclusions

The present study used docking and (un)binding simulation analyses to identify MK-2206 interacting residues of human AKT isoforms. The MK-2206 is an allosteric inhibitor of AKT1 and exerts its inhibitory mechanism by binding to the allosteric site of AKT1 and engaging the functionally important residues in various

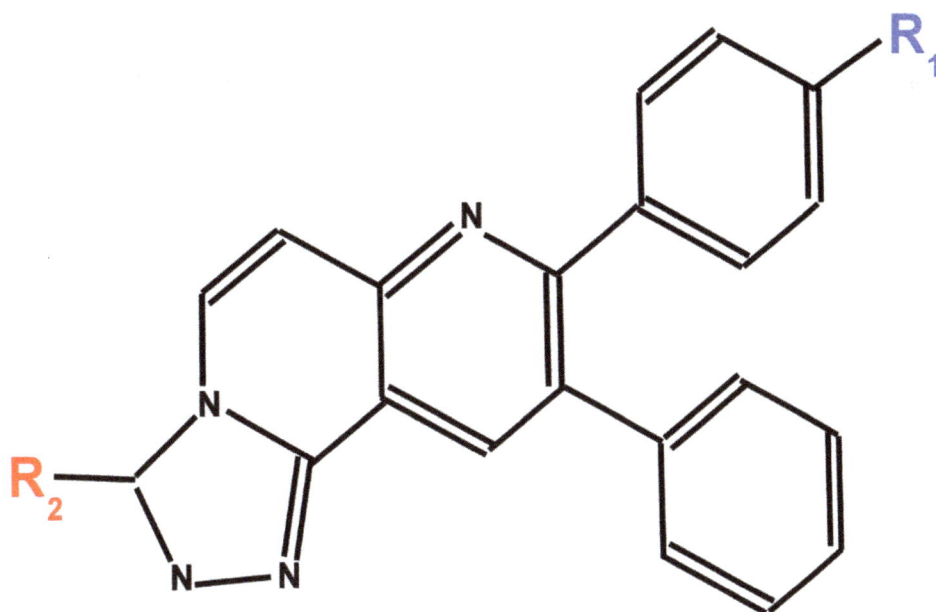

Figure 7. Structural analogs of MK-2206 can be derived by varying R1 and R2 group on common scaffold according to devised rules on visual analyses. For the drug MK-2206, R_1 is 1-amino cyclo-butyl group and R_2 is ketonic oxygen.

interactions. The exact binding mode of MK-2206 based on computational approach is presented and various interacting residues within the allosteric site of this protein were identified and characterized. The quality of docking was assured by the negative dock score with high absolute value and the identified various molecular interactions between the protein and the ligand. Additionally, the extent of involvement of the residues in ligand binding was calculated by ASA analyses and the residues were ranked on the basis of ΔASA score. In the docking and (un)binding simulation analyses, the Trp-80 was the key residue among various important identified residues, and showed highest decrease in the solvent accessibility, highest number of hydrophobic interactions, and the most consistent involvement in all (un)binding simulation phases. The AKT1 residues interacting with MK-2206 were also compared with those of other AKT isoforms. The lowered binding affinity of AKT3 to MK-2206 is attributed to decreased number of molecular interactions and lowered calculated- binding energy and dissociation constants. The (un)binding simulation analyses identified various Additional residues which despite being away from the binding site play important role in initial binding of the ligand and its recruitment to the binding site of the AKT1 protein. The molecular docking analyses of MK-2206 structural analogs identified one structural analog proposed as better inhibitor of AKT1 than MK-2206. Thus, the aforementioned docking and (un)binding analyses provide the structural insights into the binding mechanism of MK-2206 to the isoforms of key cancer signaling protein, AKT. The docked MK-2206–protein confor-mation is expected to serve as a suitable model for understanding

Table 6. The AKT1 residues interacting with MK-2206 analog are listed with the number of non-bonding contacts and the loss in Accessible Surface Area (ASA).

Interacting residues	No. of hydrophobic contacts	ΔASA (Å^2)
Asn-53	6	51.38[2]
Asn-54	2	13.83[9]
Gln-79	9	37.29[4]
Trp-80	7	66.43[1]
Leu-264	1	14.21[8]
Val-270	5	37.86[3]
Val-271	2	6.11[10]
Tyr-272	4	36.18[5]
Arg-273	1	31.28[6]
Asp-292	3	15.67[7]

The ranking of residues on the basis of loss in solvent accessibility is indicated by superscripts with the value of ΔASA.

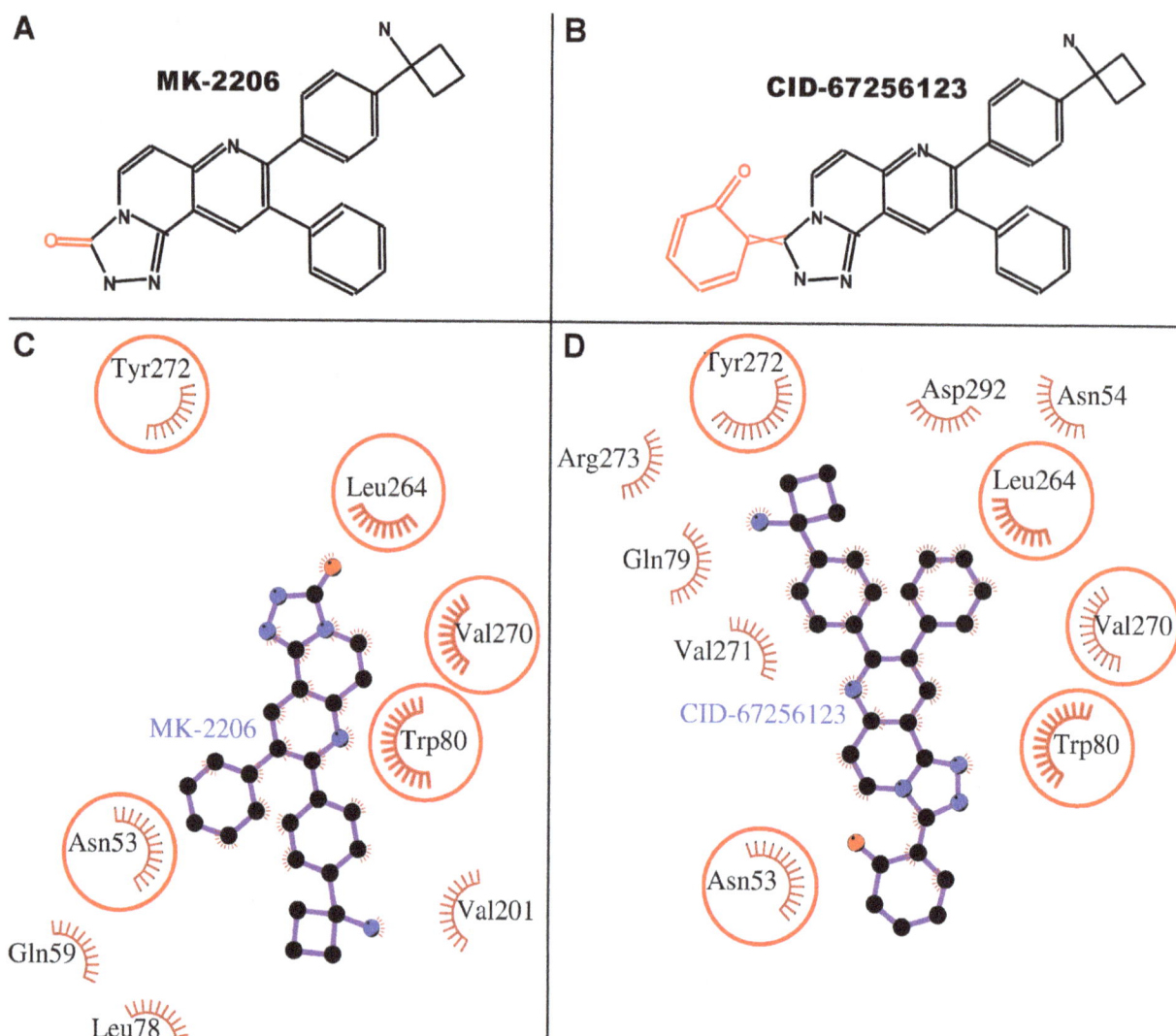

Figure 8. Comparative binding analyses of MK-2206 and its selected analog. Panels A–B: Schematic structure of MK-2206 and its analog are shown. The only difference between the compounds is change in R_2 group, shown in red on black compound-scaffold. Panel C–D: The binding of MK-2206 and its selected analog are displayed. The residues involved in hydrophobic interactions are shown as red arcs. The interacting residues which are common for both the ligands are encircled.

the drug protein interplay or more specifically the amino-acid environment mediating molecular-interactions and thus, providing electrostatic and surface complementary details for the inhibitory mechanism.

Supporting Information

File S1 R1-analogs of MK-2206.

File S2 R2-analogs of MK-2206.

File S3 R1R2-analogs of MK-2206.

Table S1 The binding strength of structural analogs of MK-2206 to human AKT1 given by various scores are listed in the table. These are classified as R1-, R2- and

R1R2- structural analogs of MK-2206 as described in Materials and Method section.

Acknowledgments

The authors are thankful to Dr. J. Cortés for providing stand-alone version of MoMA-LigPath and for his valuable suggestions. Thanks are also due to M. S. Gazdar, head of library, KFMRC for easy access of books and journals.

Author Contributions

Conceived and designed the experiments: MR SP. Performed the experiments: MR SP. Analyzed the data: MR SP. Contributed reagents/materials/analysis tools: MR SP. Contributed to the writing of the manuscript: MR MAB SP GAD GFZ.

References

1. Martelli AM, Evangelisti C, Chappell W, Abrams SL, Basecke J, et al. (2011) Targeting the translational apparatus to improve leukemia therapy: roles of the PI3K/PTEN/Akt/mTOR pathway. Leukemia 25: 1064–1079.
2. Liu P, Cheng H, Roberts TM, Zhao JJ (2009) Targeting the phosphoinositide 3-kinase pathway in cancer. Nat Rev Drug Discov 8: 627–644.
3. Kannan N, Haste N, Taylor SS, Neuwald AF (2007) The hallmark of AGC kinase functional divergence is its C-terminal tail, a cis-acting regulatory module. Proc Natl Acad Sci USA 104: 1272–1277.
4. Yang J, Cron P, Thompson V, Good VM, Hess D, et al. (2002) Molecular mechanism for the regulation of protein kinase B/Akt by hydrophobic motif phosphorylation. Mol Cell 9: 1227–1240.
5. Brazil DP, Yang ZZ, Hemmings BA (2004) Advances in protein kinase B signalling: AKTion on multiple fronts. Trends Biochem Sci 29: 233–242.
6. Carnero A (2010) The PKB/AKT pathway in cancer. Curr Pharm Des 16: 34–44.
7. Winograd-Katz SE, Levitzki A (2006) Cisplatin induces PKB/Akt activation and p38 (MAPK) phosphorylation of the EGF receptor. Oncogene 25: 7381–7390.
8. Rao E, Jiang C, Ji M, Huang X, Iqbal J, et al. (2012) The miRNA-17~92 cluster mediates chemoresistance and enhances tumor growth in mantle cell lymphoma via PI3K/AKT pathway activation. Leukemia 26: 1064–1072.
9. Wang P, Zhang L, Hao Q, Zhao G (2011) Developments in selective small molecule ATP-targeting the serine/threonine kinase Akt/PKB. Mini Rev Med Chem 11: 1093–1107.
10. Polak R, Buitenhuis M (2012) The PI3K/PKB signaling module as key regulator of hematopoiesis: implications for therapeutic strategies in leukemia. Blood 119: 911–923.
11. Barnett SF, Defeo-Jones D, Fu S, Hancock PJ, Haskell KM, et al. (2005) Identification and characterization of pleckstrin-homology-domain-dependent and isoenzyme-specific Akt inhibitors. Biochem J 385: 399–408.
12. Lindsley CW, Zhao Z, Leister WH, Robinson RG, Barnett SF, et al. (2005) Allosteric Akt (PKB) inhibitors: discovery and SAR of isozyme selective inhibitors. Bioorg Med Chem Lett 15: 761–764.
13. Wu W-I, Voegtli WC, Sturgis HL, Dizon FP, Vigers GPA, et al. (2010) Crystal Structure of Human AKT1 with an Allosteric Inhibitor Reveals a New Mode of Kinase Inhibition. PLoS ONE 5: e12913.
14. Hirai H, Sootome H, Nakatsuru Y, Miyama K, Taguchi S, et al. (2010) MK-2206, an allosteric Akt inhibitor, enhances antitumor efficacy by standard chemotherapeutic agents or molecular targeted drugs in vitro and in vivo. Mol Cancer Ther 9: 1956–1967.
15. Cheng Y, Zhang Y, Zhang L, Ren X, Huber-Keener KJ, et al. (2011) MK-2206, a novel allosteric inhibitor of Akt, synergizes with gefitinib against malignant glioma via modulating both autophagy and apoptosis. Mol Cancer Ther 11: 154–164.
16. Tan S, Ng Y, James DE (2011) Next-generation Akt inhibitors provide greater specificity: effects on glucose metabolism in adipocytes. Biochem J 435: 539–544.
17. Yan L (2009) MK-2206: a potent oral allosteric AKT inhibitor. Proceedings of the 100th Annual Meeting of the American Association for Cancer Research (April 18–22, 2009, Denver, CO). Abstract Number: DDT01-1.
18. Simioni C, Neri LM, Tabellini G, Ricci F, Bressanin D, et al. (2012) Cytotoxic activity of the novel Akt inhibitor, MK-2206, in T-cell acute lymphoblastic leukemia. Leukemia 26: 2336–2342.
19. Pant A, Irene IL, Lu Z, Rueda BR, Schink J, Kim JJ (2012) Inhibition of AKT with the Orally Active Allosteric AKT Inhibitor, MK-2206, Sensitizes Endometrial Cancer Cells to Progestin. PLoS ONE 7: e41593.
20. Knowles JA, Golden B, Yan L, Carroll WR, Helman EE, et al. (2011) Disruption of the AKT pathway inhibits metastasis in an orthotopic model of head and neck squamous cell carcinoma. The Laryngoscope 121: 2359–2365.
21. Liu R, Liu D, Trink E, Bojdani E, Ning G, et al. (2011) The Akt-specific inhibitor MK2206 selectively inhibits thyroid cancer cells harboring mutations that can activate the PI3K/Akt pathway. J Clin Endocrinol and metab 96: E577–585.
22. Pal SK, Reckamp K, Yu H, Figlin RA (2010) Akt inhibitors in clinical development for the treatment of cancer. Expert Opin Investig Drugs 19: 1355–1366.
23. Tolcher AW, Yap TA, Fearen I, Taylor A, Carpenter C, et al. (2009) A phase I study of MK-2206, an oral potent allosteric Akt inhibitor (Akti), in patients (pts) with advanced solid tumor (ST). J Clin Oncol 27: 3503.
24. Yap TA, Yan L, Patnaik A, Fearen I, Olmos D, et al. (2011) First-in-Man Clinical Trial of the Oral Pan-AKT Inhibitor MK-2206 in Patients With Advanced Solid Tumors. J Clin Oncol 29: 4688–4695.
25. Chen S-F, Cao Y, Han S, Chen J-Z (2014) Insight into the structural mechanism for PKBα allosteric inhibition by molecular dynamics simulations and free energy calculations. J Mol Graph Model 48: 36–46.
26. Sali A, Blundell TL (1993) Comparative protein modelling by satisfaction of spatial restraints. J Mol Biol 234: 779–815.
27. Laskowski RA, MacArthur MW, Moss DS, Thornton JM (1993) PROCHECK – a program to check the stereochemical quality of protein structures. J Appl Cryst 26: 283–291.
28. Ewing TJ, Makino S, Skillman AG, Kuntz ID (2001) DOCK 4.0: search strategies for automated molecular docking of flexible molecule databases. J Comput Aided Mol Des 15: 411–428.
29. Pettersen EF, Goddard TD, Huang CC, Couch GS, Greenblatt DM, et al. (2004) UCSF Chimera-a visualization system for exploratory research and analysis. J Comput Chem 25: 1605–1612.
30. DeLano WL (2002) The PyMOL Molecular Graphics System. San Carlos, CA. DeLano Scientific.
31. Laskowski RA, Swindells MB (2011) LigPlot+: multiple ligand-protein interaction diagrams for drug discovery. J Chem Inf Model 51: 2778–2786.
32. Wallace AC, Laskowski RA, Thornton JM (1995) LIGPLOT: a program to generate schematic diagrams of protein-ligand interactions. Protein Eng 8: 127–134.
33. Ghosh KS, Sen S, Sahoo BK, Dasgupta S (2009) A spectroscopic investigation into the interactions of 3'-O-carboxy esters of thymidine with bovine serum albumin. Biopolymers 91: 737–744.
34. Hubbard SJ, Thornton JM (1993) 'Naccess', computer program. Technical Report. Department of Biochemistry and Molecular Biology, University College London.
35. Wang R, Lai L, Wang S (2002) Further development and validation of empirical scoring functions for structure-based binding affinity prediction. J Comput Aided Mol Des 16: 11–26.
36. Wang R, Lai L, Wang S (2003) Comparative evaluation of 11 scoring functions for molecular docking. J Med Chem 46: 2287–2303.
37. Devaurs D, Bouard L, Vaisset M, Zanon C, Al-Bluwi I, et al. (2013) MoMA-LigPath: a web server to simulate protein-ligand unbinding. Nucleic Acids Res, 41(W1): W297–302.
38. Cortés J, Siméon T, Ruiz de Angulo V, Guieysse D, et al. (2005) A path planning approach for computing large-amplitude motions of flexible molecules. Bioinformatics 21(Suppl.1): i116–i125.
39. Robert CE (2004) MUSCLE: multiple sequence alignment with high accuracy and high throughput. Nucleic Acids Res 32 (5): 1792–1797.
40. Clamp M, Cuff J, Searle SM, Barton GJ (2004) The Jalview Java alignment editor. Bioinformatics 20: 426–427.
41. Waterhouse AM, Procter JB, Martin DMA, Clamp M, Barton GJ (2009) Jalview Version 2 – a multiple sequence alignment editor and analysis workbench. Bioinformatics 25 (9): 1189–1191.
42. Green CJ, Göransson O, Kular GS, Leslie NR, Gray A, et al. (2008) Use of Akt inhibitor and a drug-resistant mutant validates a critical role for protein kinase B/Akt in the insulin-dependent regulation of glucose and system A amino acid uptake. J Biol Chem 283: 27653–27667.

An Integrated Model of Transcription Factor Diffusion Shows the Importance of Intersegmental Transfer and Quaternary Protein Structure for Target Site Finding

Hugo G. Schmidt[1]*, Sven Sewitz[1,2], Steven S. Andrews[3], Karen Lipkow[1,2]*

1 Department of Biochemistry & Cambridge Systems Biology Centre, University of Cambridge, Cambridge, United Kingdom, 2 Nuclear Dynamics Programme, The Babraham Institute, Cambridge, United Kingdom, 3 Fred Hutchinson Cancer Research Center, Seattle, Washington, United States of America

Abstract

We present a computational model of transcription factor motion that explains both the observed rapid target finding of transcription factors, and how this motion influences protein and genome structure. Using the Smoldyn software, we modelled transcription factor motion arising from a combination of unrestricted 3D diffusion in the nucleoplasm, sliding along the DNA filament, and transferring directly between filament sections by intersegmental transfer. This presents a fine-grain picture of the way in which transcription factors find their targets two orders of magnitude faster than 3D diffusion alone allows. Eukaryotic genomes contain sections of nucleosome free regions (NFRs) around the promoters; our model shows that the presence and size of these NFRs can be explained as their acting as antennas on which transcription factors slide to reach their targets. Additionally, our model shows that intersegmental transfer may have shaped the quaternary structure of transcription factors: sequence specific DNA binding proteins are unusually enriched in dimers and tetramers, perhaps because these allow intersegmental transfer, which accelerates target site finding. Finally, our model shows that a 'hopping' motion can emerge from 3D diffusion on small scales. This explains the apparently long sliding lengths that have been observed for some DNA binding proteins observed *in vitro*. Together, these results suggest that transcription factor diffusion dynamics help drive the evolution of protein and genome structure.

Editor: Attila Csikász-Nagy, Fondazione Edmund Mach, Research and Innovation Centre, Italy

Funding: This work was funded by: a BBSRC Doctoral Training Grant (http://www.bbsrc.ac.uk/home/home.aspx) (HGS), the University of Cambridge (http://www.cam.ac.uk) (HGS), MITRE (http://www.mitre.org) subcontract 91687 awarded to R. Brent (SSA), NIGMS grant R01GM086615-01 (http://www.nigms.nih.gov/Pages/default.aspx) awarded to R. Yu and R. Brent (SSA), the EU-IndiaGrid1 & 2 (http://www.euindiagrid.eu) (KL), a Royal Society University Research Fellowship (https://royalsociety.org) (KL), a Microsoft Research Faculty Fellowship (http://research.microsoft.com/en-us/) (KL), and Apple Research & Technology Support (http://www.apple.com/uk/education/hed/arts/ - funding programme discontinued) (KL). The funders had no role in study design, data collection and analysis, decision to publish, or preparation of the manuscript.

Competing Interests: Microsoft Research Ltd. and Apple Inc. provided funding towards this study, as listed in the Funding section. There are no patents, products in development or marketed products to declare.

* Email: hugo.schmidt@st-hughs.oxon.org (HS); karen.lipkow@babraham.ac.uk (KL)

Introduction

Control of gene regulation and cellular development relies on the ability of transcription factors (TFs), a subset of the sequence-specific DNA binding proteins (ssDBP), to activate or repress selected genes in response to internal cues or changes in the environment. To perform their function, TFs must first reach relatively small regulatory sequences within much larger genomes.

Eukaryotic chromosomes are hierarchical macro-structures of DNA and proteins, of which the DNA ranges in length from hundreds of kilobases to multiple gigabases. The basic unit is the nucleosome, in which 147 bp of DNA wraps nearly twice around the protein core of a histone octamer [1]. The resulting chromatin is further compacted into higher order structures [2]. These compact structures exist in parallel with more open domains, which have highly variable structures and topologies. Recent DNA-DNA contact maps show that chromatin is segregated into territories, in which DNA loci mainly contact regions on the same chromosome. Examples of such organisation are a fractal globular arrangement [3] and multiple solenoidal structures [2] within the nucleus, depending on the species.

Transcription factors and many other proteins interact with DNA. Their sequence-specific interactions are mediated primarily by hydrogen bonds and van der Waals interactions, while their non-specific interactions are largely based on electrostatic forces [4]. In the latter case, the negatively charged DNA filament creates an electrostatic field that attracts positively charged patches on the proteins. The radius over which the electrostatic field is effective is reduced by positive ions in the nucleoplasm, which counteract the effect of the DNA's negative charge. The resulting "Manning radius," which is effectively the Debye length for protein-DNA interactions, typically extends 1–2 nm from the surface of a DNA filament [5,6]. Non-specifically bound proteins are attracted weakly enough that they can typically slide reasonably freely along the DNA filament.

Transcription factor motion, which takes place within this complex nuclear environment, has been investigated for several decades. Seminal work by Berg *et al.* [7–9] defined the four basic forms of transcription factor motion in the presence of DNA, which are collectively known as 'facilitated diffusion' (Figure 1; see [13,42] for reviews). These are: (a) 3D Brownian diffusion in the nucleoplasm, (b) 1D sliding along the DNA, facilitated by non-specific TF-DNA binding, (c) intersegmental transfer (IST), where proteins transfer directly between two DNA segments that are in close proximity to each other, and (d) hopping, in which a TF unbinds from a DNA segment, diffuses briefly in 3D space, and rebinds to a nearby section of the same DNA segment. An additional form of motion is (e) intersegmental jumping, in which a TF unbinds from a DNA segment, diffuses briefly in 3D space, and rebinds to a different DNA segment nearby [10–12]. An important criterion of all of these search mechanisms is whether they are distance dependent or independent. This refers to the time a TF takes to find its target gene (TG), called the finding time, in relation to the distance of the TF to the TG. A mechanism is called distance dependent if the mean finding time is a function of the distance between TF and target gene, and distance independent if there is no correlation between the distance and the finding time.

3D diffusion is effectively distance independent within the bounded system of the nucleus, and so is referred to distance independent throughout this paper. It is typically the slowest of the above mechanisms [7]. However, it allows access to all of the nucleus and genome [7] [14,15].

1D sliding is distance dependent with the finding time scaling with the square of the distance [16,17]. This implies that target finding is rapid over short distances but very slow for long distances [16]. 1D sliding only allows searching on uninterrupted

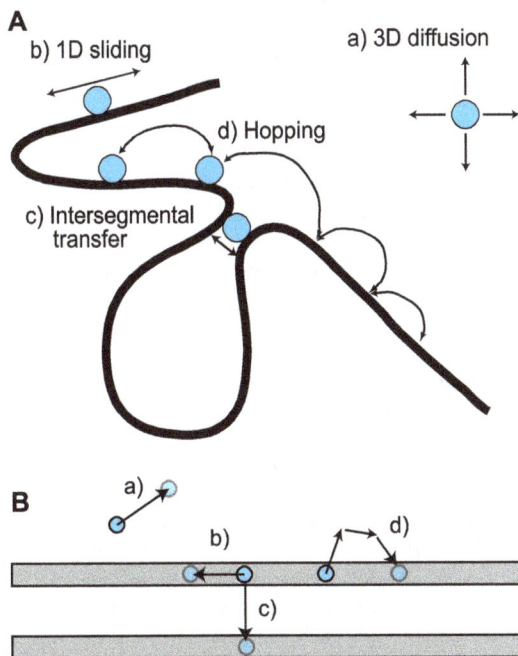

Figure 1. Modes of transcription factor motion. A) Schematic of the four modes of transcription factor (TF) motion (modified from [117]). B) Schematic of their implementation in the Smoldyn model. Modes: a) 3D diffusion within solution by Brownian motion, b) 1D sliding of a TF non-specifically bound to DNA, c) intersegmental transfer, where a TF binds two DNA segments and moves from one to the other, and d) hopping, in which a TF makes short excursions away from DNA (simulated as a sequence of elementary unbinding, diffusion, and binding processes).

stretches of DNA. Chen *et al.* (2014) recently imaged both 3D diffusion and 1D sliding in single, living cells [18].

Intersegmental transfer finding times do not obey a simple scaling law but depend on DNA conformation and concentration [19,20]. IST is generally an extremely rapid form of searching, as it effectively converts the DNA structure into a diffusion lattice. IST is a prevalent and important form of TF motion. For example, experimental work by Elf *et al.* [21] showed that finding times increased if IST was abrogated. This was consistent with experiments that selectively allowed IST by altering DNA conformation [11,12] and with the observation that TFs spend a high fraction of their time bound to DNA, as opposed to being in solution [22] [21]. The importance of IST was further emphasized by recent modelling work by Bauer *et al.* [23], by fluorescent analytic work by Esadze and Iwahara [22], and by mathematical and experimental investigation of 1D sliding and IST of the Egr-1 protein [22]. Recent advances in microscopy have enabled direct imaging of IST [10,24].

The domain structure of chromatin in a subregion of the genome affects the way in which IST may operate. If a chromatin domain approximates a solenoidal organisation or certain fractal organisations, an individual TF will be able to move by IST only between the segment it is bound to and the two that are adjacent [25–28]. On the other hand, if a chromatin domain approximates a more globular structure, with multiple contacts between DNA segments from different chromosomes, the TF would be able to move between different chromosomes, thereby extending its reach. This would be particularly the case if the 'chromatin globule' were dynamic [3,29,30]. We also note that some computational models specifically avoid IST and as a result find that DNA conformation is irrelevant to TF finding acceleration [31].

Hopping arises when a TF unbinds from DNA, diffuses in 3D, and rebinds to a *nearby* site on the same DNA segment, where this length scale is set by the definition that hopping is a distance dependent mechanism [13]. This contrasts with the distance independence of 3D diffusion with delayed rebinding, in which the rebinding location is essentially uncorrelated with the location where the TF unbound [13]. The diffusion length that marks the difference between 3D diffusion and hopping is not precisely defined [16] but generally accepted to be shorter than the DNA persistence length [16,32] (about 50 nm or 150 bp). We further distinguish hopping from the situation when a protein simply returns to the DNA because it is still electrostatically attracted to it (i.e. it does not leave the Manning radius). The distance dependence of hopping is close to that of 1D sliding, occurring with an appreciable rate only over very short distances. Hopping has been difficult to verify experimentally. However, its existence is supported by recent NMR work on the HOXD9 domain [33] and by work on the Oct1 protein [34]. Hopping has also been suggested to explain the observations that interlinked plasmids are more readily cleaved [12], and that supercoiling increases restriction enzyme motion on DNA [11]. On the other hand, some recent evidence suggests that these may be incorrectly classified cases of IST [35].

Intersegmental jumping is similar to hopping, but is between separate DNA segments rather than DNA regions on the same segment [10–12]. We define a segment as a stretch of continuous DNA on which uninterrupted 1D sliding is possible. Intersegmental jumping is also similar to IST, with the difference that intersegmental jumping requires 3D diffusion whereas IST relies on the TF binding to two DNA segments simultaneously. However, it has proven difficult to distinguish these experimentally. In particular, the *Eco*RI protein has only one DNA binding site, and therefore should not be able to perform IST, so was assumed to move by intersegmental jumping. However, it has recently been found that its transfer kinetics are closer to those of IST. This may be facilitated by a hitherto unrecognized positive patch on the protein surface opposite to the DNA binding site of the *Eco*RI protein, which may provide the necessary structure for IST [35]. This suggests that some of the other previously observed intersegmental jumps may in fact be intersegmental transfers.

In 1970, Riggs and colleagues laid the foundations of this field by investigating the binding kinetics in the *lac* operon, finding that transcription factors locate their target genes (TGs) nearly two orders of magnitude faster than the maximum speed allowed by 3D diffusion alone $(7\times10^9\ \mathrm{M}^{-1}\ \mathrm{s}^{-1}$ versus $1\times10^8\ \mathrm{M}^{-1}\ \mathrm{s}^{-1}$ respectively) [36,37]. These observations have been substantiated more recently [38] and have attracted much interest. However, they come with the caveat that the fastest finding time results arose from experiments in which the salt concentration was significantly lower than in physiological conditions, which would have effectively confined transcription factor motion to 1D sliding and intersegmental transfer, thus completely avoiding slow 3D diffusion [39] [16]. This is because low salt concentrations increase the Manning radius, which effectively prevent the TF from unbinding.

These results and others showed that facilitated diffusion increases the speed of TF target finding above that of 3D diffusion alone [36,37,40–47]. See [32] [42] [48] for excellent reviews of the field, and [49] for an emphasis on computational methods of analysis. Furthermore, experimental and theoretical work are in agreement that 1D sliding is essential to faster transcription factor finding [16,50,51]. One of the principal conceptual problems that has emerged from this work is the speed-stability paradox for ssDBPs. It states that these proteins must bind DNA sufficiently weakly to allow for sliding but also sufficiently tightly to produce the stability required to drive gene activation, which are mutually contradictory requirements [31] [23], as it implies that these proteins must bind DNA with both weak and tight modes.

Here we present a computational model of TF motion (Figure 2). It allowed us to analyse the contributions of the individual TF motion components and enables us to explain the large increase in speed of TF-TG finding first reported by Riggs *et al.* We simulated TF movements using the computer program Smoldyn, previously used for modelling signal transduction and related phenomena within or between individual cells [52–56]. Building on the established observation that 1D sliding leads to shorter finding times, we find that 1D sliding has an upper maximum search distance determined by diffusion dynamics. This maximum turns out to be a good predictor of both empirically observed TF-DNA unbinding constants and a good fit to the length of nucleosome free regions (NFRs) in eukaryotes. Our work further confirms the importance of IST and also shows how IST can lead to two forms of searching. One is distance independent and most likely to occur in areas of high concentration of chromatin, while the other is distance dependent and most likely to occur in areas of low chromatin concentration. From published data, we found that TFs and other sequence specific DNA binding proteins are enriched in dimers and tetramers, both of which promote IST.

Model

Model description

We based our model of TF motion on the yeast nucleus. The aspects of yeast nuclei that are relevant to this work are similar to those of other eukaryotes, so our model will equally apply to sequence-specific protein-DNA interactions more generally, and

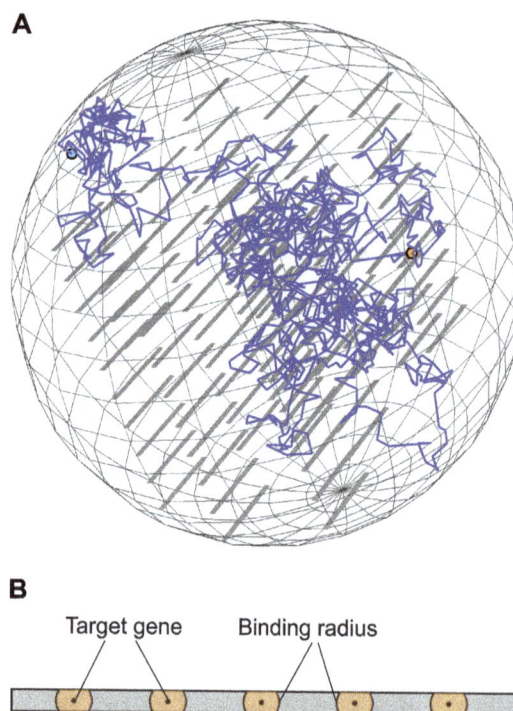

Figure 2. Smoldyn model of the yeast nucleus. A) Example of a Smoldyn simulation. The nuclear envelope is modelled as a perfect sphere containing one or more TFs (light blue dot), multiple stacks of DNA segments (grey bars), and one or more target genes (TG) on the DNA (orange hexagon). Individual DNA segments may be part of the same chromosome, but are separated so that 1D sliding is not possible between them. B) Target genes along a DNA filament, shown with their centres as black dots and their binding radii as orange regions. TF-TG complex formation occurs when a DNA-bound TF diffuses to within a binding radius of a TG.

not just to those in yeast. Our model consists of a 1.5 μm diameter sphere, modelled on the size of the yeast nucleus [57], which contains one or more stacks of virtual DNA segments. These DNA segments are 2.6 nm wide and vary from 10 to 4,200 bp in length, depending on the specific analysis. These are much shorter than yeast chromosomes (diploid cells have 32 chromosomes, ranging from 200 kb to 1.5 Mb) but, from the point of view of TF searching by 1D sliding, they are taken to be appropriate because chromosome lengths that are accessible to sliding are limited by heterochromatinisation and other obstructions. Also for this reason, we only included stretches of DNA that are accessible to the diffusing TF and are not heterochromatic or otherwise occluded. This is analogous to the way that *in vitro* studies of ssDBPs interacting with masses of discontinuous DNA have been argued to provide results that are generalisable to *in vivo* ssDBP-genome interactions [33–35]. We modelled these DNA sections as long narrow rectangles with each rectangle side representing a DNA groove. Rectangle widths were 2.6 nm, corresponding to the DNA double helix width [58].

We modelled TFs as dimensionless points. They diffused on DNA surfaces with a 1D diffusion coefficient of 0.0262 μm²/s, which was based on our analysis of fluorescence recovery after photobleaching (FRAP) data for the Ace1p yeast transcriptional activator from [59] and was comparable to other measurements [21,38,45,60–64]. We simulated this as 2D diffusion, but the long narrow DNA rectangles made it effectively 1D. TFs did not diffuse off the ends of DNA segments, but instead reflected back towards where they came from. When dissociated from DNA, TFs diffused within the 3D nuclear volume with a diffusion coefficient of $2.72\ \mu m^2/s$. This is 104 times faster than 1D diffusion, which we based on theoretical work by Berg et al. (1982) [8] and experimental work on bacterial transcription factor diffusion [45,63–65]. Our assumed 3D diffusion coefficient is comparable to that for nuclear FITC-dextrans [14].

TFs bound non-specifically to the two DNA rectangle faces, which acted as uniform binding areas, as adsorption processes. We varied the adsorption coefficient, k_{on}, from $1.7\ \mu m/s$ for chromatinised DNA to $10\ \mu m/s$ for protein-free DNA. TFs dissociated from these non-specifically bound states at a rate, k_{off}, of $11.6\ s^{-1}$, which we again based on the FRAP measurements of Ace1p [59], verifying it by replicating the recovery curve given in that work (Figure S1, compare to Figure S5 in [59]). This dissociation rate also compares well to similar studies of transcription factor binding [41,66].

We simulated IST as random TF transfers from one DNA segment to an adjacent one, with a rate constant of $11.6\ s^{-1}$. We ignored the physical proximity of these segments, which accurately captured the *effect* of concentrated and organised DNA but without needing to accurately reconstruct the DNA conformation (cf. [44]; and [23]). Because hopping occurs on a size scale below that of the DNA persistence length [16,32], our use of straight DNA segments in the model had minimal impact on hopping motions.

We represented each TF-specific target gene (TG) as a point on the centreline of the DNA rectangles. TFs diffusing in 3D space did not interact with TGs (with some exceptions, noted below). On the other hand, TFs that were already non-specifically bound to DNA could bind to TGs, which happened when a TF diffused to within one "binding radius" of a TG. We used a binding radius, σ_b, of 2.0 nm (about 6 bp), which Smoldyn computed using our assumption of a TF-TG binding rate constant of $10^5\ M^{-1}\ s^{-1}$ (comparable to ssDBP binding measurements [67–77]). This binding radius is comparable to the size of recognition sequences. More importantly though, it is substantially wider than the 1.3 nm DNA half-width; this meant that when a DNA-bound TFs diffused towards a TG, it nearly always bound the TG and did not diffuse around it (the TF diffused in discrete Gaussian-distributed displacements with 2.3 nm rms step lengths, so it was possible but unlikely for a TF to step completely over a TG). For this reason, the values used for the TF-TG binding rate constant had essentially no effect on our simulations. We modelled TF binding as a change of species, converting it from a rapidly diffusing TF to a bound TF-TG complex. This is much like the change from the fast 'search mode' to the immobile 'recognition mode' of the speed/stability paradox [48]. In some simulations, we also enabled TF-TG dissociation. In these cases, TF dissociated directly to 3D space at a rate of $0.025\ s^{-1}$, once again based on the FRAP measurements of Ace1p [59]. All simulation parameters are summarised in Supplementary Table S1.

We did not explicitly represent heterochromatinised DNA, chromosomal packaging material, RNAs, or other proteins in addition to the modelled TFs, all of which create macromolecular crowding effects in the cell [78]. This was because the values for the parameters we used in our simulation were taken from experimental measurements. These *in vivo* values already account for any crowding effects, representing the diffusion characteristics of proteins within the crowded, intranuclear environment. Additionally, preliminary work showed that simulating crowding with up to 30% volume exclusion with 100 impenetrable spheres did not affect results significantly [79]. This agreed with our prior modelling work, with similar temporal and spatial scales, where we

found that the primary effect of macromolecular crowding was to reduce diffusion coefficients [80]. It also agreed with fluorescence imaging results which showed that protein motion on the scale of transcription activators is not significantly affected by chromatin or other crowding agents [15] and is in accord with the observed high diffusion rates within the nucleus [14,21].

Model simulation and validation

We simulated our model in Smoldyn (versions 2.09 to 2.31), which is a particle-based simulator of diffusion, reactions, and surface interactions [53]. Whereas many mathematical models of cell biology are deterministic, specifying constant rates for processes (e.g. specifying the number of times hopping occurs in a given unit of time [44,81]), Smoldyn accounts for stochastic behaviour accurately, which means that complex processes emerge naturally from the fundamental diffusion dynamics. Unless otherwise specified, we ran each simulation for 1 virtual hour and used time steps of 0.1 ms. This time step caused the simulation spatial accuracy to be about 7 nm for TFs in 3D and 0.07 nm for TFs bound to DNA. This detail was fine enough to capture hopping and other TF motions, but also coarse enough, in contrast to simulators with single DNA basepair resolution such as GRiP [49], that we were able to run simulations quickly. We ran single simulations using a Mac Pro computer (2×2.8 GHz quad-core Intel Xeon), which typically completed in seconds to minutes, and batches of hundreds of simulations overnight using the Cambridge University computing grid. We analysed Smoldyn output using MATLAB version 2013a. Smoldyn input files, MATLAB scripts and simulation parameters are included as Supplementary Information.

We validated our simulations in several ways. First, we found in prior work that Smoldyn's simulations of diffusion, binding and unbinding reactions, and surface reactions all have kinetics that differ from theoretical predictions by less than 2.5% [53,82,83]. Secondly, we tested for expected behaviours and robustness as we varied simulation parameters (see [79]). As expected, simulations that investigated just reversible TF-TG complex formation, without virtual DNA, showed that the final number of TF-TG complexes increased nearly linearly with both TF and TG counts (not shown). Also, varying 3D diffusion coefficients showed few final complexes with 10^5 fold slower diffusion than our typical assumption due to slow equilibration, and the equilibrium number of complexes when diffusion coefficients were $0.027\ \mu m^2\ s^{-1}$ or greater, again as expected. Varying TG locations and the extent of TG clustering showed that these had minimal effects on results. Finally, we tested the distance dependence of finding times in different dimensionalities. We placed a cluster of 5 TGs (a) onto a long narrow rectangle that represented DNA to test 1D finding times, (b) onto a square flat surface to test 2D finding times, or (c) freely into space within the 1.5 μm diameter nuclear envelope to test 3D finding times; the outer dimensions of each system was 2 μm long on each axis. Then, we released 5 diffusing TFs a fixed distance away from the 5 TGs and recorded the mean finding time (Figure S2). As expected, the average finding time increased continuously with the distance for the 1D and 2D cases, but was essentially distance-independent for the 3D case. Together, these tests suggested that our simulations worked as intended.

Results

1D Sliding has a limited range that is only moderately influenced by the unbinding constant

The 'antenna effect' is defined as the process by which ssDBPs find their targets more readily if the recognition sequence is embedded in nonspecific DNA on which 1D sliding (and/or hopping) is possible [84–86]. We investigated this effect by placing 20 TGs at the centres of 20 DNA segments and starting 50 TFs at random locations in the 3D nuclear volume (Figure 3A). We ran each simulation 20 times, and counted the number of steady state complexes formed. They confirmed the antenna effect: the steady state number of complexes was a function of the antenna length (total DNA segment length), increasing sharply up to about 300 bp and more gradually for longer antennas (Figure 3B). Longer antennas conferred minimal additional advantage toward forming TG-TF complexes, which arose from the increasing likelihood of TF dissociation from the DNA. Presumably, the number of steady-state complexes would have turned around and started decreasing at some point if we had investigated extremely long antennas, due to TF sequestration, but these effects did not appear for the DNA lengths that we used.

Figure 3C shows that the target finding time was also a function of the antenna size, both for the first target bound (cyan line) and for the half-maximal number of targets to bind (red line, half-maximal values were computed from the steady state values plotted in panel B). In both cases, finding times decreased sharply for antenna lengths up to about 300 bp and hardly at all for longer lengths. Again, short antennas were highly effective but longer ones conferred minimal additional advantages.

Because the effectiveness of long antennas is largely limited by TF dissociation from the DNA, we explored the effect of varying the TF-DNA dissociation constant (k_{off}) on the number of TF-TG complexes at steady-state (Figure 3D). As expected, faster dissociation resulted in both fewer complexes and shorter lengths over which antennas were effective. At the other extreme, the dissociation rate had minimal effect on the number of complexes when it was very slow (compare upper curves of Figure 3D). In these cases, TFs typically did not dissociate from the DNA until after forming TF-TG complexes. We thus find that antennas do not gain substantial effectiveness for lengths that are greater than about 300 bp or for k_{off} rates that are below about 1 s^{-1}.

Hopping on 'naked' DNA as an emergent property

The hopping mode of ssDBP motion can be difficult to identify because of its short range. Because TF diffusion is not biased to move towards a particular DNA end, hopping is equally likely to carry the TF in either direction along the filament. Thus, the mean displacement over time of hopping is 0 bp. However, by measuring distance as an absolute value, the average 'hop' has been estimated to cover between one and several bp [44,87–90]. We simulated TF hopping by placing 6 individually identifiable TFs at varying distances from a single TG on each of 20 DNA segments, each 930 bp long. We allowed multiple TFs to bind to each TG and used the individual TF labels to determine which ones bound. We prevented TF-TG complex dissociation and 1D sliding, but allowed TF-DNA unbinding and rebinding. At the end of the 60 minute simulation, we counted the number of TF-TG complexes for each of the TF labels (Figure 4). Using our standard simulation parameters, including a a DNA adsorption coefficient of 1.7 μm/s, the number of complexes formed was independent of the initial TF distance away from TGs (not shown). This indicated that hopping did not occur appreciably in this case, but that TF-TG complex formation occurred via distance independent 3D diffusion. However, increasing the DNA adsorption coefficient to 10 μm/s produced complex counts that did depend on the distances (Figure 4), thus implying the presence of hopping [7,48]. The necessity of this larger adsorption coefficient suggests that hopping may only occur to an appreciable extent on "naked" DNA, meaning DNA that is not bound to histones or other

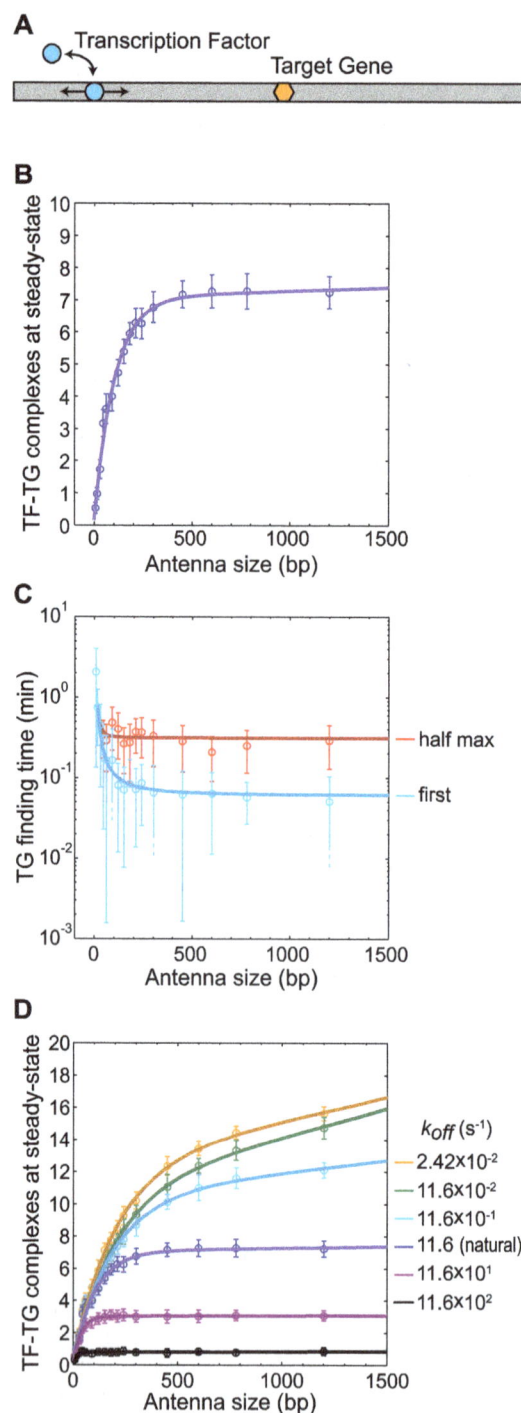

Figure 3. Antenna effect. A) Illustration of the 'antenna effect': target gene finding times are reduced when TFs can get to their targets by diffusing along the DNA. The TF (light blue circle) diffuses along the antenna DNA (grey bar) to reach the TG (orange hexagon). B) Effect of antenna length on the number of TF-TG complexes at steady-state. C) Effect of antenna length on the time for the first TF to bind to the first TG in the simulation (blue) and on the time required for half of the steady-state number of TF-TG complexes (from panel B) to form (red). D) Effect of the DNA dissociation rate (k_{off}) and antenna length on the number of steady-state TF-TG complexes. Simulation parameters: $D_{3D} = 2.72$ μm^2 s^{-1}, $D_{1D} = 0.0262$ μm^2 s^{-1}, $k_{on} = 1.7$ μm/s, $k_{off} = 11.6$ s^{-1} unless otherwise noted, $\sigma_b = 2$ nm, IST rate = 0, and specific binding was reversible with dissociation rate 0.025 s^{-1}; 50 TFs were started at random 3D locations and there were 20 TGs, each at the centre of a

DNA segment. Error bars represent one standard deviation, determined from 20 repeated simulations.

obstructions. Additionally, Figure 4 shows that hopping can be effective on distances of 360 bp or more, but its effectiveness drops off rapidly after about 60 bp. Conceptually, hopping increases the effective range of 1D sliding.

Intersegmental transfer provides two modes of motion

We investigated the effect that varying forms of chromatin organisation have on IST with three sets of simulations. The first set did not include IST; here, we placed a TG at one end of a 4,200 bp DNA segment, started a TF at a fixed distance away from the TG, and restricted TF motion to 1D sliding. The second set used "sequential IST," which might occur in solenoidal DNA structures or in regions of low chromatin concentration. Here, we created a stack of ten DNA segments, each 420 bp in length, placed a TG at one end of the top segment, and started a TF at a fixed position on one of the other segments. We restricted TF motion to 1D sliding and IST between adjacent segments. The third set used "concurrent IST," which might occur in a fractal globular DNA structure. We simulated this in the same way as for the sequential IST, but allowed IST between all segments. In the latter two cases, the plotted distance is the total DNA length between the TF and the TG. As expected, finding times were strongly distance dependent for 1D sliding (Figure 5A). On the other hand, they were less strongly distance dependent for sequential IST and essentially distance independent for concurrent IST. These results agree with prior work that has shown that IST generally accelerates finding times [11]. They also emphasize the point that the effect of IST depends strongly on the local chromatin concentration and structure [19,20].

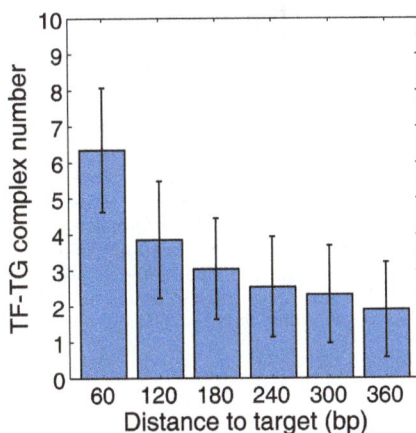

Figure 4. Transcription factor hopping. The number of TF-TG complexes that formed for TFs that started at various distances away from their targets and that could not slide along the DNA. The distance dependence shown here is indicative of hopping motion, in which TFs repeatedly unbound from the DNA, diffused briefly in 3D space, and rebound to the DNA at a location close to the unbinding location. Simulation parameters: $D_{3D}=2.72$ μm² s⁻¹, $D_{1D}=0$, $k_{on}=1.7$ μm/s, $k_{off}=11.6$ s⁻¹, $\sigma_b=2$ nm, IST rate = 0, specific binding was irreversible, and multiple TFs binding to a single TG was allowed; on each of 20 DNA segments, 6 labeled TFs were started at 60 bp distance increments away from a single TG. Bar heights represent the number of TF-TG binding events, out of 20 possible, for each TF location after 60 minutes. Error bars represent one standard deviation, determined from 20 replicate simulations.

Combining intersegmental transfer and 1D diffusion accelerates TF-TG finding

We investigated the relative effectiveness of the different types of TF motion using simulations that included them in various combinations and with various rate constants. These simulations allowed one or more of 1D sliding, 3D diffusion, and IST, and varied the *koff* rate constants. We did not specifically prevent hopping, so this undoubtedly occurred whenever 3D diffusion could. In all cases, our simulations included a stack of 50 DNA segments of 1.5 kb each, for a total of 75 kb (Figure 5B). The end of each segment linked to the start of the next, such that a TF could slide from one segment to the next. Each simulation included a single TF that started at one end of the DNA stack and a single TG that was located at the opposite end of the stack. Simulations treated TF-TG binding irreversibly and ran for 2 hours. Whereas we typically only allowed TF-TG complex formation for TFs that had already bound to the DNA, we relaxed that requirement here, also allowing TFs in 3D space to bind to TGs, using the same 2 nm binding radius. We repeated each simulation 100 times.

Figures 5C and 5D show the final number of complexes from the 100 simulations and their average finding times. Simulations that only allowed 1D sliding produced no complexes within 30 minutes because 1D sliding is very slow over long distances. IST alone was also completely ineffective, in this case because the TF couldn't search along the DNA, but was restricted to the relatively few sites where it randomly landed. 3D diffusion alone did produce complexes, but relatively slowly. In this case, a complex only formed if the TF diffused into the 2 nm binding radius of the TG, by 3D diffusion, which made this an infrequent occurrence. Combining 3D diffusion with either 1D sliding or IST alone accelerated the complex formation rate by less than a factor of 10. On the other hand, when 1D sliding and IST were combined together, then complexes were formed several orders of magnitude faster. This combination was most effective when the TF could not diffuse significantly in 3D space, and thus stayed confined to the DNA filament (Figure 5D, compare first two bars with the third bar).

Notably, when we used our default parameters for all types of motion, which we estimated as well as possible from experimental data (shown in orange, [+ + +] in Figure 5C and 5D), the finding time fell in the range of 2–3 minutes, similar to that observed by FRAP measurements [59]. This finding time is faster than we found in simulations that only combined 1D and 3D diffusion (including hopping). This suggests that IST plays a strong role *in vivo* and is essential for explaining experimentally measured finding times.

Transcription regulators are enriched for dimeric and tetrameric structures

Based on observations by us and others that IST acts as a strong accelerator of TF-TG complex formation, we wondered whether this would be reflected in the way that TF structures have been shaped by evolution. This would be reasonable because the time for a TF to find its recognition sequence has been shown to be the rate-limiting step in transcription [91], so it should be subject to selection. In particular, we investigated whether TFs and other ssDBPs are more likely to form dimers and tetramers than other proteins. This is based on the logic that dimers and tetramers allow two DNA segments to be bound simultaneously, which is the essential requirement for IST [9,92], as seen with the *lac* repressor and other TFs [93]. Additionally, transcription activator multi-merisation has been shown to aid IST [21]. On the other hand

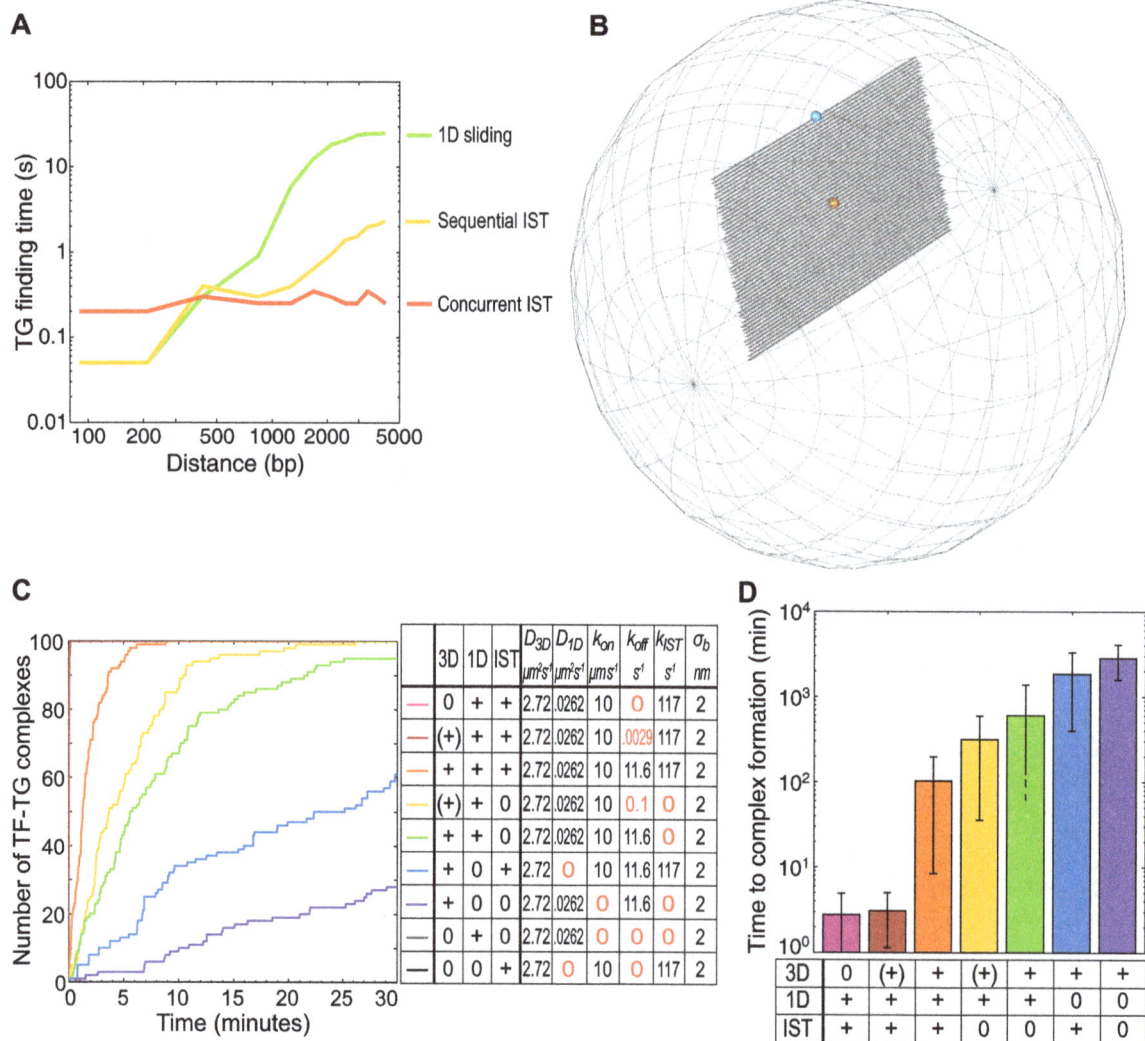

Figure 5. Intersegmental transfer. A) Mean finding time for a single transcription factor placed at varying distances away from its target. Searching was possible by 1D sliding (green), "sequential" intersegmental transfer between adjacent DNA sections (yellow) or "concurrent" intersegmental transfer between all DNA sections in a group (red). B–D) Simulations testing the effect of the three major modes of motion. B) Arrangement of the 50 DNA segments within the simulated nucleus: shown 10× wider than in the simulation for clarity. The TF is placed at the position shown in light blue, and the time is measured until it binds the target gene (orange sphere). C) Total number of complexes formed over time for 100 different simulations, each with one transcription factor and one target, using nine different combinations of 3D diffusion, 1D sliding, and intersegmental transfer (IST). 3D diffusion was varied between low (+), standard +, and zero, 0. 1D sliding and IST was either present, +, or absent, 0. Shown adjacent are the corresponding values for the simulation parameters: σ_b is the binding radius for TFs in 3D space and also bound to DNA. 1D sliding or IST alone (grey or black line) did not achieve a single binding event in the time shown. D) Mean target finding time, in the same simulations as shown in C).

though, ssDBPs may also use unstructured protein tails for IST [19,20,39], showing that IST can also happen without multi-merisation [35,94].

We used the 3D protein complex database produced by Levy *et al.* (2006) [95], who curated the set of structures deposited in the RCSB protein databank (www.rcsb.org, [96]), for a non-redundant set of protein structures. From this dataset, we counted the numbers of dimers and tetramers in the classes listed as *transcription*, *DNA binding*, *transferase*, and *oxidoreductase*, as well as in the complete curated set. The protein databases might be biased towards multimers and homodimers for technical reasons of crystallisation, but this bias should apply equally across all protein families. We are hence basing our analysis on the comparison of percentages between different protein families.

Figure 6 shows these values, as percentages, by class. The *total* column shows that the complete list of curated protein structures comprises ca. 32% dimers, ca. 12% tetramers, and ca. 55% others (mostly monomers). The *oxidoreductase* column represents our control class because these enzymes, which are widely used in metabolism, typically do not bind to DNA. They exhibit essentially the same fractions of dimers and tetramers as the total class. The three other classes, all of which represent sequence specific DNA binding proteins, have substantially higher fractions of dimers and especially tetramers. In particular, transcription factors, the focus of this work, are 79% dimeric or tetrameric, compared to ca. 45% in the set of all proteins. Transferases also play central roles in gene regulation, including for example methyl- and acetyltransferases that perform sequence specific DNA or histone modification; they are 85% dimeric or tetrameric, with the highest percentages for tetramers of the classes analysed. IST has been observed for this class of regulators [16,97–99]. Finally, the class of other DNA binding proteins is 63% dimeric or tetrameric. Thus, all three classes of sequence specific DNA binding proteins exhibit strong enrichment for dimers and tetramers. This is consistent with the possibility that evolution favoured their multimerisation in order to facilitate IST and thus reduce target finding times.

Discussion

Our model of transcription factor motion describes known and emergent methods of how TFs find their target genes. This question has been intensively studied in the past, and our model is based on extensively validated experimental data. Using Smoldyn, we show the degree to which intersegmental transfer (IST) accelerates TG finding, and that this feature has significant impact on the evolution of the quarternary structure of ssDBPs. We are also able to give an explanation for an observation that has caused some disquiet within the community, namely the fact that TFs are able to find their targets at speeds exceeding the 3D diffusion limit. In our analysis, this can be explained by the effect of limiting TFs to 1D diffusion and IST, as is the case at low salt concentrations (see details below).

Intersegmental transfer explains high acceleration of transcription factor search in low salt conditions

Lowering salt concentrations decreases the electrostatic shielding around DNA filaments, which increases the Manning radius and hence increases non-specific binding between ssDBP and DNA. This can largely confine TFs to the DNA, making them only able to move by 1D sliding and IST [16]. Lowering salt concentrations has also been shown to increase ssDBP searching speeds [100]. However, this acceleration only occurs if IST can take place; this has been shown with *lac* repressors that have been

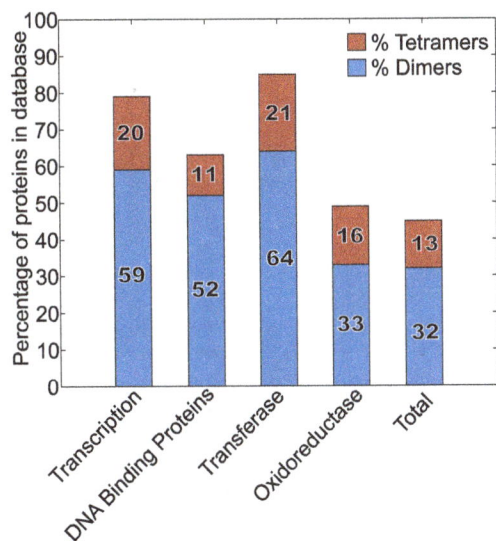

Figure 6. Multimericity of protein structures by PDB class. Bars show the percentage of proteins in different protein classes that form dimers (blue) and tetramers (red). The protein list, which is non-redundant, was obtained from the 3D complex database [118]. The first three bars represent sequence specific DNA binding proteins, the oxidoreductase bar is a control group that does not bind to DNA, and the final bar represents all proteins in the database.

mutated to make them unable to form tetramers and thus unable to bind two DNA segments simultaneously and undergo IST [21].

Our simulations help explain these experimental results, as well as the original *in vitro* low-salt experiments by Riggs *et al.* [36,37]. In particular, Figures 5C and 5D show that finding times decrease by several orders of magnitude when TF motion is restricted to 1D sliding and IST. Thus, the very fast finding times observed by Riggs *et al.* can be explained by assuming that their low salt conditions restricted TF motion to 1D sliding and IST.

The 'antenna effect' may determine the length of nucleosome free regions

We showed that the 'antenna effect' has an effective upper limit of about 300 bp. This length is largely determined by the dissociation rate of non-specific TF-DNA interactions, which sets the residence time of TF being bound to a DNA filament, and thus determines how far the TF is likely to slide. Our dissociation rate constant value, $k_{off} = 11.6$ s^{-1}, which we derived from FRAP measurements of nonspecific TF-DNA interactions, is comparable with dissociation rates that have been observed for several other transcription factors [66]. In further support of our k_{off} value, it also appears to be reasonably optimal for efficient ssDBP-DNA interaction, based on our observation (Figure 3D) that larger k_{off} values led to many fewer TF-TG complexes, but that smaller k_{off} values did not produce substantial further increases. In addition, the 300 bp antenna effect range that we found is consistent with the finding that the restriction enzyme *Eco*RV covers a similar distance by 1D sliding on DNA before dissociation [101] (longer sliding distances have also been observed for other restriction enzymes [102] and ssDBPs [103], but we speculate that those observations may have included hopping or IST).

Nucleosome free regions (NFRs) are sections of nucleosome free DNA typically associated with gene promoters. In yeast, NFRs are typically ~100–200 bp in length [104–107] and can be found at the 5' end or the 3' end of a gene. The 5'-NFRs contain a marked enrichment for TF binding sites [107–110]. Similarly, 3'-NFRs are enriched for sequences responsible for transcription termination sites (TTS) [111]. These sites are actively maintained free of nucleosomes by the action of the ATP-driven chromatin remodelers, such as Isw2 [104,112,113]. The activation of transcription is accompanied by the eviction of one or two nucleosomes, which in effect extends the usual NFR of to a length of 400–600 bp in these genes [114,115]. The similarity between the lengths of NFRs and the length of DNA traversed during effective searching is striking. We suggest that NFR lengths may be actively adjusted so as to maintain the amount of free DNA that ssDBPs require to locate their sequences efficiently. Additionally, of course, active modification of NFR lengths would create a transcriptional regulatory mechanism through modulation of the antenna effect.

Predictions made by our model

Our work enables several predictions. First, it predicts that abrogation of multiple DNA binding domains on ssDBPs should make them unable to undergo IST and thus have increased finding times. This was already shown in studies of the *lac* repressor [21]. It should also apply to other multimeric transcription regulatory proteins and to proteins that bind DNA in other ways, such as with the disordered N-terminal tails exhibited by *Hox* proteins and others [20], or the positive patches seen on the non-DNA binding side of *Eco*RI [35]. Secondly, and conversely, our model predicts that favouring IST, such as by lowering salt concentrations, should generally decrease finding times. This was our explanation for the Riggs experiments, but should also apply to other proteins that are capable of IST. Third, based on the fact that IST appears to be important for ssDBPs to locate their target sequences, we argue that most ssDBP proteins will be enriched in structures allowing for IST. Indeed, we observed a substantial enrichment for dimers and tetramers for ssDBPs. Dimerisation of TFs has other known benefits, such as being able to recognise a longer stretch of DNA, bringing with it increased specificity, which in itself has evolutionary advantages. The effects we see are expected to be additive to all other mechanisms favouring the formation of dimers and tetramers. The differences may be even more marked if positive patches and unstructured tails are analysed as well, as has already been partly shown [116].

Finally, we suggested that the size of nucleosome free regions is influenced by the upper limit of the effective antenna length. If this is the case, then shortening the NFRs around a reporter gene should substantially reduce expression but lengthening them should increase expression only minimally. Our work suggests that these effects should be reasonably independent of the sequence within the NFR, so long as the NFR remains protein free.

Conclusions

We present a computational model of transcription factor motion within cell nuclei that is based on only a few transcription factor elementary processes: diffusion in 3D space and along DNA filaments, non-specific binding and dissociation with DNA, specific binding and dissociation with target genes, and intersegmental transfer between DNA segments. This model exhibited the main modes of transcription factor motion, which are 3D diffusion, 1D sliding, hopping, and intersegmental transfer (we did not investigate intersegmental jumping). It showed that the antenna effect, in which transcription factors find their targets more quickly if the targets are embedded in DNA on which 1D sliding is possible, is extremely effective for antenna lengths up to about 300 bp but is not improved substantially with even longer antennas. From this result, we speculated that cells maintain nucleosome free regions about genes in part to accelerate expression through the antenna effect. Our model also showed that transcription factor hopping, defined as alternating DNA binding and 3D diffusion that has a distance dependent finding time, emerged naturally from our simulations; however, it required relatively rapid DNA binding and was only effective over a short distance. Additionally, our model reiterated the importance of intersegmental transfer. It showed that intersegmental transfer is essential for efficient target finding, and that this has likely led to a

substantial enrichment of dimers and tetramers for sequence specific DNA binding proteins through evolution.

Supporting Information

Figure S1   Reconstitution of FRAP recovery curve. As part of our methods verification, we ensured that we could replicate the FRAP recovery studies of Karpova *et al.* (2008) [59]. This figure shows one example, a test of the unbinding constant of a TF from its TG. As in the work of Karpova *et al.* (2008), half unbinding is reached at c. 40 seconds, and full unbinding at c. 2 minutes.

Figure S2   Distance dependence of finding times in 1D, 2D, and 3D systems. The mean time for a single TF to locate and bind a single TG is shown as a function of their initial separation for 1D, 2D, and 3D systems, in (A), (B), and (C), respectively. As expected, results are strongly distance dependent in 1D, moderately distance dependent in 2D, and nearly distance independent in 3D. Simulation parameters: $D = 2.72 \ \mu m^2/s$, $\sigma_b = 2$ nm, binding was irreversible, and the system was 2 μm wide in each dimension; 5 TFs were started at fixed distances away from 5 TGs that were all located in the centre of the system. For each simulation, we computed the mean binding time for these 5 TFs. We repeated simulations 100 times each and computed the means (solid line) and standard deviations (error bars) of the mean binding times.

Table S1   Summary of the simulation parameters used.

Code S1   Archive of Smoldyn configuration files, MATLAB scripts and Python code for simulations and data analysis. See enclosed README file for details.

Acknowledgments

We thank Steve Oliver and Boris Adryan for valuable discussions and insightful comments on the manuscript, and Mark Calleja and Jenny Barna for maintaining and helping with CamGrid.

Author Contributions

Conceived and designed the experiments: HGS KL. Performed the experiments: HGS KL. Analyzed the data: HGS SS SSA KL. Contributed reagents/materials/analysis tools: SSA KL. Wrote the paper: HGS SS SSA KL.

References

1. Luger K, Mäder AW, Richmond RK, Sargent DF, Richmond TJ (1997) Crystal structure of the nucleosome core particle at 2.8 A resolution. Nature 389: 251–260. doi:10.1038/38444.
2. Schlick T, Hayes J, Grigoryev S (2012) Toward convergence of experimental studies and theoretical modeling of the chromatin fiber. Journal of Biological Chemistry 287: 5183–5191. doi:10.1074/jbc.R111.305763.
3. Lieberman-Aiden E, Van Berkum NL, Williams L, Imakaev M, Ragoczy T, et al. (2009) Comprehensive Mapping of Long-Range Interactions Reveals Folding Principles of the Human Genome. Science 326: 289–293. doi:10.1126/science.1181369.
4. Hirayama H, Okita O (2004) Computation of diffusion limited controlled actions for gene regulating repressor particles. J Biol Phys 30: 1–31. doi:10.1023/B:JOBP.0000016448.26081.01.
5. Jayaram B, Das A, Aneja N (1996) Energetic and kinetic aspects of macromolecular association: a computational study of λ repressor-operator complexation. Journal of Molecular Structure: THEOCHEM.
6. Young MA, Jayaram B, Beveridge DL (1997) Intrusion of Counterions into the Spine of Hydration in the Minor Groove of B-DNA: Fractional Occupancy of Electronegative Pockets. J Am Chem Soc 119: 59–69. doi:10.1021/ja960459m.

7. Berg OG, Winter RB, Hippel von PH (1981) Diffusion-driven mechanisms of protein translocation on nucleic acids. 1. Models and theory. Biochemistry 20: 6929–6948.
8. Berg OG, Winter RB, Hippel von PH (1982) How do genome-regulatory proteins locate their DNA target sites? Trends in Biochemical Sciences 7: 52–55. doi:10.1016/0968-0004(82)90075-5.
9. Berg OG, Hippel von PH (1985) Diffusion-controlled macromolecular interactions. Annu Rev Biophys Biophys Chem 14: 131–160. doi:10.1146/annurev.bb.14.060185.001023.
10. Gowers DM, Halford SE (2003) Protein motion from non-specific to specific DNA by three-dimensional routes aided by supercoiling. EMBO J 22: 1410–1418. doi:10.1093/emboj/cdg125.
11. van den Broek B, Lomholt MA, Kalisch S-MJ, Metzler R, Wuite GJL (2008) How DNA coiling enhances target localization by proteins. Proc Natl Acad Sci USA 105: 15738–15742. doi:10.1073/pnas.0804248105.
12. Lomholt MA, van den Broek B, Kalisch S-MJ, Wuite GJL, Metzler R (2009) Facilitated diffusion with DNA coiling. Proc Natl Acad Sci USA 106: 8204–8208. doi:10.1073/pnas.0903293106.
13. Hippel von PH, Berg OG. Facilitated target location in biological systems. J Biol Chem 1989 Jan 15;264(2):675–8.

14. Seksek O, Biwersi J, Verkman AS (1997) Translational diffusion of macromolecule-sized solutes in cytoplasm and nucleus. J Cell Biol 138: 131–142.

15. Dross N, Spriet C, Zwerger M, Müller G, Waldeck W, et al. (2009) Mapping eGFP oligomer mobility in living cell nuclei. PLoS ONE 4: e5041. doi:10.1371/journal.pone.0005041.

16. Halford SE (2009) An end to 40 years of mistakes in DNA–protein association kinetics? Biochem Soc Trans 37: 343. doi:10.1042/BST0370343.

17. Veksler A, Kolomeisky AB (2013) Speed-Selectivity Paradox in the Protein Search for Targets on DNA: Is It Real or Not? J Phys Chem B 117: 12695–12701. doi:10.1021/jp311466f.

18. Chen J, Zhang Z, Li L, Chen B-C, Revyakin A, et al. (2014) Single-molecule dynamics of enhanceosome assembly in embryonic stem cells. Cell 156: 1274–1285. doi:10.1016/j.cell.2014.01.062.

19. Vuzman D, Levy Y (2010) DNA search efficiency is modulated by charge composition and distribution in the intrinsically disordered tail. Proc Natl Acad Sci USA 107: 21004–21009. doi:10.1073/pnas.1011775107.

20. Vuzman D, Azia A, Levy Y (2010) Searching DNA via a "Monkey Bar" mechanism: the significance of disordered tails. J Mol Biol 396: 674–684. doi:10.1016/j.jmb.2009.11.056.

21. Elf J, Li G-W, Xie XS (2007) Probing transcription factor dynamics at the single-molecule level in a living cell. Science 316: 1191–1194. doi:10.1126/science.1141967.

22. Esadze A, Iwahara J (2014) Stopped-Flow Fluorescence Kinetic Study of Protein Sliding and Intersegment Transfer in the Target DNA Search Process. J Mol Biol 426: 230–244. doi:10.1016/j.jmb.2013.09.019.

23. Bauer M, Metzler R (2013) In vivo facilitated diffusion model. PLoS ONE 8: e53956. doi:10.1371/journal.pone.0053956.

24. Gowers DM, Wilson GG, Halford SE (2005) Measurement of the contributions of 1D and 3D pathways to the translocation of a protein along DNA. Proc Natl Acad Sci USA 102: 15883–15888. doi:10.1073/pnas.0505378102.

25. Kepes F (2003) Periodic epi-organization of the yeast genome revealed by the distribution of promoter sites. J Mol Biol 329: 859–865.

26. Kepes F, Vaillant C (2003) Transcription-Based Solenoidal Model of Chromosomes. Complexus 1: 171–180. doi:10.1159/000082184.

27. Kepes F (2004) Periodic transcriptional organization of the E.coli genome. J Mol Biol 340: 957–964. doi:10.1016/j.jmb.2004.05.039.

28. Junier I, Martin O, Kepes F (2010) Spatial and topological organization of DNA chains induced by gene co-localization. PLoS Computational biology 6: e1000678. doi:10.1371/journal.pcbi.1000678.

29. Shinde UP, Sharma A, Kulkarni BD (1992) Can mobile shapes of fractals cause rate enhancement? Chaos, Solitons & Fractals 1: 401–412.

30. van Berkum NL, Lieberman-Aiden E, Williams L, Imakaev M, Gnirke A, et al. (2010) Hi-C: a method to study the three-dimensional architecture of genomes. J Vis Exp. doi:10.3791/1869.

31. Koslover EF, Díaz de la Rosa MA, Spakowitz AJ (2011) Theoretical and computational modeling of target-site search kinetics in vitro and in vivo. Biophys J 101: 856–865. doi:10.1016/j.bpj.2011.06.066.

32. Halford SE (2004) How do site-specific DNA-binding proteins find their targets? Nucleic Acids Research 32: 3040–3052. doi:10.1093/nar/gkh624.

33. Iwahara J, Clore GM (2006) Detecting transient intermediates in macromolecular binding by paramagnetic NMR. Nature 440: 1227–1230. doi:10.1038/nature04673.

34. Doucleff M, Clore GM (2008) Global jumping and domain-specific intersegment transfer between DNA cognate sites of the multidomain transcription factor Oct-1. Proc Natl Acad Sci USA 105: 13871–13876. doi:10.1073/pnas.0805050105.

35. Sidorova NY, Scott T, Rau DC (2013) DNA concentration-dependent dissociation of EcoRI: direct transfer or reaction during hopping. Biophys J 104: 1296–1303. doi:10.1016/j.bpj.2013.01.041.

36. Riggs AD, Suzuki H, Bourgeois S (1970) Lac repressor-operator interaction. I. Equilibrium studies. J Mol Biol 48: 67–83.

37. Riggs AD, Bourgeois S, Cohn M (1970) The lac repressor-operator interaction. 3. Kinetic studies. J Mol Biol 53: 401–417.

38. Wang YM, Austin RH, Cox EC (2006) Single molecule measurements of repressor protein 1D diffusion on DNA. Phys Rev Lett 97: 048302.

39. Vuzman D, Polonsky M, Levy Y (2010) Facilitated DNA search by multidomain transcription factors: cross talk via a flexible linker. Biophys J 99: 1202–1211. doi:10.1016/j.bpj.2010.06.007.

40. Ruusala T, Crothers DM (1992) Sliding and intermolecular transfer of the lac repressor: kinetic perturbation of a reaction intermediate by a distant DNA sequence. Proc Natl Acad Sci USA.; 89:4903–7.

41. Zawel L, Kumar KP, Reinberg D (1995) Recycling of the general transcription factors during RNA polymerase II transcription. Genes Dev 9: 1479–1490.

42. Halford SE, Szczelkun MD (2002) How to get from A to B: strategies for analysing protein motion on DNA. Eur Biophys J 31: 257–267. doi:10.1007/s00249-002-0224-4.

43. Slutsky M, Mirny LA (2004) Kinetics of protein-DNA interaction: facilitated target location in sequence-dependent potential. Biophys J 87: 4021–35.

44. Wunderlich Z, Mirny LA (2008) Spatial effects on the speed and reliability of protein-DNA search. Nucleic Acids Research 36: 3570–3578. doi:10.1093/nar/gkn173.

45. Hu L, Grosberg AY, Bruinsma R (2008) Are DNA transcription factor proteins maxwellian demons? Biophys J 95: 1151–1156. doi:10.1529/biophysj.108.129825.

46. Dostie J, Richmond TA, Arnaout RA, Selzer RR, Lee WL, et al. (2006) Chromosome Conformation Capture Carbon Copy (5C): A massively parallel solution for mapping interactions between genomic elements. Genome Res 16: 1299–309.

47. Das RK, Kolomeisky AB (2010) Facilitated search of proteins on DNA: correlations are important. Phys Chem Chem Phys;12: 2999.

48. Mirny L, Slutsky M, Wunderlich Z, Tafvizi A, Leith J, et al. (2009) How a protein searches for its site on DNA: the mechanism of facilitated diffusion. J Phys A: Math Theor 42: 434013. doi:10.1088/1751-8113/42/43/434013.

49. Zabet NR, Adryan B (2012) Computational models for large-scale simulations of facilitated diffusion. Mol Biosyst 8: 2815–27 doi:10.1039/c2mb25201e.

50. Ricchetti M, Metzger W, Heumann H (1988) One-dimensional diffusion of Escherichia coli DNA-dependent RNA polymerase: a mechanism to facilitate promoter location. Proc Natl Acad Sci USA 85: 4610–4.

51. Shimamoto N (1999) One-dimensional Diffusion of Proteins along DNA. J Biol Chem 274: 15293–6.

52. DePristo MA, Chang L, Vale RD, Khan SM, Lipkow K (2009) Introducing simulated cellular architecture to the quantitative analysis of fluorescent microscopy. Prog Biophys Mol Biol 100: 25–32. doi:10.1016/j.pbiomolbio.2009.07.002.

53. Andrews SS, Addy NJ, Brent R, Arkin AP (2010) Detailed simulations of cell biology with Smoldyn 2.1. PLoS Comput Biol 6: e1000705. doi:10.1371/journal.pcbi.1000705.

54. Singh P, Hockenberry AJ, Tiruvadi VR, Meaney DF (2011) Computational investigation of the changing patterns of subtype specific NMDA receptor activation during physiological glutamatergic neurotransmission. PLoS Comput Biol 7: e1002106. doi:10.1371/journal.pcbi.1002106.

55. Khan S, Zou Y, Amjad A, Gardezi A, Smith CL, et al. (2011) Sequestration of CaMKII in dendritic spines in silico. J Comput Neurosci 31: 581–594. doi:10.1007/s10827-011-0323-2.

56. Sewitz S, Lipkow K (2013) Simulating bacterial chemotaxis at high spatiotemporal detail. Current Chemical Biology 7:214–223. doi: 10.2174/2212798070314050810 1810.

57. Jorgensen P, Edgington NP, Schneider BL, Rupes I, Tyers M, et al. (2007) The size of the nucleus increases as yeast cells grow. Mol Biol Cell 18: 3523–3532. doi:10.1091/mbc.E06-10-0973.

58. Mandelkern M, Elias JG, Eden D, Crothers DM (1981) The dimensions of DNA in solution. J Mol Biol 152: 153–161.

59. Karpova TS, Kim MJ, Spriet C, Nalley K, Stasevich TJ, et al. (2008) Concurrent fast and slow cycling of a transcriptional activator at an endogenous promoter. Science 319: 466–469. doi:10.1126/science.1150559.

60. Kim JH, Larson RG (2007) Single-molecule analysis of 1D diffusion and transcription elongation of T7 RNA polymerase along individual stretched DNA molecules. Nucleic Acids Research 35: 3848–3858. doi:10.1093/nar/gkm332.

61. Biebricher A, Wende W, Escudé C, Pingoud A, Desbiolles P (2009) Tracking of single quantum dot labeled EcoRV sliding along DNA manipulated by double optical tweezers. Biophys J 96: L50–L52. doi:10.1016/j.bpj.2009.01.035.

62. Tafvizi A, Huang F, Leith JS, Fersht AR, Mirny LA, et al. (2008) Tumor suppressor p53 slides on DNA with low friction and high stability. Biophys J 95: L01–L03. doi:10.1529/biophysj.108.134122.

63. Schurr JM (1979) The one-dimensional diffusion coefficient of proteins absorbed on DNA. Hydrodynamic considerations. Biophys Chem 9: 413–414.

64. de la Rosa MAD, Koslover EF, Mulligan PJ, Spakowitz AJ (2010) Dynamic strategies for target-site search by DNA-binding proteins. Biophys J 98: 2943–2953. doi:10.1016/j.bpj.2010.02.055.

65. Etson CM, Hamdan SM, Richardson CC, van Oijen AM (2010) Thioredoxin suppresses microscopic hopping of T7 DNA polymerase on duplex DNA. Proc Natl Acad Sci USA 107: 1900–1905. doi:10.1073/pnas.0912664107.

66. Phair RD, Scaffidi P, Elbi C, Vecerová J, Dey A, et al. (2004) Global nature of dynamic protein-chromatin interactions in vivo: three-dimensional genome scanning and dynamic interaction networks of chromatin proteins. Mol Cell Biol 24: 6393–6402. doi:10.1128/MCB.24.14.6393-6402.2004.

67. Bordelon T, Wilkinson SP, Grove A, Newcomer ME (2006) The Crystal Structure of the Transcriptional Regulator HucR from Deinococcus radiodurans Reveals a Repressor Preconfigured for DNA Binding. J Mol Biol 360: 168–177. doi:10.1016/j.jmb.2006.05.005.

68. Wilkinson SP, Grove A (2006) Ligand-responsive transcriptional regulation by members of the MarR family of winged helix proteins. Curr Issues Mol Biol 8: 51–62.

69. Poon KK, Chen CL, Wong SL (2001) Roles of glucitol in the GutR-mediated transcription activation process in Bacillus subtilis: tight binding of GutR to tis binding site. J Biol Chem 276: 9620–9625. doi:10.1074/jbc.M009864200.

70. Brantl S, Licht A (2010) Characterisation of Bacillus subtilis transcriptional regulators involved in metabolic pathways. Curr Protein Pept Sci 11: 274–291.

71. Majka J, Speck C (2007) Analysis of protein-DNA interactions using surface plasmon resonance. Adv Biochem Eng Biotechnol 104: 13–36.

72. Bryan D, Aylwin SJ, Newman DJ, Burrin JM (1999) Steroidogenic factor 1-DNA binding: a kinetic analysis using surface plasmon resonance. J Mol Endocrinol 22: 241–249.

73. Jeong EH, Son YM, Hah Y-S, Choi YJ, Lee KH, et al. (2006) RshA mimetic peptides inhibiting the transcription driven by a Mycobacterium tuberculosis sigma factor SigH. Biochem Biophys Res Commun 339: 392–398. doi:10.1016/j.bbrc.2005.11.032.

74. Vainus D (2004) Investigation of Interactions between Homeodomain Proteins and DNA. PhD thesis, Mathematical Biology, University of Göttingen.

75. Kumar N, Patowary A, Sivasubbu S, Petersen M, Maiti S (2008) Silencing c-MYC expression by targeting quadruplex in P1 promoter using locked nucleic acid trap. Biochemistry 47: 13179–13188. doi:10.1021/bi801064j.

76. Patikoglou G, Burley SK (1997) Eukaryotic transcription factor-DNA complexes. Annu Rev Biophys Biomol Struct 26: 289–325. doi:10.1146/annurev.biophys.26.1.289.

77. Ujvári A, Martin CT (1996) Thermodynamic and kinetic measurements of promoter binding by T7 RNA polymerase. Biochemistry 35: 14574–14582. doi:10.1021/bi961165g.

78. Minton AP (2001) The Influence of Macromolecular Crowding and Macromolecular Confinement on Biochemical Reactions in Physiological Media. J Biol Chem 276: 10577–10580. doi:10.1074/jbc.R100005200.

79. Schmidt H (2013) Does Transcription Activator Diffusion Drive Gene Clustering in Eukaryotes? PhD Thesis, Department of Biochemistry, University of Cambridge.

80. Lipkow K, Andrews SS, Bray D (2005) Simulated diffusion of phosphorylated CheY through the cytoplasm of Escherichia coli. J Bacteriol 187: 45–53. doi:10.1128/JB.187.1.45-53.2005.

81. Rezania V, Tuszynski J, Hendzel M (2007) Modeling transcription factor binding events to DNA using a random walker/jumper representation on a 1D/2D lattice with different affinity sites. Phys Biol 4: 256–267. doi:10.1088/1478-3975/4/4/003.

82. Andrews SS, Bray D (2004) Stochastic simulation of chemical reactions with spatial resolution and single molecule detail. Phys Biol 1: 137–151. doi:10.1088/1478-3967/1/3/001.

83. Andrews SS (2009) Accurate particle-based simulation of adsorption, desorption and partial transmission. Phys Biol 6: 046015. doi:10.1088/1478-3975/6/4/046015.

84. Ricchetti M, Metzger W, Heumann H (1988) One-dimensional diffusion of Escherichia coli DNA-dependent RNA polymerase: a mechanism to facilitate promoter location. Proc Natl Acad Sci USA 85: 4610–4614.

85. Shimamoto N (1999) One-dimensional diffusion of proteins along DNA. Its biological and chemical significance revealed by single-molecule measurements. J Biol Chem 274: 15293–15296.

86. Mirny LA (2010) Nucleosome-mediated cooperativity between transcription factors. Proc Natl Acad Sci USA 107: 22534–22539. doi:10.1073/pnas.0913805107.

87. Loverdo C, Bénichou O, Voituriez R, Biebricher A, Bonnet I, et al. (2009) Quantifying hopping and jumping in facilitated diffusion of DNA-binding proteins. Phys Rev Lett 102: 188101.

88. Givaty O, Levy Y (2009) Protein sliding along DNA: dynamics and structural characterization. J Mol Biol 385: 1087–1097. doi:10.1016/j.jmb.2008.11.016.

89. DeSantis MC, Li J-L, Wang YM (2011) Protein sliding and hopping kinetics on DNA. Phys Rev E Stat Nonlin Soft Matter Phys 83: 021907.

90. Schonhoft JD, Stivers JT (2012) Timing facilitated site transfer of an enzyme on DNA. Nat Chem Biol 8: 205–210. doi:10.1038/nchembio.764.

91. Larson DR, Zenklusen D, Wu B, Chao JA, Singer RH. Real-Time Observation of Transcription Initiation and Elongation on an Endogenous Yeast Gene. Science 2011 Apr 21;332(6028):475–8.

92. Gorman J, Plys AJ, Visnapuu M-L, Alani E, Greene EC (2010) Visualizing one-dimensional diffusion of eukaryotic DNA repair factors along a chromatin lattice. Nat Struct Mol Biol 17: 932–938. doi:10.1038/nsmb.1858.

93. Gorman J, Chowdhury A, Surtees JA, Shimada J, Reichman DR, et al. (2007) Dynamic basis for one-dimensional DNA scanning by the mismatch repair complex Msh2-Msh6. Mol Cell 28: 359–370. doi:10.1016/j.molcel.2007.09.008.

94. Xia Y, Chen E, Liang D (2010) Recognition of Single- and Double-Stranded Oligonucleotides by Bovine Serum Albumin via Nonspecific Interactions. Biomacromolecules 11: 3158–3166. doi:10.1021/bm100969z.

95. Levy ED, Pereira-Leal JB, Chothia C, Teichmann SA (2006) 3D complex: a structural classification of protein complexes. PLoS Computational biology 2: e155. doi:10.1371/journal.pcbi.0020155.

96. Berman HM, Westbrook J, Feng Z, Gilliland G, Bhat TN, et al. (2000) The Protein Data Bank. Nucleic Acids Research 28: 235–242.

97. Surby MA, Reich NO (1996) Contribution of facilitated diffusion and processive catalysis to enzyme efficiency: implications for the EcoRI restriction-modification system. Biochemistry 35: 2201–2208. doi:10.1021/bi951883n.

98. Ryazanova AY, Kubareva EA, Grman I, Lavrova NV, Ryazanova EM, et al. (2011) The study of the interaction of (cytosine-5)-DNA methyltransferase SsoII with DNA by acoustic method. Analyst 136: 1227–1233. doi:10.1039/c0an00545b.

99. Zharkov DO, Shoham G, Grollman AP (2003) Structural characterization of the Fpg family of DNA glycosylases. DNA Repair (Amst) 2: 839–862.

100. Schreiber G, Shaul Y, Gottschalk KE (2006) Electrostatic design of protein-protein association rates. Methods Mol Biol 340: 235–249. doi:10.1385/1-59745-116-9:235.

101. Coppey M, Bénichou O, Voituriez R, Moreau M (2004) Kinetics of target site localization of a protein on DNA: a stochastic approach. Biophysj 87: 1640–1649. doi:10.1529/biophysj.104.045773.

102. Pingoud A, Jeltsch A (1997) Recognition and cleavage of DNA by type-II restriction endonucleases. Eur J Biochem 246: 1–22.

103. Hsieh M, Brenowitz M (1997) Comparison of the DNA association kinetics of the Lac repressor tetramer, its dimeric mutant LacIadi, and the native dimeric Gal repressor. J Biol Chem 272: 22092–22096.

104. Yadon AN, Van de Mark D, Basom R, Delrow J, Whitehouse I, et al. (2010) Chromatin Remodeling around Nucleosome-Free Regions Leads to Repression of Noncoding RNA Transcription. Mol Cell Biol 30: 5110–5122. doi:10.1128/MCB.00602-10.

105. Sun W, Xie W, Xu F, Grunstein M, Li K-C (2009) Dissecting nucleosome free regions by a segmental semi-Markov model. PLoS ONE 4: e4721. doi:10.1371/journal.pone.0004721.

106. Ioshikhes IP, Albert I, Zanton SJ, Pugh BF (2006) Nucleosome positions predicted through comparative genomics. Nat Genet 38: 1210–1215. doi:10.1038/ng1878.

107. Bernstein BE, Liu CL, Humphrey EL, Perlstein EO, Schreiber SL (2004) Global nucleosome occupancy in yeast. Genome Biol 5: R62. doi:10.1186/gb-2004-5-9-r62.

108. Lee W, Tillo D, Bray N, Morse RH, Davis RW, et al. (2007) A high-resolution atlas of nucleosome occupancy in yeast. Nat Genet 39: 1235–1244. doi:10.1038/ng2117.

109. Whitehouse I, Rando OJ, Delrow J, Tsukiyama T (2007) Chromatin remodelling at promoters suppresses antisense transcription. Nature 450: 1031–1035. doi:10.1038/nature06391.

110. Yuan G-C, Liu Y-J, Dion MF, Slack MD, Wu LF, et al. (2005) Genome-scale identification of nucleosome positions in S. cerevisiae. Science 309: 626–630. doi:10.1126/science.1112178.

111. Mavrich TN, Ioshikhes IP, Venters BJ, Jiang C, Tomsho LP, et al. (2008) A barrier nucleosome model for statistical positioning of nucleosomes throughout the yeast genome. Genome Res 18: 1073–1083. doi:10.1101/gr.078261.108.

112. Rando OJ, Ahmad K (2007) Rules and regulation in the primary structure of chromatin. Curr Opin Cell Biol 19: 250–256. doi:10.1016/j.ceb.2007.04.006.

113. Whitehouse I, Tsukiyama T (2006) Antagonistic forces that position nucleosomes in vivo. Nat Struct Mol Biol 13: 633–640. doi:10.1038/nsmb1111.

114. Shivaswamy S, Bhinge A, Zhao Y, Jones S, Hirst M, et al. (2008) Dynamic remodeling of individual nucleosomes across a eukaryotic genome in response to transcriptional perturbation. PLoS Biol 6: e65. doi:10.1371/journal.pbio.0060065.

115. Shen CH, Clark DJ (2001) DNA sequence plays a major role in determining nucleosome positions in yeast CUP1 chromatin. J Biol Chem 276: 35209–35216. doi:10.1074/jbc.M104733200.

116. Vuzman D, Hoffman Y, Levy Y (2012) Modulating protein-DNA interactions by post-translational modifications at disordered regions. Pac Symp Biocomput: 188–199.

117. Sokolov IM, Metzler R, Pant K, Williams MC (2005) Target search of N sliding proteins on a DNA. Biophys J 89: 895–902. doi:10.1529/biophysj.104.057612.

118. Levy ED, Pereira-Leal JB, Chothia C, Teichmann SA (2006) 3D Complex: a Structural Classification of Protein Complexes. PLoS Comput Biol 2: e155.

A Polyketide Synthase Acyltransferase Domain Structure Suggests a Recognition Mechanism for Its Hydroxymalonyl-Acyl Carrier Protein Substrate

Hyunjun Park, Brian M. Kevany[¤a], David H. Dyer[¤b], Michael G. Thomas*, Katrina T. Forest*

Department of Bacteriology, University of Wisconsin-Madison, Madison, Wisconsin, United States of America

Abstract

We have previously shown that the acyl transferase domain of ZmaA (ZmaA-AT) is involved in the biosynthesis of the aminopolyol polyketide/nonribosomal peptide hybrid molecule zwittermicin A from *cereus* UW85, and that it specifically recognizes the precursor hydroxymalonyl-acyl carrier protein (ACP) and transfers the hydroxymalonyl extender unit to a downstream second ACP via a transacylated AT domain intermediate. We now present the X-ray crystal structure of ZmaA-AT at a resolution of 1.7 Å. The structure shows a patch of solvent-exposed hydrophobic residues in the area where the AT is proposed to interact with the precursor ACP. We addressed the significance of the AT/ACP interaction in precursor specificity of the AT by testing whether malonyl- or methylmalonyl-ACP can be recognized by ZmaA-AT. We found that the ACP itself biases extender unit selection. Until now, structural information for ATs has been limited to ATs specific for the CoA-linked precursors malonyl-CoA and (2S)-methylmalonyl-CoA. This work contributes to polyketide synthase engineering efforts by expanding our knowledge of AT/substrate interactions with the structure of an AT domain that recognizes an ACP-linked substrate, the rare hydroxymalonate. Our structure suggests a model in which ACP interaction with a hydrophobic motif promotes secondary structure formation at the binding site, and opening of the adjacent substrate pocket lid to allow extender unit binding in the AT active site.

Editor: Avadhesha Surolia, Indian Institute of science, India

Funding: This work was funded by the National Institutes of Health grants GM100346 (to MGT and KTF), AI065850 (to MGT) and GM59721 (to KTF). Use of the Advanced Photon Source, an Office of Science User Facility operated for the U.S. Department of Energy (DOE) Office of Science by Argonne National Laboratory, was supported by the U.S. DOE under Contract No. DE-AC02-06CH11357. Use of the LS-CAT Sector 21 was supported by the Michigan Economic Development Corporation and the Michigan Technology Tri-Corridor (Grant 085P1000817). The funders had no role in study design, data collection and analysis, decision to publish, or preparation of the manuscript.

Competing Interests: Funding provided to support the LS-CAT data collection facility is provided in part by the Michigan Economic Development Corporation and the Michigan Technology Tri-Corridor. However, there are absolutely no constraints on the use of data collected at the facility, on patent rights, o n employment opportunities, etc. because of this funding.

* Email: thomas@bact.wisc.edu (MGT); forest@bact.wisc.edu (KTF)

¤a Current address: Department of Pharmacology, School of Medicine, Case Western Reserve University, Cleveland, Ohio, United States of America
¤b Current address: Department of Biochemistry, University of Wisconsin-Madison, Madison, Wisconsin, United States of America

Introduction

Fatty acids of various lengths and oxidation states are biosynthesized from malonyl-CoA and (2S)-methylmalonyl-CoA by fatty acid synthases (FASs). In contrast to FASs, the evolutionarily related polyketide synthases (PKSs), which catalyze the biosynthesis of the pharmaceutically important class of natural products called polyketides [1], are able to use a far greater repertoire of substrates [2]. The acquisition of this extended biosynthetic vocabulary by PKSs enables these enzymes to catalyze the formation of molecules with great structural and functional diversity. This diverse group includes molecules with antibacterial, antifungal, antitumor, and anticholesterol properties.

Given that PKSs descended from FASs, it is reasonable to assume that the substrates initially utilized by PKSs were limited to malonyl-CoA and (2S)-methylmalonyl-CoA. Coincidently, the PKSs that were first analyzed, and have therefore served as model

systems for PKS research, only used these two molecules as substrates. However, the evolution of PKSs resulted in the inclusion of many more molecules as polyketide substrates, and in recent years our understanding of PKSs has also progressed past relatively simple systems to include PKSs that use this expanded substrate repertory to form highly specialized structures. Engineering previously characterized PKSs to incorporate non-cognate substrates containing unique functional groups, just as nature has done, is a significant goal in natural products research.

The effort to rationally reprogram PKSs to generate useful natural product analogs must begin with a solid foundation of basic PKS enzymology. PKSs are megasynthases that catalyze the decarboxylative Claisen condensation of various short carboxylic acid precursors, the first one referred to as the starter unit, and then extender units thereafter. Despite the vast structural diversity of polyketide molecules, PKSs (like FASs) comprise highly conserved discrete functional domains and linkers; each element plays a specific role such as recognition and incorporation, condensation, or modification of extender units [3]. The acyltransferase (AT) domain in PKSs is considered the gatekeeper domain because its function is to recognize a particular thioesterified extender unit with high specificity and to transacylate it onto a downstream acyl carrier protein (ACP) domain. This transacylation reaction proceeds via a ping-pong mechanism. The first half of the reaction consists of the AT receiving the extender unit from the carrier portion of the substrate, resulting in the esterification of the moiety on the side chain of the active site serine residue [1]. In the second half of the reaction, the extender unit is transferred from the active site serine residue of the AT onto the 4′-phosphopantetheinyl arm of the downstream ACP. This second step requires all ATs to make protein-protein interactions with their partner downstream ACP domains.

The majority of AT domains characterized so far are either malonyl-CoA or (2S)-methylmalonyl-CoA specific. More rarely, AT domains are specific for an ACP-tethered extender unit, such as methoxymalonyl-ACP, hydroxymalonyl-ACP, and aminomalonyl-ACP, the final two having been identified during our analysis of zwittermicin A (ZMA) biosynthesis (Figure 1A) [2]. For ATs that are specific for extender units carried by CoA, the substrate recognition step requires a protein-small molecule interaction between the AT and CoA, whereas for ATs specific for extender units with ACP carriers, this involves an additional protein-protein interaction.

ZMA is a polyketide/nonribosomal peptide hybrid antibiotic produced by *Bacillus cereus* strains UW85 and AH1134 that exhibits activity against a variety of Gram-negative and Gram-positive bacteria, as well as certain protists and plant pathogenic fungi [4–6]. Structural analysis of ZMA [7–9] revealed an aminopolyol structure with ethanolamine and glycolyl moieties that are rare in natural products, leading our group to focus on this biosynthetic aspect of ZMA. Based on our genetic and biochemical analyses (Figure 1B) [10–13] we proposed that ZMA biosynthesis involves the synthesis of an inactive larger molecule that is processed at both its amino and carboxy termini, releasing an amino-terminal acyl-D-aspartate (Figure 1C; metabolite A), the central ZMA molecule, and a carboxyl-terminal pyruvyl-L-leucyl-L-methionine (Figure 1C; metabolite B). Our proposed mechanism of ZMA activation by a D-amino acid peptidase, which cleaves the amino-terminal acyl-D-aspartate metabolite to release the active form of ZMA, was the first example of a natural prodrug biosynthetic scheme, also found to be involved in colibactin activation (Figure 1) [14–16]. Our analyses also revealed the existence of two rare PKS extender units aminomalonyl-ACP and hydroxymalonyl-ACP [12], and the AT domains that are specific for them, ZmaF and ZmaA-AT [13], respectively.

We have focused much of our analysis on the formation of hydroxymalonyl-ACP and aminomalonyl-ACP and the subsequent incorporation of the extender units by AT domains because the hydroxyl- and amino-groups originating from the C2 position of these extender units protrude away from the polyketide backbone, potentially serving critical functions or providing useful handles for downstream semi-synthetic modifications. For these reasons it is desirable to harness the ability to place these extender units in non-natural PKS settings. To do this, it is essential to understand how the respective AT domains recognize and incorporate these rare polyketide precursors.

Previous studies have identified four conserved regions that contribute to the molecular basis of AT substrate specificity [17]. In primary sequence order these are the RVDVVQ motif, the GHSXG motif centered on the active site serine residue, the YASH motif containing the histidine that is part of the catalytic

dyad, and the last ∼30 residues of the AT domain (∼L378-S407). In addition to the four motifs that are implicated in extender unit recognition, RXR(X)₅YASH has been implicated in the AT/substrate carrier recognition [18–22].

To further shed light on the substrate selection mechanism of these AT domains, we have solved the crystal structure of the hydroxymalonyl-ACP-specific ZmaA-AT domain. The structures of AT domains published to date include PKS AT domains involved in the biosynthesis of erythromycin A, pikromycin, dynemicin, and disorazole, as well as FAS AT domain homologs (malonyl-CoA:ACP transacylases) from *Escherichia coli* and *Streptomyces coelicolor* [18,19,21,23–26]. Although these structures in the database reflect a considerable phylogenetic diversity, they are limited to recognizing malonyl-CoA or (2S)-methylmalonyl-CoA as their substrate. The structure presented here of a hydroxymalonyl-ACP-specific AT expands our understanding of AT domain recognition of ACP-linked extender units. The crystal structure of ZmaA-AT reveals an unusual solvent-exposed patch of hydrophobic residues in the proposed AT-ACP interaction surface. *In vitro* assays confirmed that this interaction plays a significant role in substrate recognition. The three-dimensional coordinates for ZmaA-AT allow us to compare the structure of an AT that is specific for an ACP-linked extender unit to the previously published structures of AT domains specific for CoA linked extender units that are involved in both polyketide and fatty acid biosynthesis. The crystal structure will be critical to the achievement of future PKS reprogramming efforts, where different substituents at the C2 position of the extender unit are desired for improved function or semi-synthetic amenability of the final PKS product.

Materials and Methods

Cloning of *zmaA* fragment *zmaA-AT*

The fragment of *zmaA* coding for the AT domain was cloned into *E. coli* expression vector pET-30a(+) (Novagen), using standard PCR-based cloning techniques, as described previously [13]. The following primers were used to introduce the gene fragment into the vector, resulting in the production of a protein containing an N-terminal histidine tag: 5′-GCACCAACCATG-GAAGCAACATCAAATAGT-3′ and 5′-TATTTTCTCGAGA-GACTACATTGGTAATGGGA-3′.

Overproduction and purification of ZmaA-AT

pET-30a(+) containing *zmaA-AT* was introduced into *E. coli* Rosetta(DE3) (Novagen) and grown to an OD₆₀₀ of 0.5 at 30°C, in lysogeny broth containing 50 μg/mL kanamycin and 15 μg/mL chloramphenicol. The temperature was reduced to 15°C and after 1 h overexpression was induced with IPTG at a final concentration of 60 μM. After 16 h, cells were harvested by centrifugation and the cell pellet was frozen at −20°C. The cell pellet was resuspended in buffer containing 20 mM Tris-HCl (pH 8.0), 300 mM NaCl and sonicated (Fisher 550 Sonic Dismembrator, power = 5, 15 min sonication with 1 s on, 1 s off). Sonicated cells were centrifuged for 30 min at 4°C to remove cell debris and insoluble protein (15,000 rpm, Beckman Model J221 centrifuge, JA-25.5 rotor, 4°C, for 30 min). ZmaA-AT was purified from the cell-free extract by nickel-affinity chromatography as previously described [27]. To enzymatically cleave the N-terminal histidine tag, the protein was concentrated to 15.5 mg/mL and dialyzed against buffer containing 20 mM Tris-HCl (pH8.0), 50 mM NaCl, and 2 mM CaCl₂. Enterokinase (New England Biolabs) was added to dialyzed protein and incubated at room temperature (22°C) for 16 h. Enterokinase was removed by benzamidine-affinity chro-

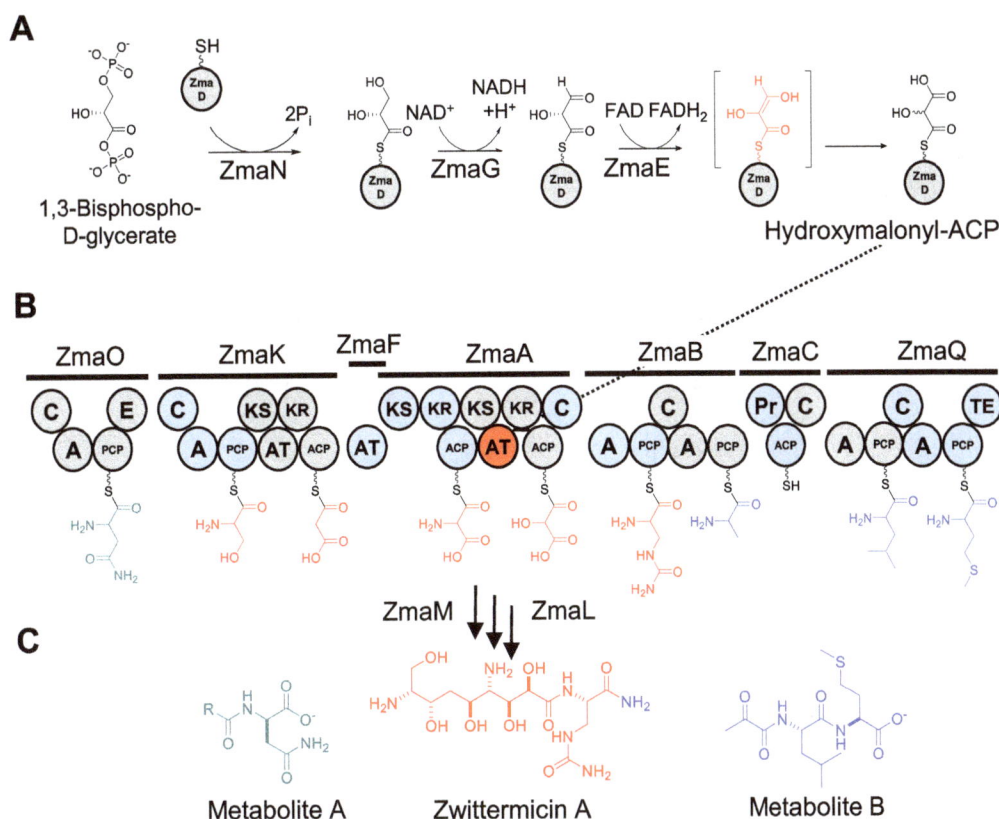

Figure 1. Biosynthesis of ZMA. (A) Biosynthetic pathway of hydroxymalonyl-ACP. The final FAD dependent oxidation step catalyzed by ZmaE may proceed through an endiol intermediate (red), resulting in the loss of stereospecificity at C2 of the final product, hydroxymalonyl-ACP. (B) ZMA PKS/NRPS. Nine extender units are utilized to form the precursor of metabolite A (green), zwittermicin A (red), and metabolite B (blue). Hydroxymalonyl-ACP is recognized by ZmaA (dotted line). Each circle represents a catalytic domain of the PKS/NRPS: C, condensation; A, adenylation; PCP, peptide carrier protein; E, epimerization; KS, ketosynthase; AT, acyltransferase; KR, ketoreductase; ACP, acyl carrier protein; Pr, protease; TE, thioesterase. (C) Natural prodrug activation. ZmaL is proposed to catalyze the cleavage of the ZMA precursor molecule from ZmaB-bound alanine, which is further condensed to leucine and methionine to form metabolite B (blue). ZmaM is proposed to catalyze the separation of metabolite A (green) from ZMA (red).

matography (HiTrap Benzamidine FF, Amersham Biosciences). Fractions containing ZmaA-AT were collected and dialyzed against buffer containing 50 mM Tris-HCl (pH8.0) and 50 mM NaCl then concentrated to 6.7 mg/mL. ZmaA-AT was further purified by size-exclusion chromatography (Superdex 75, Amersham Biosciences). Fractions containing ZmaA-AT were pooled and concentrated to 7 mg/mL.

Crystallization and Data Collection for ZmaA-AT

Initial crystallization conditions were obtained using vapour diffusion of protein (7 mg/ml) diluted with equal volume of mother liquor against the JCSG+ Suite screen (Qiagen). Crystal growth optimization resulted in final mother liquor of 100 mM BisTris pH 5.5, 200 mM $MgCl_2$, 20% PEG 4000 and 800 mM sodium formate. Cryoprotection was achieved by soaking crystals in mother liquor plus 30% glycerol.

Structure Determination

A 2.3 Å resolution native data set collected in house and processed with HKL-2000 [28] provided a highly significant molecular replacement solution using pdb code 2QO3 [23]. However, refinement was unsatisfactory. To overcome this issue

ZmaA-AT was overproduced in the *E. coli* methionine auxotroph B834 (DE3) under conditions that led to incorporation of exogenously provided selenomethionine. The resulting selenomethionine-containing ZmaA-AT was purified to homogeneity. Crystals of this protein were obtained under similar conditions as the protein lacking selenomethionine. A 1.8 Å resolution data set was collected on the MAR 300 detector on beamline 21-ID-D at LS-CAT and processed with HKL-2000 [28]. The peak wavelength provided a strong anomalous signal, and Autorickshaw [29] was used to generate SAD phases. These were combined with a new partial molecular replacement model. Finally, a higher resolution native dataset was obtained, again on beamline 21-ID-D, and used for the ultimate refinement (Table 1). Refinement and fitting were carried out iteratively using REFMAC5 [30] and Coot [31] for final R_{work} and R_{free} values of 17.3 and 20.0%, respectively. Four amino acids at the N-terminus, one at the C-terminus, and three in an internal flexible loop were not observed. All structural images were generated using PyMOL [32].

While preparing figures, we noticed the side chain of Leu192 of PDB entry 2G2Z [18] in an impossible orientation relative to the main chain and thus refitted the side chain to the publicly

Table 1. Data Collection and Refinement Statistics.

Data Collection Statistics

Data Set	Native	SeMet Peak
Space Group	$P2_12_12_1$	$P2_12_12_1$
Unit Cell (Å)	a, b, c = 43.1, 73.2, 123.5	a, b, c = 47.0, 74.3, 125.5
Wavelength (Å)	0.97872	0.97934
Resolution* (Å)	25–1.65 (1.68–1.65)	30.0–1.84 (1.89–1.84)
Unique Reflections	54447 (2669)	37946 (2321)
Total Reflections	375883	514407
Completeness (%)	99.9 (100)	98.1 (85.3)
Redundancy	6.9 (6.0)	14.0 (8.3)
Average I/σ	27.7 (4.1)	46.9 (3.5)
R_{sym} [a] (%)	4.9 (36.3)	10.5 (41.7)
Wilson B value (Å²)	20.8	21.5
Refinement Statistics		
Resolution (Å)	25–1.70 (1.74–1.70)	
Reflections	46276 (3387)	
R_{cryst} [b] (%)	17.3 (19.5)	
R_{free} [c] (%)	20.0 (24.1)	
Average B value (Å²)	22.4	
Protein	21.6	
Ligand	42.5	
Water	28.2	
Est. Coord. Error (Å)	0.10	
Rmsd bonds (Å)	0.016	
Rmsd angles (°)	1.56	
Ramachandran plot		
Favored (%)	98.4	
Allowed (%)	99.8	
Outliers (%)	0.2	

*Highest resolution shells in parentheses.

[a] $R_{sym}(I) = \sum_{hkl}\sum_i |I_i(hkl) - <I(hkl)>|/\sum_{hkl}\sum_i I_i(hkl)$ where $I(i)$ is the intensity of the ith observation of the hkl reflection and $<I(hkl)>$ is the mean intensity from multiple measurements of the h, k, l reflection.

[b] $R_{cryst}(F) = \sum_{hkl}|F_{obs}(hkl)-F_{calc}(hkl)|/\sum_{hkl}F_{obs}(hkl)$, where $F_{obs}(hkl)$ and $F_{calc}(hkl)$ are the observed and calculated structure factor amplitudes for the h, k, l reflection.

[c] R_{free} is R_{cryst} calculated for a randomly selected test set of reflections (5%) not included in the refinement.

available data followed by a single round of real space refinement of residues 191–193 against these data using Coot.

Radioactive Assays of ZmaA-AT with Malonyl-CoA, Methylmalonyl-CoA, Malonyl-ZmaD, and Methylmalonyl-ZmaD

To address the effect of the extender unit carrier on substrate recognition by the AT, ZmaA-AT was incubated with extender units with malonyl- or methylmalonyl- acyl groups on the carriers CoA or ZmaD (the ACP partner for ZmaA-AT) (Figure 1A). The reaction mixtures contained the following: 75mM Tris (pH 7.5), 10 mM MgCl$_2$, 1 mM TCEP, 5 μM ZmaA-AT, and 1 μM Sfp (*Bacillus subtilis* phosphopantetheinyl transferase). 40 μM [^{14}C-C2]malonyl-CoA (*malonyl-CoA) or 40 μM [^{14}C-C2](2-*RS*)-methylmalonyl-CoA (*(2-*RS*)-methylmalonyl-CoA) was added to the reaction mixture either in the presence or absence of 5 μM ZmaD. Reaction mixtures were incubated for 1 h at 22°C and stopped with 50 μL of 2X cracking buffer [120 mM Tris-HCl (pH 6.8), 2% (v/v) β-mercaptoethanol, 1% (w/v) sodium dodecyl

sulfate (SDS), 25% (v/v) glycerol, and 0.02% (w/v) bromophenol blue]. 30 μL was loaded onto a 15% polyacrylamide-SDS gel. The gel was stained with Coomassie Brilliant Blue, destained, dried, and exposed to a phosphorimaging screen and scanned with a Typhoon imager following a 4 day exposure. The scanned image was quantified using ImageJ [33] to determine the relative band intensities.

Results

Overall Structure of Zma-AT

The crystal structure of the ZmaA-AT domain, along with its N-terminal ketosynthase (KS)-AT linker, and 20 residue C-terminal post-AT linker, was refined against 1.7 Å resolution X-ray diffraction data (Table 1). The overall structure is similar to the analogous regions of the KS-AT domain pairs in modules 3 and 5 of the 6-deoxyerythronolide B (DEB) PKS from *Saccharopolyspora erythraea* (RMSD of 1.7 Å and 1.9 Å, respectively, for 308 and 301 C$_\alpha$ atom alignments against structures from PDB files 2QO3 and 2HG4) [23,26]. Searches on 3-D BLAST [34] and the Dali Server [35] return the same two DEB PKS structures as the most significant structural matches. The ZmaA-AT domain (residues P93-S407) forms an α/β-hydrolase core into which a small subdomain is inserted (residues A226– I292) (Figure 2). As is the case with the two KS-AT didomain structures from the DEB PKS, the post-AT linker of ZmaA (residues D408-P443) wraps around the AT domain and makes extensive contacts with the highly ordered N-terminal KS-AT linker (residues T1-H92) (Figure 2).

We note that the overwhelming majority of AT domains that partner with CoA-bound extender units have a complete ferredoxin (βαββαβ) fold as the small subdomain [36]. In our formate-bound ZmaA-AT structure, those residues which would form the final β-strand do not make the required main chain hydrogen bonds to rigorously classify them as such (Figure 2). These amino acids, roughly spanning residues 286–291, immedi-

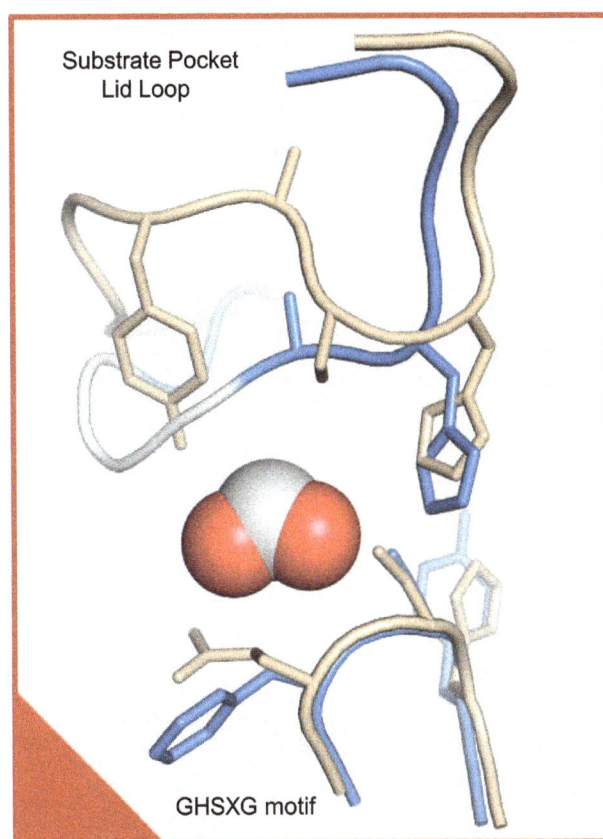

Figure 3. YASH and GHSXG motifs of ZmaA-AT compared to a methylmalonyl-CoA specific AT. The substrate binding-pocket amino acid residues (290–300 and 194) of ZmaA-AT (blue, with white span for disordered 293–295) are superimposed on those of AT from the DEB PKS module 3 (wheat). Bulky F193 is found next to the active site S192 in ZmaA-AT, instead of the glutamine residue found in methylmalonyl-CoA specific ATs. The catalytic H297 is positioned similarly to other ATs, despite its proposed steric hindrance to extender units with (2R) conformations. Despite high mobility for the substrate pocket lid YASH motif, we conclude based on the positions of well-ordered flanking residues that they must wander within the substrate binding pocket of ZmaA-AT, which holds co-crystallized formate (spheres). The red box, with its marked corner, can be compared to the same box in Figure 2 in order to orient the reader.

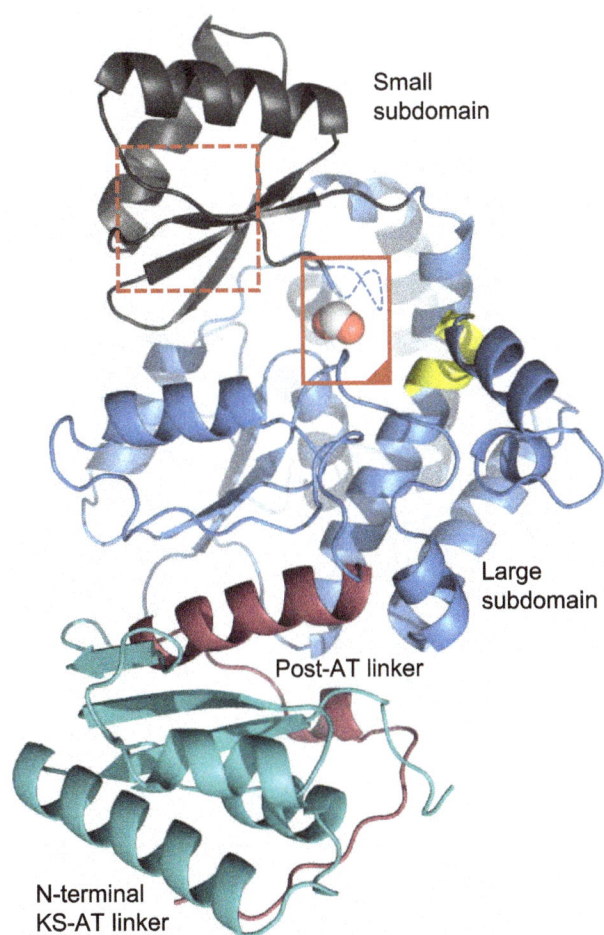

Figure 2. Overall structure of ZmaA-AT. The N-terminal KS-AT linker (green), α/β-hydrolase large subdomain (blue), small subdomain (gray), and post-AT linker (red) make up the complete asymmetric unit. The active site of ZmaA-AT (inside solid red box) is bounded on the left by the substrate pocket lid (containing the YASH motif, which in ZmaA-AT is GAAH) and on the top by the RVDVVQ motif (yellow) and is occupied by formate (spheres); cf. Figure 3. Residues E293-G294-A295 are not observed and are indicated with a dashed line. The proposed substrate ACP binding surface M286-E293 contains the methionine residues of the RXR motif (MCM in ZmaAT) (dotted red box) which correspond to the inchoate β-strand of the ferredoxin fold in the smaller subdomain of other ATs; cf. Figure 4.

ately precede the residues that form a lid over the substrate pocket, both in primary sequence and 3D space (Figure 2).

Motifs Implicated in Substrate Recognition

Structurally, the GHSXG (G190-YSF-G194 in ZmaA-AT), and the YASH (G294-AA-H297 in ZmaA-AT), motifs line the active site cleft formed between the two subdomains (Figure 3), while the RVDVVQ (R159-MEFS-Q164 in ZmaA-AT) motif forms a third wall of the active site and is positioned very close to the substrate pocket lid (Table 2, Figure 2). The structure does not inform how the C-terminal region of the AT domain indirectly influences substrate specificity and thus it will not be discussed in this report.

In ZmaA-AT, the region implicated in substrate carrier recognition is very similar to methoxymalonyl-ACP specific ATs and contains an MXW(X)$_5$YASH motif (MXM(X)$_5$GAAH in ZmaA-AT, Table 3) instead of the RXR(X)$_5$YASH motif as ATs specific for CoA-tethered substrates [22]. ZmaA-AT structure features a hydrophobic patch in this region, instead of the

Table 2. The GHSXG and the YASH Motifs of Select Acyltransferases are Responsible for ACP *vs* CoA Discrimination.

AT Domain	Substrate	GHSXG Motif					YASH Motif			
ZmaA-AT	Hydroxymalonyl-ACP	G_{190}	Y	Y	F	G	G_{294}	A	A	H
ZmaF	Aminomalonyl-ACP	G_{90}	Y	S	L	G	G_{198}	P	F	H
ZmaK-AT	Malonyl-CoA	G_{1644}	H	S	I	G	H_{1742}	A	F	H
FabD (E. coli)	Malonyl-CoA	G_{90}	H	S	L	G	V_{198}	P	S	H
FabD (S. coelicolor)	Malonyl-CoA	G_{95}	H	S	V	G	G_{198}	A	F	H
DEB PKS-AT3	Methylmalonyl-CoA	G_{649}	H	S	Q	G	Y_{751}	A	S	H
DEB PKS-AT5	Methylmalonyl-CoA	G_{640}	H	S	Q	G	Y_{742}	A	S	H

positively charged surface as on the malonyl-CoA specific ATs (Figure 4).

Alternative Substrate Recognition

We have previously established that ZmaA-AT recognizes hydroxymalonyl-ACP as its natural substrate, but it will also recognize aminomalonyl-ACP *in vitro*, when the AT is incubated with high concentrations of the latter [13]. The reduced activity of ZmaA-AT with aminomalonyl-ACP could be due to its specificity for the correct extender unit (aminomalonyl instead of hydroxymalonyl), the correct ACP (ZmaH instead of ZmaD), or a combination of both. Unfortunately the specificity of the enzymes that form aminomalonyl-ZmaH or hydroxymalonyl-ZmaD did not allow for the synthesis of hybrid precursors (e.g. aminomalonyl-ZmaD), thereby eliminating our ability to use these systems to test our hypothesis. Instead, we addressed the role of AT-ACP interaction by testing whether the AT domain can recognize the [^{14}C-C2] labeled substrates *malonyl-CoA, *(2-*RS*)-methylmalonyl-CoA, *malonyl-ZmaD, and *(2-*RS*)-methylmalonyl-ZmaD *in vitro*, using Sfp (*Bacillus subtilis* phosphopantetheinyl transferase) to generate *malonyl-ZmaD and *(2-*RS*)-methylmalonyl-ZmaD from *malonyl-CoA, *(2-*RS*)-methylmalonyl-CoA, and apo-ZmaD.

Neither *malonyl-CoA nor *(2-*RS*)-methylmalonyl-CoA was used by ZmaA-AT, whereas both *malonyl-ZmaD and *methyl-malonyl-ZmaD were used to some extent, highlighting the importance of the AT-ACP interaction (Figure 5). Quantitative analysis revealed an average of ~6 fold preference of *malonyl-ZmaD over *methylmalonyl-ZmaD, betraying an additional layer of substrate specificity at the AT-acyl unit interface. As a racemic mixture of *(2-*RS*)-methylmalonyl-CoA was used to generate methylmalonyl-ACP, it can be assumed that a racemic mixture of *(2-*RS*)-methylmalonyl-ZmaD was available to the AT. Since bacterial ATs associated with modular PKSs are known to be stereospecific [17], it is reasonable to estimate the difference in utilization of malonyl- and methylmalonyl-ACP by ZmaA-AT to be ~3 fold.

Discussion

Overall Structure of ZmaA-AT

A notable difference between ZmaA-AT and previously reported AT domain structures is the positioning of the loop reconnecting the small subdomain to the large subdomain (residues I292-S298) (Figure 3). In all other AT structures published to date, this loop is positioned away from the substrate-binding pocket of the AT, whereas in the ZmaA-AT structure, it extends into the substrate-binding pocket. There aren't any crystal packing interactions holding the lid in place. It is therefore reasonable to propose that binding of the substrate to the AT, especially the ACP portion of the substrate to the smaller subdomain, influences the positioning of this substrate pocket lid so that it moves out of the binding pocket to make room for the atoms of the extender unit. This hypothesis is supported by the high mobility of this region in ZmaA-AT. Indeed three residues (293–295) were poorly ordered and were omitted from the final model. The substrate pocket lid contains the YASH motif, which has been implicated in the extender unit specificity of AT domains (Table 2) [36].

We propose this motion is induced by the substrate carrier ACP binding to the RXR motif at the N-terminal end of this span of residues (M286-C-M288 in ZmaA-AT) (Figure 6). To date no substrate carrier ACP:AT co-crystal structures are available. Such a complex structure will be needed to validate this model of

Table 3. The RXR Motifs of Select Acyltransferases Control Extender Unit Specificity.

AT Domain	Substrate	(X)RXR			
ZmaA-AT	Hydroxymalonyl-ACP	(L)	M_{286}	C	M
ZmaF	Aminomalonyl-ACP	(G)	I_{190}	A	I
ZmaK-AT	Malonyl-CoA	(V)	K_{1734}	T	T
FabD (E. coli)	Malonyl-CoA	(K)	R_{190}	A	L
FabD (S. coelicolor)	Malonyl-CoA	(R)	K_{190}	V	V
DEB PKS-AT3	Methylmalonyl-CoA	(I)	R_{743}	V	R
DEB PKS-AT5	Methylmalonyl-CoA	(H)	K_{734}	A	R

structural rearrangement upon ACP binding. The RXR motif is proposed to be involved in substrate carrier recognition and is discussed in detail below.

AT Recognition of the Extender Unit

In ZmaA-AT, the highly conserved histidine in the GHSXG motif that includes the catalytic S192 is replaced with Y191 to form GYSFG (Figure 3, Table 2). However, the relative positioning of the phenol of Y191 to the catalytic S192 in ZmaA-AT matches that of the imidazole of the histidine and the catalytic serine in structures of FabD, DEB PKS-AT3 and DEB PKS-AT5, suggesting that they have a similar function [18,23,26]. The X following the catalytic serine in this motif is usually a bulky branched hydrophobic amino acid in ATs that recognize malonyl-CoA, whereas it is a glutamine in (2S)-methylmalonyl-CoA specific ATs (Table 2) [37]. It has been proposed that in (2S)-methylmalonyl-CoA specific ATs, the side chain of this glutamine may orient the incoming extender unit so that the α-methyl group is able to make a hydrophobic interaction with the tyrosine of the YASH motif [26]. In ZmaA-AT and in methoxymalonyl-ACP specific ATs FkbA-AT1 and FkbA-AT2 (involved in FK520 biosynthesis [38]), bulky hydrophobic amino acids such as

phenylalanine (F193, Figure 3) or leucine are found instead of glutamine in the X of the GHSXG motif, respectively, similar to malonyl-CoA specific ATs (Table 2). The side chain of F193 in the ZmaA-AT structure is pointing away from the substrate-binding pocket (Figure 3), as is the side chain of L93 in the FabD structure. Without a change in side chain rotamer compared to these crystal coordinates, F193 would not affect the orientation of the incoming substrate.

The YASH motif, which is located about 100 residues beyond the GHSXG motif, contains the histidine residue of the catalytic dyad. The side chain of the residue has been proposed to play an important part in substrate specificity in addition to its catalytic role [39]. Along with the tyrosine residue (Y742 in DEB PKS-AT5), mentioned above in methylmalonyl-CoA specificity, the imidazole ring of the histidine residue in the YASH motif is proposed to sterically hinder the α-methyl group of a (2R)-methylmalonyl-CoA as it enters the active site, providing stereo selectivity for the (2S) stereoisomer [26]. This histidine residue is part of the catalytic dyad involved in the AT mechanism [40], a fact which leads to an interesting issue regarding the orientation of the extender unit α-substituent that can be utilized in polyketide metabolism.

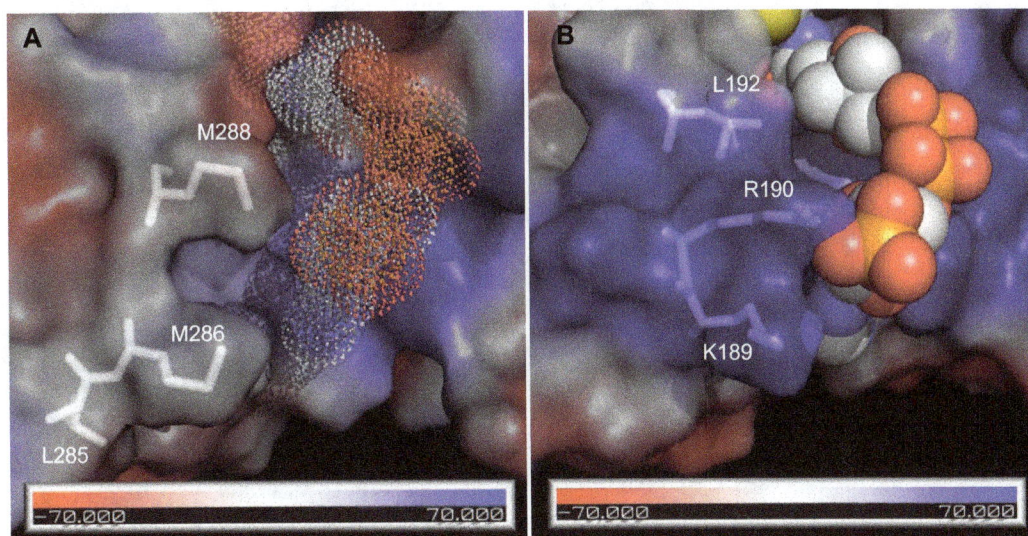

Figure 4. Proposed AT-Domain Interaction with ACP Substrate Carrier. Approximate protein contact potential calculated using PyMOL vacuum electrostatics function. The colors represent potentials ranging from −70 mV (red) to +70 mV (blue). (A) Proposed AT/ACP interface of ZmaA-AT. FabD was aligned to the structure of ZmaA-AT to show the relative position of CoA (dots, FabD not shown). (B) AT/CoA interface of E. coli FabD (PBD ID: 2G2Z, see Methods [18]). CoA is shown as spheres.

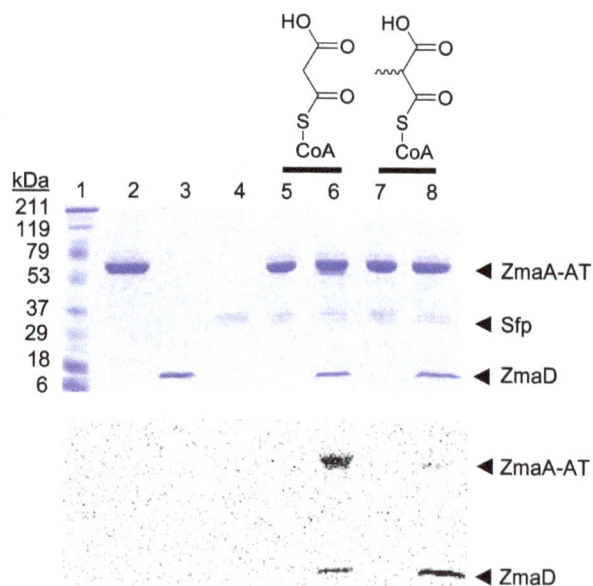

Figure 5. Transacylase assay of ZmaA-AT Distinguishes ACP from Acyl Unit Recognition. SDS-PAGE of reaction mixtures and corresponding phosphorimage. Lane 1: Molecular mass markers (Prestained Broad-range, Biorad). Lane 2: ZmaA-AT. Lane 3: ZmaD (ACP). Lane 4: Sfp (4′-phosphopantetheinyl transferase). Lane 5: ZmaA-AT, Sfp, and *Malonyl-CoA. Lane 6: ZmaA-AT, Sfp, ZmaD, and *Malonyl-CoA. Lane 7: ZmaA-AT, Sfp, and *(2-RS)-methylmalonyl-CoA. Lane 8: ZmaA-AT, Sfp, ZmaD, and *(2-RS)-methylmalonyl-CoA.

If ZmaA-AT were able to recognize (2R)-hydroxymalonyl-ACP, a hypothesis that is in keeping with the biosynthetic derivation from D-glycolytic intermediates, the stereochemistry of the extender unit would need to be reflected in the final (2R) product (Figure 1). In this case, the acyl moiety would be required to enter the substrate binding pocket at a significantly different angle than what is modeled for the (2S)-methylmalonyl-CoA entering the active site of DEB PKS-AT5 [26]. A different entrance pathway can be imagined because F193 in the GHSXG motif (GYSFG in ZmaA-AT) is positioned not to constrict the orientation of the substrate, and the α-hydroxyl group of the extender may not clash with the imidazole ring of H297 in the YASH motif (GAAH in ZmaA-AT) (Figure 7).

If the extender unit starts out in the (2R) conformation, after inversion from the condensation, there must be an additional epimerization event by the ketoreductase (KR) domain in ZmaA (ZmaA-KR2). ZmaA-KR2 does not contain an LDD motif (LGG in ZmaA-KR2) and its reduction reaction yields a hydroxyl group in the S conformation, suggesting it could be an A-type KR, by Caffrey classification [41]. However, because it also lacks an important tryptophan residue that is conserved in A type KRs, it belongs neither to the A1 nor the A2 KR type by Keatinge-Clay classification [42].

Alternatively, it is possible that ZmaA-AT recognizes the (2S)-isomer of hydroxymalonyl-ACP. The final step of hydroxymalonyl-ACP biosynthesis is an FAD-dependent oxidation of C3 by ZmaE [12]. Mechanistically, this step may proceed through an endiol intermediate (Figure 1A), which can then be re-protonated at C2 to form either the (2S) or (2R) stereoisomer of hydroxymalonyl-ACP. No epimerization would be required by the ZmaA-KR2 domain. In the ZMA molecule, the hydroxyl group at C8 is proposed to originate from the hydroxymalonyl-ACP extender unit incorporated by ZmaA-AT [11]. This hydroxyl group is in the same orientation as it would be on (2R)-hydroxymalonyl-ACP (Figure 1).

In the structure of FabD in complex with its substrate malonyl-CoA, the guanidine group of R117 is observed to stabilize the C3

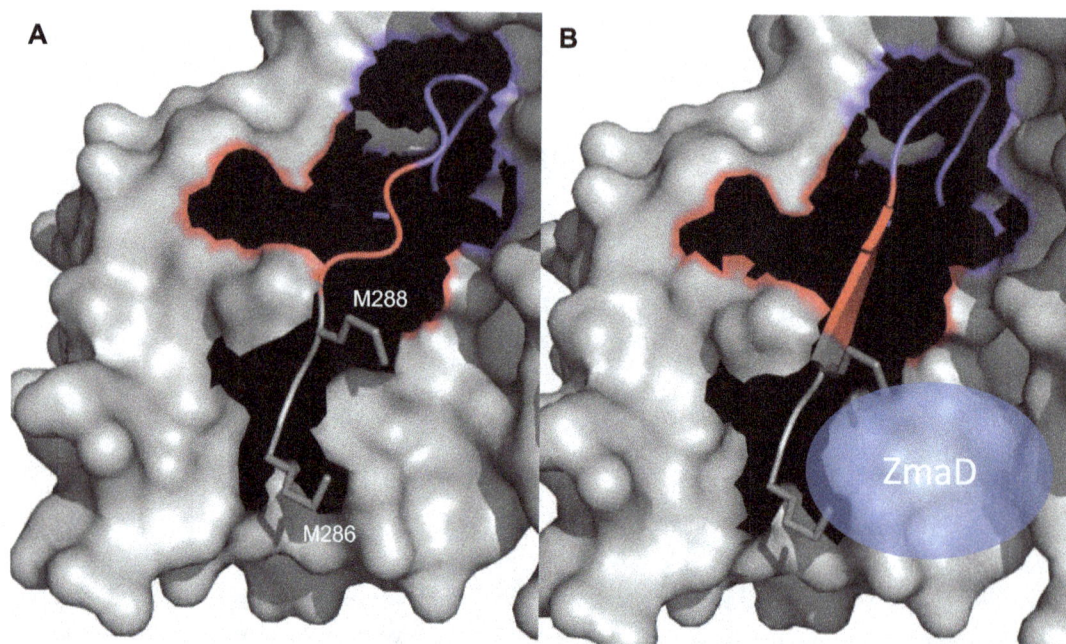

Figure 6. Proposed movement of the substrate pocket lid induced by ZmaD binding. (A) Based on the crystal structure of ZmaA-AT, the substrate pocket lid (blue) is shown in the closed position, restricting the entry of the extender unit, in the absence of substrate carrier protein. (B) Model structure of ZmaA-AT bound to substrate carrier protein, ZmaD (blue spheroid). The binding of the substrate carrier protein to the RXR motif (M286-C-M288 in ZmaA-AT; gray sticks) in the small subdomain of ZmaA-AT is proposed to cause the formation of the β-strand (red), resulting in the opening of the substrate pocket lid (blue).

A DEB PKS-AT5   **B** ZmaA-AT

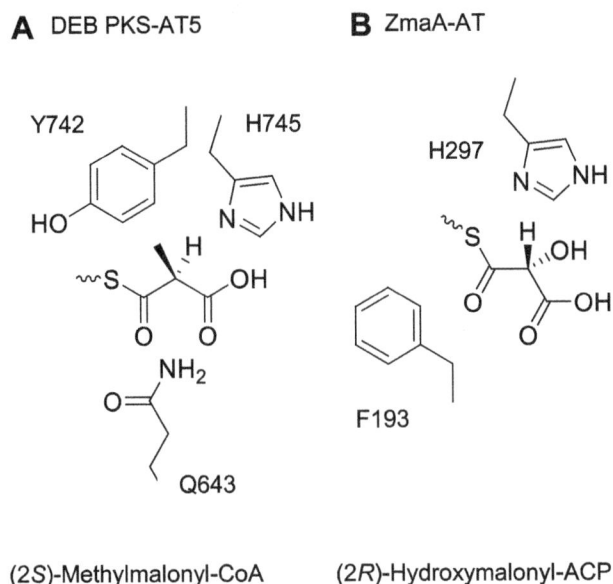

(2S)-Methylmalonyl-CoA   (2R)-Hydroxymalonyl-ACP

Figure 7. Possible difference in substrate entry angles between DEB PKS AT-5 and ZmaA-AT. (A) In DEB PKS-AT5, Q643 has been proposed to orient the incoming (2S)-methylmalonyl-CoA so that Y742 makes a hydrophobic interaction with the methyl-group and H745 sterically hinders the entry of (2R)-methylmalonyl-CoA [26]. (B) In ZmaA-AT, F193 is not positioned to orient the incoming substrate, which may allow hydroxymalonyl-ACP with (2R)-stereochemistry to enter the substrate pocket unhindered.

carboxyl group of the acylated malonate through a salt bridge. The corresponding residue in our ZmaA-AT structure, R217, is positioned close to a molecule of formate, which co-crystallized with the protein and presumably mimics the coordinates of the C3 of hydroxymalonate. We note that the relative positioning of this R217 and the catalytic S192 of ZmaA-AT in solution would require less deviation from the crystal structure to accommodate the (2S) stereoisomer than the (2R) stereoisomer of hydroxymalonyl-ACP.

We conclude that there is presently not enough evidence to support the preference for one stereoisomer over the other in the incorporation of hydroxymalonyl-ACP by ZmaA-AT.

Finally, while the residues in the RVDVVQ motif would be too far away from the extender unit to contribute to substrate specificity directly, the structure suggests that amino acid substitutions in this motif may influence positioning of the YASH-motif in the substrate pocket lid, resulting in altered specificity [37].

AT/Substrate Carrier Recognition

FabD is a malonyl-CoA specific AT in *E. coli*, involved in fatty acid synthesis. It must first interact with CoA to receive the malonyl group, then again with its partner downstream ACP to complete the transacylation reaction. Insight on the nature of the interaction between FabD and CoA was gained from the structure of FabD in complex with malonyl-CoA [18]. Later, the structure of a FabD homolog in *S. coelicolor* [19] was used for docking simulations using the structure of its partner downstream ACP [20]. These reports suggest that the arginine residues in FabD (R190 in *E. coli* and R189 in *S. coelicolor*) interact with and properly orient both the CoA and the downstream ACP. Similar results were obtained more recently, when crosslinking studies with

the AT from the disorazole PKS and its partner ACP found that K179 on the AT is important for AT-ACP interaction [21]. K179 of the disorazole PKS-AT aligns with R189 of the *S. coelicolor* FabD. Interestingly, this region of the AT has also independently been implicated in substrate selectivity between methylmalonyl-CoA and methoxymalonyl-ACP [22]. In their work, Haydock *et al.* identified the sequence RXR(X)$_5$YASH (the first Arg corresponds to R190 of *E. coli* FabD, Table 3) for methylmalonyl-CoA specific ATs, and MXW(X)$_5$YASH for methoxymalonyl-ACP specific ATs within the concanamycin PKS. They noted that the methionine and tryptophan residues in MXW can be other hydrophobic residues in methoxymalonyl-ACP specific ATs, whereas these residues are usually replaced by positively charged ones in malonyl-CoA specific ATs. Using this sequence motif in a BLAST search, they were able to locate more ATs that are proposed to be methoxymalonyl-ACP specific.

Based on the similarity of ZmaA-AT to methoxymalony-ACP specific ATs in the region that is implicated in substrate carrier recognition, we propose that the signature motif MXW is indicative of not only methoxymalonyl-ACP specific ATs, but more generally, ATs that recognize ACP tethered extender units. This hypothesis is further supported by the fact that another AT domain involved in ZMA biosynthesis, ZmaF, recognizes an ACP tethered extender unit and contains hydrophobic residues in the MXW motif, while ZmaK-AT recognizes malonyl-CoA and has a positively charged residue in that motif (Figure 1 and Table 3). The exposed hydrophobic patch may facilitate an as yet uncharacterized binding of the AT with the extender unit ACP. This binding scheme would be distinct from the previously proposed transient electrostatic mode of interaction between the AT and the downstream ACP [20,21]. Furthermore, as these hydrophobic residues lie in the region corresponding to the final β-strand of the ferredoxin fold in other ATs, we hypothesize that the binding of the substrate ACP to this region results in the formation of β-strand conformation in residues R284-T291 of the small subdomain, resulting in the displacement of the connected substrate pocket lid I292-S298 from inside the substrate binding pocket to accommodate the entry of the extender unit (Figure 6). Validation of this model awaits additional crystal structures of ACP specific ATs both alone and in complex with their ACP substrates.

Alternative Substrate Recognition

The preference for malonyl-ACP over methylmalonyl-ACP as substrate by ZmaA-AT may be explained by the C2 methyl group of (2-RS)-methylmalonyl-CoA, which has a significantly larger radius than a hydroxyl group and may be sterically hindered by two tandem alanine residues (A295 and A296 in the GAAH). In addition, the methyl substituent restricts the bond angles of the backbone carbons of methylmalonate to be significantly different from those of hydroxymalonate. Therefore, when acylated on the active S192 of the AT, the C3 carboxyl group of the methylmalonyl extender unit would not be in the optimal position to form a salt bridge with R217, which is hypothesized to stabilize the C3 carboxyl group of the hydroxymalonyl extender unit. Unfortunately, efforts to substitute the Met residues in the RXR(X)$_5$YASH of ZmaA-AT to positively charged residues resulted in insoluble protein, eliminating our ability to test whether such changes alter precursor recognition.

Conclusion

PKS extender units that are biosynthesized on ACP carriers instead of CoA molecules include aminomalonyl and hydroxymalonyl moieties. These are of particular interest in combinatorial biosynthesis of polyketides, because their incorporation results in amino- and hydroxyl- functional groups, respectively, to be present at unique positions within the product, which can further be utilized in semi-synthetic derivatizations. Our bioinformatic and crystal structural analyses of ZmaA-AT as well as published structures of other AT domains have led to our hypotheses that (1) the presence of hydrophobic residues in the RXR motif of ATs indicates specificity for ACP tethered extender units and that (2) binding of the ACP to the hydrophobic patch promotes secondary structure formation of the β-strand that leads from the ACP binding site to the extender unit binding site, and opens the latter for substrate entry. *In-vitro* biochemical analysis of ZmaA-AT has shown that the AT/substrate ACP interaction plays a significant role in substrate specificity. Taken as a whole, this work establishes an important foundation for the engineering of ATs involving the utilization of ACP linked substrates.

Accession Codes

The coordinates and structure factors have been deposited in the Protein Data Bank with accession code 4QBU.

Author Contributions

Conceived and designed the experiments: MGT KTF. Performed the experiments: HP BMK DHD MGT KTF. Analyzed the data: HP BMK DHD MGT KTF. Contributed to the writing of the manuscript: HP MGT KTF.

References

1. Staunton J, Weissman KJ (2001) Polyketide biosynthesis: a millennium review. Nat Prod Rep 18: 380–416.
2. Chan YA, Podevels AM, Kevany BM, Thomas MG (2009) Biosynthesis of polyketide synthase extender units. Nat Prod Rep 26: 90–114.
3. Walsh CT (2004) Polyketide and nonribosomal peptide antibiotics: modularity and versatility. Science 303: 1805–1810.
4. Handelsman J, Raffel S, Mester EH, Wunderlich L, Grau CR (1990) Biological Control of Damping-Off of Alfalfa Seedlings with Bacillus cereus UW85. Appl Envir Microbiol 56: 713–718.
5. Silo-Suh LA, Lethbridge BJ, Raffel SJ, He H, Clardy J, et al. (1994) Biological activities of two fungistatic antibiotics produced by Bacillus cereus UW85. Appl Env Microbiol 60: 2023–2030.
6. Silo-Suh LA, Stabb EV, Raffel SJ, Handelsman J (1998) Target range of zwittermicin A, an aminopolyol antibiotic from Bacillus cereus. Curr Microbiol 37: 6–11.
7. He H, Silo-Suh LA, Handelsman J, Clardy J (1994) Zwittermicin A, an antifungal and plant protection agent from Bacillus cereus. Tetrahedron Lett 35: 2499–2502.
8. Rogers EW, Dalisay DS, Molinski TF (2008) (+)-Zwittermicin A: assignment of its complete configuration by total synthesis of the enantiomer and implication of D-serine in its biosynthesis. Angew Chem Int Ed Engl 47: 8086–8089.
9. Rogers EW, Molinski TF (2007) Asymmetric synthesis of diastereomeric diaminoheptanetetraols. A proposal for the configuration of (+)-zwittermicin a. Org Lett 9: 437–440.
10. Emmert EA, Klimowicz AK, Thomas MG, Handelsman J (2004) Genetics of zwittermicin a production by Bacillus cereus. Appl Env Microbiol 70: 104–113.
11. Kevany BM, Rasko DA, Thomas MG (2009) Characterization of the complete zwittermicin A biosynthesis gene cluster from Bacillus cereus. Appl Env Microbiol 75: 1144–1155.
12. Chan YA, Boyne MT 2nd, Podevels AM, Klimowicz AK, Handelsman J, et al. (2006) Hydroxymalonyl-acyl carrier protein (ACP) and aminomalonyl-ACP are two additional type I polyketide synthase extender units. Proc Natl Acad Sci U S A 103: 14349–14354.
13. Chan YA, Thomas MG (2010) Recognition of (2S)-aminomalonyl-acyl carrier protein (ACP) and (2R)-hydroxymalonyl-ACP by acyltransferases in zwittermicin A biosynthesis. Biochemistry 49: 3667–3677.
14. Reimer D, Pos KM, Thines M, Grün P, Bode HB (2011) A natural prodrug activation mechanism in nonribosomal peptide synthesis. Nat Chem Biol 7: 888–890.
15. Brotherton CA, Balskus EP (2013) A prodrug resistance mechanism is involved in colibactin biosynthesis and cytotoxicity. J Am Chem Soc 135: 3359–3362.
16. Bian X, Fu J, Plaza A, Herrmann J, Pistorius D, et al. (2013) In vivo evidence for a prodrug activation mechanism during colibactin maturation. Chembiochem 14: 1194–1197.
17. Smith S, Tsai S-CC (2007) The type I fatty acid and polyketide synthases: a tale of two megasynthases. Nat Prod Rep 24: 1041–1072.
18. Oefner C, Schulz H, D'Arcy A, Dale GE (2006) Mapping the active site of Escherichia coli malonyl-CoA-acyl carrier protein transacylase (FabD) by protein crystallography. Acta Crystallogr D62: 613–618.
19. Keatinge-Clay AT, Shelat AA, Savage DF, Tsai SC, Miercke LJ, et al. (2003) Catalysis, specificity, and ACP docking site of Streptomyces coelicolor malonyl-CoA:ACP transacylase. Structure 11: 147–154.
20. Arthur CJ, Williams C, Pottage K, Ploskon E, Findlow SC, et al. (2009) Structure and malonyl CoA-ACP transacylase binding of streptomyces coelicolor fatty acid synthase acyl carrier protein. ACS Chem Biol 4: 625–636.
21. Wong FT, Jin X, Mathews II, Cane DE, Khosla C (2011) Structure and mechanism of the trans-acting acyltransferase from the disorazole synthase. Biochemistry 50: 6539–6548.

22. Haydock SF, Appleyard AN, Mironenko T, Lester J, Scott N, et al. (2005) Organization of the biosynthetic gene cluster for the macrolide concanamycin A in Streptomyces neyagawaensis ATCC 27449. Microbiology 151: 3161–3169.
23. Tang Y, Chen AY, Kim CY, Cane DE, Khosla C (2007) Structural and mechanistic analysis of protein interactions in module 3 of the 6-deoxyerythronolide B synthase. Chem Biol 14: 931–943.
24. Dutta S, Whicher JR, Hansen DA, Hale WA, Chemler JA, et al. (2014) Structure of a modular polyketide synthase. Nature 510: 512–517.
25. Liew CW, Nilsson M, Chen MW, Sun H, Cornvik T, et al. (2012) Crystal structure of the acyltransferase domain of the iterative polyketide synthase in enediyne biosynthesis. J Biol Chem 287: 23203–23215.
26. Tang Y, Kim CY, Mathews II, Cane DE, Khosla C (2006) The 2.7-Angstrom crystal structure of a 194-kDa homodimeric fragment of the 6-deoxyerythronolide B synthase. Proc Natl Acad Sci U S A 103: 11124–11129.
27. Chan YA, Thomas MG (2009) Formation and characterization of acyl carrier protein-linked polyketide synthase extender units. Methods Enzym 459: 143–163.
28. Otwinowski Z, Minor W (1997) Processing of X-ray Diffraction Data Collected in Oscillation Mode. Methods Enzym 276: 307–326.
29. Panjikar S, Parthasarathy V, Lamzin VS, Weiss MS, Tucker PA (2009) On the combination of molecular replacement and single-wavelength anomalous diffraction phasing for automated structure determination. Acta Crystallogr D65: 1089–1097.
30. Murshudov GN, Skubák P, Lebedev AA, Pannu NS, Steiner RA, et al. (2011) REFMAC5 for the refinement of macromolecular crystal structures. Acta Crystallogr D67: 355–367.
31. Emsley P, Lohkamp B, Scott WG, Cowtan K (2010) Features and development of Coot. Acta Crystallogr D66: 486–501.
32. Schrodinger LLC (2010) The PyMOL Molecular Graphics System, Version 1.3r1.
33. Abramoff MD, Magelhaes PJ, Ram SJ (2004) Image Processing with ImageJ. Biophotonics Int 11: 36–42.
34. Tung C-H, Huang J-W, Yang J-M (2007) Kappa-alpha plot derived structural alphabet and BLOSUM-like substitution matrix for rapid search of protein structure database. Genome Biol 8: R31.
35. Holm L, Rosenström P (2010) Dali server: conservation mapping in 3D. Nucleic Acids Res 38: W545–9.
36. Tsai S-CS, Ames BD (2009) Structural enzymology of polyketide synthases. Methods Enzymol 459: 17–47.
37. Haydock SF, Aparicio JF, Molnar I, Schwecke T, Khaw LE, et al. (1995) Divergent sequence motifs correlated with the substrate specificity of (methyl)malonyl-CoA:acyl carrier protein transacylase domains in modular polyketide synthases. FEBS Lett 374: 246–248.
38. Wu K, Chung L, Revill WP, Katz L, Reeves CD (2000) The FK520 gene cluster of Streptomyces hygroscopicus var. ascomyceticus (ATCC 14891) contains genes for biosynthesis of unusual polyketide extender units. Gene 251: 81–90.
39. Reeves CD, Murli S, Ashley GW, Piagentini M, Hutchinson CR, et al. (2001) Alteration of the substrate specificity of a modular polyketide synthase acyltransferase domain through site-specific mutations. Biochemistry 40: 15464–15470.
40. Röttig A, Steinbüchel A (2013) Acyltransferases in bacteria. Microbiol Mol Biol Rev 77: 277–321.
41. Caffrey P (2003) Conserved amino acid residues correlating with ketoreductase stereospecificity in modular polyketide synthases. Chembiochem 4: 654–657.
42. Keatinge-Clay AT (2007) A tylosin ketoreductase reveals how chirality is determined in polyketides. Chem Biol 14: 898–908.

Understanding the Mechanism of Atovaquone Drug Resistance in *Plasmodium falciparum* Cytochrome b Mutation Y268S Using Computational Methods

Bashir A. Akhoon[1][9], Krishna P. Singh[1][9], Megha Varshney[2], Shishir K. Gupta[3], Yogeshwar Shukla[4,5], Shailendra K. Gupta[1,5]*[¤]

1 Department of Bioinformatics, Systems Toxicology Group, CSIR-Indian Institute of Toxicology Research, Lucknow, India, 2 Interdisciplinary Biotechnology Unit, Aligarh Muslim University, Aligarh, India, 3 Department of Bioinformatics, Biocenter, Am Hubland, University of Würzburg, Würzburg, Germany, 4 Department of Proteomics, CSIR-Indian Institute of Toxicology Research, Lucknow, India, 5 Academy of Scientific and Innovative Research (AcSIR), New Delhi, India

Abstract

The rapid appearance of resistant malarial parasites after introduction of atovaquone (ATQ) drug has prompted the search for new drugs as even single point mutations in the active site of Cytochrome b protein can rapidly render ATQ ineffective. The presence of Y268 mutations in the Cytochrome b (Cyt b) protein is previously suggested to be responsible for the ATQ resistance in *Plasmodium falciparum* (*P. falciparum*). In this study, we examined the resistance mechanism against ATQ in *P. falciparum* through computational methods. Here, we reported a reliable protein model of Cyt bc1 complex containing Cyt b and the Iron-Sulphur Protein (ISP) of *P. falciparum* using composite modeling method by combining threading, *ab initio* modeling and atomic-level structure refinement approaches. The molecular dynamics simulations suggest that Y268S mutation causes ATQ resistance by reducing hydrophobic interactions between Cyt bc1 protein complex and ATQ. Moreover, the important histidine contact of ATQ with the ISP chain is also lost due to Y268S mutation. We noticed the induced mutation alters the arrangement of active site residues in a fashion that enforces ATQ to find its new stable binding site far away from the wild-type binding pocket. The MM-PBSA calculations also shows that the binding affinity of ATQ with Cyt bc1 complex is enough to hold it at this new site that ultimately leads to the ATQ resistance.

Editor: Adrian J.F. Luty, Institut de Recherche pour le Développement, France

Funding: This work was supported by the Council of Scientific & Industrial Research (CSIR) - network project GENESIS (BSC0121) and INDEPTH (BSC0111). The funder had no role in study design, data collection and analysis, decision to publish, or preparation of the manuscript.

Competing Interests: The authors have declared that no competing interests exist.

* Email: skgupta@iitr.res.in

¤ Current address: Department of Systems Biology and Bioinformatics, University of Rostock, Rostock, Germany

9 These authors contributed equally to this work.

Introduction

Studies revealed that human malaria is caused by protozoan parasites of the genus *Plasmodium*. The four most common *Plasmodium* species that infect human are *P. vivax*, *P. ovale*, *P. malariae*, and *P. falciparum*. Additionally, a fifth one *P. knowlesi* has also been identified as responsible for infection in human [1] often in many countries of Southeast Asia [2]. According to the latest World malaria report (2012) by World health organization, there were about 219 million cases of malaria in 2010 and an estimated 660 000 deaths. *P. falciparum* predominates in Africa and is the most deadly form leading to death due to malaria. 90% of malaria occurs in Africa and among which 85% deaths happen in children under the age of 5 [3].

Plasmodium species can acquire drug resistance through several mechanisms, like change in drug permeability, increased expression of the drug target, or changes in the enzyme target [4]. ATQ drug acts against malarial parasites by inhibiting mitochondrial electron transport [5] and collapsing mitochondrial membrane potential [6]. Based on its structural similarity to ubiquinol, it has been postulated that ATQ binds to parasite Cyt b protein [7]. It is

supported by experimental findings that mutations at 268th position in *P. falciparum* Cyt b are unambiguously associated with acquired ATQ resistance [8]. However, the mechanism of ATQ resistance is still not well understood. Thus, there is an urgent need to develop novel disease management strategies against various *Plasmodium sp* induced malaria.

In several studies [9–11], researchers have modeled some *P. falciparum* mutations including Y268S using *in silico* methods, however none of them have completely modeled the *P. falciparum* Cyt bc1 complex rather they rely on the ATQ-bound yeast Cyt bc1 complex. Moreover, none of the study has examined the dynamics of the Cyt bc1 ATQ-bound complex. Therefore in the present study, we exploited the *in silico* approaches to identify molecular basis of ATQ drug resistance in the Y286S mutation model of Cyt b protein of *P. falciparum*. To best of our knowledge, this is the first study to report the modeling and molecular dynamics simulation of ATQ-bound *P. falciparum* Cyt bc1 complex in both wild and mutant-type models for nanoseconds time scale.

Materials and Methods

Computational model building and quality assessment

The ubiquinol oxidation (Qo) site of the Cytochrome bc1 complex serves as a pocket for ATQ binding [11] and two subunits of the complex (Cyt b and ISP) are involved in ATQ binding [11,12]. To model the whole complex, amino acid sequences of *P. falciparum* Cyt b (Genbank accession no: NP_059668.1) and ubiquinol-Cyt C reductase ISP subunit (Genbank accession no: XP_001348547.1) were retrieved from the Entrez protein database available at NCBI (http://www.ncbi.nlm.nih.gov). In the process of protein modeling, we observed that no single template was able to satisfy ~100% query coverage. Hence, the composite modeling which combines various techniques such as threading, *ab initio* modeling and atomic-level structure refinement approaches [13–16] implemented in the iterative threading assembly refinement (I-TASSER) server was preferred to build the full-length protein structure of both the protein chains. I-TASSER generates 3D atomic models from multiple threading alignments and iterative structural assembly simulations. The full methodology of the server has been described elsewhere [17]. The template modeling score (TM-score) calculation [18] was used to assess the structural similarity of model and template protein structures [Eq. i].

$$TM\text{-}score = Max \left[\frac{1}{L} \sum_{i=1}^{L_{ali}} \frac{1}{1 + d_i^2/d_0^2} \right] \quad (1)$$

where L is the length of the target protein, L_{ali} is the number of the equivalent residues in two proteins, di is the distance of the ith pair of the equivalent residues between the two structures, which depends on the superposition matrix; the 'max' means the procedure to identify the optimal superposition matrix that superposition matrix that maximizes the sum in Eq. i. The scale $d_0 = \sqrt[3]{(L-15)} - 1.8d$ is defined to normalize the TM-score in a way that the magnitude of the average TM-score for random protein pairs is independent on the size of the proteins.

Confidence score (C-score) was taken into consideration to determine the accuracy of the predicted structure. The score is defined based on the quality of the threading alignments and the convergence of the I-TASSER's structural assembly refinement simulations [Eq. ii].

$$C\text{-}score = \ln \left[\frac{M}{M_{tot}} \times \frac{1}{\langle RMSD \rangle} \times \frac{1}{7} \sum_{i=1}^{7} \frac{Z(i)}{Z_0(i)} \right] \quad (2)$$

Where M is the number of structure decoys in the cluster and M_{tot} is the total number of decoys generated during the I-TASSER simulations. <RMSD> is the average RMSD of the decoys to the cluster centroid. $Z(i)$ is the Z-score of the best template generated by i^{th} threading in the seven LOMETS programs and $Z_0(i)$ is a program-specified Z-score cutoff for distinguishing between good and bad templates.

The geometry of the theoretical model was improved by side-chain geometry optimization using the ChiRotor algorithm [19]. The modeled structures were further subjected to energy minimization followed by model quality estimation. In order to further design the Cyt b-ISP complex from the individually modeled structures of Cyt b and ISP subunits from *P. falciparum*, we superimposed them to the *S. cerevisiae* bc1 complex already available in Protein Data Bank (pdb entry: 3CX5). From this superimposed structure, we got the coordinates of modelled Cyt b and ISP subunits of *P. falciparum* in the orientation similar to the one found in Cyt bc1 complex of *S. cerevisiae*. Since the yeast bc1 complex also strongly interacts with water molecules in the vicinity of Glu 272 [20], the water molecule was also added to the modeled *P. falciparum* Cyt b-ISP complex before performing the docking experiments.

Mutation mapping of Cyt b protein at 268th position

All the known point mutations observed at position 268 of *P. falciparum* Cyt b were individually incorporated in the modeled 3D protein structure to scan their impact on ATQ binding. For this, all the mutant models of the Cyt b protein of *P. falciparum* were generated using the mutational modeling protocol of DS3.1. The Build Mutants protocol mutates selected residues to specified types and optimizes the conformation of the mutated residues and their neighbors using MODELER program.

Retrieval of ATQ structure and modeling of Cyt bc1 complex

PubChem Compound, one of the linked databases within the NCBI's Entrez information retrieval system was accessed for the retrieval of ATQ structure. Both the protein and ligand molecules were prepared before being subjected to docking analysis using Prepare Protein and Prepare Ligand protocols of DS3.1 respectively. Prepare Protein protocol rectify the protein for various problems, such as missing atoms in incomplete residues; missing loop regions; alternate conformations (disorder); nonstandard atom names; incorrect protonation state of titratable residues etc. The generation of 3D conformation of ATQ was attained by the Prepare Ligands tool.

Molecular docking experiments of the ATQ into the Qo site of both mutated and non-mutated variants of the Cyt b protein was performed by CDOCKER, a molecular dynamics (MD) simulated-annealing-based algorithm [22]. The ATQ was assumed to bind in the same binding pocket to that of ligand stigmatellin, a known inhibitor of Qo site of the Cyt bc1 complex. Water molecules were removed, except HOH7187 which has been reported to play important role in the observed hydrogen bonding network [9]. General-purpose all-atom force field (CHARMm) with a wide coverage for proteins, nucleic acids and general organic molecules was included in the random structure generation. 10 orientations were generated for the ATQ, improved by performing simulated annealing method and finally refined by applying low, but most accurate full potential as a refined pose minimization method.

Molecular dynamics simulations

Molecular dynamics simulations were performed with Gromacs ver. 4.5.3. The ligand topology and parameterization was attained with SwissParam (http://www.swissparam.ch/), an automatic tool that generates topology and parameters based on the Merck molecular force field. The wild and mutant-type Cytbc1 complexes (Cyt b-ISP/ATQ) were subjected to molecular dynamics simulation with explicit TIP3P water solvation model in the Isothermal–isobaric (NPT) ensemble using the AMBER99SB force field. Each system was minimized with the steepest descent method to relax unfavorable contacts between molecules and equilibrated for 150 ps before production runs to achieve stability during production dynamics. Simulations were performed at a constant temperature of 310 K and pressure of 1 atm, using Particle Mesh Ewald method [23] for long-range electrostatic [24] and van der Waals (vdW) [25,26] interactions with a cut-off of 1.4 nm while constraints were applied on all bonds using the LINCS [27] algorithm. All systems were simulated in the NPT (fixed number of atoms N, pressure P, and temperature T) ensemble using the v-rescale coupling algorithm [28] and the Parrinello–Rahman coupling algorithm [29] for 90 ns and time step of 2 fs without any position restraints.

MM/PBSA calculations

The MM/PBSA approach [30] was applied to perform the binding free energy calculations. The binding free energy of a protein to a ligand (ΔG_{bind}) is defined from the complex, protein and ligand free energies ($G_{complex}$, $G_{protein}$ and G_{ligand}, respectively) as [Eq. iii]

$$\Delta G_{bind} = G_{complex} - (G_{protein} + G_{ligand}) \qquad (3)$$

Each free energy term is obtained from a MD-derived ensemble of structures as the sum of six terms as mentioned in [Eq. iv].

$$\begin{aligned} <G> \; = \; &<G_{int}> + <G_{vdW}> + <G_{coul}> \\ &+ <G_{ps}> + <G_{nps}> - T<SMM> \end{aligned} \qquad (4)$$

Where G_{int}, G_{vdW} and G_{coul} indicate the internal (including bond, angle, and torsional angle energies), van der Waals and coulombic energy terms, respectively, collectively defined "gas phase terms". $<G_{ps}>$ and $<G_{nps}>$ are the polar and nonpolar solvation energy terms, respectively. $<SMM>$ is the entropic term. Angle brackets denote the average along the structures.

The single trajectory method (STM) has been used for both the wild-type and the mutant systems. The STM requires the trajectory of the complex to be run only. The structures of the free forms of the protein and ligand species were obtained by stripping the partner molecule from the structure of the complex. Thus, zeroing out the $<G_{int}>$ term in the STM analysis.

For the MM/PBSA calculations, the GMXAPBS tool was used [31]. In particular: (1) the van der Waals term was calculated with Gromacs, (2) the coulombic term was calculated using the APBS accessory program coulomb, (3) the polar solvation term was calculated via APBS [32], using the non-linearized Poisson Boltzmann equation. Internal and external dielectric constants were set to 1 and 80, respectively; temperature was set to 310 K; the salt concentration was defined as 0.15 M; grid spacing was set to an upper limit of 0.5 Å, (4) The nonpolar solvation term was considered proportional to the solvent accessible surface area (SASA) as shown in [Eq. v]

$$<G_{nps}> \; = \; <SASA> \gamma + \beta \qquad (5)$$

where $\gamma = 0.0227$ kJ mol^{-1} Å^{-2} and $\beta = 0$ kJ mol^{-1}[33]. The dielectric boundary was defined using a probe of radius 1.4 Å.

The equilibrium phase (70–90 ns) of the two molecular dynamics simulations (i.e., 151 equally time-distant frames for each system) was considered for MM/PBSA calculations. The standard errors (SE) were calculated as [Eq. vi]

$$SE = SD/sqrt \, (151) \qquad (6)$$

where SD is the standard deviation.

Principal component analysis

The Principal Component Analysis (PCA) was used to characterize and compare the overall motions of the two complexes. We calculated the principal components on the converged simulation and focused on the movement of the 731 Cα atoms of the protein that resulted in 2193 dimensional displacement vectors. The 2193×2193 covariance matrix was then diagonalized to obtain its eigenvalues and eigenvectors. The PCA method decomposes the overall protein motion into a set of modes (eigenvectors) that are ordered from largest to smallest contributions to the protein fluctuations. The contribution of atom j to the ith mode's fluctuation was obtained using the following equation [Eq. vii]:

$$|m_i^j| = \sqrt{\left(m_i^{jx}\right)^2 + \left(m_i^{jy}\right)^2 + \left(m_i^{jz}\right)^2} = component_i^j \quad (7)$$

The $m_i^j = \left(m_i^{jx}, m_i^{jy}, m_i^{jz}\right)$ term represents the component vectors of the jth atom for the ith mode.

Each of the eigenvectors depicts a collective motion of particles and their respective amount of participation is represented by eigenvalues. Usually, the first ten eigenvectors are sufficient to describe almost all of the conformational subspace accessible to the protein.

Results and Discussion

The Y268 residue of Cyt b in *Plasmodium* is known to play a key role in ATQ drug resistance [8] and thus, can be used as a potential resistance marker [34]. Studies of *P. falciparum* resistance to ATQ revealed 3 point mutations at 268th position (Y268N, Y268S, Y268C) [8,21,35]. The substitution of one or several amino acid residues in a protein often lead to substantial changes in properties such as thermodynamic stability, catalytic activity, or binding affinity [36–38]. As point mutations at Y268 have already been identified for ATQ resistance, it is obvious that this substitution should affect the fitting and binding of the drug. ATQ drug, an analogue of coenzyme Q (ubiquinone), interrupts electron transport and leads to loss of the mitochondrial membrane potential [6].

Molecular modeling of Cyt b-ISP complex of *P. falciparum*

The 3D structure details of proteins are of major importance in providing insights into their molecular functions. Since computational methods not only help in directing the selection of key experiments, but also in the formulation of new testable hypotheses [39]. Therefore, in the absence of X-ray structures, the 3D theoretical models were built using the I-TASSER. I-TASSER [12,13], a hierarchical protein structure modeling approach based on 2 protein structure prediction methods i.e., threading and *ab initio* prediction, was used to build 3D models of the Cyt b and ISP sequences from *P. falciparum*. I-TASSER uses restraints from templates identified by multiple threading programs to build full length model using replica-exchange Monte-carlo simulations. Cyt b chain was modeled by I-TASSER using restraints from PDB templates 2IBZ, 3CX5, 1EZV, 1BCC and 3H1J while the ISP subunit used restrains from PDBs 1KB9, 3CX5, 1EZV, 1BCC, 3CWB, 1BGY and 3L72. No significant similarity was observed with any of the PDB templates against N-terminal sequence of ISP, therefore, this part was modeled by I-TASSER using ab-initio approach. Although, including the long

N-terminal sequence, modeled by ab-initio method can compromise the overall model accuracy, in the present study, we used full length model because of multiple reasons, (i) N-terminal 158 residues (modeled) of ISP subunit may also contribute in the reliable folding of tertiary structure as we observed several secondary structure elements in that region (Figure S1). Several studies have shown that the removal of N-terminal building blocks from the structure may contribute in error during protein folding [40–42] as the protein may acquire a non-native stable conformation due to mis-association of the adjunct building blocks. (ii) After modeling of the Cyt b-ISP complex, we observed the N-terminal sequence of chain moves back and forth over the active-site cleft (Figure S2) therefore this part was also taken into consideration. (iii) Since, even a single amino acid residue may significantly affect the conformation of binding site if present around 4.5 Å radius of the ligand, and may also alter the binding efficacy [43], thus we preferred to include this region in our model, so that the possible impact of N-terminal sequence in ATQ binding may be evaluated during simulations.

The computation of a structural alignment of 2 protein structures is critical in modeling, as in contrast to sequence alignment methods, structure alignment methods aim directly on optimizing the structural similarity of the input proteins [44]. The implemented TM-score in I-TASSER is a sensitive scale to the global topology for measuring the structural similarity between 2 proteins. Statistically, a TM-score <0.17 means a randomly selected protein pair with the gapless alignment taken from PDB. The better TM-score of our Cyt b model (0.99±0.04) and ISP subunit model (0.35±0.12) indicates much better structural match of the target sequence with the templates. I-TASSER provides C-score to estimate the quality of the predicted models, and is calculated based on the significance (i.e. Z-score) of the threading alignments in LOMETS and the convergence parameters (i.e. cluster density) of the I-TASSER structure assembly simulations. The C-score scheme has been extensively tested in large-scale benchmarking tests [13,18] and is typically in the range (−5, 2), where a higher score reflects a model of better quality. The C-score of the best predicted model of Cyt b and ISP subunit model was +2.00 and −3.25.

The Side-Chain Refinement protocol of DS was used to optimize the protein side-chain conformations. This protocol uses ChiRotor algorithm and CHARMm force field to systematically search for optimal side-chain conformation of all residues and generates a model structure with the best side-chain conformation. Model quality estimation is critically important in computational protein modeling, since the accuracy of a model determines its suitability for specific biological and biochemical experimental design [45]. The fitness of a protein sequence in its current 3D environment before and after side chain refinement was evaluated by Verify Protein (Profiles-3D). The Verify score of the protein is the sum of the scores of all residues in the protein and has been used by several researchers for structural assessment of theoretical models [46–48]. If the overall quality is lower than the expected low score, the structure is certainly misfolded. The verify score of the Cyt b model before and after energy minimization was 89.2 and 93.5, with the expected low score of 77.1, showing that the structure after side chain refinement was much better than the non-refined one. Similarly, we observed the verify score of the ISP model as 89.4, with the expected low score of 72.7906. The exact percentage of amino acids located in the core region was calculated by Procheck program. Generally, the atomic resolution structures have over 90% of their residues in the most favorable regions and for lower resolution structures resolved at 3.0–4.0 Å, the core percentage is around 70% [49]. The Ramachandran plot

showed that 90.3% residues were located in the core region of Cyt b chain and 83.33% residues were in the core region of ISP chain, indicating the reliability of models for further studies (Figure S3).

Moreover, we also looked into the RMSD of the modelled Qo site and the templates chosen for modeling of wild-type protein of *P. falciparum*. The calculated residuals for 1EZV, 2IBZ, 3CX5, 3H1J and 1BCC templates from our modeled protein were 0.48 Å, 0.44 Å, 0.46 Å, 0.56 Å and 1.02 Å respectively. The insignificant RMSD of Qo site from the respective templates further supports the model accuracy.

Mutagenesis of Cyt b protein

Most of the reported mutations in Cyt b either destabilize the important hydrophobic interactions between ATQ and the amino acid residues in the binding site of the protein, or are responsible for the change of pocket volume [9,50]. The conserved bulky Tyrosine (T) residue at 268th position forms hydrophobic contact with the ATQ drug in the Qo region of the ubiquinol oxidation site. Substitution of the hydrophilic and less bulky asparagine (N) at position 268 not only reduces the volume of the binding pocket but it also decreases the affinity and binding of ATQ [51]. Besides, substitution of serine (S), a hydrophilic amino acid, limits hydrophobic contact with ATQ resulting in marked decrement of ATQ susceptibility in mutated malaria parasites [21,35]. Moreover, a role for cysteine (C) in impairment of ATQ binding has also been observed [8]. Hence, *P. falciparum* Cyt b protein mutations (Y268S, Y268N and Y268C) were implemented in the 3D structure of the parent model using mutational modeling protocol of DS.

Active site selection and docking calculations of ATQ

ATQ is very likely to bind in a manner similar to stigmatellin, a known inhibitor of Qo site of the Cyt bc1 complex [10] and hence the potential binding site for stigmatellin as proposed by Solmaz and Hunte (PDB acquisition code 3CX5) [20] was chosen as the biologically favorable site for ATQ docking. After checking the conservancy of the active site residues of Cyt bc1 complex of *P. falciparum* (CYTB: MET116, ILE119, VAL120, PHE123, VAL124, MET133, TRP136, GLY137, VAL140, ILE141, THR142, LEU144, LEU145, ILE155, PHE169, LEU172, ILE258, VAL259, PRO260, GLU261, PHE264, PHE267, TYR268, LEU271, VAL284, LEU285; ISP: HIS104, LEU302, CYS319) with the Cyt b protein of *S. cerevisiae*, we observed that with the exception of 8 amino acid residues, all other residues were conserved between these 2 species (Figure 1). Moreover, we observed the non-identical residues of *P. falciparum* were also showing strongly similar properties (scoring>0.5 in the Gonnet PAM 250 matrix) with the amino acid residues present in *S. cerevisiae*. *Plasmodium* Cyt b is unusual in the sense that cd2 helix (a critical structural component of the catalytic Qo site) contains a 4-residue deletion that is not found in non-Apicomplexan sequences. It would seem very likely that this would alter the fold of the Qo site when compared to the Cyt b structural data available in the PDB. Therefore, we aligned the 3D structures of Cyt b from *P. falciparum* and *S. cerevisiae*. The structural overlay of the homology model of the *P. falciparum* Cyt b with the yeast Cyt b has been presented for comparison in Figure 2. Our model suggests that the 4 residue deletion in the cd2 helix results in a 0.83 Å displacement of this structural element compared with the yeast Cyt b (Figure 2). Likewise catalytically essential 'PEWY' motif of the ef helix was observed to be displaced by 0.35 Å from the yeast enzyme. To perform docking analysis, the ATQ structure was modeled properly using the Prepare/Filter Ligand tool of DS. Docking is a potentially powerful and inexpensive method for the

```
Cyt b_P. falciparum    MNFYS----INLVKAHLINYPCPLNINFLWNYGFLLGIIFFIQIITGVFLASRYTPDVSY  56
Cyt b_S. cerevisiae    MAFRKSNVYLSLVNSYIIDSPQPSSINYWWNMGSLLGLCLVIQIVTGIFMAMHYSSNIEL  60

Cyt b_P. falciparum    AYYSIQHILRELWSGWCFRYMHATGASLVFLLTYLHILRGLNYSYMYLP--LSWISGLIL  114
Cyt b_S. cerevisiae    AFSSVEHIMRDVHNGYILRYLHANGASFFFMVMFMHMAKGLYYGSYRSPRVTLWNVGVII  120

Cyt b_P. falciparum    FMIFIVTAFVGYVLPWGQMSYWGATVITNLLSSIPVA----VIWICGGYTVSDPTIKRFF  170
Cyt b_S. cerevisiae    FILTIATAFLGYCCVYGQMSHWGATVITNLFSAIPFVGNDIVSWLWGGFSVSNPTIQRFF  180

Cyt b_P. falciparum    VLHFILPFIGLCIVFIHIFFLHLHGSTNPLGYDTALK-IPFYPNLLSLDVKGFNNVIILF  229
Cyt b_S. cerevisiae    ALHYLVPFIIAAMVIMHLMALHIHGSSNPLGITGNLDRIPMHSYFIFKDLVTVFLFMLIL  240

Cyt b_P. falciparum    LIQSLFGIIPLSHPDNAIVVNTYVTPSQIVPEWYFLPFYAMLKTVPSKPAGLVIVLLSLQ  289
Cyt b_S. cerevisiae    ALFVFYSPNTLGHPDNYIPGNPLVTPASIVPEWYLLPFYAILRSIPDKLLGVITMFAAIL  300

Cyt b_P. falciparum    LLFLLAE-QRSLTTIIQFKMIFGARDYSVPIIWFMCAFYALLWIGCQLPQDIFILYGRLF  348
Cyt b_S. cerevisiae    VLLVLPFTDRSVVRGNTFKVLS-----KFFFFIFVFNFVLLGQIGACHVEVPYVLMGQIA  355

Cyt b_P. falciparum    IVLFFCSGLFVL--VHYRRTHYDYSSQANI  376
Cyt b_S. cerevisiae    TFIYFAYFLIIVPVISTIENVLFYIGRVNK  385
```

```
ISP_P. falciparum    MNNIKYVELFYKCKIFRKNGLNRIIRRNGGTFNHNIKENERIPPASEDPS  50
ISP_S. cerevisiae    MLGIR------------------------------------SSVKTC  11

ISP_P. falciparum    YKNLFDHAEDIKLWEIEEKQNVSHKKVEDLSELVEPSNHPHQYEGIFART  100
ISP_S. cerevisiae    FK---------------PMSLTSKRLISQSLLASKS-------------  32

ISP_P. falciparum    RYAHYNQTAEPVFPRKPDLEKGELASGANVTRTDVWHNPKEPAIVSIGKF  150
ISP_S. cerevisiae    --TYRTPNFDDVLKENNDADKG--------------------------  52

ISP_P. falciparum    EPRNFRPAGYAENCPNPESINSDHHPDFREYRLRSGNEDRRSFMYFISAS  200
ISP_S. cerevisiae    ----------------------------------------RSYAYFMVGA  62

ISP_P. falciparum    YFFIMSSIMRSAICKSVHFFWISKDLVAGGTTELDMRTVNPGEHVVIKWR  250
ISP_S. cerevisiae    MGLLSSAGAKSTVETFISSMTATADVLAMAKVEVNLAAIPLGKNVVVKWQ  112

ISP_P. falciparum    GKPVFVKHRTPEDIQRAKEDDKLIQTMRDPQLDSDRTIKPEWLVNIGICT  300
ISP_S. cerevisiae    GKPVFIRHRTPHEIQEANSVD--MSALKDPQTDADRVKDPQWLIMLGICT  160

ISP_P. falciparum    HLGCVPA-QGGNYSGYFCPCHGSHYDNSGRIRQGPAPSNLEVPPYEFVDE  349
ISP_S. cerevisiae    HLGCVPIGEAGDFGGWFCPCHGSHYDISGRIRKGPAPLNLEIPAYEFDGD  210

ISP_P. falciparum    NTIKIG  355
ISP_S. cerevisiae    -KVIVG  215
```

Figure 1. Sequence alignment of Cyt b protein and ISP chain of *P. falciparum* **and** *S. cerevisiae.* Amino acid residues involved in the formation of Qo site are highlighted in yellow tinted color.

discovery of binary interactions. The Qo site residues were chosen to define the binding site in our modeled Cyt bc1 complex of *P. falciparum* based on known Qo site inhibitor interactions for *S. cerevisiae* available in protein data bank (pdb id: 3CX5, 2IBZ). A total of 10 random ligand conformations were generated from the ATQ structure through high temperature molecular dynamics, followed by random rotations. These conformations were refined by grid-based (GRID 1) simulated annealing and a final full force field minimization method. We observed that ATQ was showing less binding affinity towards all the mutant variants when compared to the wild-type.

'PEWY' motif of ef helix
(Blue: *P. falciparum*; Cyan: *S. cerevisiae*)

cd2 helix
(Red: *P. falciparum*; Green: *S. cerevisiae*)

Figure 2. Structural overlay of the homology model of Cyt b protein of *P. falciparum* (blue) with the Cyt b unit of *S. cerevisiae* (golden) (PDB ID: 3CX5). A total of 4 amino acid residues deletion in cd2 helix (red) of *P. falciparum* resulted in structural displacement when compared with the same domain of *S. cerevisiae* (green). Also the structural changes in 'PEWY' motif of ef helix are shown.

While examining all the contact amino acid residues within 5 Å of the ATQ, as shown in the Figure 3, we observed the presence of Y268 within the ATQ binding site in the wild-type protein. Surprisingly when Y268 was mutated to any of the 3 possible amino acid mutations, i.e.Y268N, Y268S, Y268C, this position shifted far away from ATQ binding site. We feel the shift of amino acid residue at 268th position after point mutation might be the main reason of ATQ resistance in the mutant models. In order to understand the mechanistic insight of ATQ resistance in the mutant models, we further performed detailed molecular dynamics simulation studies by considering only the most prevalent

mutant variant (Y268S) identified in various experimental settings [9].

Dynamic insights into the Cyt bc1 modeled complexes

Several simulation studies have already shown nice correlation between computational and experimental measurements of macromolecular dynamics [52–55]. As molecular dynamics based techniques can provide more precise protein–ligand models in the state close to natural conditions therefore to get detailed insights into the molecular basis of ATQ resistance in malaria, we individually simulated ATQ-bound wild-type and the most prevalent mutant variant (Y268S) of *P. falciparum* Cyt b protein

Figure 3. Two dimensional contact plots of amino acid residues from the wild and all screened mutant models of Cyt bc1 complex from P. falciparum in the vicinity of 5 Å radius around ATQ. It may be noted that in the mutant models the position 268 shifted away from the 5 Å radius of ATQ Binding site. Whereas green color indicates that the particular amino acid residue is present in the wild as well as in all mutant models in the observed area; yellow, red, blue, cyan color shows amino acid residues present only in wild type, Y268S, Y268N and Y268C mutant models respectively.

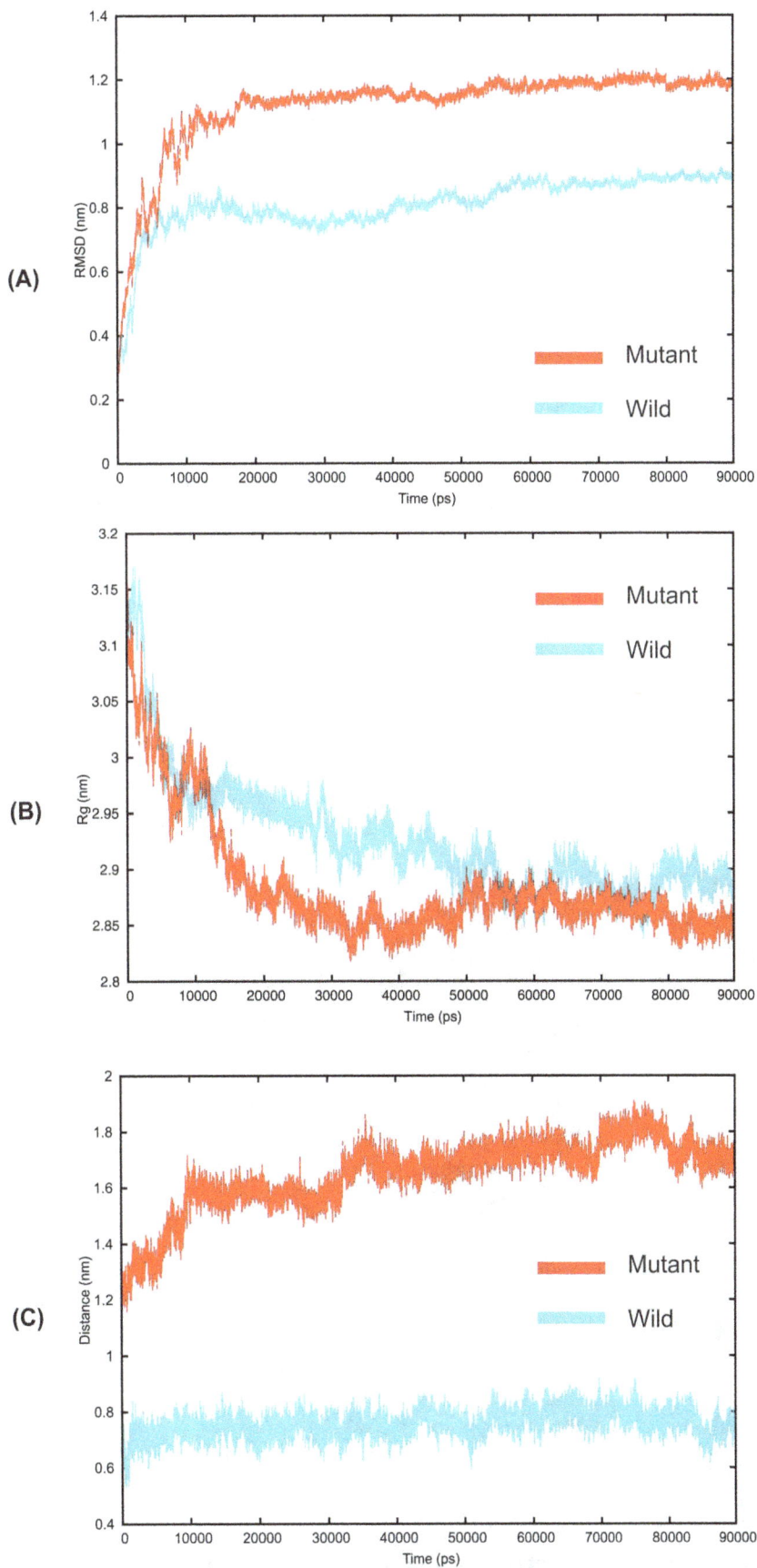

Figure 4. Molecular dynamics simulation graph of wild (Y268) and mutant (Y268S) Cyt bc1 complexes of *P. falciparum* in solution. (A) Root mean square deviation (RMSD) of backbone atoms with respect to their initial complexes over a period of 90 ns simulation time. (B) Radius of gyration (Rg) graph and (C) Distance of ATQ from the Qo site over the whole simulation in wild as well as mutant type.

[9] till we attained the convergence of MD simulation around 70–90 ns of production run. To explore the dynamic stability of the systems, root mean square deviation (RMSD) of the Cyt b-ISP backbone atoms (both wild-type and mutant complexes) was computed with reference to their respective initial structures as a function of simulation run time. MD simulation shows that the wild-type complex undergoes less structural changes when compared with the mutant complex over a period of 90 ns simulation time (Figure 4A, in Cyan) and RMSD of both the complexes remain almost stagnant after ~65 ns time period. The plotted graph in Figure 4A (red) shows that although the RMSD of mutant complex was initially in agreement with the wild-type complex but the complex underwent significant deviation after 7ns and reaches at ~1.1 nm RMSD (consistent throughout the whole dynamics run), showing the comparatively unstable behavior of the mutant complex. We further extended our study to analyze the compactness (radius of gyration) of both the wild-type and mutant complexes. Though we did not find much significant difference in Rg values, however we noticed that at the equilibrium state mutant structure was more compact than the wild-type (Figure 4B). We were also interested to check the ATQ distance from the Qo site during the entire dynamics run. After analyzing the results, we noticed that ATQ retained its position throughout the whole dynamics run in the wild-type however it showed significant fluctuations during the entire 90ns simulation in mutant case (Figure 4C). Even as clear from the Figure 4C, the distance between the Qo site and ATQ was increasing with respect to time. We may attribute this behavior to the change in binding pocket configuration and also because of some steric clashes of ATQ near the binding site.

In order to understand how the resistant mutations affect the interaction of ATQ in mutant protein, we calculated root mean square inner product (RMSIP) from both the complexes to ascertain the convergence of conformational sampling instep-wise

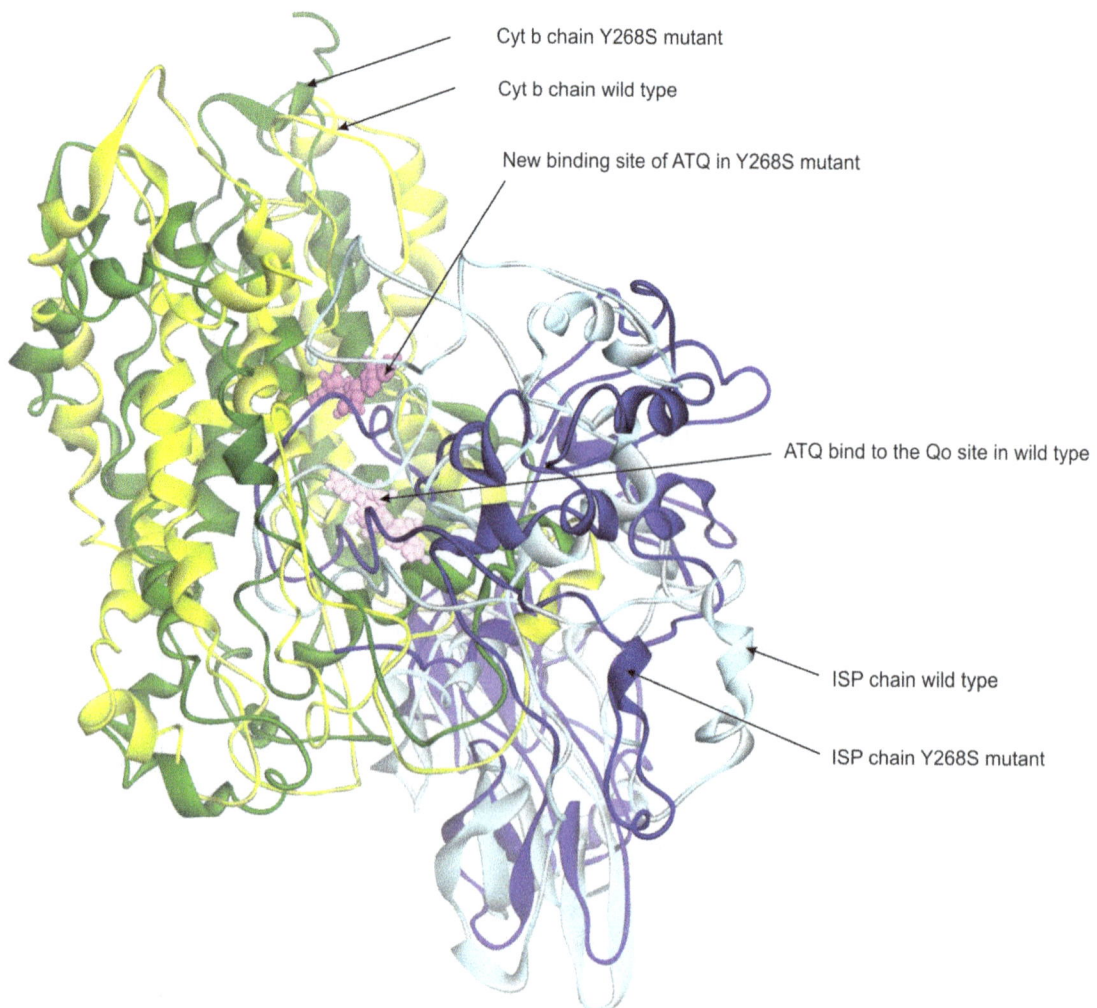

Figure 5. ATQ binding site in the wild and mutant (Y268S) Cyt bc1 complex of *P. falciparum*. The figure indicates that ATQ binds to a new site in the mutant model which is around 12 Å distant from the Qo site. The structure was captured from the average structure of the Cyt bc1 complex of *P. falciparum* over 70–90 ns (converged part of the trajectory).

Table 1. MM/PBSA binding free energies (kJ/mol) of wild-type and mutant Cyt bc1/ATQ complexes.

Binding free energies (kJ/mol)	Y268S	Y268
ΔG_{coul}[1]	-71.438 ± 1.993	-49.334 ± 1.527
ΔG_{vdw}[2]	-188.381 ± 1.118	-209.546 ± 1.023
ΔG_{ps}[3]	190.650 ± 2.357	246.728 ± 2.430
ΔG_{nps}[4]	-17.924 ± 0.070	-19.962 ± 0.037
ΔG_{polar}[5]	119.212	197.394
$\Delta G_{nonpolar}$[6]	-206.305	-229.508
ΔG_{bind}[7]	-87.094 ± 2.178	-32.113 ± 2.653

[1]: coulombic term;
[2]: van der Waals term;
[3]: polar solvation term;
[4]: nonpolar solvation term;
[5]: polar term (sum of coulombic and polar solvation terms);
[6]: nonpolar term (sum of van der Waals and nonpolar solvation terms);
[7]: computational binding free energy.

manner till we find the convergence. In this process, we increased the production run from initial 20 to 90 ns in the block of 10 ns. In general, RMSIP values between 0.5–0.7 represent adequate convergence [56] and here in our case, we observed an acceptable convergence measure of 0.55 from 70–90 ns trajectory. Therefore, this part of simulation was used for the computation of average structure of the complex. To remove the crudeness of the average structure, the structure was subjected to 1000 steps of energy minimization using the Smart Minimizer (SM) available in DS. SM begins with the Steepest Descent method, followed by the Conjugate Gradient method for faster convergence towards a local minimum. We observed the ATQ interactions were predominantly hydrophobic, although certain hydrophilic interactions exist temporarily during the MD simulation. Histidine residue 181 in the ISP (Yeast) is reported to form a strong hydrogen-bond with certain classes of Qo-bound inhibitors such as stigmatellin ('b-distal' inhibitors). We noticed that in *P. falciparum*, HIS104 of ISP chain forms such stable interaction with the ATQ. However, we did not find the stability of the ATQ hydrogen bond that was supposed to be formed via a water molecule with Glu (Glu-272 in yeast) of Cyt b. In mutant case, both these interactions were altogether absent. In the mutant model irrespective of ATQ binding at Qo site, it was found to get stabilized at a new site (site II) which is around 12 Å apart from

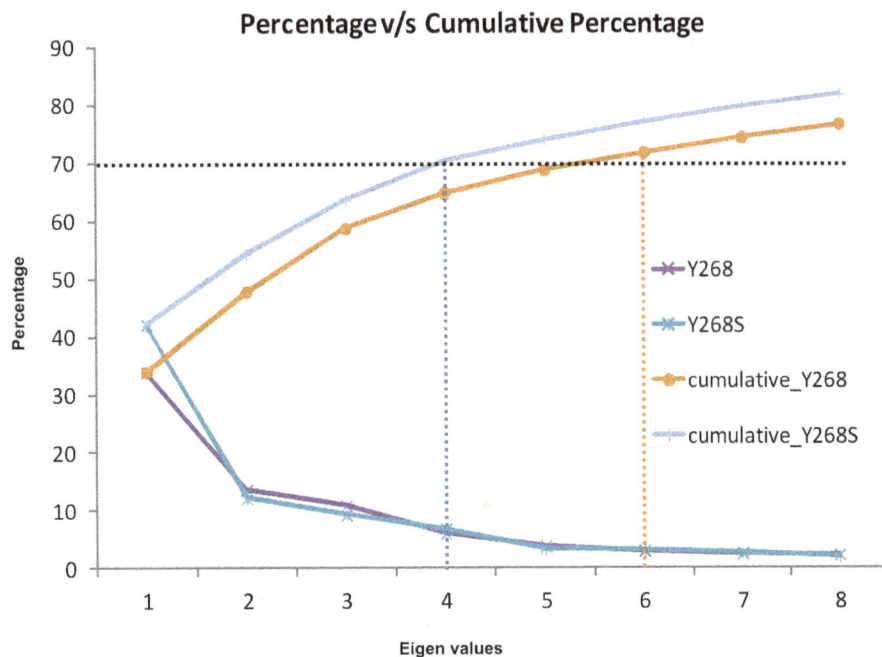

Figure 6. Proportion of variance and cumulative proportion of total variance of first ten eigenvalues of the wild (Y268) and mutant variant (Y268S) of *P. falciparum* proteins.

the Qo site. The two different binding modes of ATQ in wild and mutant-type are shown in Figure 5. Moreover, we also noticed that the induced mutation makes the active site to undergo significant conformational changes that reduce both the active site volume (7174.6 Å^3) and its surface area (4783.4 Å^2) than the wild-type (11388 Å^3; 7269.3 Å^2). The shrinkage in both volume and surface area (also reflected by Rg values, Figure 4C) might be also one of the probable reasons for different binding site selectivity of ATQ in mutant-type. Overall, all the above consequences might be the causing agents for ATQ to have different selectivity in the mutant protein than the wild-type.

MD-based binding free energy calculations of ATQ-bound Cyt bc1 complex

To acquire an estimate of the binding free energies in the two systems (i.e., Y268 and Y268S) and to inspect the differences in terms of polar and non-polar interactions, we performed binding free energy calculations on the converged MD trajectories (70–90 ns) using the MM/PBSA approach. For this purpose, we took advantage of the GMXAPBS tool [30]. Here we would like to mention that in this investigation we were interested in highlighting the differences of binding free energies of similar complexes. As already reported for several other investigations [30,57,58], we will thus ignore the entropic term and focus only on the binding enthalpy, defined as the sum of 4 terms, namely coulombic, van der Waals, polar solvation and non-polar solvation. The results of our calculations are shown in Table 1.

Both systems present a similar pattern: the van der Waals, coulombic and non-polar solvation terms are negative, meaning that they favor the formation of the complex. On the contrary, the polar solvation term is positive, which indicates that it antagonizes the binding process. This observation might be due to the cost of desolvating the polar moieties in the protein residues and in the ligand. It is noteworthy that in both cases the polar contribution, defined as the sum of the coulombic and the polar solvation terms, is positive. Overall, the MM/PBSA analysis reveals that the complex stabilization is promoted by the non-polar contribution only. Such a behavior has been described in the past both for protein-ligand [59] and for protein-protein [31] interactions. The Y268S mutation significantly affects all terms.

Our study demonstrates that the binding of ATQ with the site II (newly identified binding site) of *P. falciparum* Cyt bc1 complex was energetically favorable compare to the site I (Qo site). We assume the sufficient binding energy of ATQ at site II might be preventing ATQ to bind its native site (Qo) thereby resulting in the loss of the anti-malarial efficiency of ATQ in the Y268S mutant-type.

Essential dynamics analyses of the Cyt bc1 complexes

To support the MD results, we performed essential dynamics study of both the simulated complexes and our results shows that the cumulative variance captured by the first few eigenvectors or principal components of the wild-type complex was comparatively lower than the mutant-type (Figure 6).

The analysis indicates that the Y268S/ATQ complex pertain more motions than the wild-type complex. This is in consistent with the earlier dynamic results where we observed some fluctuations and far away displacement of ATQ from the Qo site in the Y268S/ATQ complex.

Conclusion

The emerging acquired drug resistance because of mutations has presented a challenge to follow the traditional drug discovery pipelines. Even a single point mutation has potential to produce the drug resistance. In this study, we evaluated the drug-resistance mechanism of ATQ in the mutated (Y268S) Cyt b protein of *P. falciparum* using the potential of *in silico* methods. We observed that the interaction between ATQ and Cyt b is mainly stabilized by the hydrophobic contacts and after mutation of Y268S ATQ contacts with the Qo site are greatly reduced. Such findings have also been reported by other authors [9]. We presume that this reduction in Qo contacts and also change in the volume and surface area of the binding pocket (similar observations have been made by Kessl et al. 2005 [9] in I269M mutation of ATQ-bound yeast bc1 complex) enforce ATQ to find its desirable contacts at distant location from its wild active site. Moreover, the MM-PBSA calculations firmly proved the tighter binding of ATQ with the mutant-type at additional binding site (first time observation), present at ~12 Å faraway from the active site, thereby raising no choice for ATQ to bind its native site. This might be the probable reason for ATQ anti-malarial efficacy loss in Y268S mutants. We hope the structural details presented in this study would aid the experimental plan to design new suitable selective ligands that could have correct size to fit properly in the active site even after mutation. Such ligands might be able to resist the mutation effect and can be used as future effective drugs against malaria.

Supporting Information

Figure S1 ISP subunit of Cyt bc1 complex of *P. falciparum* with predicted secondary structure elements. It is important to note that the initial 158 N-terminal residues, which were not present in the *S. cerevisiae* Cyt bc1 complex subunit in PDB file 3CX5, are also involved in critical secondary structure confirmations. This is the reason why we consider full length ISP subunit in our analysis.

Figure S2 The Qo site (ATQ binding site) of Cyt bc1 complex of *P. falciparum* is shown. The N-terminal residues of ISP chain are involved in the formation of active site (Qo) cleft (red color and shown as surface model). Cyt b subunit is shown in green color and ISP subunit as blue.

Figure S3 The Ramachandran plots of modeled Cyt b protein and ISP subunit of *P. falciparum* in Cyt bc1 complex are shown. The plots indicate the quality of the modeled structure was satisfactory.

Acknowledgments

We would like to thank Giovanna Musco, Cristina Paissoni and Dimitrios Spiliotopoulos from S. Raffaele Scientific Institute, Biomolecular NMR Laboratory, Milano, Italy for their support in MM/PBSA calculations and fruitful discussion to improve this manuscript. KPS and SKG acknowledge Council of scientific and industrial Research (CSIR) India network project GENESIS (BSC 0121) and INDEPTH (BSC 0111). BAA is recipient of Senior Research Fellowship from CSIR India. CSIR-IITR Manuscript Communication number 3246.

Author Contributions

Conceived and designed the experiments: Shailendra Gupta BAA. Performed the experiments: BAA KPS MV Shishir Gupta. Analyzed the data: Shailendra Gupta BAA Shishir Gupta KPS YS. Contributed reagents/materials/analysis tools: Shailendra Gupta. Wrote the paper: Shailendra Gupta BAA KPS YS.

References

1. Singh B, Kim Sung L, Matusop A, Radhakrishnan A, Shamsul SS, et al. (2004) A large focus of naturally acquired Plasmodium knowlesi infections in human beings. Lancet 363: 1017–1024.

2. Daneshvar C, Davis TM, Cox-Singh J, Rafa'ee MZ, Zakaria SK, et al. (2009) Clinical and laboratory features of human Plasmodium knowlesi infection. Clin Infect Dis 49: 852–860.

3. Crowther GJ, Napuli AJ, Gilligan JH, Gagaring K, Borboa R, et al. (2011) Identification of inhibitors for putative malaria drug targets among novel antimalarial compounds. Mol Biochem Parasitol 175: 21–29.

4. Garcia LS (2010) Malaria. Clin Lab Med 30: 93–129.

5. Fry M, Pudney M (1992) Site of action of the antimalarial hydroxynaphtho-quinone, 2-[trans-4-(4'-chlorophenyl) cyclohexyl]-3-hydroxy-1,4-naphthoqui-none (566C80). Biochem Pharmacol 43: 1545–1553.

6. Srivastava IK, Rottenberg H, Vaidya AB (1997) Atovaquone, a broad spectrum antiparasitic drug, collapses mitochondrial membrane potential in a malarial parasite. J BiolChem 272: 3961–3966.

7. Vaidya AB, Lashgari MS, Pologe LG, Morrisey J (1993) Structural features of Plasmodium cytochrome b that may underlie susceptibility to 8-aminoquinolines and hydroxynaphthoquinones. Mol Biochem Parasitol 58: 33–42.

8. Musset L, Bouchaud O, Matheron S, Massias L, Le Bras J (2006) Clinical atovaquone-proguanil resistance of Plasmodium falciparum associated with cytochrome b codon 268 mutations. Microbes Infect 8: 2599–2604.

9. Kessl JJ, Ha KH, Merritt AK, Lange BB, Hill P, et al. (2005) Cytochrome b mutations that modify the ubiquinol-binding pocket of the cytochrome bc1 complex and confer anti-malarial drug resistance in Saccharomyces cerevisiae. J Biol Chem 280: 17142–17148.

10. Kessl JJ, Meshnick SR, Trumpower BL (2007) Modeling the molecular basis of atovaquone resistance in parasites and pathogenic fungi. Trends Parasitol 23: 494–501.

11. Fisher N, Abd Majid R, Antoine T, Al-Helal M, Warman AJ, et al.(2012) Cytochrome b mutation Y268S conferring atovaquone resistance phenotype in malaria parasite results in reduced parasite bc1 catalytic turnover and protein expression. J Biol Chem287: 9731–41.

12. Hill P, Kessl J, Fisher N, Meshnick S, Trumpower BL, et al. (2003) Recapitulation in Saccharomyces cerevisiae of cytochrome b mutations conferring resistance to atovaquone in Pneumocystis jiroveci. Antimicrob Agents Chemother 47: 2725–31.

13. Zhang Y (2008) I-TASSER server for protein 3D structure prediction. BMC Bioinformatics 9: 40.

14. Das R, Qian B, Raman S, Vernon R, Thompson J, et al. (2007) Structure prediction for CASP7 targets using extensive all-atom refinement with Rosetta@home. Proteins 69: 118–28.

15. Zhang Y (2007) Template-based modeling and free modeling by I-TASSER in CASP7. Proteins 69: 108–117.

16. Zhou H, Pandit SB, Lee SY, Borreguero J, Chen H, et al. (2007) Analysis of TASSER-based CASP7 protein structure prediction results. Proteins 69: 90–97.

17. Roy A, Kucukural A, Zhang Y (2010) I-TASSER: a unified platform for automated protein structure and function prediction. Nat Protoc 5: 725–738.

18. Zhang Y, Skolnick J (2004) Scoring function for automated assessment of protein structure template quality. Proteins 57: 702–10.

19. Spassov VZ, Bashford D (1999) Multiple-Site Ligand Binding to Flexible Macromolecules: Separation of Global and Local Conformational Change and an Iterative Mobile Clustering Approach J Comput Chem 2: 1091–1111.

20. Solmaz SR, Hunte C (2008) Structure of complex III with bound cytochrome c in reduced state and definition of a minimal core interface for electron transfer. J Biol Chem 283: 17542–17549.

21. Korsinczky M, Chen N, Kotecka B, Saul A, Rieckmann K, et al. (2000) Mutations in Plasmodium falciparum cytochrome b that are associated with atovaquone resistance are located at a putative drug-binding site. Antimicrob Agents Chemother 44: 2100–2108.

22. Wu G, Robertson DH, Brooks CL 3rd, Vieth M (2003) Detailed analysis of grid-based molecular docking: A case study of CDOCKER-A CHARMm-based MD docking algorithm. J Comput Chem 24: 1549–1562.

23. Essmann U, Perera L, Berkowitz ML, Darden T, Lee H, et al. (1995) A smooth particle mesh Ewald method. J Chem Phys 103: 8577–8593.

24. Merlino A, Mazzarella L, Carannante A, Di Fiore A, Di Donato A, et al. (2005) The importance of dynamic effects on the enzyme activity: X-ray structure and molecular dynamics of onconase mutants. J Biol Chem 280: 17953–1760.

25. Taly JF, Marin A, Gibrat JF (2008) Can molecular dynamics simulations help in discriminating correct from erroneous protein 3D models? BMC Bioinformatics 9: 6.

26. Malek K, Odijk T, Coppens MO (2005) Diffusion of water and sodium counter-ions in nanopores of a β-lactoglobulin crystal: a molecular dynamics study. Nanotechnology 16: S522–530.

27. Hess B, Bekker H, Berendsen HJC, Fraaije JGEM (1997) LINCS: A linear constraint solver for molecular simulations. J Comput Chem 18: 1463–1472.

28. Berendsen HJC, Postma JPM, Van Gunsteren WF, DiNola A, Haak JR (1984) Molecular dynamics with coupling to an external bath. J Chem Phys 81: 3684–3690.

29. Parrinello M, Rahman A (1980) Crystal Structure and Pair Potentials: A Molecular-Dynamics Study. Phys Rev Lett 45: 1196–1199.

30. Massova I, Kollman PA (1999) Computational alanine scanning to probe protein-protein interactions: A novel approach to evaluate binding free energies. J Am Chem Soc 121: 8133–8143.

31. Spiliotopoulos D, Spitaleri A, Musco G (2012) Exploring PHD fingers and H3K4me0 interactions with molecular dynamics simulations and binding free energy calculations: AIRE-PHD1, a comparative study. PLoS One 7: e46902.

32. Baker NA, Sept D, Joseph S, Holst MJ, McCammon JA (2001) Electrostatics of nanosystems: Application to microtubules and the ribosome. Proc Natl Acad Sci U S A 98: 10037–1004.

33. Brown SP, Muchmore SW (2009) Large-scale application of high-throughput molecular mechanics with poisson-boltzmann surface area for routine physics-based scoring of protein-ligand complexes. J Med Chem 52: 3159–3165.

34. Schwöbel B, Alifrangis M, Salanti A, Jelinek T (2003) Different mutation patterns of atovaquone resistance to Plasmodium falciparum in vitro and in vivo: rapid detection of codon 268 polymorphisms in the cytochrome b as potential in vivo resistance marker. Malar J 2: 5.

35. Fivelman QL, Butcher GA, Adagu IS, Warhurst DC, Pasvol G (2002) Malarone treatment failure and in vitro confirmation of resistance of Plasmodium falciparum isolate from Lagos, Nigeria. Malar J 1: 1.

36. Wells JA (1990) Additivity of mutational effects in proteins. Biochemistry 29: 8509–17.

37. Lo TP, Komar-Panicucci S, Sherman F, McLendon G, Brayer GD (1995) Structural and functional effects of multiple mutations at distal sites in cytochrome c. Biochemistry 34: 5259–5268.

38. Schreiber G, Fersht AR (1995) Energetics of protein-protein interactions: analysis of the barnase-barstar interface by single mutations and double mutant cycles. J Mol Biol 248: 478–486.

39. Baloria U, Akhoon BA, Gupta SK, Sharma S, Verma V (2012) In silico proteomic characterization of human epidermal growth factor receptor 2 (HER-2) for the mapping of high affinity antigenic determinants against breast cancer. Amino Acids 42: 1349–60.

40. Ma B, Tsai CJ, Nussinov R (2000) Binding and folding: in search of intramolecular chaperone-like building block fragments. Protein Eng 13: 617–27.

41. Sham YY, Ma B, Tsai CJ, Nussinov R (2001) Molecular dynamics simulation of Escherichia coli dihydrofolate reductase and its protein fragments: relative stabilities in experiment and simulations. Protein Sci 10: 135–148.

42. Kumar S, Sham YY, Tsai C-J, Nussinov R (2001) Protein folding and function: the N-terminal fragment in adenylate kinase. Biophys J 80: 2439–2454.

43. Feyfant E, Sali A, Fiser A (2007) Modeling mutations in protein structures. ProtSci 16: 2030–2041.

44. Gupta SK, Srivastava M, Akhoon BA, Gupta SK, Grabe N (2012) Insilico accelerated identification of structurally conserved CD8+ and CD4+ T-cell epitopes in high-risk HPV types. Infect Genet Evol 12: 1513–8.

45. Srivastava M, Gupta SK, Abhilash PC, Singh N (2012) Structure prediction and binding sites analysis of curcin protein of Jatrophacurcas using computational approaches. J Mol Model. 18: 2971–9.

46. Akhoon BA, Gupta SK, Verma V, Dhaliwal G, Srivastava M, et al. (2010) In silico designing and optimization of anti-breast cancer antibody mimetic oligopeptide targeting HER-2 in women. J Mol Graph Model 28: 664–9.

47. Cherkis KA, Temple BR, Chung EH, Sondek J, Dangl JL (2012) AvrRpm1 missense mutations weakly activate RPS2-mediated immune response in Arabidopsis thaliana. PLoS One 7:e42633.

48. Frauer C, Rottach A, Meilinger D, Bultmann S, Fellinger K, et al. (2012) Different binding properties and function of CXXC zinc finger domains in Dnmt1 and Tet1. PLoS One 6:e16627.

49. Laskowski RA (2003) Structural quality assurance. In Structural Bioinformatics. Edited by Bourne P, Weissig H. Wiley-Liss, Inc. 292 p.

50. Fisher N, Meunier B (2008) Molecular basis of resistance to cytochrome bc1 inhibitors. FEMS Yeast Res 8: 183–192.

51. Khositnithikul R, Tan-Ariya P, Mungthin M (2008) In vitro atovaquone/proguanil susceptibility and characterization of the cytochrome b gene of Plasmodium falciparum from different endemic regions of Thailand. Malar J 28: 7: 23.

52. Friedman SH, DeCamp DL, Sijbesma RP, Srdanov G, Wudl F, et al. (1993) Inhibition of the HIV-1 protease by fullerene derivatives: model building studies and experimental verification. J Am Chem Soc 115: 6506–6509.

53. Friedman SH, Ganapathi PS, Rubin Y, Kenyon GL (1998) Optimizing the binding of fullerene inhibitors of the HIV-1 protease through predicted increases in hydrophobic desolvation. J Med Chem 41: 2424–9.

54. Cheng Y, Li D, Ji B, Shi X, Gao H (2010) Structure-based design of carbon nanotubes as HIV-1 protease inhibitors: atomistic and coarse-grained simulations. J Mol Graph Model 29: 171–7.

55. Lee VS, Nimmanpipug P, Aruksakunwong O, Promsri S, Sompornpisut P, et al. (2007) Structural analysis of lead fullerene-based inhibitor bound to human immunodeficiency virus type 1 protease in solution from molecular dynamics simulations. J Mol Graph Model 26: 558–70.

56. Laberge M, Yonetani T (2008) Molecular dynamics simulations of hemoglobin A in different states and bound to DPG: Effector-linked perturbation of tertiary conformations and HbA concerted dynamics. Biophys J 94: 2737–2751.

57. Huo S, Massova I, Kollman PA (2002) Computational alanine scanning of the 1:1 human growth hormone-receptor complex. J Comput Chem 23: 15–27.

58. Bradshaw RT, Patel BH, Tate EW, Leatherbarrow RJ, Gould IR (2011) Comparing experimental and computational alanine scanning techniques for probing a prototypical protein-protein interaction. Protein Eng Des Sel 24: 197–207.

59. Chiappori F, Merelli I, Milanesi L, Marabotti A (2013) Static and dynamic interactions between GALK enzyme and known inhibitors: Guidelines to design new drugs for galactosemic patients. Eur J Med Chem 63C: 423–434.

Exploring the Genes of Yerba Mate (*Ilex paraguariensis* A. St.-Hil.) by NGS and *De Novo* Transcriptome Assembly

Humberto J. Debat[1,9], Mauro Grabiele[2,4,9], Patricia M. Aguilera[2,4], Rosana E. Bubillo[3], Mónica B. Otegui[4], Daniel A. Ducasse[1], Pedro D. Zapata[4], Dardo A. Marti[2,4]*

1 Instituto de Patología Vegetal, Centro de Investigaciones Agropecuarias, Instituto Nacional de Tecnología Agropecuaria (IPAVE-CIAP-INTA), Córdoba, Argentina, 2 Instituto de Biología Subtropical, Universidad Nacional de Misiones (IBS-UNaM-CONICET), Posadas, Misiones, Argentina, 3 Estación Experimental Cerro Azul, Instituto Nacional de Tecnología Agropecuaria (EEA Cerro Azul-INTA), Misiones, Argentina, 4 Instituto de Biotecnología de Misiones, Facultad de Ciencias Exactas Químicas y Naturales, Universidad Nacional de Misiones (INBIOMIS-FCEQyN-UNaM), Misiones, Argentina

Abstract

Yerba mate (*Ilex paraguariensis* A. St.-Hil.) is an important subtropical tree crop cultivated on 326,000 ha in Argentina, Brazil and Paraguay, with a total yield production of more than 1,000,000 t. Yerba mate presents a strong limitation regarding sequence information. The NCBI GenBank lacks an EST database of yerba mate and depicts only 80 DNA sequences, mostly uncharacterized. In this scenario, in order to elucidate the yerba mate gene landscape by means of NGS, we explored and discovered a vast collection of *I. paraguariensis* transcripts. Total RNA from *I. paraguariensis* was sequenced by Illumina HiSeq-2000 obtaining 72,031,388 pair-end 100 bp sequences. High quality reads were *de novo* assembled into 44,907 transcripts encompassing 40 million bases with an estimated coverage of 180X. Multiple sequence analysis allowed us to predict that yerba mate contains ~32,355 genes and 12,551 gene variants or isoforms. We identified and categorized members of more than 100 metabolic pathways. Overall, we have identified ~1,000 putative transcription factors, genes involved in heat and oxidative stress, pathogen response, as well as disease resistance and hormone response. We have also identified, based in sequence homology searches, novel transcripts related to osmotic, drought, salinity and cold stress, senescence and early flowering. We have also pinpointed several members of the gene silencing pathway, and characterized the silencing effector Argonaute1. We predicted a diverse supply of putative microRNA precursors involved in developmental processes. We present here the first draft of the transcribed genomes of the yerba mate chloroplast and mitochondrion. The putative sequence and predicted structure of the caffeine synthase of yerba mate is presented. Moreover, we provide a collection of over 10,800 SSR accessible to the scientific community interested in yerba mate genetic improvement. This contribution broadly expands the limited knowledge of yerba mate genes, and is presented as the first genomic resource of this important crop.

Editor: Jin-Song Zhang, Institute of Genetics and Developmental Biology, Chinese Academy of Sciences, China

Funding: The authors have no support or funding to report.

Competing Interests: The authors have declared that no competing interests exist.

* Email: darmarti@gmail.com

9 These authors contributed equally to this work.

Introduction

Ilex paraguariensis (Aquifoliaceae) is a dioecious crop tree native to the subtropical rainforest of Northeastern Argentina, Southwestern Brazil and Eastern Paraguay, where it is widely cultivated [1]. This evergreen holly is colloquially known as "yerba mate" or "erva mate" as it is mainly consumed as a nutritional and stimulant beverage named "mate", a type of hot infusion made from dried milled leaves and twigs of *I. paraguariensis*. Yerba mate is also extensively used to prepare infusions, concoctions and quenchers with similar purposes and, more recently as admixture in ice creams, candies and energy drinks [2], as well as in dyes, cosmetics and spa ingredients [3]. Antioxidant, anti-inflammatory, antimutagenic and lipid-lowering properties have been reported in *I. paraguariensis* [2], leading to an increasing interest in this tree. Yerba mate is an economically

important crop cultivated and produced on a total area of more than 326,000 ha [4,5]. Argentina is the main producer with a total yield of over 880,000 t, representing ~85% of world-wide yerba mate production [6]. About 15% of total yerba mate production is exported to South American, European and Asian markets [7]. Besides the agricultural and economic importance of yerba mate, it is worth noting its profound and omnipresent influence in Latin American socio-cultural dynamics. Yerba mate widespread consumption embraces and extends ubiquitously, pervasively reaching every economic and cultural niche in South America [8–11]. To emphasize the relevance of yerba mate in South American tradition and its introduction and dissemination particularly in Argentinean culture, in 2009 a 5,000 people survey projected that while 81% of the Argentinean population consumes coffee, a striking 98% of the population consumes yerba mate [12].

Genetic improvement of *I. paraguariensis* has been limited by several factors. Agronomic evaluation and selection programs have been performed essentially to improve yield in this crop [13,1]. However, apart from this trait, very little is known about agronomically important loci on the limited available germplasm of this species. Moreover, yerba mate plants cannot be recognized as male or female prior to their first blooming, which takes 3 to 10 years post seed germination [14], delaying the selection of parentals for breeding purposes. Likewise, knowledge of sequences of interesting genes is needed to achieve genetic improvement based on molecular tools, a valuable information that is lacking in yerba mate. Currently, merely ~80 sequences originated from *I. paraguariensis* are available in the National Center for Biotechnology Information (NCBI) database, most of them corresponding to microsatellites. In addition, genetic information in the genus is scarce, annotated sequences are virtually inexistent and no expressed sequence tags (ESTs) libraries have been generated so far.

Massively parallel sequencing of RNA (RNA-Seq) is an efficient way to characterize the transcriptional landscape of a species and reveal its complexity [15]. It allows to investigate the transcriptome composition and expression and, in this direction, to explore and reveal the expressed profile of a defined organism [16]. This next-generation sequencing technology (NGS) is a simple and fast tool to analyze the transcriptome since it requires neither cloning library of the cDNAs nor any *a priori* knowledge of the species. Instead of this, RNA-Seq technology generates millions of short direct cDNA reads which are subsequently assembled to construct transcripts [15]. *De novo* transcriptome assembly is suitable in order to reconstruct full length transcripts from these short reads in organisms without a sequenced genome as reference. The most advanced algorithms to achieve this strategy consists in efficiently constructing and analyzing sets of *de Bruijn* graphs to construct and assemble transcripts and requires a great amount of parallel sequence short reads provided by high throughput sequencing technology [17].

This study presents the first analysis of the *I. paraguariensis* transcriptome. We employed the Illumina total RNA-Seq sequencing method to generate 72,031,388 pair-end 100 bp sequence reads. The obtained high quality reads were *de novo* assembled into 44,907 primary transcripts encompassing 40 million bases with an estimated coverage of 180X. Multiple sequence analysis allowed us to predict that yerba mate contains about 32,355 genes and 12,551 gene variants or isoforms. An initial analysis of these genes allowed us to identify and categorize members of more than 100 metabolic pathways. The transcriptome characterization of *I. paraguariensis* generated from our study is a very useful tool derived from a convenient and exhaustive approach of annotation and discovery of genes of several major metabolic pathways. The vast amount of information obtained would encourage and serve as reliable source in the path to the discovery of biologically and agronomically important traits, as well as for molecular markers development, gene mapping, analysis of genetic diversity and selective breeding in yerba mate.

Results and Discussion

RNA sequencing analysis and transcriptome *de novo* assembly

In order to shed light on the transcriptional landscape of yerba mate, total RNA was extracted from pooled leaves of *I. paraguariensis* breeding line Pg538 from INTA EEA-Cerro Azul, Misiones, Argentina. After initial quality controls the isolated RNA

was sequenced by the Illumina HiSeq-2000 platform. A total of 72,031,388 100 bp pair-end reads were obtained (**Table 1**). An analysis of the sequencing run indicated an absence of cycle-wise multiplied calls of the same nt, an average high quality of Q = 36.3, a lack of positional biases in the call frequency for each base and a typical unimodal distribution of quality average and Kmer enrichment frequency (**Figure S1 a-e**).

Recently, towards the identification of phosphate starvation-responsive genes in wheat (*Triticum aestivum*), a similar NGS approach was employed based in *de novo* assembly of 73,8 million reads from RNA-seq libraries [18]. The extension of this sequencing process, similar to that of our study, was effective and sufficient to generate a comprehensive transcriptome in wheat in the absence of reference genome information.

After quality filtering, the sequencing reads were assembled with the Trinity software [17] and a transcriptome of 44,907 assembled sequences was obtained. A quality analysis of the assembly suggested a typical distribution and coverage of GC content, a lack of positional kmer enrichment, a high percentage coverage consistent with the sequence length distribution and a regular positional nucleotide contribution in the assembled transcripts (**Figure S2 a-f**). The transcriptome covers 39,969,375 bp, with a mean contig length of 890 bp, N50 of 1,430 bp and 8,353 sequences with a length of over the N50. Our *de novo* assembly utterly indicates that yerba mate presents an estimate of 32,356 genes and 12,551 gene variants or isoforms (**Table 2**).

Evaluating the yerba mate transcriptome by DEG analysis

It has been proposed that a comprehensive catalog of essential genes may constitute a minimal genome, forming a set of functional modules, which play key roles in eukaryotic metabolism [19]. In that direction, a catalog of over 356 genes has been assigned as essential in the cruciferous plant *Arabidopsis thaliana* [20]. To assess and estimate the "completeness" of the assembled yerba mate transcriptome, a DEG (Database of Essential Genes) analysis was performed (**Data S1**). Exploring the genes of yerba mate we observed that the orthologs of 97.2% of the *A. thaliana* essential genes were present in our assembled transcriptome. In a recent study [21] a highly representative *Nicotiana benthamiana* transcriptome was evaluated under a similar platform depicted as "Core eukaryotic genes dataset" (CEGMA) [22], which includes a widely conserved set of 248 eukaryotic proteins. In this CEGMA analysis, 95% of the core proteins were identified in the *N. benthamiana* transcriptome. In this scenario we consider that the yerba mate DEG score is indicative of an overall representative status of the transcript library produced.

Characterization and functional annotation of yerba mate transcripts

The assembled transcriptome of *I. paraguariensis* was sequentially subjected to homology searches using the BLASTX platform against the UniProt *viridiplantae* database (**Table S3**) and TAIR. BLASTX hits E-value distribution of assembled transcripts to TAIR *A. thaliana* proteins is presented in **Figure 1**. Using a cut-off value of 10E-05, over 77% of transcripts (31,787) attained a blast hit based on identity conservation (**Table S4**).

The obtained BLASTX hits were subsequently imported into the Blast2GO software, the KAAS server, and the agriGO platform, where gene ontology annotation, metabolic pathway profiling and GO categorization were performed. Over 217,655 GO term tags were identified in the yerba mate transcriptome, of which 4,341 were associated to KEGG ids corresponding to 315 KEGG maps (**Table S2**). An initial sorting of the yerba mate sequences based on GO terms is presented in **Figure 2** using the

Table 1. Yerba mate Illumina HiSeq-2000 sequencing run statistics.

Sequencing stats	*Ilex paraguariensis* **RNA-seq**
Total Bases	7,275,170,188
Read Count	72,031,388
GC (%)	45,38
N (%)	0,027
Q20 (%)	98,21
Q30 (%)	94,99
Average Q	36,3
read length	100 nt×2

A. thaliana transcriptome as background. The assembled transcripts are categorized by cell component (**Figure 2a**) where an enrichment in organelle and cell structural components was observed, molecular function (**Figure 2b**) that showed an elevated percentage of catalytic and binding representatives, and biological process (**Figure 2c**) where the distribution of sequences followed the typical frequency observed in *Arabidopsis*. A closer analysis of the GO associated yerba mate sequences (**Table S5**) by semantic similarity-based scatterplots representations and tree-maps (**Figure S3** and **Figure S4**) highlighted several terms based on p-values (circle size, rectangle size) associated to the GO enriched categories, such as growth, methylation and reproductive structure development on biological process (**a**), chloroplast, membrane enclosed lumen and ubiquitin ligase complex on cellular component (**b**) and chlorophyll binding, methyltransferase activity and sequence specific DNA binding on molecular function (**c**). An exhaustive analysis of the GO terms is presented as AgriGO generated plots of GO enrichment, significance and relationships in yerba mate based in biological process (**Figure S5**), cellular component (**Figure S6**) and molecular function (**Figure S7**). A 166 catalog of KEGG drawn maps representing the gene members of the yerba mate transcriptome extensively associated with numerous metabolic pathways is presented in **Data S2**. In order to explore the yerba mate genes, we approached a categorization of transcripts based in BLASTX. Overall, we have identified over 1,000 putative transcription factors of yerba mate (**Table S1**), 50 transcripts involved in heat-stress, more than 200 oxidative stress responsive putative genes, 30 transcripts associated with pathogen response, a significant number of transcripts associated with ribosome constituents, ribosome processing, trafficking, rRNA maturation, and ribosome assembly (**Figure S8**), as well as 60 assembled transcripts involved in disease resistance and 150 transcripts probably engaged in hormone response (**Figure S9**). We have also identified nearly 100 transcripts related to osmotic, drought, salinity and cold stresses, senescence, early flowering, and biosynthesis of sugars, flavonoids, carotenoids and chlorophyll (**Table S3**).

When compared with other plant species reported by previous studies that used the Illumina system and the Trinity software, the quality of our transcriptome sequence shows to be significantly high, which is evidenced in several aspects. First, the average length of the unigenes we observed is 890 bp which is comparable to that of chili pepper (*Capsicum frutescens*, 712 bp) [23], ramie (*Boehmeria nivea*, 824 bp) [24], *Salvia splendens* (779 bp and 812 bp for two different strains) [25] and peanut (*Arachis hypogaea*, 751 bp) [26] transcriptome sequences. Second, approximately 77% of the genes discovered in this study were successfully annotated for their putative functions. Previous reports of

Table 2. Yerba mate Trinity *de novo* assembled transcriptome statistics.

Assembly	*Ilex paraguariensis*
method	Trinity k25
assembled seq.	44,907
unigenes	44,906
gene families	32,355
gene variants	12,551
n: 100	44,907
n: N50	8,353
min	201 bp
median	544 pb
mean	890 bp
N50	1,430 bp
max	15,716 bp
sum	39,969,375 bp

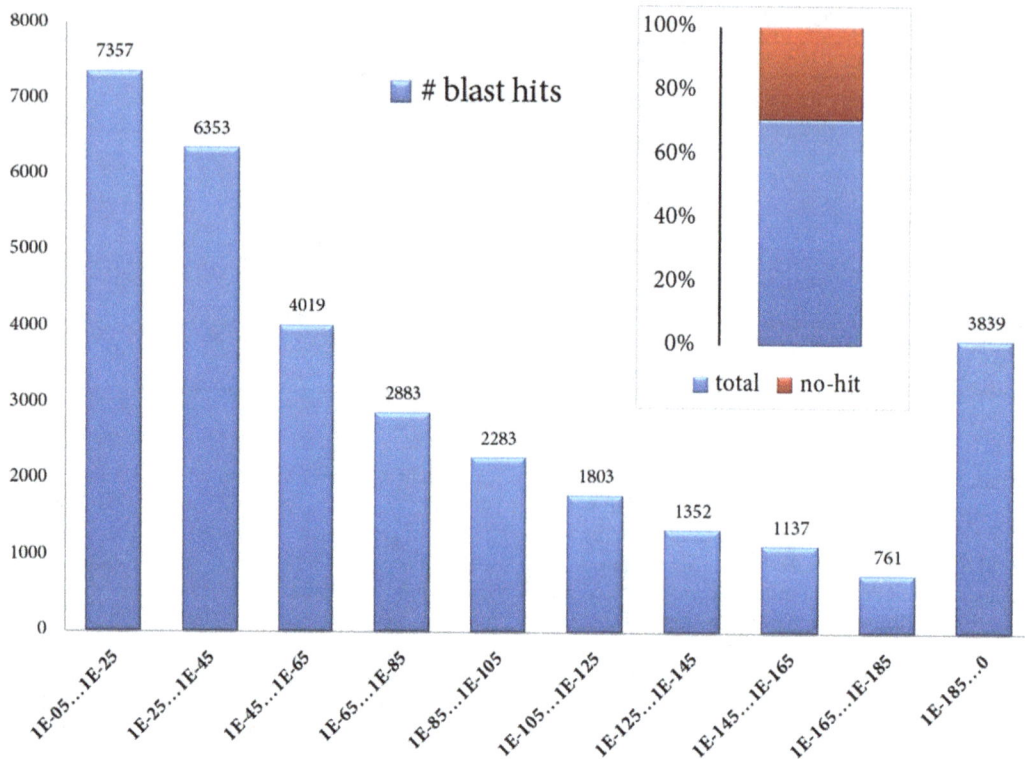

Figure 1. BLASTX hits E-value distribution of assembled transcripts to TAIR *Arabidopsis thaliana* proteins. Using a cut-off value of 10E-05, over 77% of transcripts (31,787 contigs) attained a blast hit based in identity conservation.

annotated genes in species such as *C. frutescens* (72.33%) [23], *Boehmeria nivea* (77.70%) [24], barnyardgrass (*Echinochloa crus-galli*, 57.45%) [27] and sugarcane varieties (*Saccharum offici-*

narum x *S. spontaneum*, 49.06%) [28] further support the notion of our assembly dataset being a fair representation of the yerba mate transcriptome.

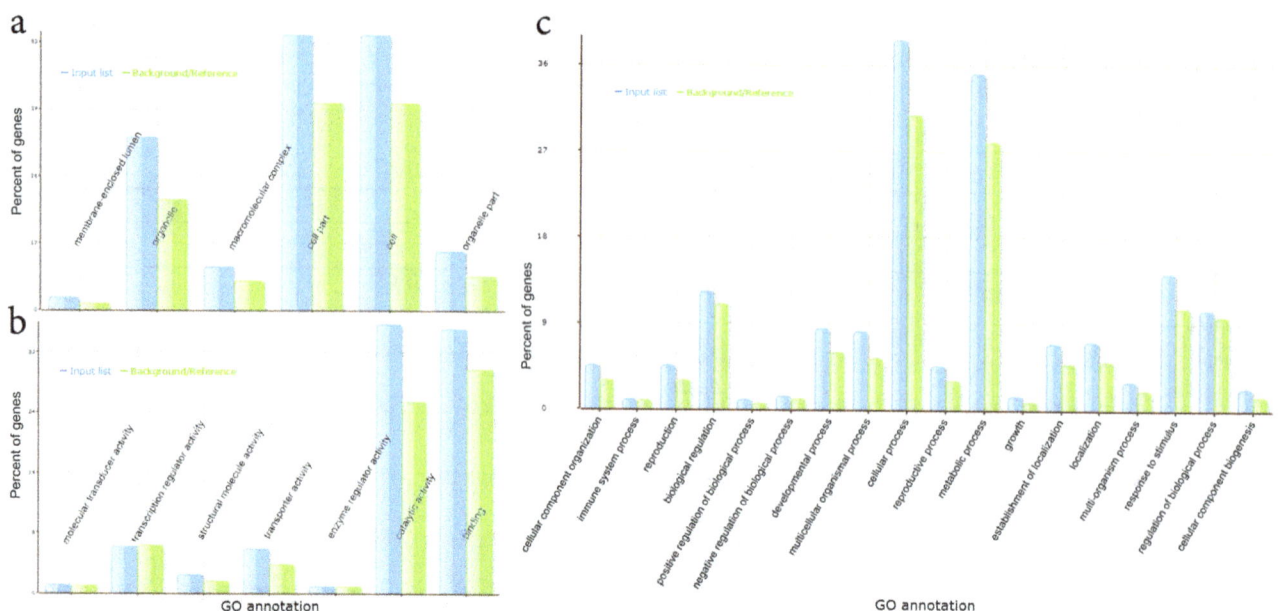

Figure 2. GO annotations obtained for the yerba mate transcriptome. Categorization by cell component (a), molecular function (b), and biological process (c). *Ilex paraguariensis* GO percentages are based on 31,787 BLASTX hits (blue), and the *Arabidopsis thaliana* transcriptome was employed as background (green).

Prediction of yerba mate SSRs

Simple sequence repeat (SSR) markers are well-known and widely used as valuable tools for assessing genetic diversity. SSRs are useful in the development of genetic maps, comparative genomics and marker-assisted selection breeding [28]. Thus, in parallel, the yerba mate transcripts library was comprehensively analyzed in search of SSRs. A total of 10,813 SSRs were identified in 8,449 sequences along the transcriptome. We analyzed our data and *in silico* predicted SSRs using 6,4,3,3,3 motifs repeats criteria for di-, tri-, tetra-, penta-, and hexa-nucleotides SSRs. In this context, the 2 nt motif repeats represented 40.9% of total SSRs found, while 3 nt motif repeats constituted roughly a 35.8% of total SSRs (**Figure 3a**). The most represented SSR corresponded to 2 nt motif ct/ag-tc/ga (**Figure 3b**) which encompassed over 84% of the 4,429 SSRs of 2 bp motif (**Table S11**). Among the tri-nucleotide motif repeats, with over 26% of the hits, aag/ctt-tct/aga-ttc/gaa are the most common SSR found in *I. paraguariensis* (**Figure 3c**). In most plant transcriptome studies, tri-nucleotide are the most frequent SSRs. However, the repeat motif abundance in plant transcriptomes is affected by the *in silico* determination of SSRs prediction criteria. For instance, several studies consider di-, tri-, tetra- penta- and hexa-nucleotides when diverse motif repeats are present, i.e. 6,5,4,4,4 in *Salvia splendens* [25], 6,5,5,5,5 in *Saccharum* spp [28], 6,5,5,4,4 in *Capsicum frutescens* [23], 6,4,3,3,3 in *Curcuma longa* [29], 4,4,4,4,4 in *Ipomoea batatas* [30]. In order to be consistent with the literature, we have *in silico* predicted SSRs using 6,4,3,3,3 motifs repeats criteria. In this background, di-nucleotides were the most representative SSR species, followed by tri-nucleotides. This non-standard distribution has also been described for *Salvia splendes* [25] with 39.9%/29.3% di- and tri-nucleotide frequencies, respectively, sweet potato with 43.3%/42.4% [30], rubber tree with 38%/34% [31] and several other plants such as cucumber [32], sesame [33], kiwi [34] and coffee [35] where di-nucleotides are also the most represented SSR species.

Transposable elements discovery

Several transposable elements (TE) were identified by sequence homology in yerba mate. In the literature, only a few sequences have been recently reported corresponding to yerba mate TE fragments obtained by DNA based methods [14,36], hence this is the first report describing actively yerba mate TEs. As expected for a transcriptional library, most of the sequences corresponded to Retro-Transposon elements, mainly Group Antigen polyprotein (GAG-Pol), reverse-transcriptase and RNAse H domain hits of Gypsy-like and Copia-like retro-elements. Also, a handful of Non-LTR retro-elements were identified and a few CACTA, En/Spm sub-class Transposons. The predicted repetitive elements were explored in detail, and representative results obtained by the NCBI conserved domain search web-service are presented as graphical summaries depicting typical TE domains such as Reverse Transcriptase domain in Gypsy-like elements and Transposase domain in En/Spm Transposons, of several yerba mate putative TEs (**Table S12**).

Organelles draft genome *de novo* assembly and analysis

Recent NGS based studies have emphasized in the abundance and wide extension of chloroplast and mitochondrial transcripts, postulating that most of the organelles genomic DNA is actively transcribed in plants [37–40]. In this scenario we surveyed the yerba mate transcriptome in order to generate a draft genome of both organelles based in RNA transcripts and sequence similarity to reference organelle sequences of slightly related plant species. Illumina reads were relaxedly mapped to a *Lactuca sativa*

chloroplast sequence and a total of 10,798,227 reads comprised and sustained a high coverage library that was assembled into a consensus sequence draft of the yerba mate chloroplast (**Table 3**). The assembled *I. paraguariensis* chloroplast is predicted to be ~152,872 bp long, consisting in 51.6% of coding sequences, representing 83 protein coding genes, 37 transfer RNA genes and 7 ribosome RNA genes (**Figure 4**). The 83 protein coding genes included several ribosomal proteins, constituents of photosystem I & II, NADH dehydrogenases and ATP synthases among others (**Table 4** and **Table S6**). A sequence alignment of the *L. sativa* chloroplast complete sequence and yerba mate draft chloroplast shows extensive identity, in some regions exceeding 90%, particularly at gene transcripts such as 16 s rRNA, 23 s rRNA and several transfer RNA genes (**Figure S10**). Mapping of *I. paraguariensis* assembled transcripts to the chloroplast sequence draft shows an extensive and pervasive coverage (**Figure S11**). The assembled draft was subjected to microsatellite discovery (**Table S8**) and a total of 94 SSRs were identified, consisting mainly of 6 bp motif (57.5% of total SSRs).

A similar approach was employed in order to envisage a mitochondrial genome draft of yerba mate. A total of 1,265,566 Illumina reads were mapped to the *Helianthus annuus* mitochondrial genome sequence. In this case, most mapped reads corresponded mainly to the gene coding regions, and the assembled draft extended at about a third of the total predicted genome (**Table 3**). A sequence alignment of sunflower mitochondrial complete sequence and yerba mate mitochondrial sequence consensus (**Figure S12**) presented high identity at most of the 43 coding sequences that corresponded to protein coding genes such as Complex I NADH dehydrogenases, Complex V ATP synthases and ribosomal proteins (SSU and LSU), transfer RNA genes and ribosome RNA genes (**Table 5, Table S7**).

A *Mauve* alignment, which is preferred for rearranged genome sequences [41], was performed with the yerba mate and sunflower mitochondrial sequences. The higher identity, mostly confined to gene encoding regions, is represented hierarchically from white to red. The consensus *I. paraguariensis* sequence conserved most of the *Helianthus* gene annotations. As an example, the consensus sequence of *I. paraguariensis* at 76,000 bp coordinates presented high identity to the 230,000 bp coordinates of sunflower, corresponding to the ccmFn coding sequence (**Figure S13**). The assembled yerba mate consensus sequence was subjected to microsatellite discovery (**Table S9**) and a total of 69 SSRs were identified, consisting mainly of 6 bp motif (69.8% of total predicted SSRs).

The yerba mate RNA silencing and degradation pathway

A particular limited set of seventy six yerba mate transcripts identified yielded considerable similarity with several members of the RNA silencing and degradation pathway (**Table S3**). RNA interference is a post-transcriptional sequence-specific process of gene silencing that mediates resistance to both endogenous parasitic and exogenous pathogenic nucleic acids, and regulates the expression of protein-coding genes in eukaryotic organisms [42]. Among several enzymatic components of RNA interference such as Dicer-Like proteins, RNA dependent RNA polymerases, exosome members and dsRNA binding domain proteins, the family of Argonaute effectors (AGO) was also pinpointed in the yerba mate transcriptome. AGO and AGO-like proteins are the main RNA silencing effectors across kingdoms, and they mediate the cleavage of target RNAs using small RNAs of 20–24 nt as guides [43]. Argonaute 1 (AGO1) is responsible of two important circuits in plants: gene silencing of endogenous transcripts by microRNAs and virus resistance based in viral derived small

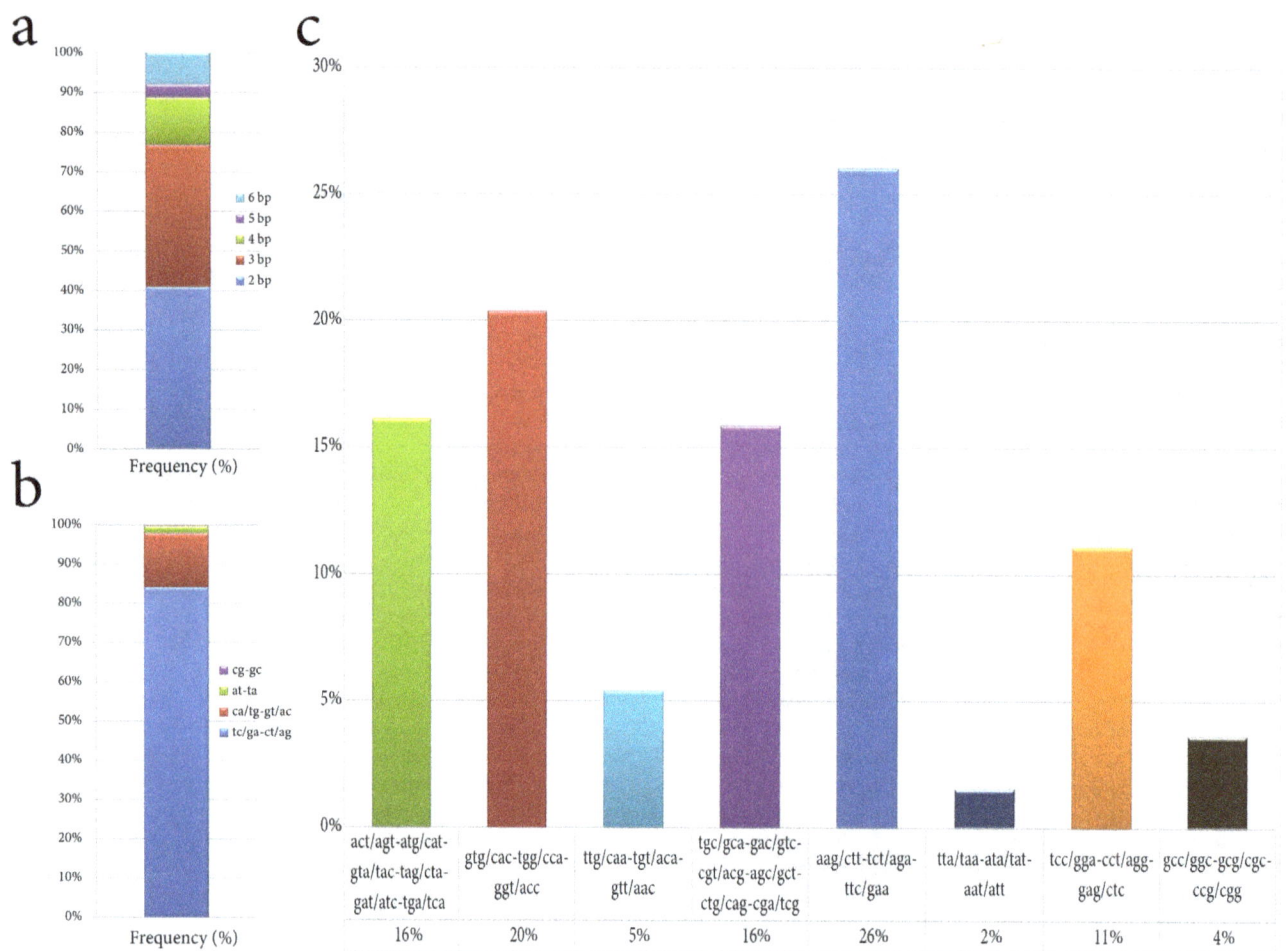

Figure 3. Proportion and frequencies of predicted SSRs in *Ilex paraguariensis* transcriptome. (a) Proportion of SSR predicted in yerba mate transcriptome categorized by k-mer length. (b) ct/ag-tc/ga account for 84% of di-nucleotide SSRs found in yerba mate. (c) Frecuency of tri-nucleotide SSRs predicted in yerba mate. With over 26% of the hits, aag/ctt-tct/aga-ttc/gaa are the most common SSR found in *Ilex paraguariensis*.

RNAS [44,45]. The yerba mate AGO1 predicted protein is estimated to be 1,062 aa in length and presented the typical AGO1 glycine rich domain (E-value = 1,8e-25), the PAZ domain (E-value = 4,3e-38) which is predicted to interact with single stranded small RNAs and the PIWI domain (E-value = 3,5e-112), responsible of the RNA-guided hydrolysis of single stranded-RNA (**Figure 5a**). Multiple protein alignment and secondary structure prediction of yerba mate, *N. benthamiana*, carrot and tomato AGO1 showed an important conservation in gene structure and domains. A phylogenetic tree based in Jukes-Cantor, neighbor-joining and 1000 bootstraps indicated that AGO1 from yerba mate is more related with carrot than *Solanaceae* AGO1 despite the basic genetic distance among them (**Figure 5 b–d**, **Figure S14**, **Figure S15**).

MicroRNAs (miRNAs) are small non-coding RNAs that modulate plant gene expression by means of gene silencing through sequence-specific inhibition of target mRNAs. MiRNAs derive from precise processing of precursor transcripts with stem-loop secondary-structure features that are recognized by a Dicer-like complex. Mature miRNAs are loaded predominantly onto AGO1 and target endogenous RNAs for their degradation or translational arrest [46]. By combining two *in silico* based approaches we engaged in an attempted characterization of putative miRNA precursors in yerba mate. A yerba mate miRNA

sequence prediction report based in UEA small RNA workbench platform and canonical relaxed mapping of conserved precursor miRNAs to the yerba mate transcriptome, indicated the presence of at least 59 pre-miRNAs corresponding to 41 of both young and ancient miRNA families (**Table S10**). The miR156 gene family has been involved in the regulation of developmental timing, vegetative phase change, flowering and sex identity in plants [47–49]. In yerba mate several mature miRNAs were predicted based in sequence homology to miRBase [50] (**Figure 6a**). In the particular case of miR156, nine isoform variants were predicted with high sequence homology and minor mismatches. An insertion of a "A" at position 10 in miR156b and c forms, slightly affected the precursors secondary structure at the miRNA/miRNA* coordinates that can be observed as a bulge in **Figure 6b**. While the homology at the mature miR156 was high, the diversity among precursors of the miRNA gene family was extensive (**Figure 6d**). A library generated of predicted Squamosa Promoter Binding Protein-Like (SPL) mRNAs of yerba mate was evaluated as a target of Ipa-miR156. A strong interaction with a high expectation score was *in silico* predicted for SPL9, SPL6 and SPL4 with Ipa-miR156 (**Figure 6c**), which are typically conserved and validated targets of miR156 in plants [51]. These SPL genes significantly differed in their nucleotide sequence, however a strong conservation of the specific miR156 target could be

Table 3. General features of Yerba mate draft assembled organelles.

Feature	Chloroplast	Mitochondrion
Genome size (bp) estimated	~150,872	~301,093
contig coverage (bp)	118,064	90,151
total mapped 100 bp reads	10,798,227	1,265,566
GC content (%)	56.06	42.06
Coding sequences	127	43
Gene content (%)	51.6	9.03
No. of protein-coding gene	83	26
No. of introns	17	0
No. of tRNA genes	37	14
No. of rRNA operons	7	3
Sequence repeat	94	69

observed in the 3 genes (**Figure 6e**). The identification of transcripts related to sex identity, such as miR156 and SPL gene families, is of special interest in yerba mate. In this diclino-dioecious crop, plants cannot be recognized as male or female prior to their first blooming, which occurs between 3 and 10 years post seedling emergence [14], delaying considerably the selection of parentals for breeding purposes. So, a cost-effective early sex determination system would be promising for the yerba mate breeding programs. It is tempting to postulate, that perhaps the determination of expression levels of these particular genes during yerba mate plant development, may be employed as a gender predictor at early stages.

The yerba mate caffeine synthase

One of the most important constituents of yerba mate extracts is caffeine [52,53]. Caffeine is responsible for the stimulant effect of mate [54], and perhaps the underlying rationale of its profound influx in Latin American culture based on its effect on the body and mind and its properties that aid in staying awake and improving mental alertness after fatigue among others [55]. Caffeine or 1,3,7-trimethylxanthine is a crystalline xanthine alkaloid. Caffeine biosynthesis involves a series of reactions that direct the conversion of xanthosine to 7-methylxanthosine, to 7-methylxanthine to theobromine which is converted into caffeine [56]. The enzyme, assigned the name caffeine synthase (EC 2.1.1.160), catalyses the last two steps of caffeine biosynthesis, the conversion of 7-methylxanthine to caffeine via theobromine [57]. The gene encoding caffeine synthase (CS) was originally cloned from young tea leaves by Kato et al. [58]. Using the sequence of *Camellia sinensis* CS we probed our library and identified a sequence corresponding to the full length of an assembled transcript of the yerba mate transcriptome. The putative yerba mate assembled CS was identified by similarity to the *C. sinensis* CS complete mRNA (E-value = 8e-168). The yerba mate 1,491 nt CS transcript was assembled based in 16,851 100 bp reads with an average coverage of 1,113X. Exploring this transcript, a single ORF was predicted encompassing 1,098 nt between coordinates 113 to 1,210, encoding a 366 aa protein with a 61% identity (E-value = 2e-157) to the corresponding 369 aa *C. sinensis* CS protein, sharing the presence of the Methyltransferase_7 SAM dependent carboxyl methyltransferase domain involved in caffeine synthesis. With the predicted CS ORF, we performed multiple MUSCLE protein alignment and secondary structure prediction.

By comparing *Coffea arabica*, *Theobroma cacao* and *C. sinensis* CS to *I. paraguariensis* predicted CS transcript an important conservation in gene structure and domains was observed (**Figure S16**). Since caffeine content is a desirable and important character in breeding programs of this crop, the preliminary and putative nature of the yerba mate predicted CS assembled transcript encourage further experimental validation, heterologous expression experiments and biochemical characterization of the full length CS coding sequence by traditional methods. After sequence annotation, we exploited the SWISS-MODEL algorithm to generate a yerba mate CS 3D prediction using *C. arabica* CS as a template (**Figure S17**). The 3D structure of the predicted *I. paraguariensis* caffeine synthase is presented in **Figure 7a** based in the X-ray crystallography solved structure of *C. arabica* CS (**Figure 7b**). A ribbon model of yerba mate CS (**Figure 7c**) and coffee CS (**Figure 7e**) suggested high conservation of secondary structure when superimposed (**Figure 7d**). A reconstruction of a mesh model of yerba mate CS is presented (**Figure 7f**) and compared with the coffee EM (**Figure 7h**), showing extensive quaternary structure similarity (**Figure 7g**).

Chlorogenic acid in yerba mate

Chlorogenic acid (CGA, caffeoyl quinate, KEGG compound C00852) is the major phenolic compound found in yerba mate [59]. CGA acts as an antioxidant in plants, is involved in resistance to insects and defense against fungal pathogens. Human CGA consumption and its antioxidant role have been associated to protection against degenerative age-related diseases [59]. The enzyme 4-coumaroyl-CoA: quinnate O-(hydroxycinnamoyl)trans-ferase (HQC, E.C.: 2.3.1.133) and p-coumaroyl quinate 3′-hydroxylase (C3′H, E.C.: 1.14.13.36), corresponding to the flavonoid and stilbenoid metabolic pathways, respectively (**Data S2**, pages 50 and 145), are responsible and limiting for CGA biosynthesis in tobacco, coffee and switchgrass [60–62]. We examined our transcriptome and, interestingly we found four putative HQC and two C3′H gene versions sharing over 75% identity with the corresponding coffee enzymes, presenting the typical condensation and Cytochrome P450 domains, respectively. We noticed that these transcripts were amongst the 1–4% most abundantly expressed, ranging from 40 to 160 FPKM. Although further studies are required, we suggest that the presence of redundant and highly expressed Chlorogenic acid related genes might be directly associated with the important production of

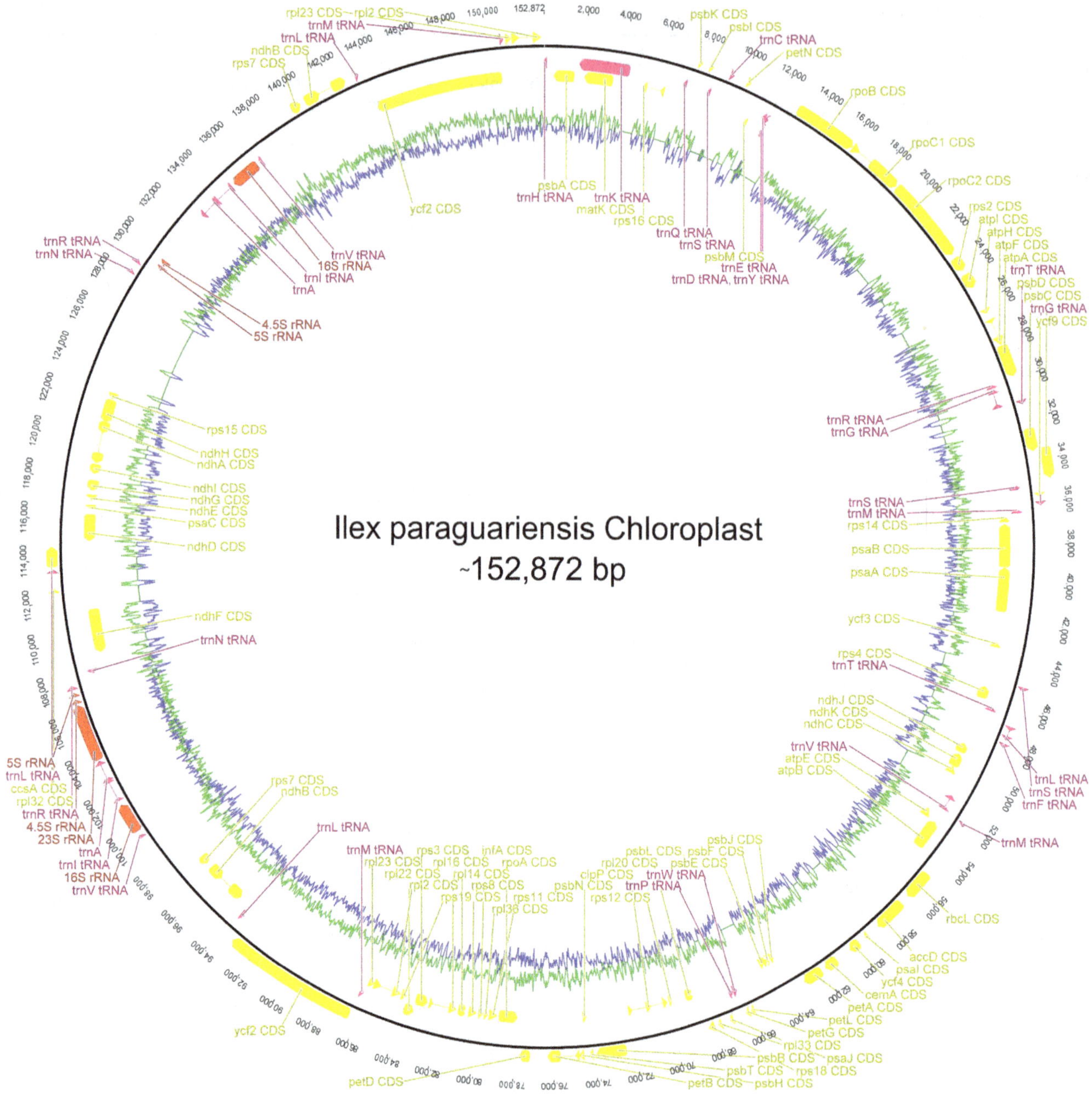

Figure 4. *Ilex paraguariensis* chloroplast is predicted to be ~152,872 bp long, consisting in 51.6% of coding sequences, representing 83 protein coding genes (yellow), 37 transfer RNA genes (pink) and 7 ribosome RNA genes (red). The 83 protein coding genes include several ribosomal proteins, constituents of photosystem I & II, NADH dehydrogenases and ATP synthases among others.

CGA in yerba mate and its significant effects on oxidative stress reduction when consumed [63].

Ilex paraguariensis PCR detection assay based in 5.8S and ITS2 assembled rRNA regions

In Argentina, *I. paraguariensis* is the only species authorized to be used to manufacture yerba mate products (Argentina law, Act 18.284, articles 1192–1193). It is interesting to note that a molecular based, standardized and cost effective method for *I. paraguariensis* detection is lacking in the literature. Moreover, *I. dumosa* is frequently used in mixtures with *I. paraguariensis*, and amounts above 1% are considered adulterants [64]. Based on our

sequencing data we were able to assemble the *I. paraguariensis* rDNA corresponding regions of ITS1, 5.8S and ITS2. Multiple alignment of *I. paraguariensis* assembled 45S consensus sequence transcripts and *I. dumosa* homologous regions was performed (**Data S4**). This analysis showed that while ITS1 is highly conserved in both species, 5.8S and ITS2 are distinctive enough to differentiate them by means of PCR with species-specific primers. PCR amplifications were performed and agarose gel electrophoresis revealed that the designed primers from *I. paraguariensis* are species-specific and able to detect unequivocally genomic *I. paraguariensis* DNA. Moreover, by using the *I. dumosa* species-specific primers we were able to detect *I. dumosa* in DNA

Table 4. Yerba mate chloroplast encoded genes by category.

Chloroplast gene category	Gene name
Photosystem I	psaA, psaB, psaC, psaI, psaJ
Photosystem II	q
Cytochrome b/f complex	petA, petB, petD, petG, petL, petN
ATP synthase	atpA, atpB, atpE, atpF, atpH, atpI
NADH dehydrogenase	ndhA, ndhB, ndhC, ndhD, ndhE, ndhF, ndhG, ndhH, ndhI, ndhJ, ndhK
RubisCO large subunit	rbcL
RNA polymerase	rpoA, rpoB, rpoC1, rpoC2
Ribosomal proteins (SSU)	rps2, rps3, rps4, rps7(2), rps8, rps11, rps12, rps14, rps15, rps16, rps18, rps19
Ribosomal proteins (LSU)	rpl2(2), rpl14, rpl16, rpl20, rpl22, rpl23(2), rpl32, rpl33, rpl36
Other genes	clpP, matK, accD, ccs1, ccsA, infA, cemA
hypothetical	ycf2(2), ycf3, ycf4, ycf9
Transfer RNAs	trnA-UGC(2), trnC-GCA, trnD-GUC, trnE-UUC, trnF-GAA, trnG-UCC(2), trnH-GUG, trnI-CAU, trnI-GAU, trnK-UUU, trnL-CAA, trnL-GAG, trnL-UAA, trnL-UAG, trnM-CAU(4), trnN-GUU(2), trnP-UGG, trnQ-UUG, trnR-ACG(3), trnS-GCU(2), trnS-UGA, trnT-GGU(2), trnV-GAC(2), trnV-UAC, trnW-CCA, trnY-GUA
Ribosomal RNAs	rRNA 4.5 s(2), rRNA 5 s(2), rRNA 16 s(2), rRNA 23 s

preparations obtained of mix leaf tissue of fractions above 1% of *I. dumosa* respective to *I. paraguariensis* as previously reported [64] (**Data S4**). PCR amplifications with both species-specific primers is a versatile, simple and cost effective method to detect *I. paraguariensis* and the typical yerba mate products adulterant.

Conclusions

This is the first publicly available *I. paraguariensis* NGS study performed to investigate the entire yerba mate transcriptome, and our data provides the unique comprehensive transcriptome resource currently existing for yerba mate. In sum, through a systematic and exhaustive process of gene analysis and annotation, we have identified ~1,000 putative transcription factors, genes involved in heat and oxidative stress, pathogen response, as well as disease resistance and hormone response. We have also identified transcripts related to osmotic, drought, salinity and cold stress, senescence and early flowering. We have also pinpointed several members of the gene silencing pathway and characterized the silencing effector Argonaute1. We predicted a diverse supply of putative microRNA precursors involved in developmental processes. We developed a draft of the transcribed genomes of the yerba mate chloroplast and mitochondrion. The putative sequence and predicted structure of the caffeine synthase of yerba mate is presented. Finally, we provide here a collection of over 10,800 SSR accessible to the community interested in yerba mate genetic improvement.

The transcriptome characterization of *I. paraguariensis* generated from our study is a very useful tool derived from a convenient and exhaustive approach of annotation and discovery of genes of several major metabolic pathways in this important crop. The vast amount of information obtained would encourage and serve as reliable source in the path to the discovery of biological and agronomic interesting traits, as well as for molecular markers development, gene mapping, analysis of genetic diversity and selection breeding in yerba mate.

Table 5. Yerba mate mitochondrial encoded genes by category.

Mitochondrial gene category	Gene name
Complex I (NADH dehydrogenase)	nad3, nad4, nad5, nad6, nad8, nad9
Complex III (cytochrome c reductase)	cob
Complex IV (cytochrome c oxidase)	coxI, coxIII
Complex V (ATP synthase)	atp1, atp4, atp6, atp8, atp9
Cytochrome c biogenesis	ccmB, ccmC, ccmFc, ccmFn(2)
Ribosomal proteins (SSU)	rps4, rps12, rps13
Ribosomal proteins (LSU)	rpl5, rpl10, rpl16
Maturases	matR
Other genes	orf873
Transfer RNAs	trnC-GCA, trnD-GUC, trnE-UUC, trnF-GAA, trnG-GCC, trnH-GUG, trnK-UUU, trnL-UAA, trnM-CAU(2), trnN-GUU, trnP-UGG, trnQ-UUG, trnS-GCU, trnW-CCA, trnY-GUA
Ribosomal RNAs	rrn5, rrn18, rrn26

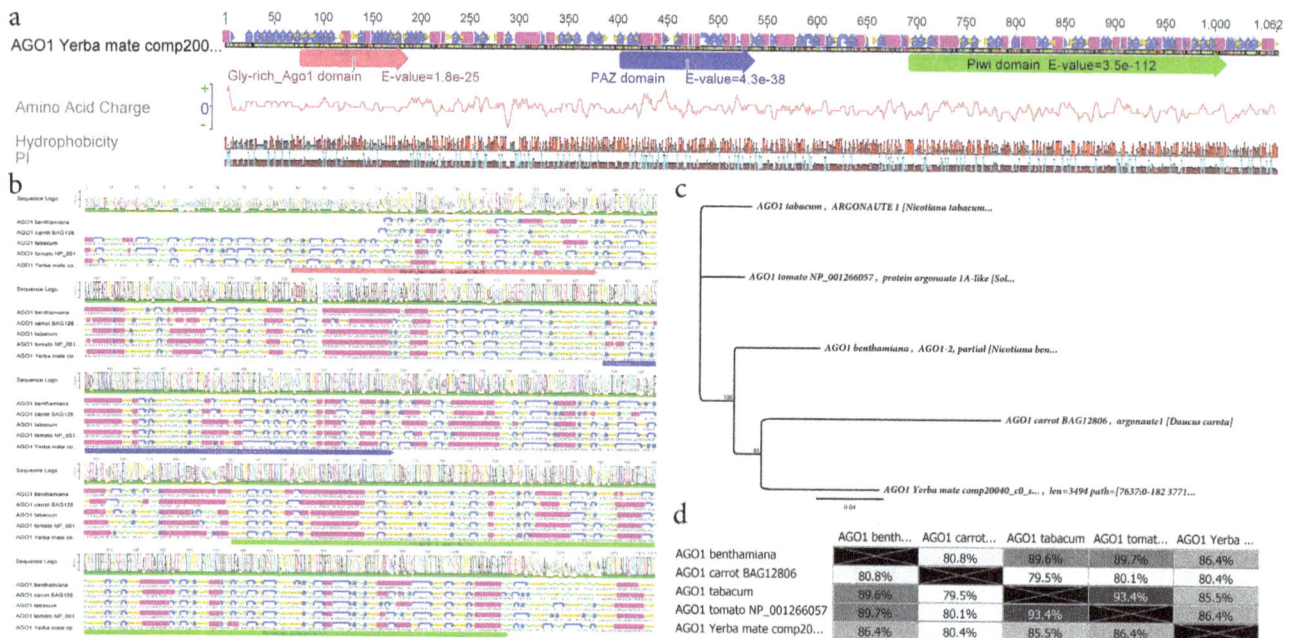

Figure 5. Yerba mate *Argonaute 1* (AGO1): characterization of the catalytic component of the miRNA pathway. (a) The predicted *Ilex paraguariensis* AGO1 protein is 1,062 aa in length and presents the typical AGO1 glycine rich domain, the PAZ domain which is predicted to interact with single stranded small RNAs and the PIWI domain, responsible of the RNA-guided hydrolysis of single stranded-RNA. (b) Multiple protein alignment and secondary structure prediction of yerba mate, *Nicotiana benthamiana*, carrot and tomato AGO1 showing an important conservation in gen structure and domains. (c) A phylogenetic tree based in Jukes-Cantor, neighbor-joining and 1000 bootstraps indicates that AGO1 from yerba mate is more related with carrot than *Solanaceae* AGO1 despite the basic genetic distance among them (d).

Materials and Methods

Plant materials and RNA extraction

Leaf samples at emerging, young, fully expanded, and early and late senescent stages from *I. paraguariensis* breeding line Pg538 from INTA EEA-Cerro Azul, Misiones, Argentina, were collected and immediately frozen in liquid Nitrogen. Total RNA was isolated from pooled leaf tissue with the RNeasy Plant Mini Kit (Qiagen Inc.) and supplemented with RNase-free DNase (Qiagen Inc.). To increase the depth of depletion of ribosomal RNA, a process with the RiboMinus Plant Kit (Life Sciences Inc.) was performed with the isolated RNA. The resulting RNA was evaluated in concentration and purity using Nanodrop 1000 (Thermo Inc.), and subsequently subjected to an integrity analysis by agar electrophoresis and by Bioanalizer 2100 (Agilent Inc.) to determine quality parameters by QC and RING.

RNA-seq library construction for Illumina sequencing

The resulting high quality RNA was employed for the generation of a cDNA library through TruSeq RNA Sample Preparation Kit (Illumina Inc.). The purified cDNA library was used for cluster generation on the Illumina Cluster Station and then sequenced on Illumina HiSeq 2000 following vendor instruction. A paired-end sequencing run with 100 nt read length for each read was performed for RNA-Seq. Raw sequencing intensities were then extracted and the bases were called using Illumina RTA software, followed by sequence quality filtering. The extracted sequencing reads were saved as a pair of fastq files for the first and second read, respectively.

Sequence data analysis and assembly

Quality reports and filtering of the sequencing run and assembly were generated using CLC Genomics Workbench v7.0.4 (http://

www.clcbio.com/) and the RobiNA v1.2.4 software (http://mapman.gabipd.org/web/guest/robin). All raw reads generated from the sequencer were quality filtered and *de novo* assembled into contigs using the Trinity program [17] with optimal parameters of 25 kmer word and group pairs distance of 500. The abundance of assembled contigs/isoforms was estimated using RSEM (http://deweylab.biostat.wisc.edu/rsem/) following the Trinity protocol. The raw sequencing data with quality scores have been deposited in the NCBI Short Read Archive database under accession number SRP043293.

Sequence annotation

The obtained contigs were bulk analyzed in homology searches by BLASTX (http://blast.ncbi.nlm.nih.gov/Blast.cgi) to protein databases nr, Swiss-Prot (http://www.uniprot.org/), KEGG (http://www.genome.jp/kegg/) and COG (http://www.ncbi.nlm.nih.gov/COG/). The yerba mate transcripts were also investigated and analyzed with the Plant Ontology database (http://www.plantontology.org/), the database of essential genes DEG (http://www.essentialgene.org), Plaza 2.5 (http://bioinformatics.psb.ugent.be/plaza/) and alternatively with TAIR (http://www.arabidopsis.org/) the batch blast tool of the Rosaceae Genome Database (http://www.rosaceae.org/tools/batch_blast) and the PlantGDB BLAST (http://www.plantgdb.org/cgi-bin/blast/PlantGDBblast) using E≤1e-5 as threshold, retrieving best hits and functional annotations. The Blast2GO program www.blast2go.com/ (E≤1e-5) was also used to obtain GO annotation of genes. The GO annotations retrieved were subjected to enrichment analysis with the WEGO software (wego.genomics.org.cn/), Revigo web server (revigo.irb.hr/) and the AgriGO platform (bioinfo.cau.edu.cn/agriGO/).

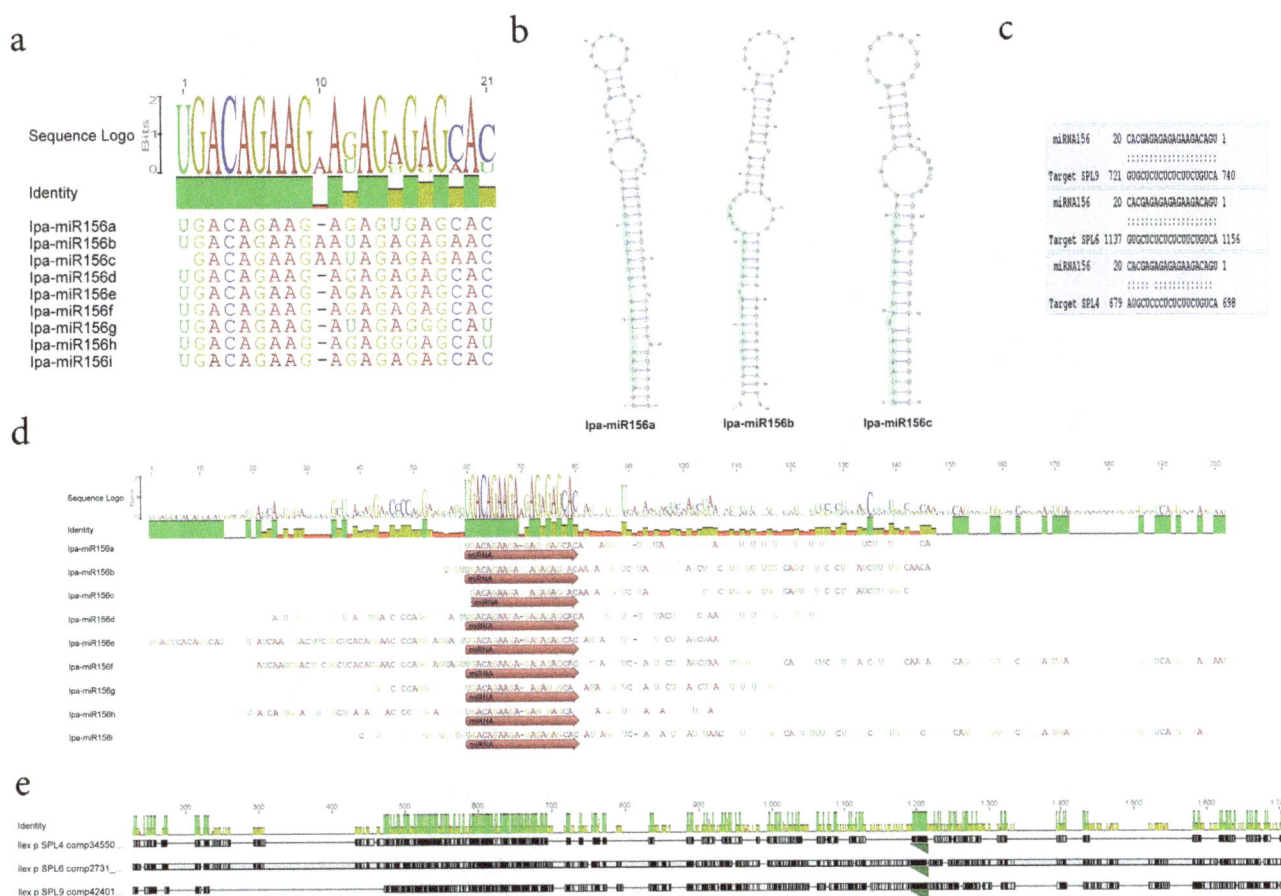

Figure 6. *MiR156* gene family in yerba mate. (a) Several mature miRNAs were predicted in yerba mate based in sequence homology to Mirbase. In the particular case of miR156, nine isoform variants were predicted with high sequence homology and minor mismatches. An insertion of a "A" at position 10 in miR156b and c forms, slightly affected the precursors secondary structure at the miRNA/miRNA* coordinates that can be observed as a bulge in (b). While the homology at the mature miR156 is high, the diversity among precursors of the miRNA gene family is extensive (d). A library generated of predicted SPL mRNAs of yerba mate was evaluated as a target of Ipa-miR156. A strong interaction with a high expectation score was *in silico* predicted for SPL9, SPL6 and SPL4 with Ipa-miR156 (c). These SPL genes significantly differ in their nucleotide sequence, however a strong conservation of the miR156 target can be observed in the 3 genes (green triangle, e).

Organelle genome assembly and annotation

To generate a draft of yerba mate transcribed chloroplast, Illumina reads were relaxedly mapped to a *Lactuca sativa* chloroplast sequence (Accession no. AP007232.1) employing the map to reference utility of the Geneious 7.0 software (Geneious assembler, medium sensitivity, iterations up to 5 times). A consensus sequence was generated with the mapped reads and aligned to the *Lactuca sativa* chloroplast complete sequence using a Geneious global aligment with free end gaps (93%, gap open penalty 12, gap extension penalty 3). The yerba mate chloroplast draft was annotated integrating *Lactuca sativa* chloroplast gene predictions and *in silico* based estimations obtained with the Dual Organellar GenoMe Annotator (DOGMA, http://dogma.ccbb. utexas.edu/). A similar process was employed to generate a yerba mate mitochondrial draft. In this latter case to compensate for significant sequence gaps a Mauve genome alignment of yerba mate and *Helianthus annuus* mitochondrial complete sequence (Accession no. KF815390.1) was generated. The annotation of the yerba mate transcribed mitochondrion draft was generated based in sequence homology to the respective sunflower predicted genes.

MicroRNA prediction

We analyzed both the assembled transcripts and the raw reads by two *in silico* based approaches in an attempted characterization of putative miRNA precursors in yerba mate. The UEA small RNA workbench platform (srna-workbench.cmp.uea.ac.uk/) with plant standard parameters and a cut-off P-value of 0.05 was employed using the yerba mate assembled transcriptome as reference. In parallel, a canonical relaxed mapping (similarity: 0.8 to 0.75, length fraction: 0.8, mismatch cost: 2, insertion cost: 2, deletion cost: 3) of conserved precursor miRNAs from miRBASE Release 20 (ftp://mirbase.org/pub/mirbase/CURRENT/miRNA.dat.gz) to the yerba mate transcriptome on the CLC Genomics Workbench v7.0.4 environment was generated and mapped reads were analyzed by eye and evaluated by proper secondary folding. The secondary structures of the predicted stem-loops secondary structures were solved using the mfold web server (http://mfold.rna.albany.edu/) using version 2.3 and adjusting folding temperature to 24°C, and Vienna type RNA structures were predicted with Context Fold (http://www.cs.bgu.ac.il/~negevcb/contextfold/).

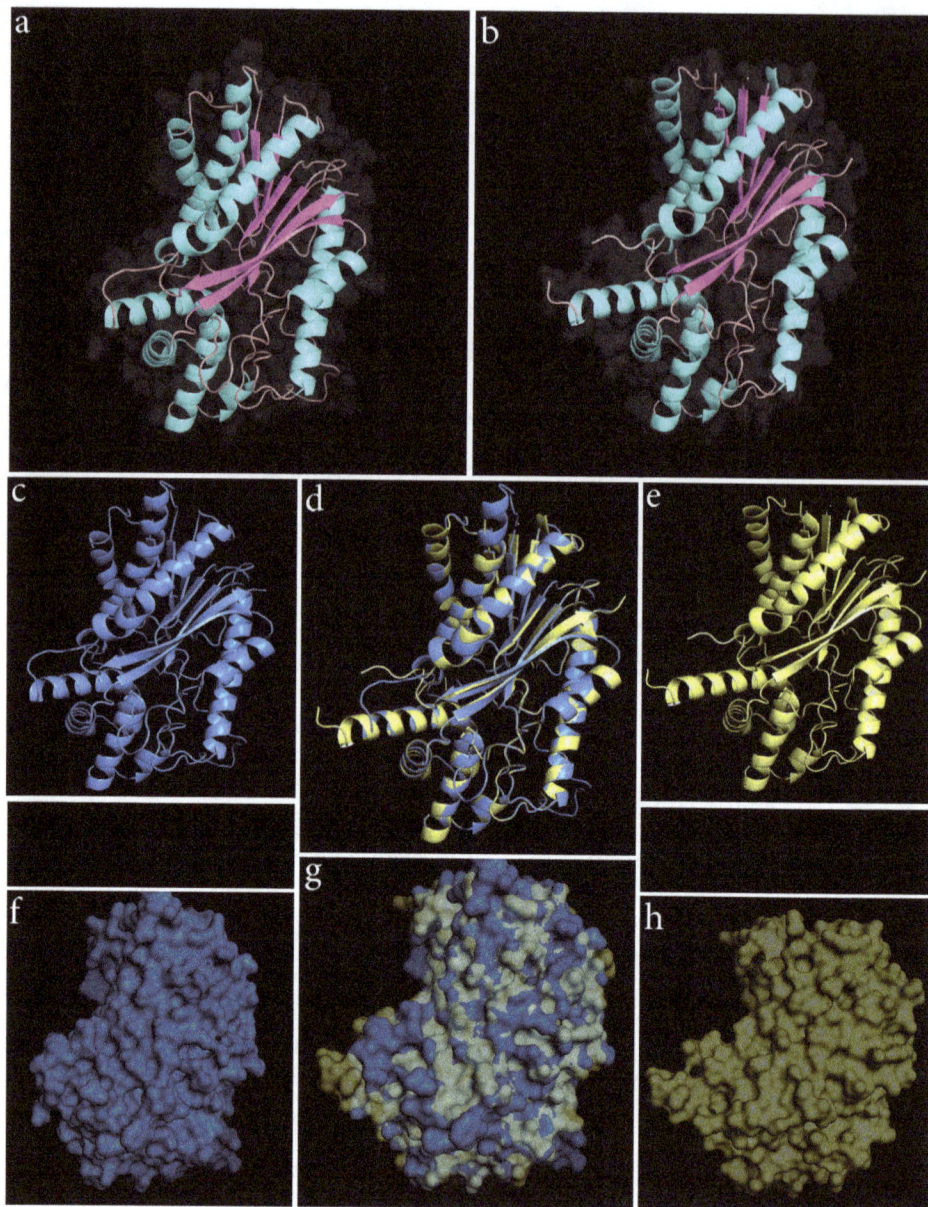

Figure 7. 3D structure of *Ilex paraguariensis* caffeine synthase. (CS). Employing the X-ray crystallography solved structure of *Coffea arabica* CS as a template (b), the 3D structure of yerba mate CS was predicted by the swiss-model algorithm (a). A ribbon model of yerba CS (c) and coffee CS (e) suggest high conservation of secondary structure when superimposed (d). A reconstruction of a mesh model of yerba CS is presented (f) and compared to the coffee EM (h), showing extensive quaternary structure similarity (g).

Caffeine synthase structure prediction

Coffea arabica (Acc. no: BAC43759.1), *Theobroma cacao* (Acc. no: BAE79730.1), *Camellia sinensis* (Acc. no: ABP98983.1) and *Ilex paraguariensis* predicted caffeine synthase coding sequences were subjected to multiple MUSCLE protein alignments (8 iterations, distance methods kmer6-6 and pctd-kimura, and clustering by UPGMB) and secondary structure prediction with the GOR method (Garnier-Osguthorpe-Robson) with Genious v7.0.

Yerba mate CS cuaternary structure was predicted with the automated protein structure homology-modelling server (swissmo-del.expasy.org/; [65]) using as a template the solved structure of *Coffea arabica* CS (2efjA (2.00 A)). The pdb obtained structure

was analyzed and rendered using the PyMol software version 1.7 (www.pymol.org/).

SSR detection

The GDR SSR Server [66,67] was employed to identify single sequence repeats in the transcriptome of yerba mate. 5 types of SSRs (di-nucleotide, tri-nucleotide, tetra-nucleotide, penta-nucleotide, hexa-nucleotide) were screened with an specific minimum number of 6, 4, 3, 3, 3 repeats respectively.

Transposable elements analysis

Transposable elements were first identified based in homology searches by NCBI-BLASTN and BLASTX. Specifically, Non-LTR Long Interspersed Elements (LINE) were pinpointed by the retrotransposase domain; LTR elements (Copia-like and Gypsy) were scanned with the LTR-FINDER software [68] and DNA transposons by domain based search. Every candidate was then explored and re-evaluated with NCBI conserved domain search (http://www.ncbi.nlm.nih.gov/Structure/cdd/wrpsb.cgi?).

Supporting Information

Figure S1   Illumina RNA sequencing analysis. (a) A lack of peaks reaching up to 100% at individual cycles in the homopolymer graph indicates an absence of a common technical artifact of cycle-wise multiplied calls of the same nt. (b) The Sequence quality plots allow an overview of the base call qualities assigned to each base by the base caller module of the sequencing pipeline. The plot shows the median (solid blue line), the 25th percentile and the 75th percentile (lower and upper bound of the light blue area) of the qualities at each position (cycle). (c) The base call frequency plot indicates a lack of positional biases in the call frequency for each base. (d) The plot shows the distribution of qualities averaged across the reads. In the yerba mate data, the average quality score is 36.3 (indicated by the red line). (e) The Kmer frequency check identifies short sequences that occur more often than expected. 10 Kmers that occur 3 times more often than expected are indicated.

Figure S2   Trinity *de novo* assembly report. (a) Distribution of GC-contents: The GC-content of a sequence is calculated as the number of GC-bases normalized to the total number of sequences. (b) Combined coverage of G and C bases: number of G and C bases observed at current position normalized to the total number of bases observed at that position. (c) The five most-overrepresented 5mers. The over-representation of a 5mer is calculated as the ratio of the observed and expected 5mer frequency. The expected frequency is calculated as product of the empirical nucleotide probabilities that make up the 5mer. (d) The number of sequences that support (cover) the individual base positions normalized to the total number of sequences. (e) Distribution of sequence lengths: x: sequence length in base-pairs y: number of sequences featuring a particular length normalized to the total number of sequences. (f) Coverages for the four DNA nucleotides.

Figure S3   Semantic similarity-based scatterplots of gene ontology categories representations of the yerba mate transcriptome based in biological process (a), cellular component (b) or molecular function (c). Circle size is estimated based in p-values associated to the GO categories. Color legend is represented as an inset in (a). The Revigo web server was used to generate the plots: http://revigo.irb.hr/.

Figure S4   Tree-map visualization of enriched GO categories of yerba mate transcriptome based in biological process (a), cellular component (b) or molecular function (c). Rectangle size is estimated based in p-values associated to the GO categories. The Revigo web server was used to generate the maps: http://revigo.irb.hr/.

Figure S5   AgriGO generated plot of GO enrichment in yerba mate based in biological process. Significance color levels and arrow types associated with GO relationships are represented as an inset at the superior left corner.

Figure S6   AgriGO generated plot of GO enrichment in yerba mate based in cellular component. Significance color levels and arrow types associated with GO relationships are represented as an inset at the superior left corner.

Figure S7   AgriGO generated plot of GO enrichment in yerba mate based in molecular function. Significance color levels and arrow types associated with GO relationships are represented as an inset at the superior left corner.

Figure S8   Profile of ribosomal proteins and ribosome biogenesis related transcripts obtained from the yerba mate transcriptome. Plant Ribosomes are constituted by 4 rRNAs and ~80 ribosomal proteins. In this study every major structural ribosome constituents (**a**), and roughly every enzyme (**b**) responsible for ribosome processing, trafficking, rRNA maturation, and ribosome assembly were identified (green). Images credits: Kanehisa Laboratories, Japan.

Figure S9   Profile of plant hormone signal transduction related transcripts obtained from the yerba mate transcriptome. Plant development is regulated by endogenous signaling molecules including plant hormones. Perception of biological cues and signal transduction involves several hormone sensing and effector pathways. The major yerba mate enzymes involved in plant growth, cell division, stem growth, seed dormancy, senescence and cell elongation were identified in the assembled transcriptome (green). Images credits: Kanehisa Laboratories, Japan.

Figure S10   Genome alignment of *Lactuca sativa* chloroplast complete sequence (Accession no. AP007232.1) and yerba mate chloroplast. Identity is obtained based in 1 nt sliding window size and represented by color and bar height from 0% (red) to 100% (green). Annotations are depicted as protein coding genes (yellow), transfer RNA genes (pink) and ribosome RNA genes (red).

Figure S11   Mapping of *Ilex paraguariensis* assembled transcripts to the chloroplast sequence draft.

Figure S12   Genome alignment of sunflower mitochondrial complete sequence and yerba mate mitochondrial sequence consensus. Identity is obtained based in 1 nt sliding window size and represented by color and bar height from 0% (red) to 100% (green). Annotations are depicted as protein coding genes (yellow), transfer RNA genes (pink) and ribosome RNA genes (red).

Figure S13   Mauve genome alignment of yerba mate and *Helianthus annuus* mitochondrial complete sequence (Accession no. KF815390.1). Identity is represented hierarchically from white to red. The consensus *Ilex paraguariensis* sequence conserves most of the *Helianthus* gene annotations (rectangles). As an example, the consensus sequence of Ilex p. at 76,000 bp coordinates presents high identity to the 230,000 bp coordinates of sunflower (transparent bars), corresponding to the ccmFn coding sequence.

Figure S14 Bayesian phylogenic tree of the *Argonaute 1* (AGO1) genes of 35 plant species and yerba mate determined by the Geneious 7.0 platform. Values at the nodes indicate bootstrap support percentage obtained for 1,000 replicates.

Figure S15 Multiple gene alignment of *Argonaute 1* (AGO1) genes of 35 plant species and yerba mate. Identity is obtained based in 1 nt sliding window size and represented by color and bar height from 0% (red) to 100% (green).

Figure S16 Multiple MUSCLE protein alignment and secondary structure prediction of *Coffea arabica*, *Theobroma cacao*, *Camellia sinensis* and *Ilex paraguariensis* caffeine synthase showing an important conservation in gene structure and domains.

Figure S17 SWISS-MODEL report of yerba mate caffeine synthase 3D prediction using *Coffea arabica* CS as a template.

Table S1 Transcription factors predicted in yerba mate based in BLASTX hits descriptions employing the *viridiplantae* UniProt, or *Arabidopsis* TAIR protein database as reference.

Table S2 Gene Ontology terms associated to the yerba mate transcriptome. The GO terms were obtained employing the KAAS (KEGG Automatic Annotation Server, http://www.genome.jp/kegg/kaas/) for ortholog assignment and pathway mapping.

Table S3 Categorized yerba mate complete transcriptome best BLASTX hits, against the *viridiplantae* UniProt database.

Table S4 Yerba mate complete transcriptome best BLASTX hits, against the TAIR protein database.

Table S5 List and description of Gene Ontology, KO terms and mapsassociated to the yerba mate transcriptome obtained by the KAAS and deduced from BLASTX hits against *Arabidopsis* TAIR protein database and *viridiplantae* UniProt.

Table S6 Complete gene catalog list of yerba mate draft chloroplast and mitochondrion.

Table S7 *Helianthus annuus* mitochondrial gene coordinates and the analogous yerba mate mitochondrial draft ortholog gene sequence coordinate list.

Table S8 Predicted SSR summary, frequencies report and designed alternative primers for the yerba mate chloroplast genome.

Table S9 Predicted SSR summary, frequencies report and designed alternative primers for the yerba mate mitochondrial genome draft.

Table S10 Yerba mate miRNA sequence prediction report based in UEA small RNA workbench platform and canonical relaxed mapping of conserved precursor miRNAs to the yerba transcriptome on the CLC Genomics environment.

Table S11 Predicted SSR summary and frequencies report for the yerba mate complete transcriptome.

Table S12 Transposable element-like sequences present in the yerba mate transcriptome.

Data S1 DEG analysis of the yerba mate transcriptome.

Data S2 166 KEGG pathways models graphs representing most of yerba mate annotated genes (in green). Images credits: Kanehisa Laboratories, Japan.

Data S3 Alternative language abstract in Spanish.

Data S4 PCR detection assay of *I. paraguariensis* vs. *I. dumosa* based in 5.8S and ITS2 distinctive regions. Designed primer features, sample preparation and PCR conditions are presented as tables. Multiple alignment of *I. paraguariensis* assembled 45S consensus sequence vs. *I. dumosa* Acc. No. AJ492657 reveals that ITS1 is highly conserved in both species; however 5.8S and ITS2 are distinctive enough to differentiate them by means of PCR with species-specific primers. Primer bind locations are annotated in blue in both species. Agarose gel electrophoresis showing differential amplification according to primers specificity and reliability of the method is presented. M1: Promega 200 bp ladder molecular DNA ruler; M2: in-house control Marker; NTC: Non Template Control; 100%, 50%, 10%, 1%, 0.1%: percentage of *I. dumosa* (I.du) fraction respective to *I. paraguariensis* (I.pa) of leaf tissue employed for DNA isolation and subjected to specific PCR.

Acknowledgments

M.G. and D.A.M. are career members and P.M.A. is a postdoctoral research fellow of the Consejo Nacional de Investigaciones Científicas y Técnicas (CONICET-Argentina). H.J.D., R.E.B. and D.A.D. are researchers at the Instituto Nacional de Tecnología Agropecuaria (INTA-Argentina). We specially acknowledge Ing. Agr. Sergio D. Prat Kricun for his 40 years of field collection and selection of yerba mate plants from South America now available at the EEA Cerro Azul-INTA, Misiones. Authors also thank Universidad Nacional de Misiones for continuous support. Finally, we would like to thank Presidencia de la Nación Argentina, Ministerio de Ciencia, Tecnología e Innovación Productiva and Ministerio de Agricultura, Ganadería y Pesca for their permanent and compelling support of science, innovation, education and inclusion in Argentina.

Alternative language abstract

An Alternative language abstract in Spanish is available as **Data S3**.

Author Contributions

Conceived and designed the experiments: MG HJD DAM. Analyzed the data: HJD MG. Contributed reagents/materials/analysis tools: MG HJD DAM PMA REB DAD MBO PDZ. Wrote the paper: HJD MG PMA DAM. Reviewed and approved the final manuscript: HJD MG PMA REB MBO DAD PDZ DAM.

References

1. Giberti GC (1999) Recursos fitogenéticos relacionados con el cultivo y explotación de la Yerba Mate (*Ilex paraguariensis* St. Hil., Aquifoliáceas) en el cono sur de América. Acta Horticulturae 500: 137–144.

2. Bracesco N, Sanchez AG, Contreras V, Menini T, Gugliucci A (2011) Recent advances on *Ilex paraguariensis* research: Minireview. Journal of Ethnopharmacology 136: 378–384.

3. Gauer L, Cavalli-Molina S (2000) Genetic variation in natural populations of maté (*Ilex paraguariensis* A. St.-Hil., Aquifoliaceae) using RAPD markers. Heredity 84: 647–656.

4. Instituto Nacional de la Yerba Mate, Argentina. INYM website. Available: http://www.inym.org.ar/inym/imagenes/Estadisticas/sup%20cultivada%20depa.pdf. Accessed 2014 Jun 4.

5. Jerke G, Horianski MA, Salvatierra KA (2009) Evaluación de géneros micotoxigénicos en yerba mate elaborada. Revista de Ciencia y Tecnología (UNaM) 12: 41–45.

6. Sistema Integrado de Información Agropecuaria. Ministerio de Agricultura, Ganadería y Pesca, Presidencia de la Nación, Argentina. SIIA website. Available: http://www.siia.gob.ar/. Accessed 2014 Jun 4.

7. Canitrot L, Grosso MJ, Méndez A (2011) Complejo Yerbatero. Serie "Producción regional por complejos productivos". Ministerio de Economía y Finanzas Públicas, Argentina. Mecon website. Available: http://www.mecon.gov.ar/peconomica/docs/Complejo_Yerbatero.pdf. Accessed 2014 Jun 4.

8. Burtnik OJ (2006) Yerba Mate: Manual de Producción. INTA, AER Santo Tomé, Corrientes, Argentina. 52p.

9. Instituto Nacional de Tecnología Agropecuaria EEA Cerro Azul, Cambio Rural (1998) Yerba mate. Biblioteca para el Productor. 60p.

10. Sosa DA, Bárbaro S, Alvarenga FA, De Coll ODR, Ohashi DV, et al. (2011) Yerba mate. Manual de campo. 51p.

11. Ricca J (2012) El mate. Random House Mondadori Press. 288p.

12. Yerba mate Argentina website. Available: http://yerbamateargentina.org.ar/zona_archivo.php?archivo=descargas_02_Consumo_Cualitativo_YM_Congreso_Sud.ppt&titulo=Investigacion%20Cualitativa. Accessed 2014 Jun 4.

13. Belingheri LD, Prat Kricun SD (1997) Programa de mejoramiento genético de la Yerba Mate en el INTA. I Congresso Sul-Americano de Erva-Mate. II Reuniao Técnica do Cone Sul Sobre a Cultura da Erva-Mate: pp 267–277.

14. Gottlieb AM, Poggio L (2010) Genomic screening in dioecious "yerba mate" tree (*Ilex paraguariensis* A. St. Hill., Aquifoliaceae) through representational difference analysis. Genetica 138: 567–578.

15. Fan H, Xiao Y, Yang Y, Xia W, Mason AS, et al. (2013) RNA-Seq analysis of *Cocos nucifera*: Transcriptome sequencing and subsequent functional genomics approaches. PLoS ONE 8(3): e59997.

16. Shi CY, Yang H, Wei CL, Yu O, Zhang ZZ, et al. (2011) Deep sequencing of the *Camellia sinensis* transcriptome revealed candidate genes for major metabolic pathways of tea-specific compounds. BMC Genomics 12: 131.

17. Grabherr MG, Haas BJ, Yassour M, Levin JZ, Thompson DA, et al. (2011) Full-length transcriptome assembly from RNA-Seq data without a reference genome. Nature Biotechnology 29: 644–652.

18. Oono Y, Kobayashi F, Kawahara Y, Yazawa T, Handa H, et al. (2013) Characterisation of the wheat (*Triticum aestivum* L.) transcriptome by de novo assembly for the discovery of phosphate starvation-responsive genes: gene expression in Pi-stressed wheat. BMC Genomics 14: 77.

19. Luo H, Lin Y, Gao F, Zhang C-T, Zhang R (2013) DEG 10, an update of the Database of Essential Genes that includes both protein-coding genes and non-coding genomic elements. Nucleic Acids Research 42: D574-D580.

20. Meinke D, Muralla R, Sweeney C, Dickerman A (2008) Identifying essential genes in *Arabidopsis thaliana*. Trends in Plant Science 13: 483–491.

21. Nakasugi K, Crowhurst RN, Bally J, Wood CC, Hellens RP, et al. (2013) De novo transcriptome sequence assembly and analysis of RNA silencing genes of *Nicotiana benthamiana*. PLoS ONE 8: e59534.

22. Parra G, Bradnam K, Ning Z, Keane T, Korf I (2009) Assessing the gene space in draft genomes. Nucleic Acids Research 37: 289–297.

23. Liu S, Li W, Wu Y, Chen C, Lei J (2013) De novo transcriptome assembly in chili pepper (*Capsicum frutescens*) to identify genes involved in the biosynthesis of capsaicinoids. PLoS ONE 8(1): e48156.

24. Touming L, Siyuan Z, Qingming T, Ping C, Yongting Y, et al. (2013) De novo assembly and characterization of transcriptome using Illumina paired-end sequencing and identification of CesA gene in ramie (*Boehmeria nivea* L. Gaud). BMC Genomics 14: 125.

25. Ge X, Chen H, Wang H, Shi A, Liu K (2014) De novo assembly and annotation of *Salvia splendens* transcriptome using the Illumina platform. PLoS ONE 9(3): e87693.

26. Yin D, Wang Y, Zhang X, Li H, Lu X, et al. (2013) De novo assembly of the peanut (*Arachis hypogaea* L.) seed transcriptome revealed candidate unigenes for oil accumulation pathways. PLoS ONE 8(9): e73767.

27. Yang Y, Yu X-Y, Li Y-F (2013) De novo assembly and characterization of the barnyardgrass (*Echinochloa crus-galli*) transcriptome using next-generation pyrosequencing. PLoS ONE 8(7): e69168.

28. Cardoso-Silva CB, Costa EA, Mancini MC, Balsalobre TWA, Canesin LEC, et al. (2014) De novo assembly and transcriptome analysis of contrasting sugarcane varieties. PLoS ONE 9(2): e88462.

29. Annadurai RS, Neethiraj R, Jayakumar V, Damodaran AC, Rao SN, et al. (2013) *De Novo* Transcriptome Assembly (NGS) of *Curcuma longa* L. Rhizome Reveals Novel Transcripts Related to Anticancer and Antimalarial Terpenoids. PLoS ONE 8(2): e56217.

30. Wang Z, Fang B, Chen J, Zhang X, Luo Z, et al. (2010) De novo assembly and characterization of root transcriptome using Illumina paired-end sequencing and development of cSSR markers in sweetpotato (*Ipomoea batatas*). BMC Genomics 11: 726.

31. Mantello CC, Cardoso-Silva CB, da Silva CC, de Souza LM, Scaloppi Junior EJ, et al. (2014) De Novo Assembly and Transcriptome Analysis of the Rubber Tree (*Hevea brasiliensis*) and SNP Markers Development for Rubber Biosynthesis Pathways. PLoS ONE 9(7): e102665.

32. Cavagnaro PF, Senalik DA, Yang L, Simon PW, Harkins TT, et al. (2010) Genome-wide characterization of simple sequence repeats in cucumber (*Cucumis sativus* L.). BMC genomics 11(1): 569.

33. Wei W, Qi X, Wang L, Zhang Y, Hua W, et al. (2011) Characterization of the sesame (*Sesamum indicum* L.) global transcriptome using Illumina paired-end sequencing and development of EST-SSR markers. BMC Genomics 12: 451.

34. Fraser LG, Harvey CF, Crowhurst RN, De Silva HN (2004) EST-derived microsatellites from *Actinidia* species and their potential for mapping. Theoretical and Applied Genetics 108: 1010–1016.

35. Aggarwal RK, Hendre PS, Varshney RK, Bhat PR, Krishnakumar V, et al. (2007) Identification, characterization and utilization of EST-derived genic microsatellite markers for genome analyses of coffee and related species. Theoretical and Applied Genetics 114: 359–372.

36. Gottlieb AM, Poggio L (2014) Quantitative and qualitative genomic characterization of cultivated *Ilex* L. species. Plant Genetic Resources: Characterization and Utilization 1–11. doi:10.1017/S1479262114000756

37. Hotto AM, Schmitz RJ, Fei Z, Ecker JR, Stern DB (2011) Unexpected diversity of chloroplast noncoding RNAs as revealed by deep sequencing of the *Arabidopsis* transcriptome. G3: Genes, Genomes, Genetics 1: 559–570.

38. Hotto AM, Germain A, Stern DB (2012) Plastid non-coding RNAs: emerging candidates for gene regulation. Trends in Plant Science 17: 737–744.

39. Zhelyazkova P, Sharma CM, Förstner KU, Liere K, Vogel J, et al. (2012) The primary transcriptome of barley chloroplasts: numerous noncoding RNAs and the dominating role of the plastid-encoded RNA polymerase. The Plant Cell Online 24: 123–136.

40. Small ID, Rackham O, Filipovska A (2013) Organelle transcriptomes: products of a deconstructed genome. Current Opinion in Microbiology 16: 652–658.

41. Darling AC, Mau B, Blattner FR, Perna NT (2004) Mauve: multiple alignment of conserved genomic sequence with rearrangements. Genome Research 14: 1394–1403.

42. Hannon GJ (2002) RNA interference. Nature 418: 244–251.

43. Bologna NG, Voinnet O (2014) Diversity, Biogenesis, and Activities of Endogenous Silencing Small RNAs In Arabidopsis. Annual Review of Plant Biology. doi: 10.1146/annurev-arplant-050213-035728

44. Vaucheret H, Vazquez F, Crété P, Bartel DP (2004) The action of ARGONAUTE1 in the miRNA pathway and its regulation by the miRNA pathway are crucial for plant development. Genes & Development 18: 1187–1197.

45. Baumberger N, Baulcombe DC (2005) Arabidopsis ARGONAUTE1 is an RNA Slicer that selectively recruits microRNAs and short interfering RNAs. Proceedings of the National Academy of Sciences of the United States of America 102: 11928–11933.

46. Debat HJ, Ducasse DA (2014) Plant microRNAs: Recent Advances and Future Challenges. Plant Molecular Biology Reporter. doi:10.1007/s11105-014-0727-z

47. Wu G, Poethig RS (2006) Temporal regulation of shoot development in *Arabidopsis thaliana* by miR156 and its target SPL3. Development 133: 3539–3547.

48. Wang JW, Czech B, Weigel D (2009) miR156-Regulated SPL transcription factors define an endogenous flowering pathway in *Arabidopsis thaliana*. Cell 138: 738–749.

49. Xing S, Salinas M, Höhmann S, Berndtgen R, Huijser P (2010) miR156-targeted and nontargeted SBP-box transcription factors act in concert to secure male fertility in *Arabidopsis*. The Plant Cell Online 22: 3935–3950.

50. Kozomara A, Griffiths-Jones S (2014) miRBase: annotating high confidence microRNAs using deep sequencing data. Nucleic Acids Research 42(D1): D68-D73.

51. Yang L, Conway SR, Poethig RS (2011) Vegetative phase change is mediated by a leaf-derived signal that represses the transcription of miR156. Development 138: 245–249.

52. Bastos DHM, Oliveira DM, Matsumoto RLT, Carvalho PO, Ribeiro ML (2007) Yerba mate: pharmacological properties, research and biotechnology. Med Aromat Plant Sci Biotechnol 1: 37–46.

53. Murakami ANN, Amboni RD, Prudêncio ES, Amante ER, Fritzen-Freire CB, et al. (2013) Concentration of biologically active compounds extracted from *Ilex paraguariensis* St. Hil. by nanofiltration. Food Chemistry 141: 60–65.

54. Silva RD, Bueno ALS, Gallon CW, Gomes LF, Kaiser S, et al. (2011) The effect of aqueous extract of gross and commercial yerba mate (*Ilex paraguariensis*) on intra-abdominal and epididymal fat and glucose levels in male wistar rats. Fitoterapia 82: 818–826.

55. Heckman MA, Weil J, Mejia D, Gonzalez E (2010) Caffeine (1, 3, 7-trimethylxanthine) in foods: a comprehensive review on consumption, functionality, safety, and regulatory matters. Journal of Food Science 75: R77–R87.

56. Ashihara H, Sano H, Crozier A (2008) Caffeine and related purine alkaloids: biosynthesis, catabolism, function and genetic engineering. Phytochemistry 69(4): 841–856.

57. Ashihara H, Monteiro AM, Gillies FM, Crozier A (1996) Biosynthesis of caffeine in leaves of coffee. Plant Physiology 111: 747–753.

58. Kato M, Mizuno K, Crozier A, Fujimura T, Ashihara A (2000) A gene encoding caffeine synthase from tea leaves. Nature 406: 956–957.

59. Deladino L, Teixeira A, Reta M, Molina García AD, Navarro AS, et al. (2013) Major phenolics in yerba mate extracts (*Ilex paraguariensis*) and their contribution to the total antioxidant capacity. Food and Nutrition 4: 154–162.

60. Niggeweg R, Michael AJ, Martin C (2004) Engineering plants with increased levels of the antioxidant chlorogenic acid. Nature Biotechnology 22(6): 746–754.

61. Lepelley M, Cheminade G, Tremillon N, Simkin A, Caillet V, et al. (2007) Chlorogenic acid synthesis in coffee: An analysis of CGA content and real-time RT-PCR expression of HCT, HQT, C3H1, and CCoAOMT1 genes during grain development in *C. canephora*. Plant Science 172(5): 978–996.

62. Escamilla-Treviño LL, Shen H, Hernandez T, Yin Y, Xu Y, et al. (2014) Early lignin pathway enzymes and routes to chlorogenic acid in switchgrass (*Panicum virgatum* L.). Plant Molecular Biology 84(4–5): 565–576.

63. Zenaro LC, Andrade LB, Santos P, Locatelli C (2014) Effects of aqueous extract of Yerba Mate (*Ilex Paraguariensis*) on the oxidative stress in rats fed a cafeteria diet. International Journal of Natural Sciences Research 2(3): 30–43.

64. Barchuk ML, Tiscornia MM, Giorgio EM, Fonseca MI, Zapata PD (2013) Diseño de un método molecular para la detección de *Ilex dumosa* en yerba mate elaborada utilizando una secuencia específca ubicada en la región ITS2 del DNA ribosómico. Revista de Ciencia y Tecnología (UNaM) 19: 28–34.

65. Kiefer F, Arnold K, Künzli M, Bordoli L, Schwede T (2009) The SWISS-MODEL Repository and associated resources. Nucleic Acids Research 37: D387–D392.

66. Jung S, Ficklin SP, Lee T, Cheng CH, Blenda A, et al. (2014) The Genome Database for Rosaceae (GDR): year 10 update. Nucleic Acids Research 42(D1): D1237–D1244.

67. Jung S, Staton M, Lee T, Blenda A, Svancara R, et al. (2008) GDR (Genome Database for Rosaceae): integrated web-database for Rosaceae genomics and genetics data. Nucleic Acids Research 36: D1034–D1040.

68. Zhao X, Wang H (2007) LTR_FINDER: an efficient tool for the prediction of full-length LTR retrotransposons. Nucleic Acids Research 35: W265–W268.

8

Bayesian Model of Protein Primary Sequence for Secondary Structure Prediction

Qiwei Li[1], David B. Dahl[2]*, Marina Vannucci[1], Hyun Joo[3], Jerry W. Tsai[3]

1 Department of Statistics, Rice University, Houston, Texas, United States of America, **2** Department of Statistics, Brigham Young University, Provo, Utah, United States of America, **3** Department of Chemistry, University of the Pacific, Stockton, California, United States of America

Abstract

Determining the primary structure (i.e., amino acid sequence) of a protein has become cheaper, faster, and more accurate. Higher order protein structure provides insight into a protein's function in the cell. Understanding a protein's secondary structure is a first step towards this goal. Therefore, a number of computational prediction methods have been developed to predict secondary structure from just the primary amino acid sequence. The most successful methods use machine learning approaches that are quite accurate, but do not directly incorporate structural information. As a step towards improving secondary structure reduction given the primary structure, we propose a Bayesian model based on the knob-socket model of protein packing in secondary structure. The method considers the packing influence of residues on the secondary structure determination, including those packed close in space but distant in sequence. By performing an assessment of our method on 2 test sets we show how incorporation of multiple sequence alignment data, similarly to PSIPRED, provides balance and improves the accuracy of the predictions. Software implementing the methods is provided as a web application and a stand-alone implementation.

Editor: Yang Zhang, University of Michigan, United States of America

Funding: This work is supported by National Institutes of Health National Institute of General Medical Sciences R01 GM104972. The funder had no role in study design, data collection and analysis, decision to publish, or preparation of the manuscript.

Competing Interests: The authors have declared that no competing interests exist.

* Email: dahl@stat.byu.edu

Introduction

For protein sequences of unknown structure and/or biological function, one of the first and quite insightful analyses of the linear sequence of amino acids (i.e., the primary structure) is a prediction of the secondary structure. Fig. 1 shows a four-state secondary structure definition of the primary amino acid sequence. Advances in genomic sequencing technologies have made obtaining protein sequences relatively cheap, accurate and fast, in comparison to the costly and involved approaches to solving a protein's structure. However, the number of protein sequences far outpaces knowledge of their structure. Improvements in secondary structure prediction would have impact across many fields of computational biology. As the basis for higher order protein structure, more accurate secondary structure predictions is a necessary step for improved modeling of a protein's fold [1,2] and identification of its function [3]. Secondary structure modeling also plays an important role in the rational design of protein structure [4] and enzymatic function [5] as well as in drug development [6].

Depending on the set of protein sequences assessed, the accuracy of secondary structure prediction methods has improved steadily to an average of upwards of 80% [7]. The most successful approaches for secondary structure prediction apply machine learning algorithms to maximize the sequence relationship between a proteins' primary sequences and their assigned secondary structure as defined by the program DSSP [8]. One of the early approaches that has become a standard in the field,

PHD [9] and its current incarnation PredictProtein [10], employs an artificial neural network and sequence profiles in identifying secondary structure from a protein sequence. Other successful servers such as Jpred [11] and PSIPRED [12] also apply artificial neural networks. As an approach, neural net based prediction methods are quite popular and accurate [13,14]. Other machine learning methods attempt to match prediction accuracy using hidden Markov models (HMM) [15–17] and support vector machines (SVM) [18,19]. Due to their consistently high accuracy of prediction, these methods have become the *de facto* standard against which other secondary structure prediction methods measure their success, many of which have been evaluated in a recent review [20]. However, the accuracy has essentially remained at 80% for many years [2,20].

Because these expert systems rely on indeterminate relationships between the primary sequence and a 3 or 4 state secondary structure classification, a potential approach to improving secondary structure predictions is to incorporate higher order structural information. The initial use of structural information to model protein secondary structure was based on the hydrophobic patterns found in amphipathic helices and sheets [21] or the hydrophilic spacing between residues and turn regions [22]. With the recent success of fragment-based structure prediction, numerous methods have incorporated structural information from local fragment prediction [23–25] to more global structural relationships [26–29] into secondary structure prediction. These recent

a	E	E	N	I	I	P	Y	I	T	N	V	L	Q	N	P	D	L	A	L	R	M	A	V	R	N	N	L	A	G
η	C			H				T		C				H										C					
λ	3			6				3		2				10										5					

Figure 1. The first 29 amino acids from the protein clathrin 1⑨/domain a [46] with our associated parameterization (η, λ) of the secondary structure. If this were the entire protein, then we would have $L = 29$, $M = 6$, and $(\eta, \lambda) = ((\eta_1, ..., \eta_6), (\lambda_1, ..., \lambda_6)) = ((C, H, T, C, H, C), (3, 6, 3, 2, 10, 5))$.

methods have been able to reproduce the 80% accuracy of the machine learning approaches. The approach tested in this paper applies sequence to structure relationships defined by packing of residues based on the knob-socket model [30,31].

Improving the previous models of packing in helix [32] and sheet [33], the knob-socket model provides a simple and general motif to describe the packing in protein structure that has been shown to relate the primary sequence to the packing structure at both the secondary and tertiary structure levels in both helices [30] and sheets [31]. Whereas the previous knob-into-holes [32] and ridges-into-grooves [33] are each limited to describing packing at defined angles within only a single type of secondary structure, the knob-socket model encompasses all packing within proteins at all angles and between all types of secondary structure. The knob-socket model simplifies the convoluted packing of side-chains into regular patterns of a single knob residue from one element of secondary structure packing into a socket formed by 3 residues from another element of secondary structure. Because the composition of both the knobs and socket exhibit preferences for certain amino acids, this knob-socket model not only relates primary sequence to tertiary packing structure, but also associates the primary sequence with secondary structure packing. At the level of secondary structure, only the local 3-residue socket plays a role in this model, since the knob residue defines tertiary packing structure. The repetitive main-chain hydrogen bonding for regular secondary structure produces a consistent arrangement of sockets. The arrangements defines the secondary structure packing motifs that provide the sequence patterns to identify secondary structure (Fig. 2). This is the case even for the irregular coil secondary structure.

In this paper we propose a Bayesian model for secondary structure prediction given the primary structure. The method considers the packing influence of residues on the structure determination, including those packed close in space but distant in sequence. Fundamentally, secondary structure is defined by the regular hydrogen bonding patterns between the main-chain polar groups of amino acids in the linear sequence. From this definition, secondary structure is usually described by three classes (α-helix, β-sheets, and other or coil) for three-state predictions. The regular hydrogen bonding patterns define the states of α-helix or β-strand, whereas the irregular coil is defined by the lack of repetitive hydrogen bonding. Instead of hydrogen bond patterns, the knob-socket model identifies the regular patterns of packing not only in α-helices and β-strand, but also in coil structure (Fig. 2). We therefore develop a probabilistic model for secondary structure prediction that is informed by the packing structure between residue defined by the knob-socket model. The local sequence relationships are incorporated, similarly to PSIPRED [12] [34], as a set of multiple sequence alignments (MSA). The incorporation of MSA information has been a standard approach for many years [35]. We compare performances of our method with TorusDBN [36] and the benchmark machine learning approach PSIPRED,

on two test sets. TorusDBN is a conformational sampling method (akin to a fragment library) and was not developed for secondary structure prediction. A side-benefit of the TorusDBN method, however, is the prediction of protein secondary structure. Our results demonstrate that adding local structural information (as defined by the knob-socket model) in the prior distribution increases our method's accuracy to just below current standards on one of the test set and above on the larger test set.

We have developed both a web application and stand-alone implementation of the methods described in this paper. The website is http://bamboo.byu.edu and the stand-alone implementation is bamboo, an open-source R package available on the Comprehensive R Archive Network (CRAN). The package implements all the methods described in the paper and it provides all the data used in our analysis. Documentation and an example are provided. The package may be installed on the latest version of R by running: install.packages ("bamboo").

Results and Discussion

We evaluate the performance of our proposed method against the 3,344 chains in the ASTRAL30 and the 203 chains in the CASP9 data sets. For each data set, the DSSP annotation [8] was used as the true secondary structure state. Over the thousands of proteins, the secondary structure prediction based on the amino acid sequence a is compared to the DSSP value at each position. Since test sets were not part of the 15,470 chain training set, the accuracy reported for our approach compared to the true secondary structure value are reliable assessments of performance. The knob-socket packing was implemented in 2 different ways with 2 posterior summaries, which amounts to 4 separate methods. We compare the prediction accuracy of our method against 2 other secondary structure prediction methods: TorusDBN [36] and PSIPRED [12]. PSIPRED was run using 2 sequence databases: sequences from the PDB for PSIPRED-PDB and a non-redundant sequence database for PSIPRED-NR, described in more detail in the methods. The purpose of the PSIPRED-PDB is to provide a more direct comparison of methods, since the MSA's used by our method were limited to sequences from the PDB [37]. However, with a deep sequence alignment as an input, PSIPRED-NR produces the best results, which coincide with the benchmark for secondary structure prediction accuracy.

Table 1 reports the percent accuracy of each method on the ASTRAL30 and CASP9 test sets in terms of classification recall, i.e., the percent of the true DSSP defined secondary states that are correctly predicted for 3 classes, or Q3. Because TorusDBN and PSIPRED are three-state models that do not predict turn residues, the turn predictions of our model were merged into the coil class to facilitate comparisons. TorusDBN and PSIPRED both performed consistently between the 2 test sets. Although secondary structure prediction is not the focus of TorusDBN, TorusDBN exhibits a high of 62% accuracy for the ASTRAL30 test set. For PSIPRED,

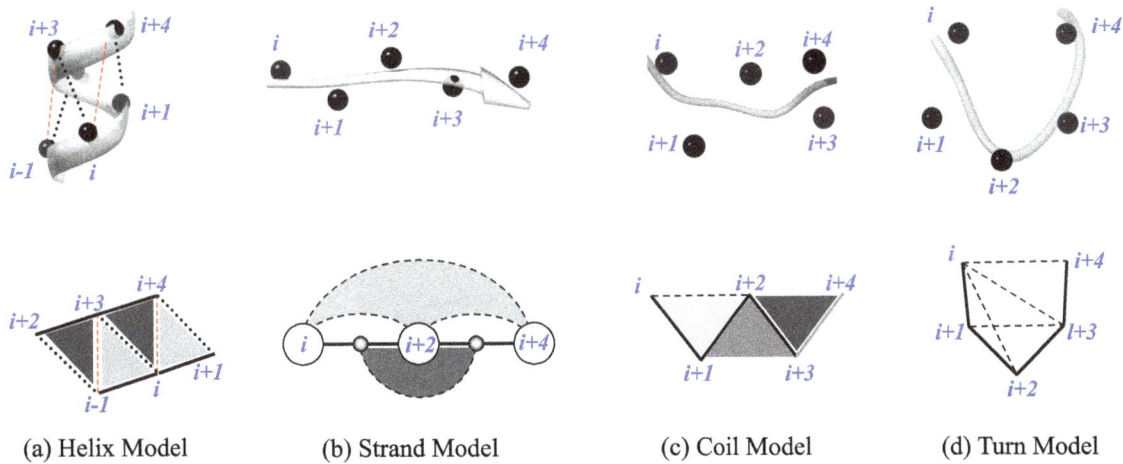

(a) Helix Model   (b) Strand Model   (c) Coil Model   (d) Turn Model

Figure 2. Local structural motifs used to model protein secondary structure as defined by the knob-socket model. On the top for each type of secondary structure, ribbon diagrams of the protein backbone with black spheres at $C\alpha$ positions are presented. On the bottom, two-dimensional lattice representations are shown of the local residue interactions that define secondary structure, where solid lines represent covalent contacts between residues and broken lines are packing interactions. Because only the local interactions are being considered to predict secondary structure, only the socket portion of the knob-socket model is used. The knob portion signifies interactions at the level of tertiary structure or packing of non-local residues distant in the protein sequence. Each of the 4 types of secondary structure are described in more detail. (a) Helix Model: Relative residue positions and interactions are shown. Two types of sockets are represented in different grey scale: $(i,i+3,i+4)$ sockets in dark grey and $(i,i+1,i+4)$ sockets in light grey. (b) Strand Model: Double-side sheet sockets are shown. Sockets $(i,i+1,i+4)$ and $(i+2,i+3,i+4)$ in white are facing one direction, a socket $(i+1,i+3,i+4)$ in dark grey faces the other side. Also, the side chain only socket $(i,i+2,i+4)$ is shown in light grey. (c) Coil Model: Three types of coil sockets are shown. The socket $(i,i+1,i+2)$ is closed socket with all three residues in contact one another, the socket $(i+1,i+2,i+3)$ is open socket with $(i+1,i+2)$ contact and $(i+2,i+3)$ contact but no contact between $i+1$ and $i+3$, and the socket $(i+2,i+3,i+4)$ is strained socket with no contact between $i+3$ and $i+4$. (d) Turn Model: Three residue sockets $(i,i+1,i+2)$, $(i+1,i+2,i+3)$, $(i,i+2,i+3)$, and $(i,i+3,i+4)$ in the 5 residue turn are shown.

using only sequences from solved structures performs up to 15 percentage points better than TorusDBN, while allowing deeper sequence alignments into the larger non-redundant sequence database increases the Q3 accuracy upwards of 20 points to the established standard of 80% on the ASTRAL30 data set and slightly higher 81% on the CASP9 data set. As a baseline for our method, the NonInfo-MAP uses a non-informative prior and maximization of the posterior probability, producing the lowest measured accuracy of 52% on the CASP9 data set. Implementing the secondary structure guided block sampling of the posterior distribution (NonInfo-MP in Table 1) provides a modest increase in prediction accuracy, although still a bit below TorusDBN for both test sets. The fact that TorusDBN and our NonInfo-MP essentially perform equally is to be expected given the fact that neither use an informative prior. Incorporating MSA information to inform our method's prior distribution results in a significant improvement in accuracy, especially with the ASTRAL30 test set where both implementations of the posterior sampling reach accuracies above the PSIPRED-NR benchmark of 80% at 84% for MSA-MAP and 88% for MSA-MP. Accuracies improve for the CASP9 test set too, although they remain lower than PSIPRED-NR at 69% for MSA-MAP and 75% for MSA-MP. This decrease in performance is likely due to the different amount of MSA information available for the two data sets. For ASTRAL30, in fact, there were 201 sequences without an MSA, which is 6%, while for CASP9, there were 109 sequences, which is 54%.

Recall and precision is broken down for each of the secondary structure states to provide a more detailed understanding of Q3 prediction accuracy. For the ASTRAL30 and CASP9 sets respectively, a comparison of Q3 classification recall is shown in Tables 2 and 3 and Q3 classification precision is shown in Tables 4 and 5. The recall tables indicate that TorusDBN best predictions are of the helix and coil states. The majority of

incorrectly predicted helix residues are assigned the coil state and the converse is true for the coil state. The strand residues are assigned essentially by chance with a uniform distribution over the 3 states. For precision, the TorusDBN results are consistently the same across all three states, where the distribution of prediction is 60% correct and then about equally mispredicted at 20% for the other 2 states. The PSIPRED program's recall in both implementations and across both test sets performs the best at identifying the coil state, then the helix state and finally the strand state. Incorrect predictions of state are consistently coil that should be strand or helix. PSIPRED does not mix up helix and strand states often. Precision of PSIPRED predictions is the best for the helix state and worst for the coil state, where the strand state is twice as likely to be predicted as coil than the helix state. Reiterating the recall results, the precision of PSIPRED helix and strand predictions consistently are incorrectly assigned coil states.

For the baseline NonInfo-MP implementation of our method, the recall results (Tables 2 and 3) indicate that inclusion of this model for coil correctly predicts a high of 75% of coil residues. This result is accomplished by over-predicting the coil state such that coil is the major error in predicting helix and strand at around 37% for each. The precision for NonInfo-MP reveals that the helix state is most precisely predicted followed by coil and then sheet for both the ASTRAL30 (Table 4) and CASP9 (Table 5) test sets. NonInfo-MP and TorusDBN do not use prior information and, therefore, do not perform particularly well, our MSA-MP predictions perform very well on the ASTRAL30 set at 88% accuracy. The recall in Table 2 is at 90% for helix, 89% for coil and 85% for strand. The predominate error is to assign coil to helix and sheet at 9% and 14%, respectively. This is especially encouraging as the MSA-MP uses essentially the same sequence database as PSIPRED-PDB. The precision for the MSA-MP method (Table 3) corroborates these results. While TorsuDBN

Table 1. Overall Q3 accuracy (%) of TorusDBN, PSIPRED, and our method under different priors and segmentations on ASTRAL30 and CASP9 test data sets.

Dataset	Other methods			Our method			
	TorusDBN	PSIPRED-PDB	PSIPRED-NR	NonInfo-MAP	NonInfo-MP	MSA-MAP	MSA-MP
ASTRAL30	62	77	80	53	60	84	88
CASP9	61	75	81	52	58	69	74

and PSIPRED accuracies are consistent between the 2 test sets, our MSA-MP method exhibits worse Q3 prediction accuracies for the CASP9 data set. However, the overall accuracy of 74% is on par with the PSIPRED-PDB value of 75%. For recall, the drop in accuracy accompanies an increase in incorrect assignment of helix and strands to coil. The precision drop with the MSA-MP on the CASP9 data set shows an increase in all of the off-diagonal misprediction states. Indeed, the NonInfo implementation of our method is used for those sequences without MSA information, resulting in the overall predictions being influenced by the over prediction of the coil state.

Fig. 3 compares the Q3 results from the different methods to the DSSP [8] defined states for the phospholipase protein 3rvc [38], where the marginal posterior probabilities from our NonInfo-MP and MSA-MP methods are also plotted. Performing the worst in this set are TorusDBN and our NonInfo-MP, and both of these are limited to primarily local information. As indicated in the recall and precision tables, TorusDBN over predicts the helix state and under predicts the sheet state, which is clearly shown by the prediction in Fig. 3. Our NonInfo-MP method is slightly better at finding regions of correct secondary structure, but the length and limits of secondary structure states are poorly predicted. As shown by Fig. 2, the longest sequence distance that the knob-socket model considers is 5 residues. Clearly, information limited to local residues is unable to accurately reproduce the native secondary structure over large segments of sequence. Yet, over all of our predictions using the NonInfo-MP implementation, the average difference between the incorrect and correct probabilities was 0.244, with a standard deviation of 0.173 and an interquartile range (that is, the difference between the 25th and the 75th percentiles) of 0.244. This implies that the probabilities are somewhat close in areas of misprediction. With a residue window of 15 residues [34], PSIPRED in both of its applications is able to better identify the sheet residues and define the transitions between the different secondary structure states. Adding in the MSA information, the MSA-MP method adds more global secondary structure state information to the local model provided but the knob-socket model. The accuracy in identification of secondary structure states is very accurate, with most of the errors in defining the ends of secondary structure segments. In the plot of the MSA-MP, the marginal probability for a certain type of structure is clearly dominant at many positions in the middle of secondary structure segments, with values close to 1, but drops at the residues that transition between secondary structure states, with values of 0.5 or a little higher. While the case shown in Fig. 3 displays a favorable prediction for our MSA-MP method, it is at these transition points where the marginal probabilities of the secondary structure states clearly show less confidence in the prediction. Plots of this type help us understanding how incorporating information about the global influences on secondary structure can aid prediction.

As noted above and shown in lower portion of Fig. 3, the posterior distribution of marginal probabilities facilitates inference and adds a level of interpretation available only through a Bayesian approach. In particular, these marginal probabilities provide an extra level of confidence in the predictions. The posterior distributions in Fig. 3 are not only higher in areas that are correctly predicted, but there is a great spread between the marginal probability of the correct state than the next closest incorrect state. For the transitions areas, the marginal probabilities are much closer in value. To quantify of this, the difference in probability between the correct and highest incorrect state have been calculated and are reported for each state in Table 6. In the ASTRAL30 data set, the difference is between 56% and 70%

Table 2. Classification Q3 recall (%) of TorusDBN, PSIPRED, and our method under different priors on ASTRAL30 test dataset.

	Other Methods								
	(a) TorusDBN			(b) PSIPRED-PDB			(c) PSIPRED-NR		
	Helix	Strand	Coil	Helix	Strand	Coil	Helix	Strand	Coil
Helix	69	30	23	77	5	9	79	3	6
Strand	7	33	6	3	66	8	1	68	6
Coil	24	37	71	20	29	83	20	29	88
Overall	62			77			80		
	Our Methods								
	(d) NonInfo-MAP			(e) NonInfo-MP			(f) MSA-MAP			(g) MSA-MP		
	Helix	Strand	Coil	Helix	Strand	Coil	Helix	Strand	Coil	Helix	Strand	Coil
Helix	42	16	17	50	12	13	84	2	8	90	1	6
Strand	17	43	14	15	51	11	1	79	6	1	85	5
Coil	41	41	69	35	37	76	15	19	86	9	14	89
Overall	53			60			84			88		

*Each column of the matrix represents the instances in an actual class, while each row represents the instances in a predicted class. Note that the sum of elements of each column equals to 100.

when the prediction is correct, whereas the difference is only between 34% and 39% when incorrect. In the CASP9 data set, the values are smaller in both cases, where the difference for correct predictions ranges from 42% to 54% and the difference for incorrect predictions is 24% to 28%. So, as a simple rule of thumb, a separation of over 50% would strongly indicate a good prediction, and values less than 50% would indicate that the prediction is potentially wrong.

Not only can we examine the marginal posterior probability at each position, but our method allows us to make inference on the number of blocks in total and the number of blocks of each type. Take, for example, protein T0622-D10 from the CASP9 data set.

These posterior distributions are plotted in Fig. 4. For this protein, the model estimated well the total number of blocks, but has over estimated the number of coil blocks and under estimated the number of turn blocks.

Our results reinforce the general concept that more context is necessary to understand the environment that induces secondary structure that in effect goes against its amino acid composition. Improvement may be possible by considering higher order packing that can be provided by the knob-socket model. As an example, instead of strand predictions, a construct for tertiary packed sheets would potentially improve the accuracy by providing a strong differentiation over coils using the knob-socket

Table 3. Classification Q3 recall (%) of TorusDBN, PSIPRED, and our method under different priors on CASP9 test dataset.

	Other Methods								
	(a) TorusDBN			(b) PSIPRED-PDB			(c) PSIPRED-NR		
	Helix	Strand	Coil	Helix	Strand	Coil	Helix	Strand	Coil
Helix	72	31	24	77	6	10	82	2	7
Strand	6	30	5	4	62	10	1	69	7
Coil	22	39	71	19	32	80	17	29	86
Overall	61			75			81		
	Our Methods								
	(d) NonInfo-MAP			(e) NonInfo-MP			(f) MSA-MAP			(g) MSA-MP		
	Helix	Strand	Coil	Helix	Strand	Coil	Helix	Strand	Coil	Helix	Strand	Coil
Helix	41	16	18	47	14	14	66	8	13	73	7	11
Strand	18	42	14	16	49	12	7	61	10	7	67	9
Coil	41	42	68	37	37	74	27	31	77	20	26	81
Overall	52			58			69			74		

*Each column of the matrix represents the instances in an actual class, while each row represents the instances in a predicted class. Note that the sum of elements of each column equals to 100.

Table 4. Classification Q3 precision (%) of TorusDBN, PSIPRED, and our method under different priors on ASTRAL30 test dataset.

Other Methods									
(a) TorusDBN			**(b) PSIPRED-PDB**			**(c) PSIPRED-NR**			
Helix	**Strand**	**Coil**	**Helix**	**Strand**	**Coil**	**Helix**	**Strand**	**Coil**	
63	16	21	86	3	11	91	2	7	Helix
22	59	19	5	78	17	3	84	13	Strand
20	18	62	17	13	70	16	13	71	Coil
62			77			80			Overall

Our Methods												
(d) NonInfo-MAP			**(e) NonInfo-MP**			**(f) MSA-MAP**			**(g) MSA-MP**			
Helix	**Strand**	**Coil**	**Helix**	**Strand**	**Coil**	**Helix**	**Strand**	**Coil**	**Helix**	**Strand**	**Coil**	
61	13	26	71	10	19	91	1	8	93	1	6	Helix
30	44	26	27	52	21	2	86	12	1	89	10	Strand
30	17	53	26	16	58	13	9	78	8	7	85	Coil
53			60			84			88			Overall

*Each column of the matrix represents the instances in an actual class, while each row represents the instances in a predicted class. Note that the sum of elements of each row equals to 100.

model. While long range residue interactions have been integrated in a general sense into previous methods [26,28], the knob-socket model provides constructs that can correlate specific residue patterns derived from packing interactions that identify secondary structure.

Materials and Methods

Data Set

The secondary structure data was derived from the ASTRAL SCOP 1.75 structure set [39] filtered at 95% sequence identity. This structure set consisted of 15,470 individual protein domains from the PDB [37] whose length range from 22 to 1,419 amino acids and total 2,751,815 amino acids. Besides the training set, we used two test sets. The first test set is the current release of SCOPe 2.03 data set [40] filtered at 30% sequence identity (ASTRAL30). In this ASTRAL30 set we included the domains that are not included in 1.75 version and only included in 2.03 version. The transmembrane proteins were also excluded and this gave 2,794 domains. The data set integrity was further tested by breaking down into the actual segments. When the structure has missing residues, the chain was split into separate sequences and omitted in this study if a chain is shorter than 25 residues. This produced 3,344 chains with 523,332 amino acids. The second test set was

Table 5. Classification Q3 precision (%) of TorusDBN, PSIPRED, and our method under different priors on CASP9 test dataset.

Other Methods									
(a) TorusDBN			**(b) PSIPRED-PDB**			**(c) PSIPRED-NR**			
Helix	**Strand**	**Coil**	**Helix**	**Strand**	**Coil**	**Helix**	**Strand**	**Coil**	
61	17	22	84	4	12	91	2	7	Helix
19	63	18	7	74	19	2	84	14	Strand
19	21	60	16	16	68	13	14	73	Coil
61			75			81			Overall

Our Methods												
(d) NonInfo-MAP			**(e) NonInfo-MP**			**(f) MSA-MAP**			**(g) MSA-MP**			
Helix	**Strand**	**Coil**	**Helix**	**Strand**	**Coil**	**Helix**	**Strand**	**Coil**	**Helix**	**Strand**	**Coil**	
58	15	27	67	12	21	77	6	17	82	5	13	Helix
30	45	25	27	52	21	12	69	19	11	73	16	Strand
30	19	51	27	17	56	22	16	62	17	14	69	Coil
52			58			69			74			Overall

*Each column of the matrix represents the instances in an actual class, while each row represents the instances in a predicted class. Note that the sum of elements of each row equals to 100.

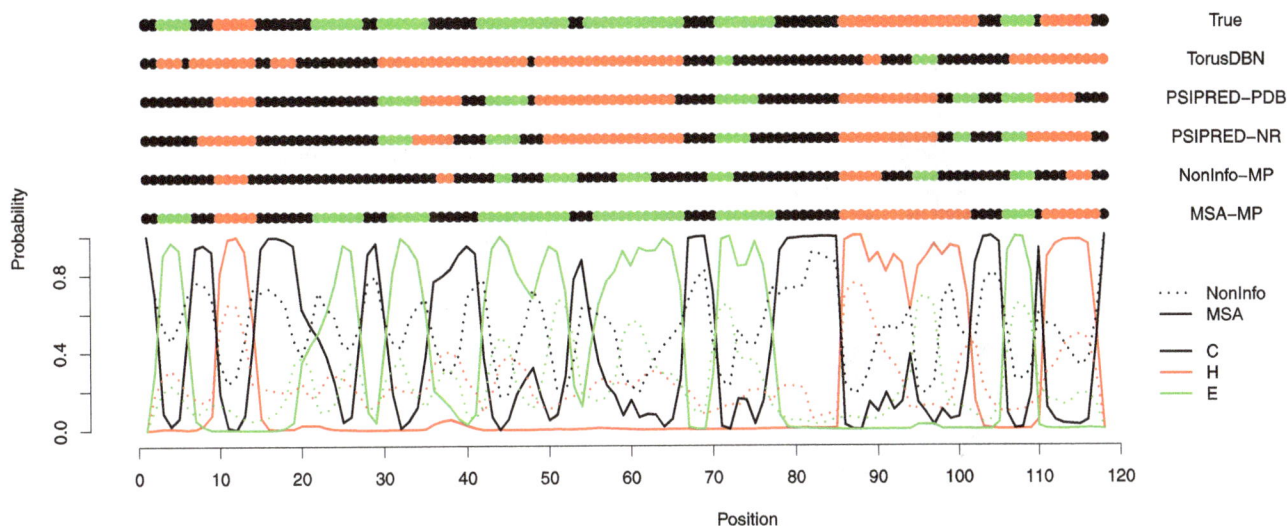

Figure 3. Marginal probability (MP) curves across positions for the phospholipase protein 3rvc [38]. Shown at the top is the true secondary structure, TorusDBN's prediction, PSIPREDs' prediction, and the prediction from our method (MP-MSA).

created from the targets used in CASP9 experiments in 2010 [41]. The CASP9 set includes 147 structures, and the same cleanup procedure produced 203 chains with 23,298 amino acids.

To compare the performance with TorusDBN [36] and PSIPRED [34], our method was trained with the older ASTRAL30 1.75 set [42,43]. The training set did not contain any chains from either the ASTRAL30 and CASP9 test sets described, and so is properly jack-knifed with regards to the test data. TorusDBN and PSIPRED predictions were carried out locally using downloaded copies of the programs. For TorusDBN, the backbone-dbn-torus predictor program was used without any additional input. The prediction with PSIPRED was carried out with a BLAST [44,45] search on two different databases. The NR is a full non-redundant sequence database with low-complexity regions filtered and the other nrPDB is the subset of sequences with determined structures in the PDB [37]. The BLAST search was also performed on a local computer with the downloaded program. Because of the large number of sequences in the NR database, the BLAST search took significantly longer than nrPDB. Multiple sequence alignments of the similar structures were obtained from the BLAST search with the nrPDB database for our MSA prediction. Also, in our prediction, the sequences in the ASTRAL30 and CASP9 were jack-knifed out of the nrPDB

database. This insured that our MSA based predictions had no information from the native sequence.

Notation

Let $\boldsymbol{a} = (a_1, \cdots, a_L)$ be an observed amino acid sequence, i.e., protein primary structure, where $a_l \in \mathcal{A} = \{A, R, N, D, C, E, Q, G, H, I, L, K, M, F, P, S, T, W, Y, V\}$ is a one-letter code denoting one of the 20 proteinogenic amino acids and L is the protein length. The secondary structure of a protein is the general form of its local segments, which we refer to as "block types". [8] proposed the Dictionary of Protein Secondary Structure (DSSP) for protein secondary structure with single letter codes. Although generalizations may be desirable, we consider the following 4 block types (in italics) from the original 8 structures defined in DSSP (in parentheses):

- Helix "*H*": 3_{10} helices (G), α-helices (H), or π-helices (I);
- Strand "*E*": extended strands in parallel or anti-parallel β-sheets (E);
- Turn "*T*": hydrogen bonded turns of length 3 or more amino acids (T);
- Coil "*C*": β-bridge residues (B), bends (S), or random coils (C).

Table 6. Means and standard deviations (in parenthesis) of differences in marginal probability between correctly predicted secondary structure (Correct) and the next highest probability, and between secondary structure predicted incorrectly (Wrong) and highest probability for ASTRAL30 and CASP9 data sets.

		Helix	Strand	Turn	Coil
ASTRAL30	Correct	0.71(0.25)	0.70(0.27)	0.59(0.29)	0.56(0.31)
	Wrong	0.34(0.26)	0.36(0.25)	0.39(0.27)	0.35(0.26)
CASP9	Correct	0.54(0.30)	0.53(0.30)	0.42(0.28)	0.45(0.34)
	Wrong	0.24(0.19)	0.25(0.19)	0.28(0.23)	0.28(0.22)

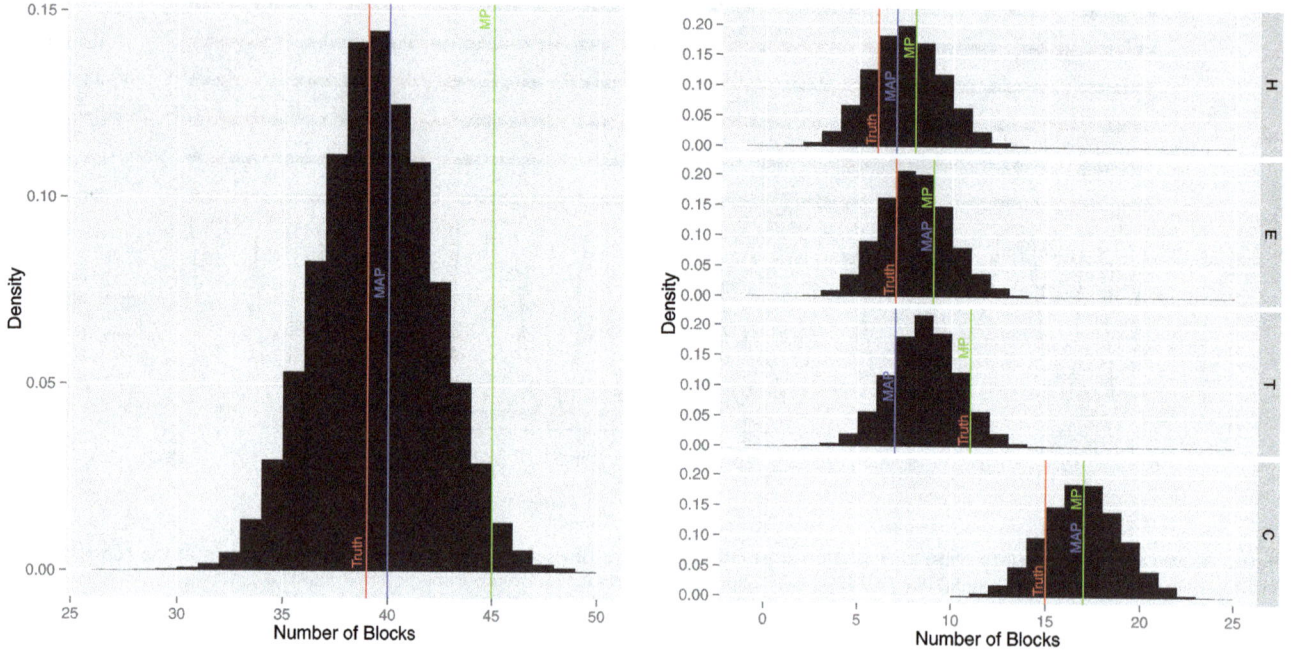

Figure 4. The posterior distribution of the number of blocks in total (left) and the number of blocks of each type (right) for protein T0622-D10 from the CASP9 data set. Also displayed is the number of the blocks in the truth, the MAP estimate, and the MP estimate in red, blue, and green color, respectively.

Let $\mathcal{S} = \{H,E,T,C\}$ denote the set of block types. The secondary structure can be encoded in a convenient fashion by representing the structural types and segment length $(\boldsymbol{\eta},\boldsymbol{\lambda}) = ((\eta_1,\lambda_1),\cdots,(\eta_M,\lambda_M))$, where $\eta_m \in \mathcal{S}$ gives the secondary structure type in the m-th block and λ_m gives the length of that block. Note that $\lambda_m \in \{1,\cdots,L\}$ and $\sum_{m=1}^{M} \lambda_m = L$. For example, Fig. 1 shows the representation of the secondary structure of the protein clathrin $1c9l$ [46].

In the case of secondary structure prediction, the quantities of interest are $\boldsymbol{\eta}$ and $\boldsymbol{\lambda}$ corresponding to the known amino acid sequence $\boldsymbol{a}$, i.e., the type and length of each secondary structural segment. The cumulative length also contains the segment location information. Thus, mathematically, the problem is to infer the values of $(\boldsymbol{\eta},\boldsymbol{\lambda})$ given the amino sequence $\boldsymbol{a}$.

Sampling Model

We start by considering the joint distribution of the data $\boldsymbol{a} = (a_1,\cdots,a_L)$ given the latent secondary structure, $(\boldsymbol{\eta},\boldsymbol{\lambda})$. We write the joint probability mass function (p.m.f.) $p(\boldsymbol{a}|\boldsymbol{\eta},\boldsymbol{\lambda})$ as a product over blocks:

$$p(\boldsymbol{a}|\boldsymbol{\eta},\boldsymbol{\lambda}) = \prod_{m=1}^{M} p_{\eta_m}(a_{i_m},\cdots,a_{j_m}), \qquad (1)$$

where $i_m = 1 + \sum_{m'<m} \lambda_{m'}$ is the starting position of the m-th block, $j_m = \sum_{m'\leq m} \lambda_{m'}$ is its ending position, and p_{η_m} is one of p_H, p_E, p_T, and p_C based on the value of $\eta_m \in \mathcal{S}$. By grouping portions of the sequence into blocks, our method leverages the natural property that secondary structure states are necessarily formed by groups of residues. Our method thus captures the local context or environment around a residue that influences its secondary structure state, which aids prediction accuracy in all three states. As described below, p_H, p_E, p_T, and p_C are designed to reflect the

protein three-dimensional local structure at the molecular level. (See Fig. 2.)

We evaluate the sampling model for each block as the product of position-specific marginal or conditional distributions estimated from the PDB. At each position, the sampling model for a single amino acid a is of the form:

$$p(a|X) = \int p(a|\boldsymbol{\theta})p(\boldsymbol{\theta}|X)d\boldsymbol{\theta}, \qquad (2)$$

where $X = (X_1,\ldots,X_{20})$ is the count vector for the number of times that each of the 20 amino acids is found in the training data (from the PDB) for the situation of interest. Specific situations are described in the following subsections and could be, for example, the start of a helical block or the third position in a strand with amino acids A then C proceeding it. When a is viewed as a vector of length 20 with all zeros except a single 1, $p(a|\boldsymbol{\theta})$ in (2) is a multinomial distribution with one trial and probability vector $\boldsymbol{\theta}$. Assume the following Bayesian model: $X|\boldsymbol{\theta} \sim \text{Multinomial}(n,\boldsymbol{\theta})$ and $\boldsymbol{\theta} \sim \text{Dirichlet}(1,\ldots,1)$, where $n = \sum_{k=1}^{20} X_k$. Due the conjugacy, the posterior distribution $p(\boldsymbol{\theta}|X)$ is $\boldsymbol{\theta}|X \sim \text{Dirichlet}(X_1+1,\ldots,X_{20}+1)$. The integration of the product of $p(a|\boldsymbol{\theta})$ and $p(\boldsymbol{\theta}|X)$ with respect to $\boldsymbol{\theta}$ makes $p(a|\boldsymbol{\theta})$ a Dirichlet-multinomial distribution. Because the number of trials is simply 1, evaluating $p(a|\boldsymbol{\theta})$ requires only that we divide one plus the number of times the amino acid a is present in the situation of interest in the training dataset by $n+20$.

Sampling Model for Helices. We propose that the sampling model for helices is defined by a product of four p.m.f.'s as follows:

$$p_H(a_i,\ldots,a_j) =$$

$$p_{H_1}(a_i)p_{H_2}(a_{i+1}|a_i)p_{H_3}(a_{i+2}|a_i,a_{i+1})p_{H_4}(a_{i+3}|a_{i+1},a_{i+2}) \times \qquad (3)$$

$$p_{H_5}(a_{i+4}|a_i,a_{i+1},a_{i+3}) \times \cdots \times p_{H_5}(a_j|a_{j-4},a_{j-3},a_{j-1}),$$

where p_{H_1} is a multinomial distribution with a category for each of the 20 amino acids, p_{H_2} is a 20-dimensional multinomial distribution conditioned on the antecedent amino acid, p_{H_3} and p_{H_4} are multinomial distributions conditioned on the two previous amino acids, and p_{H_5} is a 20-dimensional multinomial distribution conditioned on the previous amino acid, the amino acid three positions back, and the amino acid four positions back. In the case of a short helical block, terms beyond the length of the helix are simply ignored (i.e., $p_H(a_i, a_{i+1}, a_{i+2}) = p_{H_1}(a_i) p_{H_2}(a_{i+1}|a_i) p_{H_3}(a_{i+2}|a_i, a_{i+1})$ for a helix of three amino acids). This formulation for $p_H(a_i, \cdots, a_j)$ is tractable, yet still respects the biochemistry of helices, as shown in Fig. 2(a).

As previously explained, in our approach we evaluate the simpler p.m.f.'s in (3) based on training data. The 20-dimensional probability vector for p_{H_1} is taken to be the posterior mean from a Bayesian model assuming a multinomial sampling model and a non-informative Dirichlet prior with all hyperparameters equal to 1. The data for this estimation is obtained from the PDB by counting the number of helical blocks that start with each of the 20 amino acids. Similarly, since there are 20 amino acids on which to condition, there are 20 p.m.f.'s of type p_{H_2} and 20×20 p.m.f.'s of type p_{H_3}. Likewise, since there are $20 \times 20 \times 20 = 8,000$ combinations of three amino acids, there are 8,000 p.m.f.'s of type p_{H_4}. Again, these probability vectors for p_H are calculated from all the sequences in the PDB and stored for evaluating the likelihood for a helical block as in (2).

As described to this point, the sampling model for helix is a "forward" model in which the contribution of each amino acid is conditioned on previous amino acids. An important aspect of the biochemistry of each block is the existence of capping signals: the preference, through side chain-backbone hydrogen bonds or hydrophobic interactions, for particular amino acids at the N- and C-terminals of a helix. Usually, the terminal end is the first and last 3 or 4 positions in a block [47], whose effect is reflected by the amino acid distribution which significantly differ from that of internal positions. These signals have been characterized experimentally in terms of their stabilizing effect in helical peptides [47].

Whereas the forward model captures the capping signal in the N-terminus, we also consider a "backward" model. The backward model is the exact opposite of the forward model. It is built sequentially by starting at the C-terminus of the block and working backwards to the front, each time conditioning on amino acids closer to the C-terminus. Apart from the direction, the form of the conditioning is the same as the forward model. Thus, the sampling model for the helical blocks is a mixture model, composed of the forward component and the backward component as follows:

$$p_H(a_i, \cdots, a_j) = w_+ p_{H_+}(a_i, \cdots, a_j) + w_- p_{H_-}(a_j, \cdots, a_i),$$

where p_{H_+} is the forward model defined in (3) and $w_+ = w_- = 1/2$. A mixture model is not the only way to handle both capping ends and, for example, a single unified model would also be valid. We do not expect a major difference in performance among models that account for capping. As such, we propose the two-component mixture model for ease of exposition.

Sampling Model for Strands. We propose that the sampling model for strands, with joint p.m.f. $p_E(a_i, \cdots, a_j)$, is defined by a product of six simpler p.m.f.'s $p_{E_1}, p_{E_2}, p_{E_3}, p_{E_4}, p_{E_5}$, and p_{E_6} as follows:

$$
\begin{aligned}
p_E(a_i, \ldots, a_j) = & \; (p_{E1}(a_i) \, p_{E3}(a_{i+2}|a_i) \times \\
& p_{E5}(a_{i+4}|a_i, a_{i+2}) \times \cdots \times p_{E5}(a_j|a_{j-4}, a_{j-2})) \times \\
& (p_{E2}(a_{i+1}) \, p_{E4}(a_{i+3}|a_{i+1}) \times \\
& p_{E6}(a_{i+5}|a_{i+1}, a_{i+3}) \times \cdots \times p_{E6}(a_j|a_{j-4}, a_{j-2})),
\end{aligned}
\tag{4}
$$

where p_{E_1} and p_{E_2} are a multinomial distribution with a category for each of the 20 amino acids, p_{E_3} and p_{E_4} are 20-dimensional multinomial distributions conditioned on the value of the antecedent amino acid two positions back, etc. In the case of a short strand block, terms beyond the length of the strand are simply ignored (i.e., for a stand of length three, $p_E(a_i, a_{i+1}, a_{i+2}) = p_{E1}(a_i) p_{E3}(a_{i+2}|a_i) p_{E2}(a_{i+1})$. Again, this formulation $p_E(a_i, \cdots, a_j)$ is tractable, yet still respects the biochemistry of strands, as shown in Fig. 2(b). Note that $p_{E_1} \neq p_{H_1}$, despite the fact that both are marginal multinomial distributions. Likewise, $p_{E_3} \neq p_{H_2}$, despite the fact that both are conditional multinomial distributions given an amino acid. In particular, p_{E_1}, $p_{E_2}, p_{E_3}, p_{E_4}, p_{E_5}$, and p_{E_6} are estimated from PBD data involving strands, whereas $p_{H_1}, p_{H_2}, p_{H_3}, p_{H_4}$, and p_{H_5} are estimated from PBD data involving helices, but the estimation strategy is the same.

Sampling Model for Coil. We propose a sampling model for coils as the product of p.m.f.'s as follows:

$$
\begin{aligned}
p_C(a_i, \ldots, a_j) = & \; p_{C_1}(a_i) p_{C_2}(a_{i+1}|a_i) \times \\
& p_{C_3}(a_{i+2}|a_i, a_{i+1}) \times \cdots \times p_{C_3}(a_j|a_{j-2}, a_{j-1}).
\end{aligned}
\tag{5}
$$

In the case of a short coil block, terms beyond the length of the coil are simply ignored (i.e., for a coil of length two, $p_C(a_i, a_{i+1}) = p_{C_1}(a_i) p_{C_2}(a_{i+1}|a_i)$. Again the formulation respects the biochemistry of coils as shown in Fig. 2(c) and the sampling models are estimated from the PDB.

Sampling Model for Turn. According to the turn structure as shown in Fig. 2(d), we propose that the sampling model for turns be defined by a product of p.m.f.'s, as follows:

$$
\begin{aligned}
p_T(a_i, \ldots, a_j) = & \\
& p_{T_{31}}(a_i) p_{T_{32}}(a_{i+2}|a_i) p_{T_{33}}(a_{i+1}|a_i, a_{i+2}) \\
& \qquad\qquad\qquad\qquad\qquad\qquad \text{if } j-i+1=3; \\
& p_{T_{41}}(a_i) p_{T_{42}}(a_{i+3}|a_i) p_{T_{43}}(a_{i+1}|a_i, a_{i+3}) p_{T_{43}}(a_{i+2}|a_i, a_{i+3}) \\
& \qquad\qquad\qquad\qquad\qquad\qquad \text{if } j-i+1=4; \\
& p_{T_{51}}(a_i) p_{T_{52}}(a_{i+4}|a_i) p_{T_{53}}(a_{i+1}|a_i, a_{i+4}) p_{T_{53}}(a_{i+3}|a_i, a_{i+4}) p_{T_{54}}(a_{i+2}|a_{i+1}, a_{i+3}) \times \\
& p_{T_{55}}(a_{i+5}|a_{i+2}, a_{i+3}, a_{i+4}) \times \cdots \times p_{T_{55}}(a_j|a_{j-3}, a_{j-2}, a_{j-1}) \\
& \qquad\qquad\qquad\qquad\qquad\qquad \text{if } j-i+1 \geq 5,
\end{aligned}
\tag{6}
$$

where each conditional p.m.f. in the equation above is estimated based on the PDB data using hydrogen bonded turns of length 3, 4, and 5 or more amino acids, respectively.

Prior Distribution

The model is completed by specifying the prior distribution, with p.m.f. $p(\boldsymbol{\eta}, \boldsymbol{\lambda})$. First, we make the p.m.f. equal zero if the biochemistry inherent in secondary structure is violated. Specifically it is zero if, for $m = 1, \cdots, M$, any of the following conditions are met:

- $\eta_1 \neq C$ or $\eta_M \neq C$ (i.e., if it does not start and end in coil)

- $\eta_m = H, \eta_{m+1} = E$ (i.e., if helix is followed by strand)
- $\eta_m = E, \eta_{m+1} = H$ (i.e., if stand is followed by helix)
- $\lambda_m < 3$ and $\eta_m = H$ (i.e., if helix block is less than 3 positions)
- $\lambda_m < 3$ and $\eta_m = E$ (i.e., if strand block is less than 3 positions)
- $\lambda_m < 3$ and $\eta_m = T$ (i.e., if turn block is less than 3 positions).

The implies that helix, strand, coil, and turn blocks are at least 3, 3, 1, and 3 amino acids long, respectively. The first prior is a noninformative (NonInfo) prior and it provides equal weights to all the allowable secondary structures, that is,

$$p_{\text{NonInfo}}(\boldsymbol{\eta}, \boldsymbol{\lambda}) \propto 1,$$

for all $(\boldsymbol{\eta}, \boldsymbol{\lambda})$ except the above listed conditions.

We also consider an informative prior distribution which incorporates multiple sequences alignment (MSA) information. For an observed amino acid sequence $\boldsymbol{a}$, we first search for a set of proteins with similar amino acid sequences whose secondary structures is already known. The candidate database and the matching criterion are a modeling choice. For our analysis, we used a PSI-BLAST search of the nrPDB database. PSI-BLAST searches were performed on the local sever against the non-redundant protein sequence database with entries from GenPept, Swissprot, PIR, PDF, PDB and NCBI RefSeq, downloaded from NCBI website (ftp://ftp.ncbi.nih.gov/blast/db/nr.*). The low complexity sequence regions were filtered to avoid the artifactual hits. The structures with E-values better than 0.001 from the search were used in the alignments. Also, sequences already in the validation datasets (ASTRAL30 and CASP9) were excluded to insure that our MSA based predictions had no information from the native sequence. We build the prior distribution for $\boldsymbol{a}$ as the product of position-specific marginal distributions estimated from its corresponding alignment outputs. Let

$$\boldsymbol{X} = (\boldsymbol{X}_{\cdot 1} \quad \boldsymbol{X}_{\cdot 2} \quad \cdots \quad \boldsymbol{X}_{\cdot L}) = \begin{pmatrix} X_{H1} & X_{H2} & \cdots & X_{HL} \\ X_{E1} & X_{E2} & \cdots & X_{EL} \\ X_{C1} & X_{C2} & \cdots & X_{CL} \\ X_{T1} & X_{T2} & \cdots & X_{TL} \end{pmatrix},$$

where $\boldsymbol{X}_{\cdot j}$ is the count vector for the number of times that each of the four secondary structure types is found in the j-th position of the alignment output. Assume the following Bayesian model:

$$\boldsymbol{X}_{\cdot j} | \phi_{\cdot j} \sim \text{Multinomial}(n_j, \phi_{\cdot j}),$$

and

$$\phi_{\cdot j} \sim \text{Dirichlet}(\alpha_H, \alpha_E, \alpha_C, \alpha_T),$$

where n_j is the number of aligned sequences minus the number of times that gap is found in the j-th position. Due to the conjugacy, the posterior distribution is

$$\phi_{\cdot j} | \boldsymbol{X}_{\cdot j} \sim \text{Dirichlet}(X_{Hj} + \alpha_H, X_{Ej} + \alpha_E, X_{Cj} + \alpha_C, X_{Tj} + \alpha_T).$$

We suggest default values of $\alpha_H = \alpha_E = \alpha_C = \alpha_T = 1$. Let

$$\boldsymbol{\Phi} = (\phi_{\cdot 1} \quad \phi_{\cdot 2} \quad \cdots \quad \phi_{\cdot L}),$$

then we assume the secondary structure sequence follows a product of L p.m.f.'s, i.e.,

$$p_{\text{MSA}}(\boldsymbol{\eta}, \boldsymbol{\lambda}) = \prod_{m=1}^{M} \prod_{l=i_m}^{j_m} \phi_{\eta_m l},$$

where l indexes the position, $i_m = 1 + \sum_{m' < m} \lambda_{m'}$ is the starting position of the m-th block, and $j_m = \sum_{m' \le m} \lambda_{m'}$ is its ending position.

MCMC Algorithm

Our goal is to make inference on the secondary structure $(\boldsymbol{\eta}, \boldsymbol{\lambda})$ given the amino acid sequence $\boldsymbol{a}$. We use Markov chain Monte Carlo (MCMC) methods to sample from the posterior distribution:

$$p(\boldsymbol{\eta}, \boldsymbol{\lambda} | \boldsymbol{a}) \propto p(\boldsymbol{a} | \boldsymbol{\eta}, \boldsymbol{\lambda}) p(\boldsymbol{\eta}, \boldsymbol{\lambda}). \tag{7}$$

We update $(\boldsymbol{\eta}, \boldsymbol{\lambda})$ using a Metropolis algorithm. The factorization in (1) allows Hastings ratios to be evaluated locally with respect to the affected segments [48]. We note that this algorithm is sufficient to guarantee ergodicity for our model. In this algorithm, a new candidate $(\boldsymbol{\eta}^*, \boldsymbol{\lambda}^*)$ is generated according to the following scheme:

- Switch the type of a randomly chosen block: Randomly choose a number $m \in \{1, M\}$ and change the new m-th block type to $\eta_m^* \in \{H, E, T, C\} \setminus \{\eta_{m-1}, \eta_m, \eta_{m+1}\}$ with equal probability. Leave all other block types and lengths unchanged.
- Change the position of boundary between two blocks: Randomly choose a number $m \in \{1, M-1\}$ and draw the new m-th block length λ_m^* from $\text{Uniform}(1, \lambda_m + \lambda_{m+1} - 1)$ and hence the new $(m+1)-th$ block length λ_{m+1}^* equals to $\lambda_m + \lambda_{m+1} - \lambda_m^*$. Leave all block types and other lengths unchanged.
- Split a block into two adjacent blocks: Randomly choose a number $m \in \{1, M\}$. Make space for a new block to be placed between blocks m and $m+1$ as follows. Let $\eta_t^* = \eta_t$ and $\lambda_t^* = \lambda_t$ for $t = 1, \ldots, m-1$, and let $\eta_t^* = \eta_{t-1}$ and $\lambda_t^* = \lambda_{t-1}$ for $t = m+2, \ldots, M+1$. Let $\eta_m^* = \eta_m$. What remain to define are values for λ_m^*, η_{m+1}^*, and λ_{m+1}^*. Assign the new $(m+1)$-th block type to $\eta_{m+1}^* \in \{H, E, T, C\} \setminus \{\eta_m, \eta_{m+1}\}$ with equal probability. Draw the new m-th block length λ_m^* from $\text{Uniform}(1, \lambda_m - 1)$ and hence the new $(m+1)$-th block length λ_{m+1}^* equals to $\lambda_m - \lambda_m^*$.
- Merge two adjacent blocks into one block: Randomly choose a number $m \in \{1, M-1\}$. Let $\eta_t^* = \eta_t$ and $\lambda_t^* = \lambda_t$ for $t = 1, \cdots, m-1$, and let $\eta_t^* = \eta_{t+1}$ and $\lambda_t^* = \lambda_{t+1}$ for $t = m+1, \cdots, M-1$. Finally, let $\eta_m^* = \eta_m$ and $\lambda_m^* = \lambda_m + \lambda_{m+1}$.

The Hastings ratio can be written as:

$$r = \frac{p(\boldsymbol{\eta}^*, \boldsymbol{\lambda}^* | \boldsymbol{a})}{p(\boldsymbol{\eta}^{(t-1)}, \boldsymbol{\lambda}^{(t-1)} | \boldsymbol{a})} \frac{q(\boldsymbol{\eta}^{(t-1)}, \boldsymbol{\lambda}^{(t-1)}; \boldsymbol{\eta}^*, \boldsymbol{\lambda}^*)}{q(\boldsymbol{\eta}^*, \boldsymbol{\lambda}^*; \boldsymbol{\eta}^{(t-1)}, \boldsymbol{\lambda}^{(t-1)})},$$

where $q(\boldsymbol{\eta}^*, \boldsymbol{\lambda}^*; \boldsymbol{\eta}^{(t-1)}, \boldsymbol{\lambda}^{(t-1)})$ is the proposal density, the density for proposing a move to $(\boldsymbol{\eta}^*, \boldsymbol{\lambda}^*)$ given the previous state $(\boldsymbol{\eta}^{(t-1)}, \boldsymbol{\lambda}^{(t-1)})$ and $q(\boldsymbol{\eta}^{(t-1)}, \boldsymbol{\lambda}^{(t-1)}; \boldsymbol{\eta}^*, \boldsymbol{\lambda}^*)$ is the reverse case. The move is accepted $(\boldsymbol{\eta}^{(t)}, \boldsymbol{\lambda}^{(t)}) = (\boldsymbol{\eta}^*, \boldsymbol{\lambda}^*)$ with the probability min(1, r), otherwise, the move is rejected and $(\boldsymbol{\eta}^{(t)}, \boldsymbol{\lambda}^{(t)}) = (\boldsymbol{\eta}^{(t-1)}, \boldsymbol{\lambda}^{(t-1)})$.

For the results presented in this paper, 1,000,000 MCMC proposal were obtained (each time randomly selecting among the four proposal schemes described earlier). About half of the proposals lead to valid secondary structures. (For example, proposing to switch a block to helix is not valid if the block is already adjacent to an helix block.) Among the valid proposals, about 20% were accepted. The first 10,000 samples were discarded for burnin. MCMC convergence can be assessed by comparing the stability of the marginal probabilities of the states at each position across independent MCMC runs with different starting secondary structure states.

Posterior Estimation

The goal is to infer the secondary structure $(\boldsymbol{\eta}, \boldsymbol{\lambda})$. We considered two ways to summarize the posterior distribution to yield a point estimate. Among all samples obtained by the MCMC algorithm, choose the $(\boldsymbol{\eta}, \boldsymbol{\lambda})$ that maximizes the posterior probability $p(\boldsymbol{\eta}, \boldsymbol{\lambda}|\boldsymbol{a})$:

$$(\boldsymbol{\eta}, \boldsymbol{\lambda})_{\mathrm{MAP}} = \mathrm{argmax}_{\boldsymbol{\eta}, \boldsymbol{\lambda}} \; p(\boldsymbol{\eta}, \boldsymbol{\lambda}|\boldsymbol{a}).$$

We name this estimate as maximum *a posteriori* (MAP) estimate.

To describe the second posterior estimation method, it is convenient to introduce the *linear sequence* parameterization that encodes the secondary structure using a vector $\boldsymbol{\rho} = (\rho_1, \cdots, \rho_L)$, where $\rho_l \in S$ indicates the secondary structure at position l. This parameterization encodes the same information as the original parameterization $(\boldsymbol{\eta}, \boldsymbol{\lambda})$. We construct the estimate by selecting the most likely block type for each position:

$$\boldsymbol{\rho}_{\mathrm{MP}} = (\rho_1^{\mathrm{MP}}, \ldots, \rho_L^{\mathrm{MP}}),$$

where $\rho_l^{\mathrm{MP}} = Y^{\dagger}$ if $\Pr(\rho_l = Y^{\dagger}|\boldsymbol{a}) \geq \Pr(\rho_l = Y|\boldsymbol{a})$ for $Y, Y^{\dagger} \in S = \{H, E, T, C\}$ and $l = 1, \ldots, L$. We call estimates obtained in this manner marginal probability (MP) estimates.

Conclusions

A statistical model for knob-socket packing [30,31] between residues has been developed for prediction of protein secondary structure. The unique feature of this approach is that the knob-socket model provides constructs for the direct inclusion and prediction of the secondary states of coil and turn (Fig. 2(c) and (d), respectively). Other secondary structure prediction methods do not make direct prediction of coil structure and essentially apply indirect identification of coil residues as neither helix and sheet. We assess our method's Q3 prediction accuracy on 2 test sets and compare results with those obtained with the benchmark method PSIPRED [12]. From an investigation of the accuracy of prediction for each state, we found improved predictions adding context in terms of blocks of amino acids; however, our basic model over predicts the coil state. We show how incorporation of multiple sequence alignment data, similarly in spirit to PSIPRED, provides balance and improves prediction accuracy. Indeed, our method achieves slightly less accurate predictions than does PSIPRED on one test set, and almost reaches 90% on the other. Our results reinforce the general concept that more context is necessary to understand the environment that induces secondary structure.

Our work takes the initial step to enable Bayesian method to infer the secondary structure of proteins and serves as a call for participation. Many interesting and important directions are worth exploring. For example, our work is limited in the sense that only considers local dependency. We are exploring several ways of incorporating non-local information in future work. This may be especially beneficially improving strand predictions. Another interesting line of research is how to borrow information across probability vectors in the sampling models to improve the algorithm performance.

Author Contributions

Conceived and designed the experiments: QL DBD MV HJ JWT. Performed the experiments: QL DBD MV HJ JWT. Analyzed the data: QL DBD MV HJ JWT. Wrote the paper: QL DBD MV HJ JWT.

References

1. Adams PD, Baker D, Brunger AT, Das R, DiMaio F, et al. (2013) Advances, interactions, and future developments in the cns, phenix, and rosetta structural biology software systems. Biophysics 42.
2. Pirovano W, Heringa J (2010) Protein secondary structure prediction. In: Data Mining Techniques for the Life Sciences, Springer. pp. 327–348.
3. Sleator RD (2012) Prediction of protein functions. In: Functional Genomics, Springer. pp. 15–24.
4. Das R, Baker D (2008) Macromolecular modeling with rosetta. Annual Review of Biochemistry 77: 363–382.
5. Kiss G, Çelebi-Ölçüm N, Moretti R, Baker D, Houk K (2013) Computational enzyme design. Angewandte Chemie International Edition 52: 5700–5725.
6. Winter C, Henschel A, Tuukkanen A, Schroeder M (2012) Protein interactions in 3d: From interface evolution to drug discovery. Journal of Structural Biology 179: 347–358.
7. Rost B (2001) Review: protein secondary structure prediction continues to rise. Journal of Structural Biology 134: 204–218.
8. Kabsch W, Sander C (1983) Dictionary of protein secondary structure: pattern recognition of hydrogen-bonded and geometrical features. Biopolymers 22: 2577–2637.
9. Rost B, Sander C, Schneider R (1994) Phd–an automatic mail server for protein secondary structure prediction. Computer Applications in the Biosciences: CABIOS 10: 53–60.
10. Rost B, Yachdav G, Liu J (2004) The predictprotein server. Nucleic Acids Research 32: W321–W326.
11. Cole C, Barber JD, Barton GJ (2008) The jpred 3 secondary structure prediction server. Nucleic Acids Research 36: W197–W201.
12. Buchan DW, Minneci F, Nugent TC, Bryson K, Jones DT (2013) Scalable web services for the psipred protein analysis workbench. Nucleic Acids Research 41: W349–W357.

13. Bettella F, Rasinski D, Knapp EW (2012) Protein secondary structure prediction with sparrow. Journal of Chemical Information and Modeling 52: 545–556.
14. Yasee A, Li Y (2014) Context-based features enhance protein secondary structure prediction accuracy. Journal of Chemical Information and Modeling.
15. Aydin Z, Altunbasak Y, Borodovsky M (2006) Protein secondary structure prediction for a single-sequence using hidden semi-markov models. BMC Bioinformatics 7: 178.
16. Yao XQ, Zhu H, She ZS (2008) A dynamic bayesian network approach to protein secondary structure prediction. BMC Bioinformatics 9: 49.
17. Malekpour SA, Naghizadeh S, Pezeshk H, Sadeghi M, Eslahchi C (2009) A segmental semi markov model for protein secondary structure prediction. Mathematical Biosciences 221: 130–135.
18. Guo J, Chen H, Sun Z, Lin Y (2004) A novel method for protein secondary structure prediction using dual-layer svm and profiles. PROTEINS: Structure, Function, and Bioinformatics 54: 738–743.
19. Nguyen MN, Rajapakse JC (2004) Two-stage multi-class support vector machines to protein secondary structure prediction. In: Pacific Symposium on Biocomputing. Pacific Symposium on Biocomputing. pp. 346–357.
20. Zhang H, Zhang T, Chen K, Kedarisetti KD, Mizianty MJ, et al. (2011) Critical assessment of high-throughput standalone methods for secondary structure prediction. Briefings in Bioinformatics 12: 672–688.
21. Lim V (1974) Algorithms for prediction of α-helical and β-structural regions in globular proteins. Journal of Molecular Biology 88: 873–894.
22. Cohen FE, Abarbanel RM, Kuntz I, Fletterick RJ (1983) Secondary structure assignment for α/β proteins by a combinatorial approach. Biochemistry 22: 4894–4904.
23. Figureau A, Soto M, Toha J (2003) A pentapeptide-based method for protein secondary structure prediction. Protein Engineering 16: 103–107.

24. Birzele F, Kramer S (2006) A new representation for protein secondary structure prediction based on frequent patterns. Bioinformatics 22: 2628–2634.

25. Feng Y, Luo L (2008) Use of tetrapeptide signals for protein secondary-structure prediction. Amino Acids 35: 607–614.

26. Montgomerie S, Sundararaj S, Gallin WJ, Wishart DS (2006) Improving the accuracy of protein secondary structure prediction using structural alignment. BMC Bioinformatics 7: 301.

27. Mooney C, Pollastri G (2009) Beyond the twilight zone: automated prediction of structural properties of proteins by recursive neural networks and remote homology information. Proteins: Structure, Function, and Bioinformatics 77: 181–190.

28. Madera M, Calmus R, Thiltgen G, Karplus K, Gough J (2010) Improving protein secondary structure prediction using a simple k-mer model. Bioinformatics 26: 596–602.

29. Bondugula R, Wallqvist A, Lee MS (2011) Can computationally designed protein sequences improve secondary structure prediction? Protein Engineering Design and Selection 24: 455–461.

30. Joo H, Chavan AG, Phan J, Day R, Tsai J (2012) An amino acid packing code for α-helical structure and protein design. Journal of Molecular Biology 419: 234–254.

31. Joo H, Tsai J (2014) An amino acid code for β-sheet packing structure. Proteins: Structure, Function, and Bioinformatics.

32. Crick FH (1953) The packing of α-helices: simple coiled-coils. Acta Crystallographica 6: 689–697.

33. Chothia C, Levitt M, Richardson D (1977) Structure of proteins: packing of α-helices and pleated sheets. Proceedings of the National Academy of Sciences 74: 4130–4134.

34. Jones DT (1999) Protein secondary structure prediction based on position-specific scoring matrices. Journal of Molecular Biology 292: 195–202.

35. Rost B (1996) Phd: predicting 1d protein structure byprofile based neural networks. Methods Enzymol 266: 525–539.

36. Boomsma W, Mardia KV, Taylor CC, Ferkinghoff-Borg J, Krogh A, et al. (2008) A generative, probabilistic model of local protein structure. Proceedings of the National Academy of Sciences 105: 8932–8937.

37. Berman HM, Westbrook J, Feng Z, Gilliland G, Bhat T, et al. (2000) The protein data bank. Nucleic Acids Research 28: 235–242.

38. Kerry PS, Long E, Taylor MA, Russell RJ (2011) Conservation of a crystallographic interface suggests a role for-sheet augmentation in influenza virus ns1 multifunctionality. Acta Crystallographica Section F: Structural Biology and Crystallization Communications 67: 858–861.

39. Chandonia JM, Hon G, Walker NS, Conte LL, Koehl P, et al. (2004) The astral compendium in 2004. Nucleic Acids Research 32: D189–D192.

40. Fox NK, Brenner SE, Chandonia JM (2013) Scope: Structural classification of proteinsextended, integrating scop and astral data and classification of new structures. Nucleic Acids Research.

41. Moult J, Fidelis K, Kryshtafovych A, Tramontano A (2011) Critical assessment of methods of protein structure prediction (casp)round ix. Proteins: Structure, Function, and Bioinformatics 79: 1–5.

42. Andreeva A, Howorth D, Brenner SE, Hubbard TJ, Chothia C, et al. (2004) Scop database in 2004: refinements integrate structure and sequence family data. Nucleic Acids Research 32: D226–D229.

43. Andreeva A, Howorth D, Chandonia JM, Brenner SE, Hubbard TJ, et al. (2008) Data growth and its impact on the scop database: new developments. Nucleic Acids Research 36: D419–D425.

44. Altschul SF, Gish W, Miller W, Myers EW, Lipman DJ (1990) Basic local alignment search tool. Journal of Molecular Biology 215: 403–410.

45. Zhang J, Madden TL (1997) Powerblast: a new network blast application for interactive or automated sequence analysis and annotation. Genome Research 7: 649–656.

46. ter Haar E, Harrison SC, Kirchhausen T (2000) Peptide-in-groove interactions link target proteins to the β-propeller of clathrin. Proceedings of the National Academy of Sciences 97: 1096–1100.

47. Schmidler SC, Liu JS, Brutlag DL (2000) Bayesian segmentation of protein secondary structure. Journal of Computational Biology 7: 233–248.

48. Schmidler SC, Liu JS, Brutlag DL (2002) Bayesian protein structure prediction. In: Case Studies in Bayesian Statistics, Springer. pp. 363–378.

9

In Vitro, In Silico and *In Vivo* Studies of Ursolic Acid as an Anti-Filarial Agent

Komal Kalani[1,5], **Vikas Kushwaha**[2], **Pooja Sharma**[3], **Richa Verma**[2], **Mukesh Srivastava**[4], **Feroz Khan**[3,5], **P. K. Murthy**[2*], **Santosh Kumar Srivastava**[1,5*]

1 Medicinal Chemistry Department, CSIR-Central Institute of Medicinal and Aromatic Plants, Lucknow, 226015 (U.P.) India, **2** Division of Parasitology, CSIR-Central Drug Research Institute, Lucknow, 226001, UP, India, **3** Metabolic & Structural Biology Department, CSIR-Central Institute of Medicinal and Aromatic Plants, Lucknow, 226015 (U.P.) India, **4** Clinical and Experimental Medicine, Biometry section, CSIR-Central Drug Research Institute, Lucknow, 226001, UP, India, **5** Academy of Scientific and Innovative Research (AcSIR), Anusandhan Bhawan, New Delhi, 110 001, India

Abstract

As part of our drug discovery program for anti-filarial agents from Indian medicinal plants, leaves of *Eucalyptus tereticornis* were chemically investigated, which resulted in the isolation and characterization of an anti-filarial agent, ursolic acid (UA) as a major constituent. Antifilarial activity of UA against the human lymphatic filarial parasite *Brugia malayi* using *in vitro* and *in vivo* assays, and *in silico* docking search on glutathione-s-transferase (GST) parasitic enzyme were carried out. The UA was lethal to microfilariae (mf; LC$_{100}$: 50; IC$_{50}$: 8.84 μM) and female adult worms (LC$_{100}$: 100; IC50: 35.36 μM) as observed by motility assay; it exerted 86% inhibition in MTT reduction potential of the adult parasites. The selectivity index (SI) of UA for the parasites was found safe. This was supported by the molecular docking studies, which showed adequate docking (LibDock) scores for UA (−8.6) with respect to the standard antifilarial drugs, ivermectin (IVM −8.4) and diethylcarbamazine (DEC-C −4.6) on glutathione-s-transferase enzyme. Further, *in silico* pharmacokinetic and drug-likeness studies showed that UA possesses drug-like properties. Furthermore, UA was evaluated *in vivo* in *B. malayi-M. coucha* model (natural infection), which showed 54% macrofilaricidal activity, 56% female worm sterility and almost unchanged microfilaraemia maintained throughout observation period with no adverse effect on the host. Thus, in conclusion *in vitro, in silico* and *in vivo* results indicate that UA is a promising, inexpensive, widely available natural lead, which can be designed and developed into a macrofilaricidal drug. To the best of our knowledge this is the first ever report on the anti-filarial potential of UA from *E. tereticornis*, which is in full agreement with the Thomson Reuter's 'Metadrug' tool screening predictions.

Editor: Gnanasekar Munirathinam, University of Illinois, United States of America

Funding: The funding sources are CSIR-network project BSC-0121 and CSIR-SPLENDID and the funders had no role in study design, data collection and analysis, decision to publish, or preparation of the manuscript.

Competing Interests: The authors have declared that no competing interests exist.

* Email: skscimap@gmail.com (SKS); drpkmurthy@gmail.com (PKM)

Introduction

Among the six neglected tropical diseases, lymphatic filariasis (LF) is one of the major health problems in 73 tropical and subtropical countries in Africa, Asia, South and Central America and the Pacific Islands. According to the World Health Organization (WHO) global report, over 120 million people are currently infected with LF [1,2] of which about 40 million people are suffering with chronic disease manifestations: Elephantiasis and hydrocele [3], which cause permanent, long-term disability and economic loss to the nations [3,4]. The LF is caused by the nematode parasites *Brugia malayi, B. timori* and *Wuchereria bancrofti* and according to a recent report about 1 billion people (18% of the world's population) are at risk of infection (www.globalnetwork.org). Although, the World Health Organization launched a global filariasis elimination programme [5,6] using diethylcarbamazine (DEC) or ivermectin (IVM), but due to serious technical difficulties the programme is facing problem in the eradication of this endemic disease [4,7–8]. Since, DEC and IVM both are microfilaricides with poor or no activity on adult parasites

[9], the peripheral blood microfilaremia reappears in patients after a certain period of withdrawal of the drug. This depressing perspective demands, an urgent need for new molecular structures associated with macrofilaricidal activity/or sterilizing the adult worms is therefore needed [8–10] as adult parasites not only produce millions of microfilariae (mf) that are picked up by the mosquito vector and transmitted, but are also responsible for the debilitating pathological lesions. Therefore, macrofilaricidal agents are the need of hour, which not only adversely affect the target but should have also very low or no side effect [11].

As a part of our drug discovery program, we recently reported a pentacyclic triterpenoid, glycyrrhetinic acid [9] as a novel class of anti-filarial agent. This prompted us to investigate anti-filarial activity in other pentacyclic triterpenoids, widely available in Indian medicinal plants. For this purpose, in the present study leaves of *Eucalyptus tereticornis* were chemically and biologically investigated in details, which afforded an anti-filarial agent, Ursolic acid (UA, a pentacyclic triterpenoid) as a major constituent. The *in-vitro* activity of UA against the mf and adult worms, *in-silico* docking studies on glutathione-s-transferase (GST)

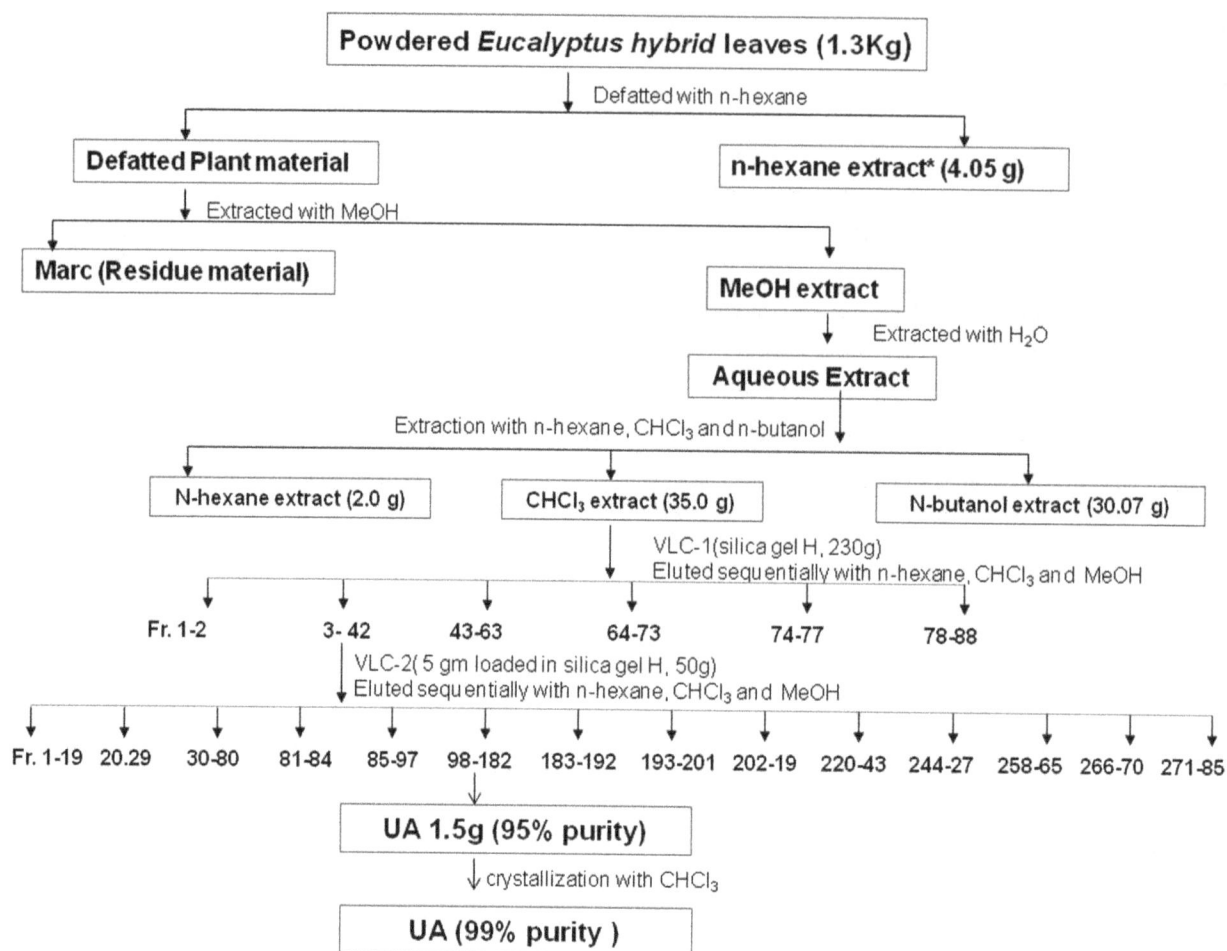

Figure 1. The schematic extraction and fractionation of UA from the leaves of *E. tereticornis*. §Washed with water and the solvent was dried over anhydrous Na_2SO_4. *Solvent was completely removed under vacuum at 35°C on a Buchi Rota vapour.

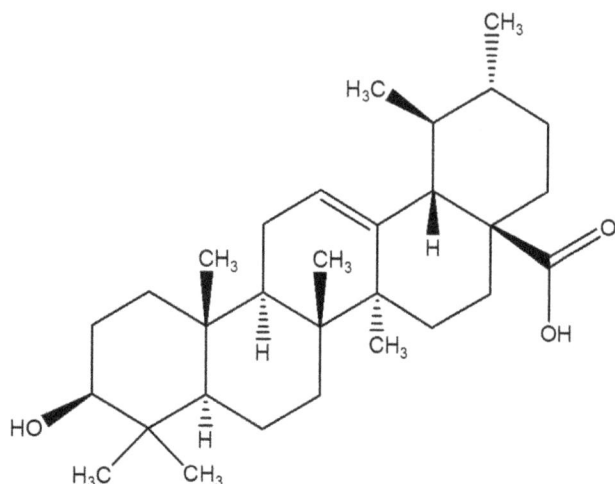

Figure 2. 2D structure of Ursolic Acid (UA).

parasitic enzyme and *in vivo* activity against *B. malayi in Meriones unguiculatus* model have been discussed here in detail.

Materials and Methods

General experimental procedure

The ^1H and ^{13}C NMR spectra were recorded on a Bruker 300 MHz spectrometer in deuterated pyridine. ESI-MS was carried out on a LCMS-2010 V (Shimadzu, Kyoto, Japan) simultaneously in positive (detector voltage 1.6 KV) ionization under scan mode. The scan speed of the mass analyzer was 2000 m/z per sec within the range of 400–1000 m/z. A positive full scan mode for screening and library assisted identification was used whereas time schedule selected-ion mode (SIM) in +ve ionization mode of the characteristic abundant adduct ions. Purity of UA was assessed by HPLC and was ≥95% [12]. Chemical shifts are in ppm with reference (internal) to tetramethylsilane (TMS) and J values are in hertz. With the Dept pulse sequence, different types of carbons (C, CH, CH_2 & CH_3) in UA were determined. The vacuum liquid chromatographic separations (VLC) were carried out on TLC grade Silica gel H (average particle size 10 μm) purchased from Merck, (Mumbai, India). All the required solvents and reagents were purchased from Spectrochem (Mumbai, India) and Thomas Baker Pvt. Ltd., India. Pre-coated Silica gel (60F) TLC plates 2.5 mm (Merck) were used to determine the

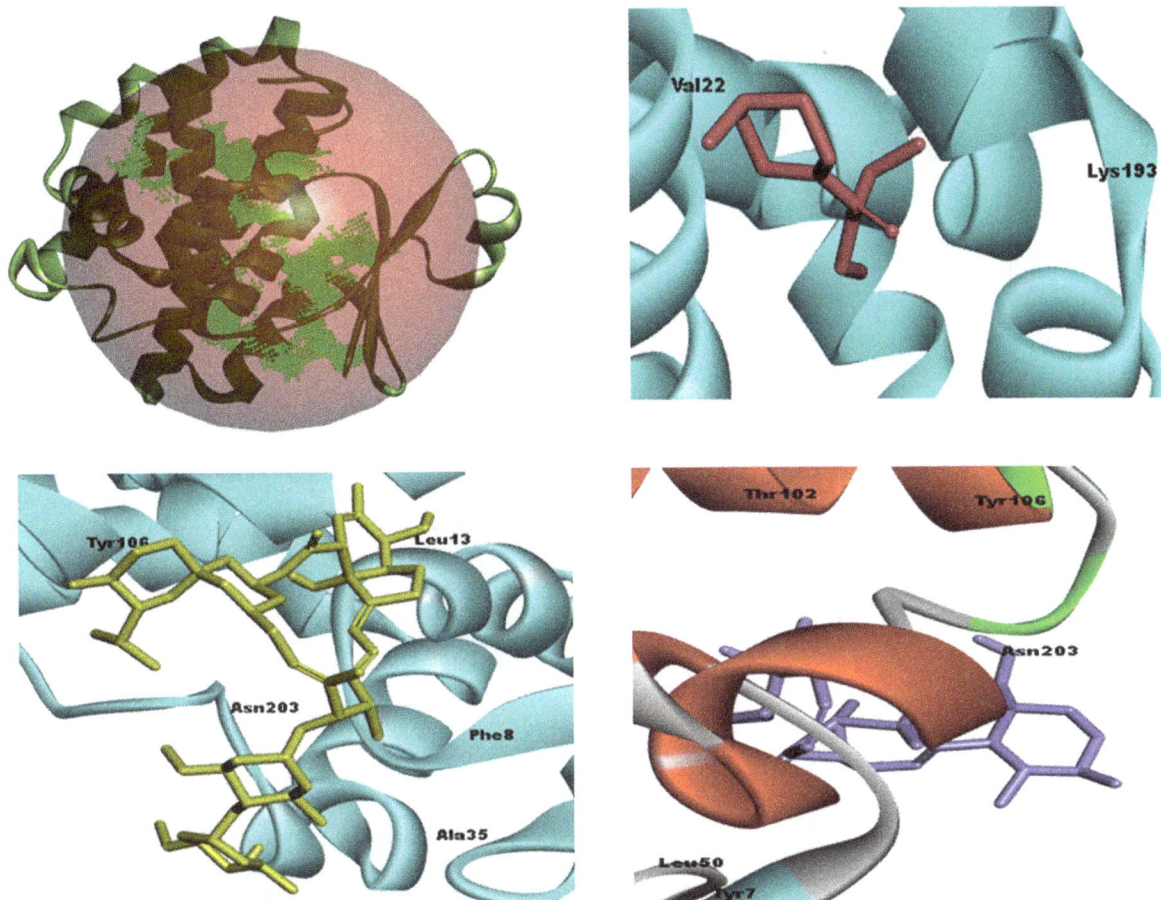

Figure 3. Docking results of studies compounds on *B. malayi* **(Filarial nematode worm) glutathione-S-transferase (***Bm***GST) homology model.** (a) docked standard drug DEC-c (control) on BmGST model active site with docking energy -4.9 kcal mol^{-1}, (b) docked another standard drug Ivermectin (control) with docking energy -8.4 kcal mol^{-1}, (c) docked UA on BmGST model with high docking energy -8.6 kcal mol^{-1}.

profiles of VLC fractions and their purity. The developed TLC plates were first observed at 254 nm in UV and then sprayed with Bacopa reagent [vanillin-ethanol sulphuric acid (1 g: 95 ml: 5 ml)] and spots were visualized after heating the TLC plate at 110°C for 5 minutes.

Plant material

The leaves of *E. tereticornis* were collected from the medicinal farm of Central Institute of Medicinal and Aromatic Plants (CIMAP), Lucknow, Uttar Pradesh, India during the month of January, 2008. A voucher specimen # 12470 was deposited in the Herbarium section of the Botany and Pharmacognosy Department of the institute.

The air dried leaves of *E. tereticornis* (1.3 kg) were powdered and defatted with n-hexane. The defatted leaves were further extracted with MeOH (4×5 L) (Figure 1).The combined MeOH extract was dried under vacuum at 40°C. The MeOH extract so obtained was dissolved in distilled water (2L) and successively fractionated with *n*-hexane, CHCl$_3$ and *n*-BuOH (saturated with H$_2$O) [12,13]. All the fractions were evaluated for anti-filarial activity, of which CHCl$_3$ fraction (35.0 g) was found active hence subjected for chromatographic separation over VLC-1 using silica gel H (260 g). The gradient elution of VLC was carried out with mixture of hexane, CHCl$_3$ and MeOH in increasing order of polarity.

Fractions 3–42 (7.2 g) eluted with hexane- CHCl$_3$ (1:1) to CHCl$_3$ – MeOH (99:1) was a complex mixture. Hence a part of it (5 g) was further chromatographed over VLC-2, using TLC grade silica gel H (50 g). Gradient elution of VLC-2 was carried out with mixture of hexane, CHCl$_3$ and MeOH in increasing order of polarity. Fractions 175–182 (1.5 g) eluted with CHCl$_3$ (100%) afforded a white amorphous compound (95% pure) which on further crystallization with CHCl$_3$ yielded UA (99% pure.). The ^{1}H and ^{13}C NMR and ESI-MS spectra of the homogenous compound (UA) were recorded and the spectroscopic data are presented as below:

ESI-MS m/z 457 [M+H]$^+$, C$_{30}$H$_{48}$O$_3$, ^{1}H NMR (300 MHz, Pyridine): δ 0.77, 0.78, 0.98, 1.09, 1.14 (3H each, all s, 5 x tert. Me) 0.92 & 0.96 (3H each, each d, J = 6.4 and 7.3 Hz, 2 x sec Me), 2.82 (1H, d, J = 9.9 Hz, H-18 β), 3.20 (1H, dd, J = 6.8 & 8.7 Hz, H-3α). ^{13}C NMR (75.5 MHz, Pyridine): C1- 39.5 (t), C2- 28.3 (t), C3- 78.7(d), C4- 39.9 (q), C5- 56.3 (d), C6- 19.1 (t), C 7- 33.9 (t), C8- 40.4 (q), C9- 47.0 (d), C10- 37.7 (q), C11- 23.9 (t), C12- 126.0 (d), C13- 139.6 (q), C14- 42.9 (q), C15- 28.9 (t), C16- 25.2 (t), C17- 48.5 (q), C18- 54.0 (d), C19- 30.5 (d), C20- 39.7 (d), C21- 31.3 (t), C22- 37.6 (t), C23- 29.0 (s), C24-15.8 (s), C25- 16.4 (s), C26- 17.5 (s), C27- 24.1 (s), C28- 179.7 (q), C29- 17.7 (s), C30- 21.4 (s) (Figure 2).

Table 1. *In vitro* activity of chloroform extract of *E. tereticornis*, its main constituent Ursolic Acid (UA) and reference drugs ivermectin and DEC on microfilariae and female adult worms of *B. malayi*.

Anti-filarial agent	Effect on female adult worm			Effect on microfilariae (Mf)		CC_{50}^c (μM)	SI	
	LC100^a (μM) in motility assay (% inhibition)	IC_{50}^b (μM) in motility assay	Mean % inhibition in MTT	LC100 (μM) in motility assay (% inhibition)	IC_{50}^b (μM) in motility assay		w.r.t. motility of Adults	w.r.t. motility of Mf
$CHCl_3$ extract	>100	-	8.55	>100	-	-	-	-
UA	100 (100)	35.36	86.12	25 (90)	8.84	300	8.48	33.94
IVM	5	3.05	5.80	2.5	1.57	250	81.96	159.23
DEC-c*	1000 (100)	314.98	62.54	500 (100)	297.30	8926	28.34	30.02

^aLC100 = 100% reduction in motility indicates death of parasite;
^bIC_{50} = 50% concentration of the agent at which 50% inhibition in motility is achieved;
^cCC_{50} = concentration at which 50% of cells are killed; SI = Selectivity Index (CC_{50}/IC_{50}); w.r.t. = with respect to; *Diethylcarbamazine citrate.

In vitro evaluation of UA/drugs against filarial parasites

Animals: The study was approved by the Institute's Animal Ethics Committee (IAEC) [approval no. 86/09/Para/IAEC; 27/4/09] of CSIR-Central Drug Research Institute, Lucknow, India, under the provisions of CPCSEA (Committee for the Purpose of Control and Supervision on Experiments on Animals), Government of India. All the experiments in animals were conducted in compliance with the IAEC guidelines for use and handling of animals. Throughout the study, jird and *M. coucha* were kept in climate ($23\pm2°C$; RH: 60%) and photoperiod (12hr light-dark cycles) controlled animal room. They were fed standard rodent chow supplemented with dried shrimps (*M. coucha*) and had free access to drinking water.

B. malayi infection in animals: The human sub-periodic strain of *B. malayi* was cyclically maintained in *M. coucha* [14] and jirds (*Meriones unguiculatus*) [15] through black-eyed susceptible strain of *Aedes aegypti* mosquitoes. Infective larvae of *B. malayi* isolated from experimentally infected *A. aegypti* mosquitoes which were fed on microfilaraemic *M. coucha* (150–200 mf/10 μl blood), were washed thoroughly with insect saline (0.6%). Each animal was inoculated with 100 (*M. coucha*) or 200 L3 (jirds), through subcutaneous (s.c.) and intraperitonial (i.p.) routes, respectively.

Isolation of parasites: Mf and adult worms (female parasites) isolated freshly from peritoneal cavity (p.c.) of jirds harboring 5–6 month old *B. malayi* infection were washed thoroughly in medium Hanks Balanced Salt Solution (HBSS; pH 7.2) containing mixture of antibiotics (penicillin: 100 U/mL; streptomycin: 100 μg/mL) and used for the present study.

In vitro anti-filarial efficacy evaluation

Primary evaluation. *In vitro* assays: Based on viability of the parasites, two *in vitro* motility and 3-(4, 5-dimethylthiazol-2-yl)-2,5 diphenyltetrazolium bromide (MTT) reduction assays [16] were carried out for UA. IVM and Diethylcarbamazine-citrate (DEC-C) were used as reference drugs. Incubation medium used was HBSS; pH 7.2 containing mixture of antibiotics as above. For incubation of mf and adult worms cell culture plate (Nunc, Denmark) were used.

The UA and IVM were dissolved in DMSO whereas DEC-C was prepared in sterile triple distilled water (STDW). The antifilarial agents were used at 2-fold serial dilutions ranged from 15.63–1000 μM (DEC), 1.56–100 μM (UA) and 0.31–20 μM (IVM). The final conc. of DMSO in the incubation medium was kept below 0.1%. DMSO (<0.1%) was used in place of test agents solution for control.

Motility assay: Efficacy of the UA and reference drugs was assessed *in vitro* on mf and adult worms of *B. malayi* (as target parasites) using motility (Mf and adult parasite) and MTT (adult parasite only) reduction assay [16,17]. Duplicate wells containing 40–50 mf/100 μl/well (of 96 well plates) and 1 female worm/ml/well (of 48-well plate) were used. UA (100 μM) or reference drugs IVM (20 μM), or DEC-C (1000 μM) were added to duplicate wells and incubated. Wells with the test compound and DEC were incubated for 24 hr and those with IVM were incubated for 24 and 48 hr as it has a slow action on the parasites. This is the standard protocol followed in our lab [16,17,18]. All incubations were at 37°C in 5% CO_2 atmosphere. The effect on motility of the parasite stages was examined under microscope and scored. The experiment was repeated twice. In case of mf, only motility assay was used.

Motility assessment: Parasite motility was assessed under a microscope after 24/48 h exposure to test substance and scored as: 0 = dead; 1–4 = loss of motility (1 = 75%; 2 = 50%; 3 = 25% and 4 = no loss of motility). Loss of motility is defined as the inability of

Table 2. Details of Docking energy, active site pocket residues and H-bonds revealed by molecular docking of DEC, IVM and UA on BmGST of *B. malayi*.

S. No.	Receptor	Anti-filarial agent	Binding Affinity (kcal/mol)	Interacting Residues	No of H-bonds
1	1SJO	DEC *	-4.9	VAL-22, ILE-26, LYS-189, GLU-190, LYS-193, ARG-195	2.9 = LYS-193
2	1SJO	IVM*	-8.4	PHE-8, LEU-13, ASN-34, ALA-35, LEU-50, TYR-106, ASN-203, ASN-205	2.7 = ASN-203 3.1 = TYR-106
3	1SJO	UA	-8.6	TYR-7, PHE-8, LEU-13, GLN-49, LEU-50, THR-102, TYR-106, ASN-203	3.0 = TYR-106

the worms to regain pretreatment level of motility even after incubating in fresh medium *minus* the test agent at 37°C for 1 h. and was expressed as percentage (%) inhibition of control.

MTT- formazan colorimetric assay for viability of worms: The same female worms used in motility were then gently blotted and transferred to 0.1 ml of 0.5% MTT in 0.01 M phosphate-buffered saline (pH 7.2) and incubated for 1 h at 37°C. The formazan formed was extracted in 1 ml of DMSO for 1 h at 37°C and its absorbance was measured at 510 nm in spectrophotometer (PowerWaveX, USA). The mean absorbance value obtained from 4 treated worms was compared with the controls. The viability of the treated worms was assessed by calculating per cent inhibition in motility and MTT reduction over DMSO control worms [16].

Criteria for assessment of *in vitro* hits: 100% inhibition in motility of female adults or mf and or ≥50% inhibition in MTT reduction ability of female parasites was considered acceptable antifilarial (microfilaricidal/adulticidal) activity and picked up as hits and subjected to further testing *in vivo* [17].

Secondary evaluation. Determination of IC$_{50}$: For IC$_{50}$ (the concentration at which the parasite motility was inhibited by 50%) determination of the parasites were incubated with two fold serial dilutions from 1.56–100 (UA), 0.31–40 (IVM) and 15.63–1000 μM (DEC-C) using triplicate wells of cell culture plate. Experiments were run in duplicate and incubations were carried out in replicates for 24/48 hr as above. After incubation, inhibition in motility (mf and female worm) and MTT reduction potential of the parasites were assessed as above. The experiment was repeated twice.

Determination of Cytotoxic concentration 50 (CC$_{50}$): The cytotoxicity assay of the test substances was carried out broadly following the method of Pagé et al. [19] with some modifications [20]. Briefly, VERO Cell line C1008 (African green monkey kidney cells) was plated in 96-well plates (Nunc, Denmark) at 0.1×10^6 cells/ml (100 μl per well) in DMEM supplemented with 10% heat inactivated FBS. A three-fold serial dilution of the test substances (starting from >20 x LC100 conc. of the test agent) in test medium was added. The plates with a final volume of 100 μl/well were incubated in 5% CO$_2$ atmosphere at 37°C. After 72 h

incubation 10 μl of 0.025% Resazurin in phosphate buffered saline (PBS; pH 7.2) was dispensed as indicator for viability followed by an additional incubation for 4 h and the plate was then read in a fluorescence reader (Synergy HT plate reader, Biotek, USA) at excitation wavelength of 530 nm and an emission wavelength of 590 nm. The assay was run in replicates in each of two independent experiments.

Data of IC$_{50}$ and CC$_{50}$ transferred to a graphic program (Excel) were calculated as described by Page et al. [19] and Mosmann [20] by linear interpolation between the two concentrations above and below 50% inhibition [21].

Selectivity Index (SI) of the UA was computed by the formula as:

$$SI = \frac{CC_{50}}{IC_{50}}$$

Molecular modeling and docking studies against glutathione-S-transferase (BmGST) enzyme

Molecular modeling and geometry cleaning of the UA was performed through ChemBioDraw-Ultra-v12.0 (Cambridge Soft, UK). The 3D structure was subjected to minimized the energy by using molecular mechanics-2 (MM2) force field until the root mean square (RMS) gradient value became smaller than 0.100 kcal mol^{-1} Å. Re-optimization was done by MOPAC (Molecular Orbital Package) method until the RMS gradient attained a value smaller than 0.0001 kcal mol^{-1} Å. The 3D chemical structure of known drugs DEC-c (CID:15432) and IVM (CID: 6321424) were retrieved from PubChem compound database (NCBI, USA). The theoretically solved structure of *B. malayi* glutathione-S-transferase (BmGST) was selected as the potential target for molecular docking simulation studies. The BmGST crystallographic protein 3D structure was retrieved from Protein Data Bank (PDB ID: 1SJO). The Ligsite program was used to identify the potential active site of BmGST model for molecular

Table 3. Predicted ADME parameters (DS v3.5, Accelrys, USA).

Anti-filarial agent	Aqueous solubility	Blood brain barrier penetration	CYP2D6 binding	Hepatotoxicity	Intestinal absorption	Plasma protein binding
DEC	4	2(Medium)	False (non-inhibitor)	True (toxic)	0 (Good)	False (Poorly bounded)
IVM	3	4 (Undefined)	False (non-inhibitor)	True (toxic)	3 (very poor)	False (Poorly bounded)
UA	1	0 (very high penetrant)	False (non-inhibitor)	False (non-toxic)	1 (moderate)	True (Highly bounded)

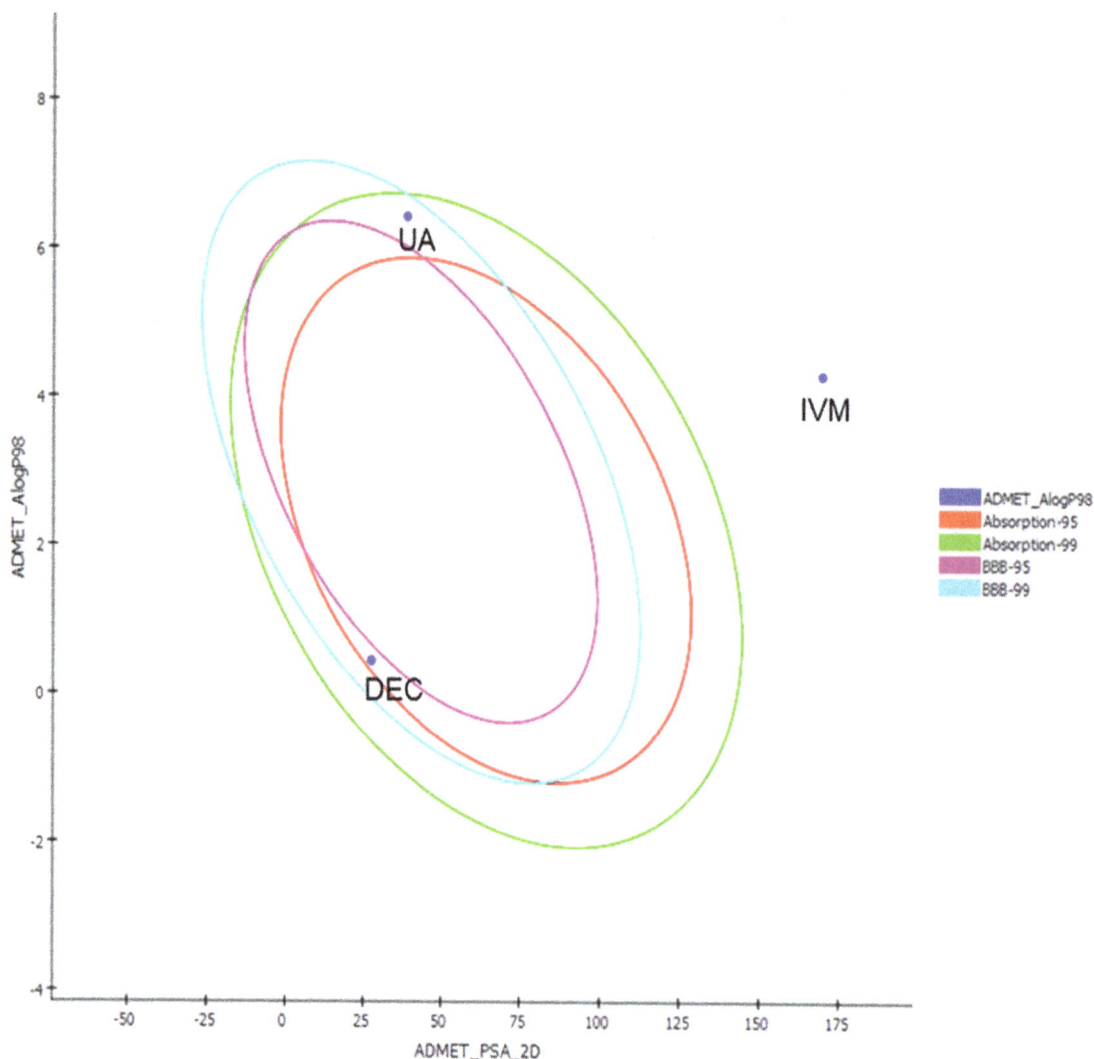

Figure 4. Adsorption model of Ursolic Acid (UA) and the standard antifilarial drugs.

docking studies and was then cross-checked with template active site as shown in Figure 3 [22]. The visualization studies were executed through Discovery Studio v3.5 (Accelrys Inc., USA, 2013).

In vivo efficacy

Administration of UA and the reference drugs: The finely powdered UA was suspended in 0.1% Tween-80 prepared in sterilized tap water. Solution of DEC was made in plain STW. *M. coucha* was administered with UA and DEC-c at 100 and 50 mg/kg body weight respectively through i.p. route for 5 consecutive days. The suspensions/solutions of UA/DEC-c were prepared daily before administration to the animals. Control animals received vehicle only.

B. malayi -M. coucha model: Animals harboring 5–7 months old *B. malayi* infection and showing progressive increase in microfilraemia were used in this study. UA and reference drug treated groups and an equal number of infected untreated animals kept as vehicle treated control, consisted of 5 animals each in two experiments were used.

Mf count in 10 μl blood drawn from tail of the animals between 12:00 noon and 1:00 PM [14] was assessed just before initiation of

treatment (day 0), on days 7/8 and 14 post initiation of treatment (p.i.t.) and thereafter at fortnightly intervals till day 84 p.i.t. [17]. The animals were killed on day 91 p.i.t.

Assessment of microfilaricidal efficacy: Microfilaricidal efficacy of UA was evaluated on day 7/8 and 14 p.i.t. and expressed as percent reduction in mf count over pretreatment level [23–25].

Assessment of macrofilaricidal and worm sterilization efficacy: Adult worms were recovered from heart, lungs and testes of treated and control animals [14]. Tissues were teased gently and the parasites recovered were then examined under microscope for status of the motility, cell adherence on their surface, dead or calcified worms [23,24]. Number of worms recovered from the treated and untreated animals was recorded. Macrofilaricidal efficacy of UA and DEC-C was assessed and expressed as percent change in adult worm recovery in treated group over control animals.

All the surviving females were teased individually in a drop of saline to examine condition of intrauterine mf stages of the parasite [23,24]. Number of sterile female worms recovered from the treated animals was compared with that of control animals and percent sterilization of female worms was determined in treated or

Table 4. Compliance of Dec, IVM & UA to the theoretical parameters of oral bioavailability and drug likeness properties.

Anti-filarial agent	Pharmacokinetic properties (ADME) dependent on chemical descriptors								
	ADM	AE	ADME	AD			H-bond acceptor		Lipinski's rule of 5 violation
	Oral bioavailability: TPSA (Å²)	MW	logP	H-bond donor					
				NH$_2$ group count	-N- group count	OH group count	N atom count	O atom count	
DEC	26.785	199	0.881	0	0	0	3	1	0
IVM	170.095	861	4.076	0	0	3	0	14	2
UA	57.527	456	6.789	0	1	1	0	3	1

Note: A = absorption, D = distribution, M = metabolism, and E = excretion; TPSA = topological polar surface area; MW = molecular weight; Log P = octanol/water partition coefficient.

control groups over total live female worms recovered from the respective groups.

Statistical analysis

Statistical analyses were carried out using Statistica version 7/ GraphPad Prism 3.0 version software. Results were expressed as mean ± S.D. of data from 5–6 animals in two experiments. The data were subjected to One-way ANOVA analysis and the significance of the difference between means were determined by Newman-Keuls Multiple Comparison Test. $P < 0.05$ was considered significant and marked as *, $P < 0.01$ as highly significant and marked as **, and $P < 0.001$ was very highly significant and marked as ***. The trend analysis was done by fitting the simple regression model ($Y = A + BX$) using the method of least squares. The slopes of the line were compared by Analysis of Variance.

Drug likeness screening studies for ADME/Tox compliance. The ADME/Toxicity parameters compliance was evaluated by screening through Metadrug^{TM}, a commercial tool of MetaDiscovery (Thomson Reuters, USA) (http://www. genego.com) [26]. MetaDrug is a system pharmacology or system chemical biology and toxicology platform designed for the assessment of would-be therapeutic indications, off-target effects and potential toxic end points of novel small molecule compounds. In the studied work, this database/tool was used to predict and evaluate the human metabolism compliance, toxicity risk assessment and mode of action by using standard experimental data.

Results

The leaves of *E. tereticornis* were extracted and fractionated, according to the scheme given in Figure 1.

Worm motility and MTT reduction assay

Of the three extracts tested *in vitro* using worm motility assay, the CHCl$_3$ extract killed adult female worms (LC$_{100}$: 400 µM) and mf (LC$_{100}$: 200 µM) (Table 1). The CHCl$_3$ extract was subjected to repeated chromatographic separations over VLC using TLC grade silica gel H, which finally resulted in the isolation of a major compound. This major compound on further crystallization with CHCl$_3$ afforded 99% pure white crystals. The ^{1}H, ^{13}C NMR and ESI-MS spectroscopic data of these crystals confirmed that this is a pentacyclic triterpene, ursolic acid (UA) (Figure 1). Finally, UA was tested for its anti-filarial activity against *B. malayi* using *in vitro* assays.

Further, UA was tested against mf and female adult worms of *B. malayi* using motility and or MTT assays and the results are summarized in Table 1. Like chloroform extract UA was also found to be more effective in killing mf (LC$_{100}$: 50 µM) than adult worms (LC100: 100 µM) and its IC$_{50}$ values were 35.36 and 8.84 µM against the respective parasite stages. UA exerted >86% inhibition in MTT reduction ability of the adult worms. It reduced the viability of female parasite in a gradual dose dependent manner as assessed by MTT reduction assay (**Figure S1**). The CC$_{50}$ (>300 µM) and SI (>10) values of UA demonstrated that it is safe for carrying out *in vivo* screening (Table 1).

The time point studied for the standard drug IVM was 24 hr and 48 hr as IVM has slow action on the parasites. After 48 hr post incubation IVM was effective in inhibiting motility of female adult worm and mf at a minimum conc. of 5.0 µM and 2.5 µM (LC$_{100}$), respectively. Its IC$_{50}$ against adult worms was 3.05 µM and that of mf was 1.49 µM. However IVM was less effective when parasites were incubated for 24 hr. IVM failed to inhibit MTT reduction ability of female worms even after 48 hr incubation (Table 1). On the other hand DEC-C required much

Table 5. Details of computational toxicity risk parameters of DEC, IVM and UA calculated by OSIRIS.

| Compound | Toxicity risk parameters | | | |
	MUT	TUMO	IRRI	REP
DEC	High Risk	No risk	No risk	High Risk
IVM	No risk	No risk	No risk	No risk
UA	No risk	No risk	No risk	No risk

Note: MUT = Mutagenicity, TUMO = Tumorogenicity, IRRI = Irritation, REP = Reproduction.

higher concentration to kill the female worms (LC_{100}: 1000 μM) and mf (LC_{100}: 500 μM); it inhibited MTT reduction skill of the adult parasite to the tune of 62.55%. The IC_{50} of DEC against the respective parasite stages were found to be 353.55 μM and 297.30 μM.

Concentration-dependent' LC_{100} and IC_{50} of UA, ivermectin and DEC for microfilariae and adult parasites of *B. malayi* in motility and MTT assays are shown in Figures S1–S3. After 24 h incubation UA (**Figure S1**) and DEC-C (**Figure S2**) caused concentration dependent decrease in viability of the parasites. However, in case of IVM the viability was time and conc. dependent (**Figure S3**).

In summary, *in vitro* findings revealed that $CHCl_3$ extract of *E. tereticornis* was microfilaricidal and macrofilaricidal, active against human filarial worm *B. malayi* and the active principal was localized to UA.

Molecular docking of UA on *Bm*GST

The current status of filaria have paved the way for investigating new lead compounds, which could be useful for the development of anti-filarial agents as there is a persistent urge for a lead to

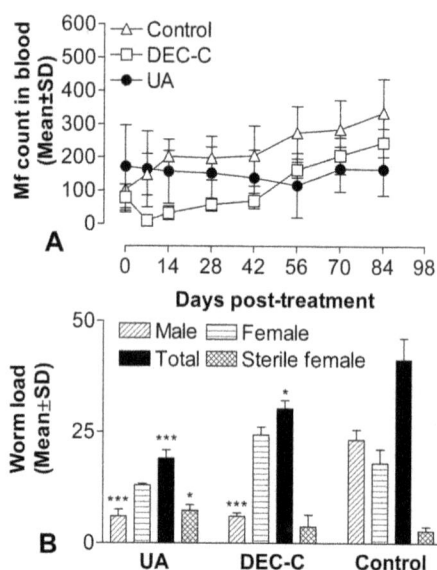

Figure 5. Micro-(A) and macrofilaricidal (B) activity of UA and reference drug diethylcarbamazine-citrate (DEC-C) against *Brugia malayi* in *Mastomys coucha*. Values are mean ± S.D. of 5 animals from two experiments. (A) No alteration in Microfilarial count in treated animals at each time point post initiation of treatment over day 0. Statistics: Student's 't' test. Significance level (B) *P<0.05 (vs sterilized female worm of control animals).

become candidate drug. The enzyme glutathione-s-transferase (GST) is playing a significant role in the long-term existence of filarial worms in mammalian host. The GST enzyme is a well known potential molecular target to inhibit filarial parasite's growth [27,28]. Therefore, the *Bm*GST theoretical protein structure 3D model was retrieved from PDB crystallographic database and later used for molecular docking simulation studies of UA, to explore the possible mechanism of action within the filarial worm (Table 2). The docking results showed high binding affinity (i.e., low docking energy; −8.6 kcal mol^{-1}) similar to that of reference drugs, DEC-c (−4.9 kcal mol^{-1}) and IVM (−8.4 kcal mol^{-1}). Docking results of UA also showed formation of H-bond (length 3.0 Å) with aromatic hydrophobic residue TYR-106, this may be the reason of high binding affinity, stability and activity of UA. The other binding site amino acid residues within a selection radius of 4 Å from the bound UA against *Bm*GST protein structure model were nucleophilic (polar, hydrophobic) e.g., threonine (THR-102), aromatic (hydrophobic) e.g., phenylalanine (PHE-8), tyrosine (TYR-7, TYR-106), polar amide e.g., asparagine (ASN-203), glutamine (GLN-49), hydrophobic e.g., leucine (LEU-13, LEU-50) (Figure 3). These results suggest that UA interacted well with the conserved hydrophobic amino acid residues of *Bm*GST. The molecular docking results showed that UA had significant similarity with respect to interacting amino acid residues and hydrogen bonds to that of the reference drug IVM, while the second reference drug Dec-c showed almost different interacting amino acid residues and hydrogen bond pattern. On the basis of docking binding affinity studies, it may be suggested that UA can be used as a potential lead against lymphatic filarial parasites by targeting GST.

ADME/Tox parameters evaluation

Since, docking results showed that UA may act as a potential anti-filarial lead, therefore *in silico* ADME/Tox parameters screening study was performed through Discovery Studio v3.5 molecular modeling & drug discovery software (Accelrys, USA). The UA, DEC and IVM were evaluated with standard descriptors and all the chemical descriptors and parameters of ADME were calculated (Table 3). The ADME results showed that there was no predictive hepatotoxicity and UA was comparable to standard range.

The ADME 2-D graph was plotted against Alogp98 versus PSA_2D (polar surface area) (Figure 4), which showed that the UA and Dec-c were inside the confidence limit ellipses of 99% for the blood brain barrier penetration and human intestinal absorption models compliance. On the other hand, IVM fallen outside the ellipse (undefined) showing very poor absorption and blood brain barrier penetration. Although, UA showed less water solubility, moderate intestinal absorption, but exhibited high plasma protein binding.

Table 6. Predicted therapeutic activity of UA against various reported diseases.

Property	Model description	Value/(TP)
Allergy	Potential antiallergic activity. Cutoff is 0.5. Values higher than 0.5 indicate potentially active compounds. Training set consists of approved drugs. Model description: Training set N = 258	0.50 (60.67)
Arthritis	Potential activity against arthritis. Cutoff is 0.5. Values higher than 0.5 indicate potentially active compounds. Training set consists of approved drugs	0.72 (58.72)
Cancer	Potential activity against cancer. Cutoff is 0.5. Values higher than 0.5 indicate potentially active compounds. Training set consists of approved drugs. Model description: Training set N = 886	0.69 (64.84)
Hyperlipidemia	Potential antihyperlipidemic activity. Cutoff is 0.5. Values higher than 0.5 indicate potentially active compounds. Training set consists of approved drugs. Model description: Training set N = 185	0.91 (66.67)
Inflammation	Potential anti-inflammatory activity. Cutoff is 0.5. Values higher than 0.5 indicate potentially active compounds. Training set consists of approved drugs. Model description: Training set N = 598	0.50 (79.31)
Migraine	Potential activity against migraine. Cutoff is 0.5. Values higher than 0.5 indicate potentially active compounds. Training set consists of approved drugs	0.62 (97.37)
Obesity	Potential activity against obesty. Cutoff is 0.5. Values higher than 0.5 indicate potentially active compounds. Training set consists of approved drugs	0.94 (97.37)
Osteoporosis	Potential anti-osteoporosis activity. Cutoff is 0.5. Values higher than 0.5 indicate potentially active compounds. Training set consists of approved drugs	0.80 (97.37)
Skin Diseases	Potential activity against skin diseases. Cutoff is 0.5. Values higher than 0.5 indicate potentially active compounds. Training set consists of approved drugs. Model description: Training set N = 255	0.94 (53.04)

Although, ADME results showed that UA violates Lipinski's rule of five due to high logP value (logP>5), hence may cause problem in absorption through biological membranes or intestinal absorption, but it still falls within the acceptable limit of rule of five, when compared with the reference drugs, DEC and IVM (Table 4).

Toxicity risk assessment

The toxicity risk assessment at high doses and/or long term use was evaluated through the OSIRIS web server for the reference drugs, DEC, IVM and the studied compound UA (Table 5). In this screening four important toxicity risk parameters *viz.*, mutagenicity, tumorogenicity, skin irritation and reproductive/developmental toxicity parameters were evaluated for high doses or long term use toxicity. The toxicity screening results showed that UA and the reference drug IVM showed no features of risk of tumorogenicity, mutagenicity, reproductive toxicity and skin irritation, therefore UA is safe for human use, whereas DEC yielded a high risk of mutagenicity and reproductive toxicity.

In vivo anti-filarial efficacy

Brugia malayi - M. coucha model. Microfilaricidal activity: Figure 5A shows anti-filarial efficacy of UA against *B. malayi* in *M. coucha* at 100 mg/kg s.c. for 5 consecutive days. UA produced 4–33% lower microfilaremia (statistically not significant) than 0 day throughout the post treatment observation period (Figure 5A). In other words, the microfilaremia in UA treated animals remained below (4–33%) the pretreatment (0 day) level throughout the observation period, while in the untreated control it was (progressively) higher than the pretreatment level and never equaled the 0 day level. This clearly shows that UA possesses considerable antifilarial efficacy. DEC-C (50 mg/kg, s.c. x 5 days), which is principally a microfilaricide, caused >85% reduction in microfilarial count on day 7 p.i.t. which progressively increased and relapsed on day 49 p.i.t.; the count further increased rapidly and crossed the pretreatment level by day 56 p.i.t. The trend of microfilaremia from day 7 to day 84 p.i.t. in the three groups (control untreated, UA treated and DEC treated) against time

(post treatment) was determined and compared among each other using linear trend analysis. The baseline of each group was subjected to equality and the observations were converted to show percent change at each time point. The baseline adjusted data was fitted to straight line. The analysis showed that while the mf count in the DEC-treated animals after an initial dramatic drop on day 7 p.i.t., increased gradually over time, it remained almost unchanged with time in UA treated group (trend not significant). Thus UA was found to be better than DEC in controlling microfilaremia. The details are given in File S1: 'In vivo antifilarial efficacy in *Brugia malayi -M. coucha* model: Microfilaricidal activity'.

Macrofilaricidal and embryostatic activity: UA (100 mg/kg, s.c. for 5 days) caused around 54% (P<0.001) adulticidal action over the untreated control. A moderate embryostatic effect of UA (56.15%; P<0.05) was also noticed in female worms (Figure 5B). DEC-C treatment (50 mg/kg, s.c. x 5 days) resulted in 26.47% reduction (P<0.05) in adult worms but did not exert any significant embryostatic effect on female worms when compared to that of untreated control animals (Figure 5B). The general behavior of the treated animals was found normal during entire observation period indicating that UA is safe.

Together, the results of UA showed promising antifilarial activity *in vitro* and *in vivo* with no adverse affect on health and general behavior of the treated animals.

ADME/Tox compliance

The compound UA was evaluated through MetaDrug tool (Thomson Reuters, USA) for compliance to the standard ADME/Tox parameters. Results showed the information of metabolites, QSAR based prediction of ADME/Tox properties, therapeutic activities, information of analogues, pathways, potential targets and signaling pathway map by leveraging an extensive database of chemical structures and pharmacological activities and visualized in the context of pathways, cell processes, toxicity and disease networks that are perturbed by the compound and its metabolites. These results for UA are briefly discussed below:

Prediction of therapeutic activities for UA: Large numbers of therapeutic activities for the compound UA were identified

Table 7. The reported interaction between UA and target.

S. No.	Target	Type	Drug	Interactions	Similarity	Effect	Pubmed/Patent ID
1	COX-2 (PTGS2)	$	Ursolic acid	€	100	Inhibition	12244669
2	OATP-C	¥	Ursolic acid	£	100	Inhibition	12871156
3	DNA ligase I	$	(1S, 2R, 4aS, 6aS, 6bR, 10S, 12aR)-10-Hydroxy-1, 2, 6a, 6b, 9, 9, 12a-heptamethyl-1, 3, 4, 5, 6, 6a, 6b, 7, 8, 8a, 9, 10, 11, 12, 12a, 12b, 13, 14b-octadecahydro-2H-picene-4a-carboxylic acid	€	100	Unspecified	15519169
4	DNA polymerase beta	$	10-Hydroxy-1, 2, 6a, 6b, 9, 9, 12a-heptamethyl-1, 3, 4, 5, 6, 6a, 6b, 7, 8, 8a, 9, 10, 11, 12, 12a, 12b, 13, 14b-octadecahydro-2H-picene-4a-carboxylic acid (1)	€	100	Inhibition	15974441
5	ACAT2	$	(1S, 2R, 4aS, 6aS, 6bR, 8aR, 10S, 12aR, 12bR, 14bS)-10-Hydroxy-1, 2, 6a, 6b, 9, 9, 12a-heptamethyl-1, 3, 4, 5, 6, 6a, 6b, 7, 8, 8a, 9, 10, 11, 12, 12a, 12b, 13, 14b-octadecahydro-2H-picene-4a-carboxylic acid	€	100	Inhibition	16462051
6	SOAT1	$	(1S, 2R, 4aS, 6aS, 6bR, 10S, 12aR)-10-Hydroxy-1, 2, 6a, 6b, 9, 9, 12a-heptamethyl-1, 3, 4, 5, 6, 6a, 6b, 7, 8, 8a, 9, 10, 11, 12, 12a, 12b, 13, 14b-octadecahydro-2H-picene-4a-carboxylic acid	€	100	Inhibition	16462051
7	SOAT2	$	(1S, 2R, 4aS, 6aS, 6bR, 10S, 12aR)-10-Hydroxy-1, 2, 6a, 6b, 9, 9, 12a-heptamethyl-1, 3, 4, 5, 6, 6a, 6b, 7, 8, 8a, 9, 10, 11, 12, 12a, 12b, 13, 14b-octadecahydro-2H-picene-4a-carboxylic acid	€	100	Inhibition	15974441
8	ACAT1	$	(1S, 2R, 4aS, 6aS, 6bR, 8aR, 10S, 12aR, 12bR, 14bS)-10-Hydroxy-1, 2, 6a, 6b, 9, 9, 12a-heptamethyl-1, 3, 4, 5, 6, 6a, 6b, 7, 8, 8a, 9, 10, 11, 12, 12a, 12b, 13, 14b-octadecahydro-2H-picene-4a-carboxylic acid	€	100	Inhibition	11794520

$ = Generic enzymes; € = Unspecified; £ = Inhibition is done with unspecified mechanism; ¥ = Transporter.

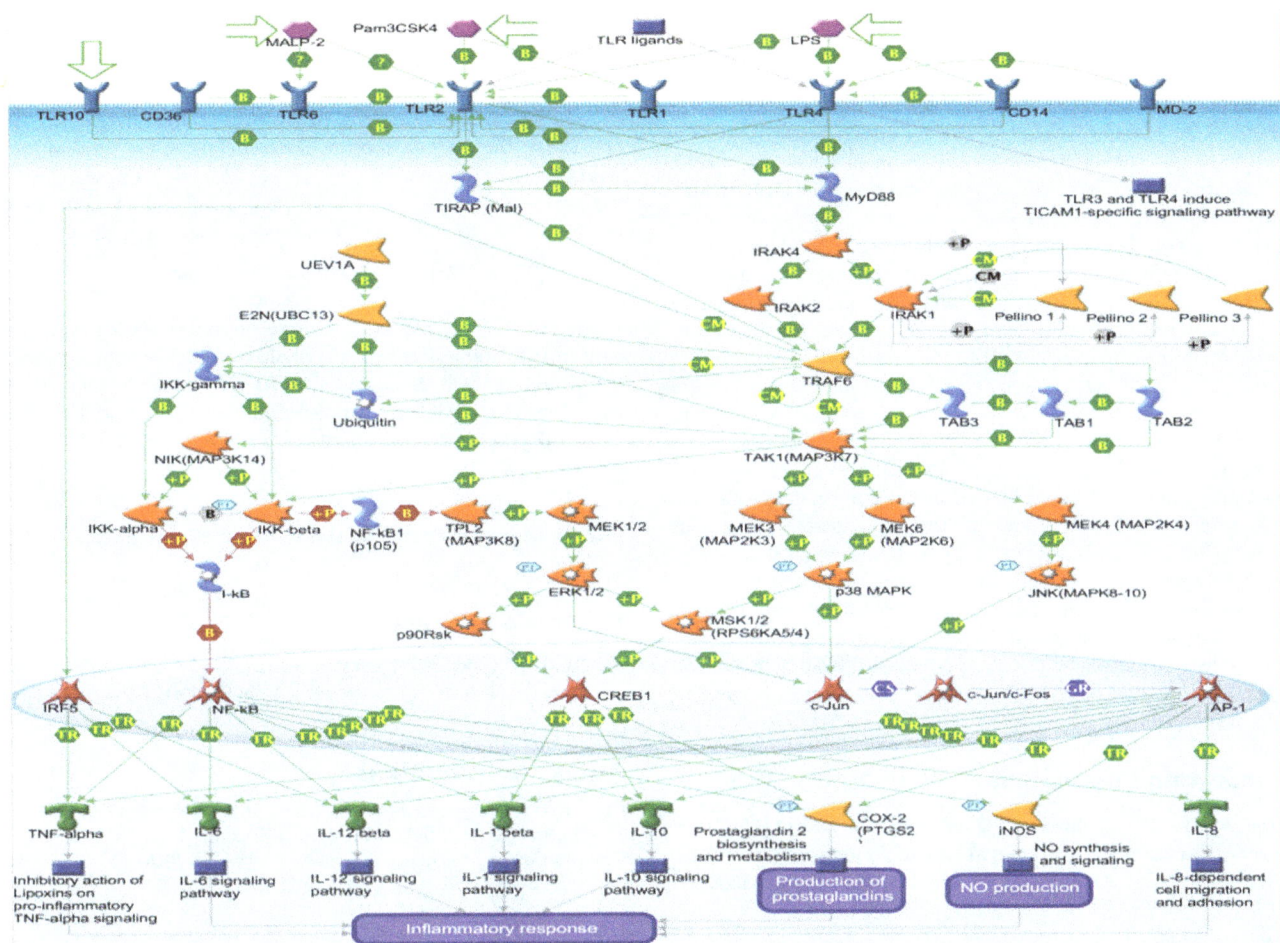

Figure 6. Signaling pathway map screened by Metadrug.

through MetaDrug tool (Thomson Reuters, USA). The evaluated therapeutic activities for UA were; allergy, Alzheimer, angina, arthritis, asthma, bacterial, cancer, depression, diabetes, HIV, heart failure, hyperlipidaemia, obesity, migraine, osteoporosis and many more. The predicted activities for UA were classified as active or non-active based on calculated values. The predicted properties of UA were calculated on the basis of Tanimoto Percentage [TP] values (standard cut-off ≥0.5) (Table 6).

Prediction of analogues, pathways and potential targets for UA: The chemical structures and the name of some known similar compounds or analogues were predicted by MetaDrug tool related to UA on the basis of structural similarity (in the range of 98–100%). MetaDrug tool also detected the potential biological pathways and the targets with experimentally known prior mode of action for UA (Table 7).

Prediction of metabolic signaling pathway map for UA: Immune response through TLR2 and TLR4 signaling pathways identified through MetaDrug tool on the basis of -lopP value *i.e.*, 1.834e-8 (7.736) with six network objects. TLR2 and TLR4 induce MyD88/IRAK/TRAF6-dependent pathway in target cells, leading to activation of transcription factors NF-kB, AP-1, CREB1 and IRF5, which induce production of various proin-flammatory mediators including cytokines, chemokines, nitric oxide (NO) and prostaglandins, leading to inflammatory response (Figure 6).

Discussion

There are only a few medicinal plant extracts and the isolated molecules, which have shown good anti-filarial activity. The literature showed that some secondary metabolites such as triterpenoids and coumarins showed significant activity against filarial parasites. Our recent finding on the antifilarial activity of pentacyclic triterpenoid, glycyrrhetinic acid (GA) has given us advantage of exploring anti-filarial activity in UA, having similar pentacyclic triterpenoid chemical structure [9,29]. The UA isolated from the leaves of *E. tereticornis* was in full agreement with the ¹H, ¹³C NMR and ESI-MS spectroscopic data with the commercially available UA (SIGMA-ALDRICH).

The *in vitro* anti-filarial activity of UA against mf and the adult worms, prompted us to carry out it's *in silico* studies to investigate its possible mechanism of action. It is well known that filarial nematode's detoxify GST enzymes, which play a significant role in the survival of the parasites inside the host's body. This enzyme has effective ability to neutralize the reactive oxygen species (ROS) attack on membrane that acts as cytotoxic products and protect the helminths inside the host [30–32]. With this background, the *in silico* molecular docking binding affinity of UA against the GST enzyme was studied. The docking experiments were performed, which showed high binding affinity of UA with *Bm*GST enzyme.

It was observed that for killing the life stages of parasites in vitro, 10 times less concentration of UA was required than the drug DEC. Similarly, in vivo, UA treatment afforded 4–33% drop in microfilaraemia over 0 day throughout the post treatment observation period. While in the untreated control it was (progressively) higher than the pretreatment level and never equaled the 0 day level. The analysis trend in the DEC-treated animals showed that the mf count after an initial dramatic drop, increased gradually over time, while in UA treated animals microfilaraemia remained almost static. This clearly shows that UA was better than DEC in controlling microfilaraemia. Further UA exhibited 54% adulticidal and 56% embryostatic effect with static microfilaraemia while DEC produced ~26% macrofilaricidal, 15% embryostatic and >85% microfilaricidal effect (on day 7 p.i.t.). These results indicate that UA is clearly superior to DEC with respect to macrofilaricidal and embryostatic effect though not with respect to microfilaricidal effect. It may be mentioned here that macrofilaricidal and embryostatic effect of UA were probably responsible for the low and static microfilaraemia. Thus, UA is better than DEC both *in vitro* and *in vivo* in its antifilarial activity. Further UA being a natural compound, has the possibility of lead optimization by QSAR approach. Thus, in view of potential antifilarial activity, absence of toxicity and favorable pharmacokinetics UA may be considered as a suitable lead for designing and development of a safe and effective antifilarial agent.

Supporting Information

Figure S1 LC$_{100}$ and IC$_{50}$ of Ursolic acid (UA) for microfilariae and adult parasites of *Brugia malayi*. After incubation with UA for 24 h the viability of parasite was assessed in motility assay using mf (**A**) and adult female worms (**B**) and in MTT reduction assay using adult female worms (**C**).

Figure S2 LC$_{100}$ and IC$_{50}$ of diethylcarbamazine citrate (DEC-C) for microfilariae and adult parasites of *B. malayi*. After incubation with DEC-C for 24 h the viability of parasite was assessed in motility assay using mf (**A**) and adult female worms (**B**) and in MTT reduction assay using adult female worms (**C**).

Figure S3 LC$_{100}$ and IC$_{50}$ of ivermectin for microfilariae and adult parasites of *B. malayi*. After incubation with ivermectin for 24 h (**A–C**) and 48 h (**D–F**) the viability of parasite was assessed in motility assay using mf (**A, D**) and adult female worms (**B, E**) and in MTT reduction assay using adult female worms (**C, F**).

File S1 Linear Trend analysis of mf count data.

Acknowledgments

The authors thank the Directors, CSIR-CIMAP and CSIR-CDRI, Lucknow, for their keen interest and encouragement during the course of this work.

Author Contributions

Conceived and designed the experiments: SKS PKM FK. Performed the experiments: KK VK RV PS MS. Analyzed the data: SKS PKM FK. Contributed reagents/materials/analysis tools: SKS PKM FK. Wrote the paper: SKS PKM FK.

References

1. WHO, 2012. Global Programme to Eliminate Lymphatic Filariasis. Weekly Epidemiological Record. Available: http://apps.who.int/iris/bitstream/10665/78611/1/WHO_HTM_NTD_PCT_2013.5_eng.pdf. Accessed 16 July 2014.
2. Molyneux DH, Zagaria N (2002) Lymphatic filariasis elimination: Progress in global programme development. Ann Trop Med Parasitol 96: 15–40.
3. WHO, 2006. Global programme to eliminate lymphatic filariasis. Weekly epidemiological record, Geneva. Available: http://www.who.int/lymphatic_filariasis/resources/wer/en/. Accessed 22 January 2013.
4. Ottesen EA (2000) The global programme to eliminate lymphatic filariasis. Trop Med Int Hlth 5: 591–594.
5. WHO, 2005. Global programme to eliminate lymphatic filariasis. Weekly epidemiological record. Available: http://www.who.int/lymphatic_filariasis/resources/wer/en/. Accessed 22 April 2014.
6. Molyneux DH, Bradley M, Hoerauf A, Kyelem D, Taylor MJ (2003) Mass drug treatment for lymphatic filariasis and onchocerciasis. Trends Parasitol 19: 516–22.
7. Dadzie Y, Neira M, Hopkins D (2003) Final report of the conference on the eradicability of Onchocerciasis. Filaria J 2: 2.
8. Burkot TR, Durrheim DN, Melrose WD, Speare R, Ichimori K (2006) The argument for integrating vector control with multiple drug administration campaigns to ensure elimination of lymphatic filariasis. Filaria J 5: 10.
9. Kalani K, Kushwaha V, Verma R, Murthy PK, Srivastava SK (2013) Glycyrrhetinic acid and its analogs: A new class of antifilarial agents. Bioorg Med Chem Lett 23: 2566–2570.
10. Murthy PK, Joseph SK, Murthy PS (2011) Plant products in the treatment and control of filariasis and other helminth infections and assay systems for antifilarial/antihelmintic activity. Planta Med 77: 647–61.
11. Kushwaha V, Saxena K, Verma SK, Lakshmi V, Sharma RK, et al. (2011) Antifilarial activity of gum from Moringa oleifera Lam. on human lymphatic filaria *Brugia Malayi*. Chronicles of Young Scientists 202–06.
12. Maurya A, Srivastava SK (2012) Determination of ursolic acid and ursolic acid lactone in the leaves of *Eucalyptus tereticornis* by HPLC. J Braz Chem Soc 23: 468–472.
13. Kalani K, Yadav DK, Khan F, Srivastava SK, Suri N (2012) Pharmacophore, QSAR, and ADME based semisynthesis and *in vitro* evaluation of ursolic acid analogs for anticancer activity. J Mol Model 18: 3389–413.
14. Murthy PK, Tyagi K, Roy Chowdhury TK, Sen AB (1983) Susceptibility of Mastomys natalensis (GRA strain) to a sub periodic strain of human *Brugia malayi*. Indian J Med Res 77: 623–630.
15. Murthy PK, Murthy PSR, Tyagi K, Chatterjee RK (1997) Fate of infective larvae of *Brugia malayi* in the peritoneal cavity of Mastomys natalensis and Meriones unguiculatus. Folia Parasitol (Praha) 44: 302–304.
16. Murthy PK, Chatterjee RK (1999) Evaluation of two *in vitro* test systems employing *Brugia malayi* parasite for screening of potential antifilarials. Curr Sci 77: 1084–1089.
17. Lakshmi V, Joseph SK, Srivastava S, Verma SK, Sahoo MK, et al. (2010) Antifilarial activity in vitro and in vivo of some flavonoids tested against *Brugia malayi*. Acta Trop 116: 127–133.
18. Sashidhara KV, Rao KB, Kushwaha V, Modukuri RK, Verma R, et al. (2014) Synthesis and antifilarial activity of chalcone-thiazole derivatives against a human lymphatic filarial parasite, *Brugia malayi*. Eur J Med Chem 81: 473–80.
19. Page C, Page M, Noel C (1993) A new fluorimetric assay for cytotoxicity measurements *in vitro*. Int J Oncol 3: 473–476.
20. Mosmann T (1983) Rapid colorimetric assay for cellular growth and survival: application to proliferation and cytotoxicity assays. J Immunol Methods 65: 55–63.
21. Huber W, Koella JC (1993) A comparison of three methods of estimating EC$_{50}$ in studies of drug resistance of malaria parasites. Acta Trop 55: 257–261.
22. Hendlich M, Rippmann F, Barnickl G (1997) LIGSITE: automatic and efficient detection of potential small molecule-binding sites in proteins. J Mol Graph Model 15: 359–63.
23. Chatterjee RK, Fatma N, Murthy PK, Sinha P, Kulshreshtha DK, et al. (1992) Macrofilaricidal activity of the stem bark of Streblus asper and its major active constituents. Drug Develop Res 26: 67–78.
24. Gaur RL, Dixit S, Sahoo MK, Khanna M, Singh S, et al. (2007) Antifilarial activity of novel formulations of albendazole against experimental brugian filariasis. Parasitology 134: 537–544.
25. Lämmler G, Wolf E (1977) Chemo prophylactic activity of filaricidal compounds on Litomosoides carinii infection of Mastomys natalensis. Tropen Med Parasitol 28: 205–225.
26. Bugrim A, Nikolskaya T, Nikolsky Y (2004) Early prediction of drug metabolism and toxicity: systems biology approach and modeling. Drug Discov Today 9: 127–135.
27. Yadav D, Singh SC, Verma RK, Saxena K, Verma R, et al. (2013) Antifilarial diarylheptanoids from Alnus nepalensis leaves growing in high altitude areas of Uttarakhand, India. Phytomedicine 20: 124–132.

28. Azeez S, Babu RO, Aykkal R, Narayanan R (2012) Virtual screening and *in vitro* assay of potential drug like inhibitors from spices against glutathione-S-transferase of filarial nematodes. J Mol Model 18: 151–63.

29. Kalani K, Yadav DK, Singh A, Khan F, Godbole MM, et al. (2014) QSAR guided semi-synthesis and *in-vitro* validation of anticancer activity in ursolic acid derivatives. Curr Top Med Chem 14: 1005–13.

30. Lanham A, Mwanri L (2013) The Curse of Lymphatic Filariasis: Would the Continual Use of Diethylcarbamazine Eliminate this Scourge in Papua New Guinea? American Journal of Infectious Diseases and Microbiology 1: 5–12.

31. Sommer A, Nimtz M, Conradt HS, Brattig N, Boettcher K, et al. (2001) Structural analysis and antibody response to the extracellular glutathione S-transferases from Onchocerca volvulus. Infect Immun 69: 7718–28.

32. Brophy PM, Pritchard DI (1994) Parasitic helminth glutathione S-transferases: an update on their potential as targets for immuno- and chemotherapy. Exp Parasitol 79: 89–96.

10

How Structure Defines Affinity in Protein-Protein Interactions

Ariel Erijman, Eran Rosenthal, Julia M. Shifman*

Department of Biological Chemistry, The Alexander Silberman Institute of Life Sciences, The Hebrew University of Jerusalem, Jerusalem, Israel

Abstract

Protein-protein interactions (PPI) in nature are conveyed by a multitude of binding modes involving various surfaces, secondary structure elements and intermolecular interactions. This diversity results in PPI binding affinities that span more than nine orders of magnitude. Several early studies attempted to correlate PPI binding affinities to various structure-derived features with limited success. The growing number of high-resolution structures, the appearance of more precise methods for measuring binding affinities and the development of new computational algorithms enable more thorough investigations in this direction. Here, we use a large dataset of PPI structures with the documented binding affinities to calculate a number of structure-based features that could potentially define binding energetics. We explore how well each calculated biophysical feature alone correlates with binding affinity and determine the features that could be used to distinguish between high-, medium- and low- affinity PPIs. Furthermore, we test how various combinations of features could be applied to predict binding affinity and observe a slow improvement in correlation as more features are incorporated into the equation. In addition, we observe a considerable improvement in predictions if we exclude from our analysis low-resolution and NMR structures, revealing the importance of capturing exact intermolecular interactions in our calculations. Our analysis should facilitate prediction of new interactions on the genome scale, better characterization of signaling networks and design of novel binding partners for various target proteins.

Editor: Bostjan Kobe, University of Queensland, Australia

Funding: This work was supported by the Israel Science Foundation (1372/10) http://www.isf.org.il/, and The Abisch-Frenkel foundation, http://ard.huji.ac.il/huard/infoPageViewer.jsp?ardNum=2033&lang=heb. The funders had no role in study design, data collection and analysis, decision to publish, or preparation of the manuscript.

Competing Interests: The authors have declared that no competing interests exist.

* Email: jshifman@mail.huji.ac.il

Introduction

Many biological processes are governed by non-covalent interactions between two or more proteins. Binding affinities of functional protein-protein interactions (PPIs) span more than nine orders of magnitude, from very weak and transient interactions observed frequently in signal transduction and membrane trafficking [1,2] to very strong interactions exhibited by several enzyme-inhibitor complexes with binding affinities (K_d) reaching 10^{-14} M [3,4]. In spite of numerous studies that analyze various PPIs [5,6], precise features that distinguish weak PPIs from high-affinity PPIs remain elusive [7]. An accurate understanding of these interactions would allow us to predict new PPIs on the genome scale, to better characterize known signaling networks, and to study PPI evolution. In addition, it would enable rational design of novel binding interactions, facilitating the discovery of therapeutic molecules that target disease-associated PPIs.

Binding interfaces of protein-protein complexes serve a dual role in protein function, since they can exist both as surfaces of monomeric proteins and as the area buried upon complex formation. They hence should differ in chemical properties from both protein surfaces and protein cores [8]. With the appearance of high-resolution structures for the first protein-protein complexes, a number of groups calculated and analyzed various structure-based features of binding interfaces such as solvent accessible surface area, packing, hydrogen bond (H bonds) and salt bridge patterns and arrived at a number of conclusions (see [9] for a review). Janin and colleagues reported that an interface covering about 1500 $Å^2$ and containing at least ten hydrogen bonds possesses sufficient enthalpy to generate binding affinities of up to 10^{-14} M [10]. Xu et al, established that hydrogen bonding geometry at interfaces is less optimal and exhibits wider distribution compared to that at the interior of globular proteins [11]. Thornton and colleagues concluded that interfaces of permanent complexes are more closely packed and contain fewer intermolecular hydrogen bonds compared to interfaces of non-obligatory complexes [5].

The growing number of high resolution structures for various PPIs as well as experiments documenting binding strengths stimulated several investigations directed at predicting PPI binding affinities from structure-derived features [12–18]. These works utilized either empirical scoring functions or knowledge–based potentials that both highly depend on the quality of the dataset. The described studies were performed using small datasets where good correlation with binding affinity was observed. However, when evaluated on larger datasets the same methods frequently fail [19]. Recent community-wide assessment of methods for predicting protein binding affinities demonstrated that even distinguishing between native PPIs and computationally designed PPIs that do not bind each other in reality is a difficult task [20].

Prediction of PPI binding affinity faces several limitations. First, binding affinity data is not always reliable and is frequently not compatible between various experiments due to different conditions and experimental techniques used for the measurement. For example, a K_d for the same protein-protein complex reported by different groups could easily range by two orders of magnitude (e.g. from 10^{-10} to 10^{-12} for the fasciculin/acetylcholinesterase complex [21–23]). An additional inaccuracy in affinity prediction might come from crystal structures that are solved at low or medium resolution and thus might misrepresent certain intermolecular interactions in the complex. Due to these uncertainties, a structure-based prediction of the binding affinity cannot exceed a certain limit of accuracy [24].

Recently, Kastritis et al reported a dataset containing 144 protein-protein complexes with their corresponding binding affinities [25]. This dataset is larger than the previously available datasets and contains only binding affinity data measured with precise experimental techniques such as SPR, ITC, and fluorescence, thus improving the accuracy of the provided K_d values. An additional advantage of this dataset is that the structural information is provided not only for the bound complex but also for the free components, allowing calculation of conformational changes associated with binding. Three recent studies used this benchmark to develop new computational methods for prediction of binding affinity and reported good correlation with experimental data, especially when utilizing a very large number of molecular descriptors [24,26,27]. However, machine-learning-based predictors are complex and thus do not provide better understanding of fundamental forces that determine protein binding affinity. In addition, such predictors are biased towards a particular dataset and are likely to perform worse on different datasets.

The goal of the present study is to examine how several types of structure-derived molecular features influence binding energetics and to define particular features that can distinguish between high-, medium- and low-affinity PPIs with statistical significance. Using the Kastritis dataset [25] we construct our own database where for each PPI we calculate a number of biophysical features from the atomic coordinates of the protein-protein complexes and the unbound structures. We explore how well each calculated biophysical feature alone correlates with binding affinity and determine the features that could be used to distinguish between high-, medium- and low- affinity PPIs. Furthermore, we test how various combinations of features could be applied to predict binding affinity and observe some improvement in correlation if more features are incorporated into the equation. In addition, we see a considerable improvement in predictions if we exclude from our analysis low-resolution and NMR structures, revealing the importance of capturing exact intermolecular interactions in our analysis.

Methods

All the data used for this work is organized in a database which includes all parameters at the atomic level. Different physical features were calculated from PDB files and included in the database.

Database preparation

144 structures of protein-protein complexes and the corresponding structures of the unbound proteins were extracted from PDB according to the list published in the benchmark database [25]. Complexes with heteroatoms closer than 1.4 Å from the

interface were discarded (PDB IDs 1BJ1, 1F34, 1JIW, 1JMO, 1S1Q, 1XD3 and 2J0T). In addition, we excluded a complex where an N-terminal tail close to the binding interface is missing from the structure but is strongly influencing the affinity measurements (PDB ID:2TGP), a complex where two different interfaces are too close to each other to analyze them independently (PDB ID:1NVU), and a complex that has been reported to exhibit an exceptionally high level of disorder-to-order transition [26] (PDB ID:2OZA).

Hydrogens were added to all files with the Reduce software [28] with histidines, asparagines and glutamines allowed to flip. The binding interface for each PPI was defined as all the atoms on one chain that are within 4.8 Å from the second chain in the complex.

Calculation of difference in accessible surface area (ΔASA)

We implemented the Lee-Richards molecular surface definition [29] using a probe radius of 1.4 Å for the calculation of the accessible surface area at the atomic level. Interface ASA is defined as the area of the atoms that make up the surface of the binding interface. ΔASA is defined as ASA that becomes inaccessible to solvent upon binding.

For each atom in the interface, we defined a *periphery index* as a distance from the closest atom on the surface of the protein that is not part of the interface. Atoms with a *periphery* index ≤3 Å were defined as peripheral atoms. Polar and non-polar area was calculated by summing up the areas of polar and non-polar interfacial atoms.

Hydrogen bonds

An angle- and hybridization-dependent 12–10 H bond potential [30] with hydrogen bond equilibrium distance of 2.8 Å and a well-depth of 8 kcal/mol was used to calculate the hydrogen bond energy for each pair of potential donor and acceptor atoms [31]. If the calculated H bond energy was lower than −0.6 kcal/mol a satisfied hydrogen bond was counted, otherwise an unsatisfied bond was counted. In the hydrogen bond analysis, we report the number of non-peripherial H bonds, involving the atoms with a periphery index of >3 Å. When an interfacial atom participates in a H bond with another atom on the same protein, an intramolecular H bond was counted. Intermolecular H bonds were counted when a H bond involves a donor and an acceptor atoms that belong to different chains.

Geometric complementarity

Van der Waals (VdW) interactions were calculated according to the Lennard-Jones 12-6 potential with softened repulsive term [31,32]. Only neighbor atoms of less than 9 Å were considered for this calculation. VdW interactions were measured between atoms of different chains in the complex. Well depth was set to 0.001 Å when atoms formed a H bond. Other methods used for geometric complementarity calculation in this work include Sc [33], implemented in the Rosetta software [34], and the Katchalsky-Katzir method [35].

Cavities

Empty spaces within the interface with sufficient volume to accommodate a water molecule were defined as cavities. We then selected only closed cavities that were at a distance of at least 2.8 Å from the surface (corresponding to the diameter of a water molecule). Cavities were detected and their volume was calculated using a simple Monte Carlo Integration technique [36].

Conformational changes

Structures of the unbound proteins were superimposed onto the structure of the complex [37]. Root Mean Square Deviation (iRMSD) between bound and unbound structures was calculated for all Cα atoms belonging to the interface.

To calculate side chain conformational changes upon binding, we assigned χ_1 and χ_2 dihedral angles to all side chains in the binding interface for the bound and the unbound structures. When at least one of these angles differed by more than 20 degrees between the unbound and the bound protein, the residue was considered to change conformation upon binding. The number of residues that do not change conformation upon binding was normalized by the total number of residues in the interface to give a percentage of residues that do not change side chain conformation. For NMR structures, only the first model was analyzed.

Hot spots

In silico alanine scanning was performed with the Robetta server that uses a fixed backbone approximation and an energy function parameterized on Ala mutations [38]. We defined hot-spot positions where mutation to Ala destabilized the complex by at least 1 kcal/mol.

Binding Interface composition

The *interface propensities* for each of the 20 natural amino-acids, for each PPI, were calculated as a ratio between ASA that the amino acid contributes to the interface and ASA that it contributes to the whole surface according to.the equation: $AA_{propensity} = (ASA\ AA_i/ASA\ interface)/(ASA\ AA_s/ASA\ surface)$. Here $ASA\ AA_i$ is the surface area of a particular amino acid in the interface and $ASA\ AA_s$ is the surface area of the same amino acid on the whole surface of the protein [5]. The results were averaged over the entire database for each amino acid. *High- and Low-affinity propensities* for each amino acid were calculated similarly as above but using only PDBs belonging to high- and low-affinity groups, respectively.

Electrostatic interactions

Salt bridges were counted according to Xu et al [11]. When distance between charged atoms is less than 4 Å and they participate in a hydrogen bond, a salt bridge is counted. Electrostatic energy was calculated using a Coulomb equation with a dielectric constant of 10 [31] and using the Delphi software [39] that utilizes linear Poisson-Boltzmann equation. For the latter calculation we used interior and exterior dielectric constants of 2 and 80, respectively and salt concentration of either 0 or 0.145 M. The electrostatic component of the binding free energy was obtained as the difference between the electrostatic free energy of the complex and those of the unbound chains.

Receiver Operator Characteristic (ROC) plots

For every tested biophysical feature we drew a ROC plot. This plot goes through all possible cut-off values that are used to distinguish between two groups, e. g. high- and low-affinity PPIs. Y-axis shows the fraction of PPIs from high-affinity PPIs that are above the cutoff values, as expected (true positive rate). X-axis shows the fraction of PPIs that are above the cut-off value in spite of belonging to low-affinity PPIs (false positive rate). Both x and y are normalized to 1 and the area under the curve (AUC) represents the performance of prediction when distinguishing between the two groups using the particular feature.

Combining biophysical features for affinity prediction

We considered all possible combinations of thirteen different biophysical features that are likely to be independent of each other (see table S2). Using a linear combination of all considered features, we optimized the weights in front of each feature to get the best correlation with experimental binding affinity. This optimization was performed starting from all combinations of two features and finishing with all possible combinations of thirteen features. Using the optimized weights, we calculated Pearson's correlation coefficient R and the AUC for ROC plots for high- vs. low-, high- vs. medium-, and medium- vs. low-affinity PPIs. Finally, the best values of R, and AUCs for each number of considered features were plotted. Models were cross-validated using the leave one out algorithm when the number of samples was below 100. This method requires fitting the data N times for a dataset of N points. In each round of fitting, one file is excluded from calculations and its affinity is predicted with the best weights in front of each feature. The N predicted values are then compared to the experimental values and the Pearson correlation coefficient R is reported.

Results and Discussion

To achieve the best possible accuracy in our analysis, we first systematically inspected all PPIs presented in the Kastritis database. We decided to exclude complexes that contain heteroatoms in the vicinity of the binding interface since the effect of heteroatoms on binding affinities cannot be accurately predicted. In addition, we excluded a complex where an N-terminal tail close to the binding interface is missing from the structure but is strongly influencing the affinity measurements (PDB ID:2TGP), a complex where two different interfaces are too close to each other to analyze them independently (PDB ID:1NVU), and a complex that has been reported to exhibit an exceptionally high level of disorder-to-order transition [26] (PDB ID:2OZA). After these exclusions we were left with 133 PPIs in our dataset. Each PPI was assigned into one of the three groups of high (Kd $\leq 10^{-9}$ M), medium (10^{-9} M <Kd $\leq 10^{-6}$ M) and low affinities (Kd>10^{-6} M) containing 43, 60, and 30 complexes in each group (see Figure S1 for distribution of datapoints). In addition, we divided the complexes into rigid and flexible according to the extent of conformational changes they exhibit upon binding. Complexes were defined flexible if the Root Mean Square Deviation between interfacial Cα atoms (iRMSD) of the bound and the unbound structures was ≥ 1 Å and were defined as rigid otherwise. For each PPI in our dataset, the binding interface was defined and was broken down into atoms with assigned coordinates and chemical properties. For each complex, we calculated a number of biophysical features that we considered important for determining binding affinity. These features included accessible surface area, inter- and intra-molecular H bonds, changes in main chain and side chain conformations between the bound and unbound structures, geometric complementarity, electrostatic interactions, the number of hot-spots, interface composition and volume of cavities. Some features were calculated with a number of different methods and the results were compared. Finally, using the Receiver Operator Characteristic (ROC) analysis, we compared how different features could be used to distinguish between low-, medium-, and high-affinity complexes.

Changes in the accessible surface area (ΔASA)

ΔASA is the total area that gets buried upon the complex formation. Earlier studies reported that protein binding affinity depends on ΔASA, with high-affinity complexes burring more surface area [5,19,40]. We found that ΔASA exhibits some correlation with binding affinity, with an R-value of 0.32 (Figure 1A). A better correlation (R = 0.41) is obtained if we normalize ΔASA by the total area of the atoms in the binding interface, providing a measure of binding interface dehydration (Figure 1B). The moderate correlation between ΔASA and affinity is not surprising since it is known that a few point mutations can produce many-fold changes in binding affinity without significant changes in ΔASA. For example, cognate and non-cognate complexes of colicin/immune proteins exhibit affinities of 10^{-14} and 10^{-8} M respectively while showing very similar ΔASA [41]. Furthermore, a single mutation in hemagglutinin from Influenza virus reduces the affinity of this protein for an antibody from 10^{-9} to 10^{-6} M [42]. Affinity maturation experiments in many antibody-antigen complexes also argue against strong correlation between ΔASA and binding affinity [43]. Thus, PPIs with larger ΔASA do not necessarily exhibit higher affinity but have a potential for containing more productive intermolecular interactions and for achieving higher affinity through mutations.

In some protein design studies, binding affinities of PPIs were enhanced by substituting polar residues with hydrophobic amino acids, implying that non-polar buried surface area should correlate with affinity [44]. We thus decided to examine whether polarity of the binding interface is correlated with binding affinity in our dataset. We found no correlation between binding interface polarity and affinity (Figure 1C) with PPIs exhibiting on average 48:52 ratio between polar and hydrophobic surface area, with a standard deviation of 7%. We conclude that binding interfaces of high-affinity complexes are not more hydrophobic than those of low-affinity complexes. This lack of correlation between polarity and affinity might be a result of the evolutionary pressure to keep protein surfaces mostly polar, preventing protein aggregation.

H bonds

H bond interactions are crucial to both protein folding and binding. While the energy of one H bond is relatively small, a large number of H bonds in proteins make them significant contributors to protein energetics. We hence calculated the number of H bonds formed across the binding interface (intermolecular H bonds), and within the interface (intramolecular H bonds) for the complexes in our dataset. Here we excluded H bonds that lie on the periphery of the binding interface since buried H bonds were expected to be more important for binding energetics compared to exposed H bonds. We also calculated the number of unsatisfied H bonds (the number of H bond donors that are buried within the interface and do not participate in H bonds). Unlike some previous studies [5] our analysis was based on calculating the energies of the H bonds according to atomic distances and angles between the donor, the hydrogen, and the acceptor atoms (see Materials and Methods). Our results show that the highest correlation (R = 0.40) is observed between affinity and the total number of H bonds in the binding interface (Figure 1D). Slightly lower correlation is observed when considering inter- or intra-molecular H bonds alone with R-values of 0.34 and 0.36, respectively (Figure 1E and 1F). Our results are in agreement with the general notion that intermolecular H bonds improve the enthalpic term of the free energy of binding, while the intramolecular bonds stabilize the interface and thus reduce the unfavorable entropic change associated with binding. Unexpectedly, no correlation with binding affinity is observed for interfacial unsatisfied H bonds for all PPIs in the dataset (Figures S2).

Geometric complementarity

Protein-protein recognition has been first described by the lock-and-key theory developed already hundred years ago [45]. This model assumes complete geometric complementarity of the two binding interfaces and does not consider any conformational change upon binding. Geometric complementarity in binding interfaces is hence a likely parameter for determining PPI binding affinity. We calculated geometric complementarity using three different methods including Katzir's molecular surface recognition [35], surface complementarity parameter Sc [33] and Van der Waals energy [30]. The best correlation with binding affinity is obtained when using VdW energy as a measure of geometric complementarity (Figure 1G). This feature shows an R-value of 0.43. Substantially lower correlation with affinity is observed when using two alternative methods for geometric complementarity calculation (Figures S3A and S3B). When separating between rigid and flexible complexes, Van der Waals energy gives a comparable correlation with binding affinity for both groups while the Katzir method gives much lower correlation for flexible complexes compared to rigid complexes and Sc parameter gives no correlation for either rigid or flexible complexes (Table S1). These results suggest that Van der Waals energy best approximates the energetic contribution of geometric complementarity to protein binding.

Cavities

Cavities could be observed as packing defects within protein cores or binding interfaces [46,47]. Cavities in protein cores have been shown to be extremely unfavorable and their removal usually produces high stabilization of the protein [48,49]. The role of cavities in protein-protein interfaces has been less studied compared to their role in protein cores. Chakrabarti and colleagues report that cavities in interfaces are on average larger compared to cavities in protein cores and their size correlates well with the size of the monomeric protein but not with the size of the interface [50].

We computed the volume of closed cavities in binding interfaces of PPIs in our dataset and examined how this feature correlates with binding affinity. In contrast to our expectations, we observed no correlation between the total cavity volume and binding affinity (Figure 1H). Interfaces of high- and low-affinity complexes exhibit an average total cavity volume of 146 and 104 A^3 respectively. The lack of correlation may be due to the fact that only empty cavities, not containing water molecules, are destabilizing for binding. However, we could not distinguish between empty and water-containing cavities in our database due to insufficient resolution of some of the crystal structures in our study.

Hot Spots

It has been observed that a relatively small number of interface residues, referred to as hot-spots, account for the majority of the binding energy [51]. Hot-spot residues are usually defined as binding interface positions where mutations to Ala produce more than 1 kcal/mol destabilization of the complex. We calculated the number of hot-spots for each PPI in the dataset using the Robetta server [52] and explored whether the calculated number of hot-spots in the interface correlates with binding affinity. We found a moderate correlation between affinity and the number of interfacial hot-spots with an R value of 0.28 (Figure 1I). One possible reason for relatively modest correlation between affinity and the number of hot-spots are errors associated with hot-spot predictions due to inaccuracies in the energy function and/or the fixed backbone approximation.

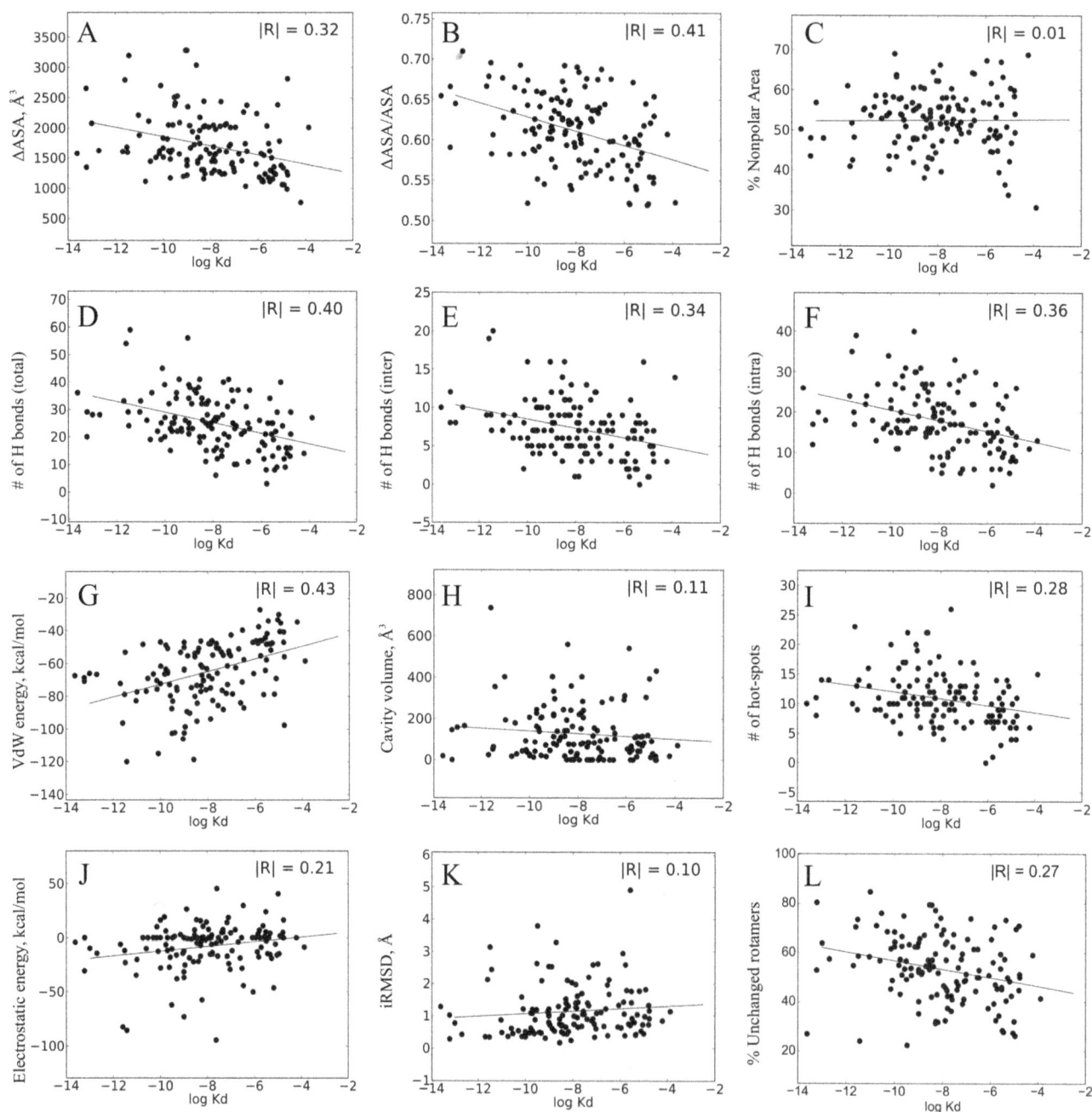

Figure 1. Dependence of Kd on various single biophysical features. (A) Change in the accessible interface surface area (ASA); (B) ΔASA normalized to the total interface area; (C) percent of non-polar change in the accessible surface area; (D) the total number of interfacial H bonds, (E) the number of intermolecular interfacial H bonds, (F) the number of intra-molecular H bonds; (G) Van der Waals energy; (H) volume of cavities; (I) number of hotspots; (J) electrostatic columbic energy; (K) RMSD between bound and unbound structures for interface Cαs; (L) percentage of rotamers that do not change conformation upon binding. Each point represents one PDB file in the database and the line corresponds to a linear fit to all data points in the database.

Electrostatic interactions

Past studies have demonstrated that electrostatic interactions, in general, are destabilizing in protein folding [39,53] but stabilizing in protein binding [54]. Such difference was explained by more favorable energetic change associated with transferring hydrophilic pairs from the surface of unbound proteins to the complex compared to the same change associated with protein folding. Here, we calculated electrostatic interactions in binding interfaces using two alternative methods, first by simply counting the number

of intermolecular salt bridges [11,55] and second by calculating the exact electrostatic interaction energy [30,39,56,57].

We found that the number of salt bridges as defined by Xu et al [11] does not correlate with binding affinity (Figure S4A). Similar results are obtained if we count repulsive same-charge intermolecular interactions (Figure S4B). Nevertheless, the number of salt bridges shows some correlation with the interfacial area (R-value of 0.44). That is, the number of salt bridges is dependent on the area of the interface, but its contribution to binding energetics is

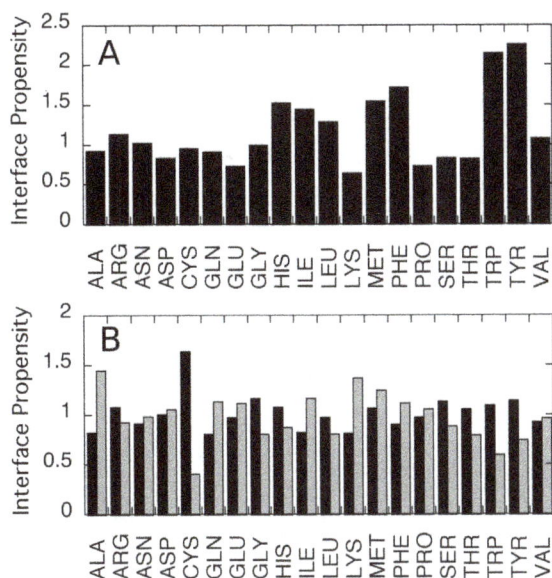

Figure 2. Amino acid interface propensities. (A) Amino acid propensities to be in an interface compared to protein surface calculated according to [5] (B) Amino acid propensities for high-affinity (black) and low-affinity (grey) complexes.

energy and binding affinity (Figure 1J and Figure S4C). These results indicate that both high- and low-affinity complexes could contain electrostatically optimized or non-optimized interfaces.

Binding interface composition

Several recent studies concluded that certain amino acids such as Tyr, Trp, Phe, Met, Val, Cys and Ile appear more frequently in PPI binding interfaces [5]. Other studies however, showed no appreciable difference in the composition of interfaces of the whole genomes [59,60]. To our knowledge, no study attempted to correlate amino acid composition to binding affinity. We hence examined whether this feature plays a role in determining binding affinity by calculating amino acid propensity to be in an interface [5], first for all complexes in our database and then separately for low- and high-affinity complexes (Figures 2A and B). For all complexes in the database, we observe that Trp, Tyr, Phe and Met are the most frequent interfacial amino acids, in agreement with previous studies. We also observe that Tyr, Trp, His, and Cys have a higher propensity to be found in high-affinity interfaces compared to low-affinity interfaces. However, due to a relatively small number of data points, these results are not statistically significant, pointing to the necessity of enlarging the dataset under study. Interestingly, Ala propensity shows an anti-correlation with affinity that is highly significant, indicating that this amino acid, within an interface, cannot provide favorable interactions. Lys also shows higher propensity to be found in low-affinity complexes compared to high-affinity complexes, in agreement with a previous study [61].

High propensity of Cys in high-affinity interactions could be partially explained by the fact that many of the high-affinity PPIs are enzyme-inhibitor complexes and inhibitors frequently contain multiple cysteins, most of them linked in intramolecular disulphide bonds. Indeed, in high-affinity complexes, Cys residues are mostly oxidized exhibiting an 8:1 oxidized/reduced ratio, while in low-

low. We thought that the lack of correlation between electrostatics and the binding affinity might be due to our simplistic approach for approximating this term. We hence performed more rigorous calculations to obtain electrostatic energy between the two protein chains involved in binding using either the Coulombic potential or a more sophisticated Poisson-Boltzmann equation [39,58]. In both cases, we found no substantial correlation between electrostatic

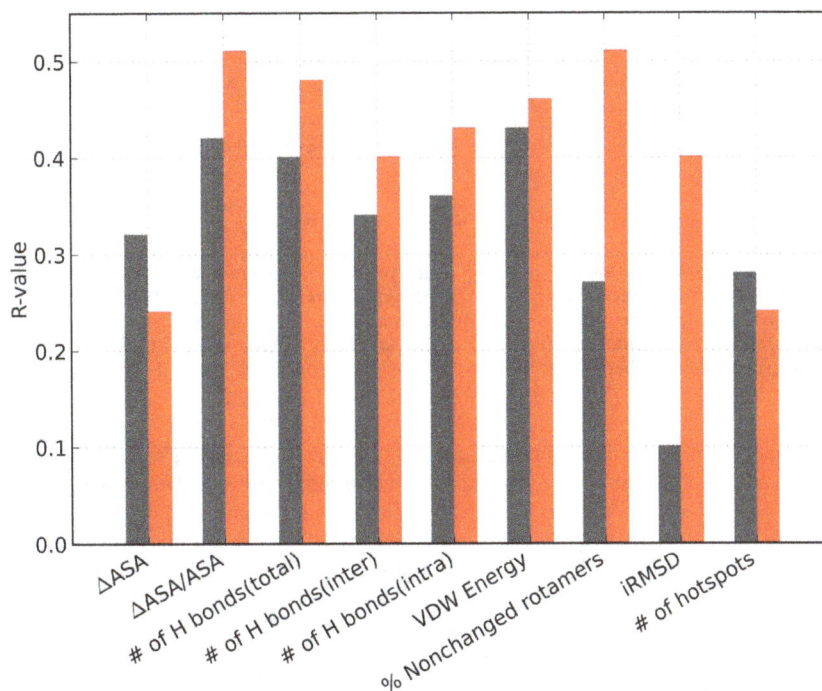

Figure 3. Improvement in R-value for high-resolution structures. Barplot displaying correlation (R-value) between different biophysical features and K_d when using only high-resolution structures (red bars) and all structures (grey bars).

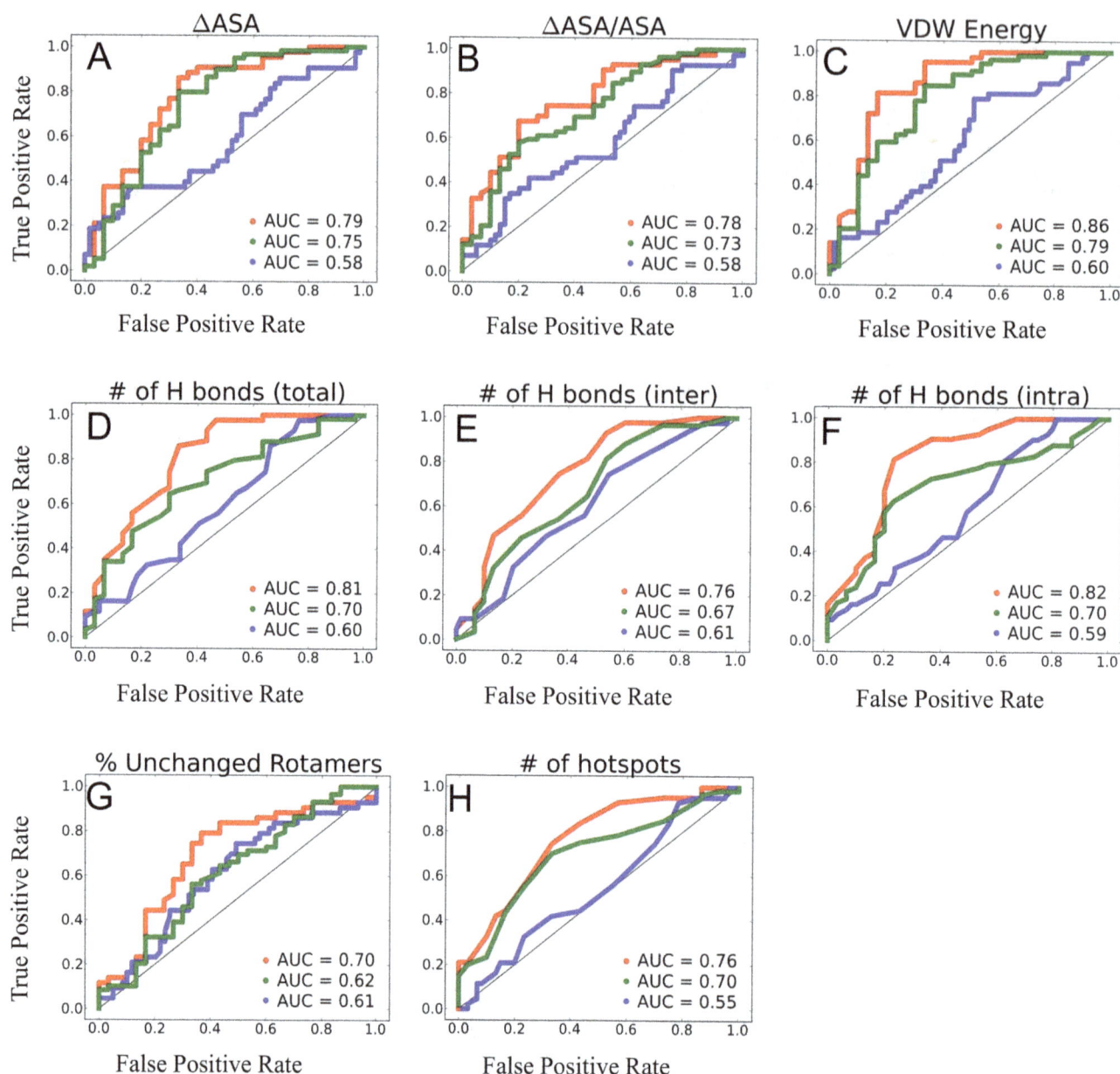

Figure 4. Receiver Operator Characteristic Analysis. The graph shows the true positive rate vs. false positive rate in discriminating high- from low-affinity PPIs (red line), medium- from low-affinity PPIs (green line) and high- from medium-affinity PPIs (blue line) for each feature. Each point represents a particular cut-off value used to discriminate between the two groups. Features included in the figure are (A) ΔASA, (B) ΔASA/ASA, (C) Van der Waals energy, (D) the total number of interfacial H bonds, (E) the number of intermolecular interfacial H bonds, (F) the number of intra-molecular H bonds; (G) Percentage of rotamers that do not change conformation upon binding; and (H) the number of hotspots.

affinity complexes this ratio reduces to 1:1. We thus conclude that cysteins bring rigidity to the main chain of the interface, lowering the entropic cost of binding.

Conformational changes upon binding

Protein-protein binding is a thermodynamic process that involves gain of favorable enthalpy and unfavorable reduction in entropy upon complex formation. However, the entropic term of binding is difficult to calculate and has been estimated only in a few studies [5,25,62,63,64]. The Kastritis database [25] allows for estimation of the entropic term by analyzing conformational changes associated with binding on both the backbone and the side chain level.

We first analyzed backbone conformational changes associated with binding by calculating iRMSD between Cα atoms of the bound and the unbound conformations. For all complexes in the database, no measurable correlation was observed between binding affinity and iRMSD (Figure 1K). However, when analyzing only rigid complexes (N = 69) we observed that iRMSD decreases with increased affinity with a correlation of R = 0.35 (Figure S5). This result indicates that iRMSD is a good measure of contribution of conformational movements to binding energetics when such movements are small. However, when large conformational changes are involved iRMSD becomes a poor measure of binding affinity.

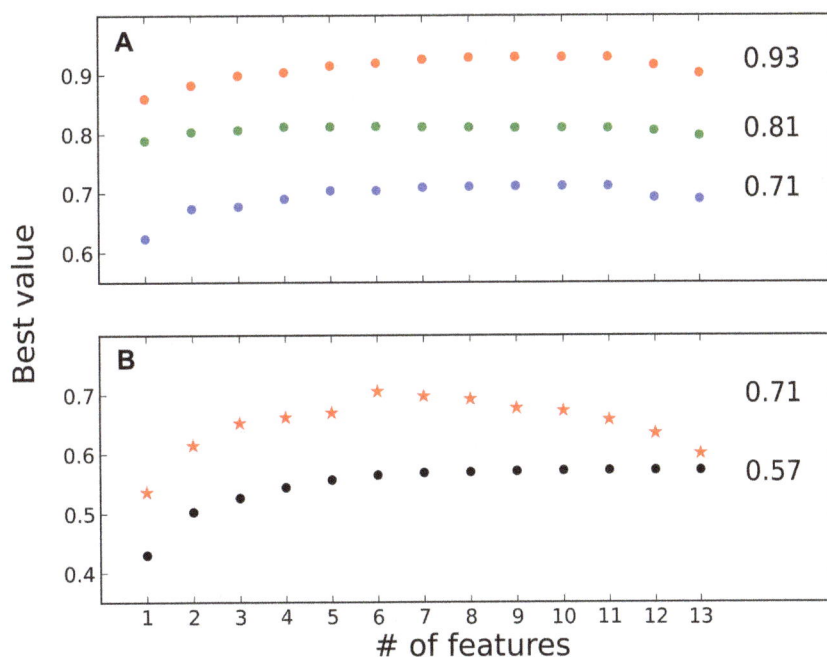

Figure 5. Incorporating more features in the prediction improves correlation with K_d and ROC analysis. The best possible weights were obtained to combine the features into one equation using a linear fit to the experimental data. X-axis shows the number of features used to predict K_d and to discriminate between the two groups. Y-axes shows the best value obtained for each number of features used in the equation. The analysis was performed on all structures in the database (filled circles) and on high-resolution structures only (red stars). (A) AUC were evaluated on high- vs low-affinity (red), medium- vs low-affinity (green) and medium- vs high-affinity (blue) PPIs (B) Pearson's correlation coefficient for all dataset (filled circles) and for high-resolution structures only (red stars).

We next measured side chain conformational rearrangements for interface residues due to binding by calculating the differences in χ1 and χ2 angles between the bound and the unbound states. We further correlated percentage of side chains that did not change conformation upon association to binding affinity. We found that when all complexes are considered, the correlation between percentage of residues that do not change conformation and affinity is moderate (R = 0.27, Figure 1L). This value however increases substantially to 0.51 if we restrict our analysis to high-resolution X-ray structures (see section on high-resolution structures). Previous reports demonstrated that some key side chains in binding interfaces are pre-oriented for binding [20,65]. Our results corroborate these findings and further reveal that high-affinity complexes are more likely to have their side chains pre-orientated for binding compared to low-affinity complexes. Such side chain pre-orientation minimizes the entropy loss upon binding and thus increases affinity. This feature however, could not be computed accurately for low-resolution structures.

High-resolution structures show better correlation to binding affinity for most of the features

We thought that modest correlation with binding affinity for most biophysical features could be in part due to the limited resolution of the structures under study. We hence compared the R-values calculated for all structures in the dataset to those calculated for only high-resolution files (X-ray structures with a resolution <2.5 Å for both the complex and the free components, 37 PPIs). To exclude possible bias coming from lowering the number of samples in our analysis, we performed the same calculation using only low-resolution structures in our dataset. Interestingly, we find that most features show a considerable improvement in R-values when only high-resolution structures are considered (Figure 3). In contrast,

decrease or no change in R-value was observed for most features for low-resolution structures (Figure S6). Very substantial improvement in R-values for high-resolution structures is shown for iRMSD and the percentage of side chains that did not change conformation upon binding (corresponding to 0.40 and 0.51, respectively), indicating that these features are most sensitive to resolution of the structure. Overall, our analysis shows that low resolution of structures is a limitation that results in inaccuracies of binding affinity predictions and some biophysical descriptors are more sensitive than others to such inaccuracies.

Distinguishing between low-, medium-, and high-affinity complexes

Using ROC analysis, we tested how each of the explored biophysical features can distinguish between high-, medium-, and low-affinity PPIs. In this analysis we included only thirteen features that showed some correlation with binding affinity (Table S2). Our results show that several single features could distinguish well between high- and low-affinity PPIs with Area Under the Curve (AUC) values reaching 0.86 and 0.81, for the VdW energy and the total number of H bonds, respectively (Figure 4, red curves). Slightly lower AUC values are obtained when distinguishing between medium- and low-affinity complexes (Figure 4, green curves). However, worse results are obtained when distinguishing high-affinity complexes from medium-affinity complexes, pointing to apparent similarity of these two groups (Figure 4, blue curves). Inability to discriminate between these two groups of PPIs is probably due to the composition of out database that contains many PPIs with a K_d near the cutoff between high- and medium-affinity groups (Figure S1).

We further explored whether our predictions could be improved by combining several features into one formula (see Methods).

Figure 5A and B shows that both AUC and R-values could be increased if several features are combined in one linear equation reaching the maximum R-value of 0.57 and maximum AUCs of 0.93, 0.81 and 0.71 for discriminating between high- and low-, medium- and low- and high- and medium-affinity PPIs, respectively. However, introduction of more than four features does not improve predictions substantially, indicating that the considered features are probably interdependent and/or overfitting is observed. Finally, when we restrict our dataset to high-resolution structures, the correlation with binding affinity significantly improves for all predictions using combinations of features reaching a maximum R-value of 0.71 (Figure 5B, red stars).

Conclusions

In this work we tested how a number of features derived from a PPI structure correlate with PPI binding affinity. The features that showed the highest correlation with binding affinity are the total number of H bonds, geometric complementarity measured by the van der Waals energy, and side chain conformational changes for high-resolution X-ray structures. In spite of moderate correlation and high variability, a number of tested single biophysical features could be used to discriminate between high- and low-affinity complexes as well as between medium- and low-affinity complexes with high significance and hence could be used to predict the range of affinities from structure. These features include not only those determining enthalpic contribution to binding but also those directly related to the entropic term such as iRMSD, change in side chain conformation, the number of disulfide bonds and intramolecular H bonds. Correlation with binding affinity for most of the studied features could be improved by restricting the analysis to high-resolution X-ray structures. Combining several biophysical features into one equation results in further improvement of our predictions and allows for unequivocal discrimination between high-, medium- and low-affinity PPIs. Finally, incorporating the information for both bound and unbound states improves the accuracy of the binding affinity predictions and could be utilized for developing new energy functions for design of PPIs.

Supporting Information

Figure S1 Density estimation of logKd. Kernel density estimation of the probability density function of the logKd for the whole dataset. Each circle represents one pdb file in the dataset.

Figure S2 Unsatisfied Interfacial H bonds. Number of unsatisfied hydrogen bonds vs. Kd. Each point represents one PDB file in the database.

Figure S3 Geometric complementarity vs. Kd. Surface complementarity calculated using the Katzir score (A) and the Sc core (B) vs. Kd. Each point represents one PDB file in the database and the line corresponds to a linear fit to all data points in the database.

Figure S4 Electrostatic interactions vs. Kd. (A) the number of interfacial salt bridges, (B) the number of repulsive electrostatic interactions and (C) the intermolecular Coulombic electrostatic energy calculated with the Poisson-Boltzmann equation vs. Kd. Each point represents one PDB file in the database and the line corresponds to a linear fit to all data points in the database.

Figure S5 iRMSD vs. Kd. iRMSD of rigid structures (iRMSD ≤1) vs. Kd. Each point represents one PDB file in the database and the line corresponds to a linear fit to all data points in the database.

Figure S6 R-values for All and low-resolution structures. Barplot displaying the correlation of different PPI features to affinity using only low-resolution structures (blue bars) compared to all structures in the database (grey bars). On the y-axis, the Pearson's correlation to affinity. On the x-axis, the different features whose correlation to affinity was measured.

Table S1 Pearson correlation coefficient R and p-values for the three methods used for surface complementarity analysis computed over the entire database, rigid, and flexible complexes.

Table S2 List of the 13 different biophysical features that were considered in a linear combination to fit experimental Kd values. linear combination to fit experimental Kd values.

Acknowledgments

We would like to thank Prof. Emil Alexov for helping us with the use of Delphi software.

Author Contributions

Conceived and designed the experiments: AE JMS. Performed the experiments: AE ER. Analyzed the data: AE ER. Contributed reagents/materials/analysis tools: AE ER. Wrote the paper: AE ER JMS.

References

1. Gao GF, Tormo J, Gerth UC, Wyer JR, McMichael AJ, et al. (1997) Crystal structure of the complex between human CD8alpha(alpha) and HLA-A2. Nature 387: 630–634.
2. Garcia KC, Scott CA, Brunmark A, Carbone FR, Peterson PA, et al. (1996) CD8 enhances formation of stable T-cell receptor/MHC class I molecule complexes. Nature 384: 577–581.
3. Kobe B, Deisenhofer J (1995) A structural basis of the interactions between leucine-rich repeats and protein ligands. Nature 374: 183–186.
4. Vicentini AM, Kieffer B, Matthies R, Meyhack B, Hemmings BA, et al. (1990) Protein chemical and kinetic characterization of recombinant porcine ribonuclease inhibitor expressed in Saccharomyces cerevisiae. Biochemistry 29: 8827–8834.
5. Jones S, Thornton JM (1996) Principles of protein-protein interactions. Proc Natl Acad Sci U S A 93: 13–20.
6. Keskin O, Nussinov R (2007) Similar binding sites and different partners: implications to shared proteins in cellular pathways. Structure 15: 341–354.

7. Bahadur RP, Chakrabarti P, Rodier F, Janin J (2004) A dissection of specific and non-specific protein-protein interfaces. J Mol Biol 336: 943–955.
8. Tsai CJ, Xu D, Nussinov R (1997) Structural motifs at protein-protein interfaces: protein cores versus two-state and three-state model complexes. Protein Sci 6: 1793–1805.
9. Janin J, Bahadur RP, Chakrabarti P (2008) Protein-protein interaction and quaternary structure. Q Rev Biophys 41: 133–180.
10. Janin J (1995) Principles of protein-protein recognition from structure to thermodynamics. Biochimie 77: 497–505.
11. Xu D, Tsai CJ, Nussinov R (1997) Hydrogen bonds and salt bridges across protein-protein interfaces. Protein Eng 10: 999–1012.
12. Audie J, Scarlata S (2007) A novel empirical free energy function that explains and predicts protein-protein binding affinities. Biophys Chem 129: 198–211.
13. Horton N, Lewis M (1992) Calculation of the free energy of association for protein complexes. Protein Sci 1: 169–181.

14. Jiang L, Gao Y, Mao F, Liu Z, Lai L (2002) Potential of mean force for protein-protein interaction studies. Proteins 46: 190–196.

15. Ma XH, Wang CX, Li CH, Chen WZ (2002) A fast empirical approach to binding free energy calculations based on protein interface information. Protein Eng 15: 677–681.

16. Qin S, Pang X, Zhou HX (2011) Automated prediction of protein association rate constants. Structure 19: 1744–1751.

17. Su Y, Zhou A, Xia X, Li W, Sun Z (2009) Quantitative prediction of protein-protein binding affinity with a potential of mean force considering volume correction. Protein Sci 18: 2550–2558.

18. Zhang C, Liu S, Zhu Q, Zhou Y (2005) A knowledge-based energy function for protein-ligand, protein-protein, and protein-DNA complexes. J Med Chem 48: 2325–2335.

19. Kastritis PL, Bonvin AM (2010) Are scoring functions in protein-protein docking ready to predict interactomes? Clues from a novel binding affinity benchmark. J Proteome Res 9: 2216–2225.

20. Fleishman SJ, Whitehead TA, Strauch E, Corn JE, Qin S, et al. (2011) Community-wide assessment of protein-interface modeling suggests improvements to design methodology. J Mol Biol 414: 289–302.

21. Eastman J, Wilson EJ, Cervenansky C, Rosenberry TL (1995) Fasciculin 2 binds to the peripheral site on acetylcholinesterase and inhibits substrate hydrolysis by slowing a step involving proton transfer during enzyme acylation. J Biol Chem 270: 19694–19701.

22. Karlsson E, Mbugua PM, Rodriguez-Ithurralde D (1984) Fasciculins, anticholinesterase toxins from the venom of the green mamba Dendroaspis angusticeps. J Physiol (Paris) 79: 232–240.

23. Radić Z, Duran R, Vellom DC, Li Y, Cervenansky C, et al. (1994) Site of fasciculin interaction with acetylcholinesterase. J Biol Chem 269: 11233–11239.

24. Tian F, Lv Y, Yang L (2012) Structure-based prediction of protein-protein binding affinity with consideration of allosteric effect. Amino Acids 43: 531–543.

25. Kastritis PL, Moal IH, Hwang H, Weng Z, Bates PA, et al. (2011) A structure-based benchmark for protein-protein binding affinity. Protein Sci 20: 482–491.

26. Moal IH, Agius R, Bates PA (2011) Protein-protein binding affinity prediction on a diverse set of structures. Bioinformatics 27: 3002–3009.

27. Vreven T, Hwang H, Pierce BG, Weng Z (2012) Prediction of protein-protein binding free energies. Protein Sci 21: 396–404.

28. Word JM, Lovell SC, Richardson JS, Richardson DC (1999) Asparagine and glutamine: using hydrogen atom contacts in the choice of side-chain amide orientation. J Mol Biol 285: 1735–1747.

29. Lee B, Richards FM (1971) The interpretation of protein structures: estimation of static accessibility. J Mol Biol 55: 379–400.

30. Gordon DB, Marshall SA, Mayo SL (1999) Energy functions for protein design. Curr Opin Struct Biol 9: 509–513.

31. Mayo SL, Olafson BD, Goddard WA (1990) DREIDING: a generic force field for molecular simulations. J. Phys. Chem. 94: 8897–8909.

32. Sharabi O, Yanover C, Dekel A, Shifman JM (2011) Optimizing energy functions for protein-protein interface design. J Comput Chem 32: 23–32.

33. Lawrence MC, Colman PM (1993) Shape complementarity at protein/protein interfaces. J Mol Biol 234: 946–950.

34. Kuhlman B, Dantas G, Ireton GC, Varani G, Stoddard BL, et al. (2003) Design of a novel globular protein fold with atomic-level accuracy. Science 302: 1364–1368.

35. Katchalski-Katzir E, Shariv I, Eisenstein M, Friesem AA, Aflalo C, et al. (1992) Molecular surface recognition: determination of geometric fit between proteins and their ligands by correlation techniques. Proc Natl Acad Sci U S A 89: 2195–2199.

36. Hammersley JM (1960) Monte Carlo Methods for Solving Multivariable Problems. Ann. New York Acad. Sci. 86: 844–874.

37. Horn BKP (1987) Closed-form solution of absolute orientation using unit quaternions. J. Opt. Soc. Am. A. 4: 629–642.

38. Kortemme T, Kim DE, Baker D (2004) Computational alanine scanning of protein-protein interfaces. Sci STKE 2004(219): pl2, 2004.

39. Honig B, Nicholls A (1995) Classical electrostatics in biology and chemistry. Science 268: 1144–1149.

40. Chen J, Sawyer N, Regan L (2013) Protein-protein interactions: general trends in the relationship between binding affinity and interfacial buried surface area. Protein Sci 22: 510–515.

41. Li W, Hamill SJ, Hemmings AM, Moore GR, James R, et al. (1998) Dual recognition and the role of specificity-determining residues in colicin E9 DNase-immunity protein interactions. Biochemistry 37: 11771–11779.

42. Fleury D, Wharton SA, Skehel JJ, Knossow M, Bizebard T (1998) Antigen distortion allows influenza virus to escape neutralization. Nat Struct Biol 5: 119–123.

43. Gilbreth RN, Koide S (2012) Structural insights for engineering binding proteins based on non-antibody scaffolds. Curr Opin Struct Biol 22: 413–420.

44. Sammond DW, Eletr ZM, Purbeck C, Kimple RJ, Siderovski DP, et al. (2007) Structure-based protocol for identifying mutations that enhance protein-protein binding affinities. J Mol Biol 371: 1392–1404.

45. Fischer E (1894) Einfluss der Configuration auf die Wirkung der Enzyme. Chem Ber 27: 2985–3198.

46. Connolly ML (1986) Atomic size packing defects in proteins. Int J Pept Protein Res 28: 360–363.

47. Hubbard SJ, Argos P (1994) Cavities and packing at protein interfaces. Protein Sci 3: 2194–2206.

48. Eriksson AE, Baase WA, Zhang XJ, Heinz DW, Blaber M, et al. (1992) Response of a protein structure to cavity-creating mutations and its relation to the hydrophobic effect. Science 255: 178–183.

49. Lee C, Park SH, Lee MY, Yu MH (2000) Regulation of protein function by native metastability. Proc Natl Acad Sci U S A 97: 7727–7731.

50. Sonavane S, Chakrabarti P (2008) Cavities and atomic packing in protein structures and interfaces. PLoS Comput Biol 4: e1000188.

51. Cunningham BC, Wells JA (1989) High-resolution epitope mapping of hGH-receptor interactions by alanine-scanning mutagenesis. Science 244: 1081–1085.

52. Kim DE, Chivian D, Baker D (2004) Protein structure prediction and analysis using the Robetta server. Nucleic Acids Res 32: W526–31.

53. Hendsch ZS, Tidor B (1994) Do salt bridges stabilize proteins? A continuum electrostatic analysis. Protein Sci 3: 211–226.

54. Xu D, Lin SL, Nussinov R (1997) Protein binding versus protein folding: the role of hydrophilic bridges in protein associations. J Mol Biol 265: 68–84.

55. Barlow DJ, Thornton JM (1983) Ion-pairs in proteins. J Mol Biol 168: 867–885.

56. Brooks BR, Bruccoleri RE, Olafson BD, States DJ, Swaminathan S, et al. (1983) CHARMM: A program for macromolecular energy, minimization, and dynamics calculations. J. Comput. Chem. 4: 187–217.

57. Cornell WD, Cieplak P, Bayly CI, Gould IR, Merz KM, et al. (1995) A Second Generation Force Field for the Simulation of Proteins, Nucleic Acids, and Organic Molecules. J. Am. Chem. Soc. 117: 5179–5197.

58. Li L, Li C, Sarkar S, Zhang J, Witham S, et al. (2012) DelPhi: a comprehensive suite for DelPhi software and associated resources. BMC Biophys 5: 9.

59. Glaser F, Steinberg DM, Vakser IA, Ben-Tal N (2001) Residue frequencies and pairing preferences at protein-protein interfaces. Proteins 43: 89–102.

60. Keskin O, Bahar I, Badretdinov AY, Ptitsyn OB, Jernigan RL (1998) Empirical solvent-mediated potentials hold for both intra-molecular and inter-molecular inter-residue interactions. Protein Sci 7: 2578–2586.

61. Levy ED (2010) A simple definition of structural regions in proteins and its use in analyzing interface evolution. J Mol Biol 403: 660–670.

62. Chothia C, Janin J (1975) Principles of protein-protein recognition. Nature 256: 705–708.

63. Goh CS, Milburn D, Gerstein M (2004) Conformational changes associated with protein-protein interactions. Curr Opin Struct Biol 14: 104–109.

64. Page MI, Jencks WP (1971) Entropic contributions to rate accelerations in enzymic and intramolecular reactions and the chelate effect. Proc Natl Acad Sci U S A 68: 1678–1683.

65. Kimura SR, Brower RC, Vajda S, Camacho CJ (2001) Dynamical view of the positions of key side chains in protein-protein recognition. Biophys J 80: 635–642.

11

On the Importance of the Distance Measures Used to Train and Test Knowledge-Based Potentials for Proteins

Martin Carlsen[1], Patrice Koehl[2], Peter Røgen[1]*

1 Department of Applied Mathematics and Computer Science, Technical University of Denmark, Kongens Lyngby, Denmark, **2** Department of Computer Science and Genome Center, University of California Davis, Davis, CA, United States of America

Abstract

Knowledge-based potentials are energy functions derived from the analysis of databases of protein structures and sequences. They can be divided into two classes. Potentials from the first class are based on a direct conversion of the distributions of some geometric properties observed in native protein structures into energy values, while potentials from the second class are trained to mimic quantitatively the geometric differences between incorrectly folded models and native structures. In this paper, we focus on the relationship between energy and geometry when training the second class of knowledge-based potentials. We assume that the difference in energy between a decoy structure and the corresponding native structure is linearly related to the distance between the two structures. We trained two distance-based knowledge-based potentials accordingly, one based on all inter-residue distances (PPD), while the other had the set of all distances filtered to reflect consistency in an ensemble of decoys (PPE). We tested four types of metric to characterize the distance between the decoy and the native structure, two based on extrinsic geometry (RMSD and GTD-TS*), and two based on intrinsic geometry (Q* and MT). The corresponding eight potentials were tested on a large collection of decoy sets. We found that it is usually better to train a potential using an intrinsic distance measure. We also found that PPE outperforms PPD, emphasizing the benefits of capturing consistent information in an ensemble. The relevance of these results for the design of knowledge-based potentials is discussed.

Editor: Yang Zhang, University of Michigan, United States of America

Funding: The first author is financed by internal funding (public, no grant number) from the Technical University of Denmark. PK acknowledges support from the NIH. The funders had no role in study design, data collection and analysis, decision to publish, or preparation of the manuscript.

Competing Interests: The authors have declared that no competing interests exist.

* Email: prog@dtu.dk

Introduction

Proteins are the essential macromolecules inside cells that perform nearly all cellular functions. Just like macroscopic tools, their shapes is a key feature for defining their functions. Structural biologists have embarked upon the challenge of finding the structures of all proteins, in hopes of unraveling this relationship between geometry and biological activity and learn in the process how cells function. Determining experimentally the structure of a protein at the atomic level however is not yet an easy task: this can be indirectly deduced from the fact that we currently know millions of protein sequences but less than hundred thousand protein structures. Predicting the structure of a protein from first principles is not much easier: direct applications of the ideas that have been used for modeling small molecules have not yet been successful on these much larger molecules. Recent reports on the advancements of *ab initio* techniques clearly show that the protein structure prediction community is making progress, but that the quality of the models they generate do not meet yet the stringent accuracy requirements to become useful to the biologists [1]. Interestingly, the series of Critical Assessment of protein Structure Prediction (CASP) meetings have highlighted that while the methods for generating models of protein structures have improved significantly [2], identifying the native-like conforma-

tions among the large collections of model structures (also called decoys) remains a significant challenge [3,4]. In this paper we focus on this problem.

Anfinsen's thermodynamics hypothesis states that the native structure of a protein is determined only by its amino acid sequence [5]. Structural and computational biologists translate this postulate into the statement, that under physiological conditions, the native state of a protein is a unique, stable minimum of the free energy. The key to solving the protein structure prediction problem amounts therefore to finding an accurate representation of this free energy function and several methods have been proposed to construct reasonable approximations of it. The two most common approaches rely on semiempirical and statistical potentials, respectively. Semiempirical methods are derived from knowledge of the basic physical principles whereas statistical potentials are based on the nonrandom statistics of known protein structures [6]. Statistical energy functions are either residue based or atom based and the most recent statistical potentials include pairwise interactions, orientations of side-chains [7], secondary structural preferences, solvent-exposure, and other geometric properties of proteins [8]. We note that there have been attempts to combine physics-based and statistics-based potentials to improve protein structure refinement [9–13].

Current protein structure prediction methods require potentials that ideally should assign "scores" to a protein structure model such that the higher the score, the less native-like the model is, where native-like is measured in terms of a distance d from the model to the native structure. If this condition is satisfied then the potential is expected to detect near native conformations even when the native conformation is not present; in addition, such an ideal potential could then be used for model refinement. In mathematical terms this can be expressed as the score function f satisfying

$$f(seq_i, \mathbf{r}_i + \mathbf{dr}) = f(seq_i, \mathbf{r}_i) + d(\mathbf{r}_i, \mathbf{r}_i + \mathbf{dr}), \tag{1}$$

for any sequence seq_i and all deformations $\mathbf{dr}$ of its native structure $\mathbf{r}_i$.

Several methods have been developed to optimize potentials towards this goal [14–17]. The choice of the distance measure d is critical to the success of these methods. The standard distance measure when comparing protein structural models is RMSD, i.e. the root mean square distance between the two models after optimal translation and rotation. RMSD however has been replaced in recent CASP experiments by the global distance test (GDT-TS [18]) due to its undesirable sensitivity towards local changes in a protein structure; GDT-TS has become one of the most commonly used distance measures in protein structure prediction. A less commonly used distance measure is the fraction of known native contacts, Q. Q quantifies the changes in the number of "contacts" found in the native structure compared to the model structure that is evaluated, where a contact corresponds to two residues being within a given threshold distance from each other. All the distance measures mentioned above identify geometric differences between two structural models but do not attempt to assess if these differences could be assigned to fluctuations due to the dynamics of the protein. Such differences would be less of a concern if they were related to geometric differences that can be explained by dynamics. As an attempt to identify the role of dynamics, Perez et al. recently introduced FlexE, a method based on a simple elastic network model that uses the deformation energy as a measure of the similarity between two structures [19]. As such, FlexE is expected to distinguish biologically relevant conformational changes from random changes.

In this work, we investigate the importance of the distance function d when optimizing an energy function f towards satisfying equation 1. We train two new $C\alpha$-based pairwise potentials, PPD and PPE, to mimic the distance between the model structure considered and its corresponding native structure, using four different definitions of the distance measure, namely RMSD, GDT-TS, Q, and MT, where MT is an anharmonic version of FlexE. These energy functions are trained and tested on sets extracted from the high resolution decoy dataset Titan-HRD [20], as well as on well known decoy datasets from DecoysRUs [21] and Rosetta [22]. We have also analyzed the performance of our potentials on the server generated Stage_1 and Stage_2 decoy sets from CASP 10 [48].

The paper is organized as follows. The next section introduces the different distance measures and describes our procedures for training and testing the potentials PPD and PPE. The following section shows the results on different decoy sets as well as a comparison between PPD, PPE, two statistical knowledge-based potentials and a semi-empirical physical potential. We conclude with a discussion of the importance of the choice of the distance measure and describe potential future work.

Materials and Methods

Geometrical distances between two structural models of the same protein

Let us consider two structural models A and B of the same protein P with N amino acids. We represent the two models as discrete sets of N points, $A = (a_1, a_2, \ldots, a_N)$ and $B = (b_1, b_2, \ldots, b_N)$ where the points a_i and b_i correspond to the positions of the $C\alpha$ atoms i in the two structures. We assume that the correspondence table between A and B is known and set such that a_i corresponds to b_i for all $i \in [1, N]$. We measure the distance between the two models either based on the Euclidean distance between the two sets of points (RMSD and GDT-TS), on differences between contact maps within each set (Q), or on an elastic network (MT).

RMSD, i.e. root mean square deviation, is the Euclidean distance between the corresponding points a_i and b_i after one of the two sets of points (usually set B) has been optimally transformed by a rigid body transformation G:

$$RMSD = \min_G \sqrt{\frac{\sum_{i=1}^{N} \|a_i - G(b_i)\|^2}{N}}. \tag{2}$$

The rigid body transformation G is a transformation that does not produce changes in the size, shape, or topology of the protein. Such transformations are compositions of rotations and translations. Many closed-form solutions to the problem of finding the optimal G have been derived [23–25]. We note that RMSD as defined above is a metric [26].

RMSD is a distance measure based on the L_2 norm; as such, it is highly sensitive to outliers, for example due to the presence of large albeit local differences between the two structures. The global distance test (GDT) was developed to decrease this sensitivity [18]. GDT focuses on the regions of the structures that can be correctly aligned by counting the number of residues that can be superimposed within a given cutoff distance. GDT-TS (where TS stands for Total Score), combines this information for multiple cutoffs:

$$GDT - TS = \frac{n_1 + n_2 + n_4 + n_8}{4n}, \tag{3}$$

where n_1, n_2, n_4, and n_8 are the numbers of aligned residues within 1, 2, 4, and 8 Ångströms, respectively, and n is the total aligned length. Note that GDT-TS is a quantity between 0 and 1 that represents similarity, with low values corresponding to bad correspondences, and high values (close to or equal to 1) indicating that the two models are highly similar. We have converted this similarity measure into a distance by considering GDT-TS* = 1-GDT-TS.

RMSD and GDT-TS* are computed after the two model structures have been optimally superposed. An alternative approach is to consider the intrinsic geometry of the two structures, as captured for example by a distance matrix that contains all $C\alpha - C\alpha$ distances internal to one structure. Q and MT are two examples of distance measures that use this alternate approach.

The fraction of native contacts, Q, is a distance measure that quantifies the changes of a contact map between two models for the same structure. A contact map is usually defined as

$$S_{i,j} = \begin{cases} 1 & \text{if residues } i \text{ and } j \text{ are in contact} \\ 0 & \text{otherwise,} \end{cases}$$

where two residues are in contact if they are within a given distance threshold. In this paper, we set this threshold to 9 Å. Q is then defined by

$$Q = \frac{sc}{sc + lc},$$

where sc is the number of shared contacts and lc is the number of lost contacts. Just like GDT-TS, Q is a measure of similarity. We convert it into a distance measure by defining $Q^* = 1 - Q$.

Q^* quantifies changes in the contact map of a structure with no consideration of what could have been the reasons for these changes. FlexE is a new measure of similarity between protein structures that was introduced as an attempt to distinguish those changes that are biologically relevant [19]. It is based on the concept of elastic network that assigns virtual isotropic springs between pairs of residues. Elastic network models are used in normal mode analysis [27,28] for example to reconstruct proteins [29], to generate decoy sets [30], or to investigate thermal fluctuations about the native or equilibrium structure [31,32]. In the formalism introduced by Perez et al [19], the distance measure FlexE between two structures N and D is assimilated to the energetic cost of deforming one of the structures into the other:

$$FlexE(N,D) = \frac{1}{N_{\text{res}}} \sum_{i,j=1}^{N_{\text{res}}} S_{i,j}^N k_{ij} \left(r_{ij}^N - r_{ij}^D \right)^2, \qquad (4)$$

where N_{res} is the number of residues in N and D, $S_{i,j}^N$ is a contact map for structure N, r_{ij}^N and r_{ij}^D are the distances between the Cα atoms of residues i and j in structures N and D, respectively, and k_{ij} is a force constant associated to the link between i and j. In our implementation of FlexE, we set all force constants to 1. We modify the quadratic term in equation 4 with a term congruent to the potential introduced by Toda [33] to study chains of particles interacting with non-linear forces.

The corresponding variant of FlexE, which we name MT, is defined as:

$$MT(N,D) = \frac{1}{N_{\text{res}}} \sum_{i,j=1}^{N_{\text{res}}} \frac{S_{i,j}^N}{b^2} \left(e^{-(r_{ij}^D - r_{ij}^N)b} + \left(r_{ij}^D - r_{ij}^N \right) b - 1 \right), \qquad (5)$$

where b is a parameter which we set to 0.5. We note that MT is equal to FlexE for small perturbations of the distances between residues; for large perturbations however, it penalizes compression more than extension. Finally the use of the fixed native contact map for all native-decoy comparisons ensures that both Flex-E(N,D) and MT(N,D) are well-defined.

Two new parametric potentials

A smooth, pairwise potential, PPD. We design a smooth knowledge based residue pair potential as done in [34]. For each of the 210 pairs of amino acids types we assume a potential that is determined by the corresponding Cα-Cα distance. We model the

interaction as a uniform cubic b-spline with compact support within 1 Å to 12 Å and 8 degrees of freedom, see e.g. [35]. With this model an interaction tends smoothly to zero energy at distances greater than 12 Å and is modeled freely within 4 Å–9 Å. The pair potential has $8 \times 210 = 1680$ parameters in total. The corresponding potential, PPD, is defined as

$$PPD = \sum_{i<j} \sum_p C_p^{aa(i)aa(j)} B_p(r_{i,j}), \qquad (6)$$

where $aa(i) \in \{1, \ldots, 20\}$ is the amino acid type of the i-th residue and $B_p(r_{i,j})$ is the p-th b-spline basis function evaluated on the distance between the i-th and j-th residues. $C_p^{aa(i)aa(j)}$ are the model parameters determined by the optimization procedure described below.

A consensus potential, PPE. We introduce a novel smooth ensemble based pair potential (PPE) that forms an artificial funnel relative to a pre-calculated contact map:

$$PPE = \sum_{i<j} S_{i,j} \sum_p C_p^{aa(i)aa(j)} B_p(r_{i,j}), \qquad (7)$$

where $S_{i,j}$ is an consensus contact map. The method to calculate the consensus contact map is described below. It is based on a similar consensus method that constructs the reference contact map from an ensemble of decoys [36].

A consensus contact map. We introduce an iterative method to compute a consensus contact map of an ensemble of decoys. The first step is to construct a contact map from the most common contacts in the ensemble. Let $M_{i,j}$ be the fraction of contacts in the ensemble for the i,j-th residue pair. The contact map is then calculated as

$$S_{i,j} = \begin{cases} 1 & \text{if } M_{i,j} > \mu \\ 0 & \text{otherwise} \end{cases} \qquad (8)$$

where μ is a cut-off fixed at 0.25. At each step, we select the 25% closest decoys to this contact map, where "closest" refers to the Hamming-distance to the contact map. This leads to a reduced ensemble from which a new contact map is computed, and the procedure is iterated. The algorithm usually converges in a few steps.

Optimizing the potentials

We design an energy landscape using a sculpting procedure. We assume that we possess a set of natives structures $\{N_i\}$ and that a set $\{D_{i,j}\}$ of decoy structures is known for each of these native structures. Let $\Delta E_{i,j}$ be the energy difference between the i-th native structure, N_i, and its j-th decoy, $D_{i,j}$, and let $d(N_i, D_{i,j})$ be the corresponding distance between N_i and $D_{i,j}$. Our method for optimizing a statistical potential [34] attempts to establish a funnel-shaped energy function by calculating the parameters that minimizes the sum of squared errors between $\Delta E_{i,j}$ and $\alpha_{N_i} d(N_i, D_{i,j})$ where α_{N_i} is a constant of proportionality. The problem can be stated as a quadratic programming (QP) problem with affine constraints,

$$\underset{X,\alpha_1\ldots\alpha_M}{\text{minimize}} \quad \sum_{i,j} \|\Delta E_{i,j}(X) - \alpha_{N_i} d(N_i, D_{i,j})\|^2 + \beta\|X\|^2$$

$$\text{subject to} \quad 0.25 \le \alpha_{N_i} \le 4, \quad \text{for} \quad i = 1\ldots M \qquad (9)$$

$$\sum_i \alpha_{N_i} = M,$$

where β is a fixed parameter used for regularization. The variables in this QP problem are X, i.e. the vector of coefficients $C^{i,j}$ introduced above, and the constants of proportionality $\alpha_{N_1}\ldots\alpha_{N_M}$, where M is the number of proteins in the training set. The last term $\beta\|X\|^2$ is a regularization term that adds a penalty onto the modulus of X. The preprocessing is trivially parallelizable since each of the terms, $\|\Delta E_{i,j}(X) - \alpha_i d(N_i, D_{i,j})\|^2$, can be calculated individually. As a consequence, the QP requires little memory and is fast to compute. We use the optimization package cplex to solve it.

Training and test sets

It is a nontrivial task to construct a "good" set of decoy structures. Any such decoy set relies on a sampling of the conformational space accessible to the protein structure of interest. The specific techniques used to generate such sampling are prone to biases [37], leading to poor sampling of the corresponding free energy surfaces. These approximate energy surfaces may not adopt a funnel like geometry in the neighborhood of the native structure and may contain many artificial potential energy barriers. To avoid the risk of learning from a specific bias introduced by one sampling technique, we have considered a variety of test sets to train and measure the performances of our energy functions. Of particular interest to us are near-native test sets since we design energy functions to mimic the neighborhoods of native structures.

We have chosen part of the Titan High Resolution Decoy set [20] as our training set. The list of proteins included in this set was originally proposed by Zhou and Skolnik [17]; it was selected on the basis that it is composed of a representative set of nonhomologous single domain proteins with maximum pairwise sequence similarity reported to be 35%. The models included in the decoy sets were generated using the torsion angle dynamics program DYANA [38] subject to distance constraints that are set to preserve the hydrophobic core of a protein. It is assumed that the hydrophobic core includes all residues within a β strand as well as all hydrophobic residues within an α-helix. The set includes 1400 proteins in total (compared to 1489 proteins in the original set of Zhou and Skolnik [17]). We eliminated all short proteins with a large radius of gyration as these proteins are overfitted by the optimization and are usually separate stretched secondary structures. We divided the remaining proteins into a training set of 1155 proteins with an average of 994 decoys per native structure (Titan-HRD*) and a test set of 142 proteins with an average of 854 decoys per native structure (Titan-HRD). The average GDT-TS distances between native and decoys over the training and test sets are 0.75 and 0.76 with a mean absolute deviation of 0.1, respectively. Note that we will use the mean absolute deviation (the l_1-norm) instead of the standard deviation (the l_2-norm) as it puts less weight on outliers.

Apart from the Titan-HRD set we use 10 freely available decoy sets that were generated using different procedures. These include 6 sets taken from DecoysRUs [21] (4 state reduced [39], hg structal [21], fisa [40], fisa casp3 [40], lmds [41] and lattice ssfit [42,43]). We also included two older versions of the Rosetta decoy

sets (Rosetta-All [44], Rosetta-Tsai [22]), the newest version Rosetta-Baker available at http://depts.washington.edu/bakerpg/decoys/ and the I-Tasser Set II [45].

The different CASP meetings have highlighted successes and failures in generating model structures that resemble the native structures of proteins. A repository of all models that have been proposed as answers to the prediction challenges that were part of these meetings is available on the CASP web page (http://predictioncenter.org). This repository provides a wealth of information on protein structure modeling, as well as useful test cases to assess the quality of new potential energy functions. We have therefore considered five CASP sets each containing models predicted by a variety of methods from the different CASP meetings (302 ensembles in total). We also generated CASP-HRD, a high resolution decoy subset of CASP 5–9, which includes models that have a TM score [46] larger than 0.5 and a RMSD less than 4 Å to the native structures. This cutoff was chosen based on the observation made by Xu and Zhang, which states that two decoys belong to the same fold when their TM-score to a native structure is higher than 0.5 [47]. CASP-HRD is constructed to have nearly the same average distance measure value as Titan-HRD but we find smaller variations of the distance measures for CASP-HRD. In that sense, it does include variations with different structural characteristics compared to Titan-HRD as it is generated by many different methods, while Titan-HRD is more homogeneous.

The total number of ensembles excluding Titan-HRD, Titan-HRD*, and CASP-HRD is 546 with an average GDT-TS between its decoys and their corresponding native structures of 0.47 with a average mean absolute deviation of 0.16. We refer to this set as "Test Set All" (TSA).

Finally, we include decoys from the latest CASP experiment, CASP10. A critical component of the CASP experiment is the assessment of the predictions that are submitted as putative models for the target proteins considered. This assessment is performed by the CASP assessors but also by the CASP community, with considerable enthusiasm, as observed in CASP10 [48]. The procedure for assessing the predictions in CASP10 differed from that of previous CASPs. The main difference was the introduction of two stages, labeled Stage_1 and Stage_2. For the former, twenty of the supposedly best predictions for each CASP target were released for assessment. Subsequently, hundred and fifty decoys were released for each target, defining Stage_2. Stage_1 ensembles are designed to survey single model assessment methods, while stage_2 allows for the survey of methods that rely on ensembles for the assessment of models. We have considered 93 targets from CASP10 for which both Stage_1 and Stage_2 test sets are available from the CASP web site (http://www.predictioncenter.org/casp10/). Compared to the other decoy sets described above, these sets contain longer protein chains. The models they include are usually as distant from their native counterparts as observed for the datasets from the previous CASP meetings. These sets however are more compact, i.e. with less diversity in distances, especially for the Stage_2 sets that resemble the CASP-HRD sets in that respect.

In table 1, we report the mean characteristics of these decoy sets (size, diversity, ...) as well as information about their availability.

Preprocessing the decoy sets. To guarantee that the decoys included in a set are consistent in length with their corresponding native structure, we performed the following two-step preprocessing. First, we removed all residues in the decoys with missing backbone atoms ($C\alpha$, N, C, and O). Second, we extracted the sequences from the decoy structure files and aligned these sequences with the native sequence of the protein of interest

Table 1. Properties of the different protein decoy sets used in this study.

Decoy set	Nprot[h]	Nres[h]	Ndecoys[h]	RMSD	MT	GDT-TS	Q
Titan-HRD[a]	142	127 (35)	854 (119)	2.4 (0.5)	2.7 (1)	0.76 (0.1)	0.85 (0.04)
Titan-HRD*[a]	1155	111 (35)	994 (138)	2.6 (0.6)	2.7 (1)	0.75 (0.1)	0.85 (0.04)
TASSER Set II[b]	55	80 (17)	438 (98)	6.3 (1.5)	9.3 (3.2)	0.54 (0.05)	0.77 (0.03)
hg Structal[c]	28	150 (7)	29 (0)	4.1 (1.2)	4.4 (1.5)	0.71 (0.07)	0.85 (0.04)
4-state[c]	7	64 (4.9)	664 (15)	5.2 (1.4)	8.3 (2.9)	0.53 (0.11)	0.75 (0.05)
fisa[c]	4	60 (10)	500 (0.4)	7.5 (1.8)	8.6 (1.7)	0.47 (0.06)	0.75 (0.06)
fisa CASP3[c]	5	88 (15)	1437 (390)	12 (1.6)	21 (4.1)	0.3 (0.03)	0.67 (0.02)
lmds[c]	10	53 (10)	433 (79)	7.7 (1.1)	12 (2.6)	0.46 (0.04)	0.72 (0.03)
lattice ssfit[c]	8	71 (10)	1997 (1.5)	9.9 (1.0)	17 (2.4)	0.3 (0.03)	0.64 (0.02)
Rosetta-All[d]	41	82 (25)	999 (0.5)	12 (1.4)	29 (5.6)	0.27 (0.03)	0.61 (0.02)
Rosetta-Tsai[d]	29	63 (9.4)	1862 (43)	7.4 (2.1)	11 (3.9)	0.46 (0.08)	0.73 (0.04)
Rosetta-Baker[d]	57	88 (20)	100 (0)	8.5 (1.4)	15 (3.3)	0.45 (0.05)	0.76 (0.03)
CASP5[e]	41	202 (78)	117 (41)	13 (3.7)	29 (14)	0.38 (0.12)	0.68 (0.08)
CASP6[e]	39	172 (71)	216 (34)	13 (4.9)	27 (16)	0.39 (0.12)	0.70 (0.08)
CASP7[e]	64	183 (80)	349 (40)	10 (3.4)	17 (10)	0.47 (0.11)	0.75 (0.07)
CASP8[e]	77	187 (81)	334 (67)	8.8 (3.1)	13 (8.6)	0.54 (0.11)	0.79 (0.06)
CASP9[e]	81	180 (81)	402 (95)	11 (4.9)	19 (14)	0.49 (0.12)	0.77 (0.07)
CASP-HRD[e]	109	188 (79)	192 (72)	2.8 (0.4)	2.2 (0.6)	0.76 (0.03)	0.89 (0.02)
CASP10-stage1[f]	93	232 (102)	18 (1.9)	13 (4.3)	20 (9.4)	0.46 (0.08)	0.76 (0.05)
CASP10-stage2[f]	93	232 (102)	132 (7.6)	11 (3.7)	17 (8.2)	0.55 (0.03)	0.80 (0.03)
TSA TM>0.5[g]	242	179 (77)	291 (119)	6.3 (2.67)	9.4 (5.5)	0.63 (0.09)	0.82 (0.05)
TSA TM <0.5[g]	303	110 (48)	602 (436)	12 (3.9)	23 (12)	0.34 (0.1)	0.68 (0.07)

[a]Training set (Titan HRD) and test set (Titan HRD*) from the Titan High resolution decoy set [20], available at http://titan.princeton.edu/2010-10-11/Decoys/.
[b]Tasser Set II is a structurally non-redundant set of protein structures and decoys derived with the program TASSER. It is available at http://zhanglab.ccmb.med.umich.edu/decoys/.
[c]Decoy sets from the Decoys 'R' us repository http://dd.compbio.washington.edu.
[d]Different decoy Rosetta-based decoy sets (see text for details), available at http://depts.washington.edu/bakerpg/decoys/.
[e]Collection of models from the successive CASP5 to CASP9 experiments, available from the CASP web site http://predictioncenter.org. CASP-HRD is a high resolution subset of the union of the five sets CASP5 to CASP9, which includes models that have a TM-score larger than 0.5 and a RMSD less than 4 Å to the native structures.
[f]The Stage_1 and Stage_2 decoy sets used in the CASP10 quality assessment category, available from the CASP web site http://predictioncenter.org. For details on how these sets are prepared, see [48].
[g]All high and low resolution targets (TSA TM-score>0.5)/(TSA TM-score <0.5) are listed in Files S1 and S2 respectively found in the supporting information.
[h]Nprot is the number of different proteins in the dataset, Nres is the average number of residues computed over all proteins in a dataset, and Ndecoys is the average number of decoys per proteins, averaged over the dataset. RMSD, MT, GDT-TS, and Q are the distance measures between the decoys and the corresponding native structures, averaged over all decoys and all proteins. We provide both the average values and the average mean absolute deviations (in parenthesis).

(where the native sequence is derived from the ATOM record in the corresponding PDB file). If these alignments include trailing unmatched residues either in the decoys or in the native structure, these residues are removed until all sequences are identical. We found that this procedure was necessary for some of the decoy sets described above.

Assessing the quality of decoy selection: R-score

Given a distance measure and an energy function, an ensemble of decoy protein conformations contains a "best" distance model, i.e. the conformation that is closest geometrically to the native structure, as well as a "best" energy model, i.e. the model whose energy is the lowest. Ideally, these two "best" models should be the same; in practice however, they are different due to shortcomings of the potential energy function. To quantify this difference we introduce the R-score as follows. Let $\mathcal{D}$ be the ensemble of decoys and let X_i be one of its elements. The corresponding native structure is N. We define the mapping S_d from $\mathcal{D}$ to $\mathbb{R}$ as $S_d(X_i) = d(X_i, N)$, i.e. the distance between the decoy X and N, where d can be any of the four distance measures defined above. We name X_E the decoy with the lowest energy, i.e. $E(X_E) \leq E(X)$ $\forall X \in \mathcal{D}$. In parallel, we name X_d the decoy closest to N with respect of the distance d, i.e. $S_d(X_d) \leq S_d(X)$ $\forall X \in \mathcal{D}$. The R score for d and E is defined as:

$$R(d,E) \equiv \begin{cases} \dfrac{S_d(X_E) - \langle S_d \rangle}{S_d(X_d) - \langle S_d \rangle} & \text{if } |S_d(X_E) - \langle S_d \rangle| \leq |S_d(X_d) - \langle S_d \rangle| \\ -1 & \text{otherwise} \end{cases}, \quad (10)$$

where $\langle S_d \rangle$ is the average value for S_d over the decoy set $\mathcal{D}$. $R(d,E)$ is designed to assess how well E mimics S in finding the best decoy. It takes values between -1 and 1 where 1 indicates that the energy has picked the best decoy. We fix the lower limit at -1 to avoid having outliers being assigned very low negative values. Note, that if an ensemble does not contain outliers then 0 is the random expectation. If we furthermore assume that the distances $S_d(X)$ are uniformly distributed then $(1 - R(d,E))/2$ is the fraction of decoys with a distance to the native structure better than $S_d(X_E)$. The R score can also be seen as the ratio between the Z-score of the best energy model, $(S_d(X_E) - \langle S_d \rangle)/\sigma(S_d)$, and the Z-score of the best distance model, $(S_d X_d - \langle S_d \rangle)/\sigma(S_d)$, where $\sigma(S_d)$ is the standard deviation for S_d over the decoy set $\mathcal{D}$.

Assessing how well the energy functions mimic a funnel in the neighborhood of the native structure

To measure how far the energy E is from the desired linear funnel shape given by Equation 1 relative to the distance measure d we report the Pearson's correlation coefficient $Corr(d,E)$ between the energy values $E(X_i)$ and distance measures $S_d(X_i)$ over all decoys X_i in the decoy set:

$$Corr(d,E) = \frac{1}{N-1} \sum_{i=1}^{N} \frac{S_d(X_i) - \langle S_d \rangle}{\sigma(S_d)} \frac{E(X_i) - \langle E \rangle}{\sigma(E)}, \quad (11)$$

where $\langle . \rangle$ and $\sigma(.)$ stand for the mean and standard deviation over the decoy set considered.

Comparing two distance measures d_1 and d_2

In the two previous subsections, we have defined a R-score $R(d,E)$ and a correlation coefficient $Corr(d,E)$ to measure how well an energy function E mimics a distance measure d. Both quantities can be used as is to compare two distance measures d_1 and d_2. Indeed, d_2 can be assimilated to a pseudo energy function, akin to the definition of FlexE given in equation 4. The R-score and correlation coefficient between d_1 and d_2 are then simply $R(d_1, d_2)$ and $Corr(d_1, d_2)$, respectively. $Corr(d1, d2)$ measures the dependence between $d1$ and $d2$ over a decoy set, while $R(d1, d2)$ checks the "quality" of the best decoy identified by d_2, as measured by d_1. Note that this R-score between distance measures may not be symmetric.

Results and Discussion

The diversity of the distance measures

There is no unique way to compare three dimensional shapes. When comparing protein structures, two main classes of distance measures have been proposed, those based on a Euclidean distance between the positions of the atoms of the two proteins (after proper translation and rotation of one of them), and those based on the intrinsic geometry of the structures. We have considered two examples in each class, namely RMSD and GDT-TS* for the former, and MT and Q* for the latter. A full description of these four distance metrics is given in Material and Methods. As these measures capture changes of different geometric properties of the protein structures, there is no reason to believe that they are equivalent. To test the degrees to which these distances differ, we have compared them on three different sets of decoys, namely Titan-HRD, CASP-HRD, and TSA, using two different report scores, $Corr$ and R, where $Corr$ is the Pearson's correlation coefficient that measures how well d_1 mimics d_2 over a large range of distance values while R measures how (metrically) wrong the best candidate of one distance measure (i.e. the decoy with the smallest distance to its corresponding native structure) is when measured by another distance (see Materials and Methods for details). Results for $Corr$ and R are given in tables 2 and 3, respectively.

The correlations between the distance measures are high on the Titan-HRD set of decoys, with values above 0.87 for the correlation coefficients. The corresponding R-scores are above 0.76. If we assume uniform distributions of the native-decoy distances over a decoy set, the best decoy by one distance measure on average is ranked within the top 5% and within the top 12% by another distance measure for R scores of 0.9 and 0.76, respectively. These high scores are expected, as the Titan-HRD decoys are high resolution, usually very close to their native structure counterparts (see Table 1). It is interesting however that the R score between RMSD and Q* is relatively low (0.76), even on this high resolution data set. This low value indicates that a "good" decoy defined by Q* may explore a range of RMSD values. In contrast, a decoy that is close to the native structure with respect to RMSD usually has a high percentage of native contacts, as highlighted by the R score between Q* and RMSD of 0.87. In fact, we observe that the best RMSD decoy is generally scored better by the three other distance measures.

While CASP-HRD also contains high resolution decoys that are close to their corresponding native structures (with RMSD <4 Å and TM scores above 0.5), the four distance measures we tested are less dependent on this dataset than on Titan-HRD, both globally as scored by correlation coefficients and locally (i.e. in picking a "best" decoy), as highlighted by the R scores. We see two possible reasons for these differences between the two groups of decoy sets. First, the decoys in Titan-HRD are homogeneous, as they all contain the same hydrophobic cores as the native structures. In contrast, the CASP decoys were derived with many different methods, leading to heterogeneity in their geometry.

Table 2. Correlations between the four distance measures.

Test set	Distance d_1	Distance d_2			
		RMSD	MT	GDT-TS*	Q*
Titan-HRD	RMSD	1[a]	0.92 (0.06)	0.92 (0.04)	0.87 (0.08)
	MT	0.92 (0.06)	1	0.92 (0.03)	0.94 (0.03)
	GDT-TS*	0.92 (0.04)	0.92 (0.03)	1	0.95 (0.03)
	Q*	0.87 (0.08)	0.94 (0.03)	0.95 (0.03)	1
CASP-HRD	RMSD	1	0.74 (0.16)	0.73 (0.14)	0.6 (0.19)
	MT	0.74 (0.16)	1	0.72 (0.13)	0.83 (0.07)
	GDT-TS*	0.73 (0.14)	0.72 (0.13)	1	0.74 (0.13)
	Q*	0.6 (0.19)	0.83 (0.07)	0.74 (0.13)	1
CASP10-stage1	RMSD	1	0.83 (0.16)	0.71 (0.24)	0.68 (0.24)
	MT	0.83 (0.16)	1	0.73 (0.2)	0.82 (0.14)
	GDT-TS*	0.71 (0.24)	0.73 (0.2)	1	0.86 (0.12)
	Q*	0.68 (0.24)	0.82 (0.14)	0.86 (0.12)	1
CASP10-stage2	RMSD	1	0.78 (0.16)	0.51 (0.22)	0.49 (0.19)
	MT	0.78 (0.16)	1	0.52 (0.2)	0.69 (0.14)
	GDT-TS*	0.51 (0.22)	0.52 (0.2)	1	0.64 (0.17)
	Q*	0.49 (0.19)	0.69 (0.14)	0.64 (0.17)	1
TSA	RMSD	1	0.92 (0.06)	0.8 (0.15)	0.82 (0.11)
	MT	0.92 (0.06)	1	0.78 (0.14)	0.85 (0.08)
TM-score> 0.5	GDT-TS*	0.8 (0.15)	0.78 (0.14)	1	0.89 (0.12)
	Q*	0.82 (0.11)	0.85 (0.08)	0.89 (0.12)	1
TSA	RMSD	1	0.8 (0.12)	0.59 (0.24)	0.56 (0.18)
	MT	0.8 (0.12)	1	0.54 (0.2)	0.68 (0.14)
TM-score <0.5	GDT-TS*	0.59 (0.24)	0.54 (0.2)	1	0.67 (0.22)
	Q*	0.56 (0.18)	0.68 (0.14)	0.67 (0.22)	1

[a]Pearson's correlation coefficient $Corr(d_1,d_2)$ between the two distance measures d_1 and d_2. We provide both the average value and the mean absolute deviation (in parenthesis) over the data set considered.

Second, we cannot exclude an effect of sample size, as on average the sets included in Titan-HRD contain four times more decoys and larger average mean absolute deviation of distance measures than the sets included in CASP-HRD (see Table 1).

TSA, which stands for "Test Sets All" is a large heterogeneous collection of decoy sets that were generated by many different techniques (see Materials and methods for details). Some of these decoy sets are high-resolution, i.e. contains mostly native-like structures, while others are more diverse, containing decoys that are very different from their corresponding native structures, both in terms of secondary structure content and three-dimensional organization. To assess the importance of this diversity, we selected within the TSA group of decoy sets two subgroups, those for which the decoys have average TM score larger than 0.5, and those with average TM score smaller than 0.5. This 0.5 cutoff was again chosen based on the observation made by Xu and Zhang that two decoys belong to the same fold when their TM-scores to a native structure is higher than 0.5 [47]. Table 1 shows that TSA TM-score> 0.5 generally contain longer chains with fewer decoys when compared to the TSA TM-score <0.5 set. The two sets are fully listed in File S1 and File S2. Tables 2 and 3 show that the distance measures behave on the high-resolution subgroup (TM> 0.5) as on the Titan-HRD test set, i.e. with high correlations and high R scores, meaning that they are very similar to each other. On the low-resolution subgroup (TM <0.5) however, the distance

measures are poorly correlated with each other, with most correlation coefficients in the range 0.5 to 0.7. Both results confirm that when two structures are very close to each other, different distance measures quantify their differences in a similar manner. When the two structures however are very different, different distance measures will focus on different geometric differences, leading to differences in their behaviors. We observe however one exception in Table 2, in that RMSD and MT clearly remains correlated (0.80) even for the diverse subgroup of TSA with TM <0.5. The reason for this exception is unclear.

The CASP 10 Stage_1 and Stage_2 test sets usually include longer proteins than the other sets considered here, with decoys that are far from their native counterparts. In the Stage_1 sets there are very few decoys per target (by construction, see Methods above) and relatively large average mean deviations of the distance measures. For the Stage_2 test sets there are more decoys per target; these decoys however are usually very similar to each other, leading to very low mean absolute deviations for the GDT-TS* and Q* distance measures, and consequently to low correlations and R scores between the measures. As an example, the correlation between RMSD and GDT-TS* for the Stage_2 decoy sets is only 0.51 and their non symmetric R scores are R(RMSD,GDT-TS*) = 0.71 and R(GDT-TS*,RMSD) = 0.73, respectively. These low values are good indicators of significant

Table 3. Comparing the best models picked by different distance measures.

Test set	Distance d_1	Distance d_2			
		RMSD	MT	GDT-TS*	Q*
Titan-HRD	RMSD	1[a]	0.88 (0.12)	0.91 (0.09)	0.76 (0.17)
	MT	0.94 (0.06)	1	0.92 (0.08)	0.91 (0.07)
	GDT-TS*	0.96 (0.04)	0.94 (0.07)	1	0.91 (0.08)
	Q*	0.87 (0.09)	0.92 (0.07)	0.89 (0.09)	1
CASP-HRD	RMSD	1	0.71 (0.26)	0.79 (0.22)	0.49 (0.38)
	MT	0.76 (0.22)	1	0.76 (0.22)	0.76 (0.23)
	GDT-TS*	0.8 (0.22)	0.68 (0.27)	1	0.48 (0.39)
	Q*	0.57 (0.33)	0.81 (0.16)	0.66 (0.24)	1
CASP10-stage1	RMSD	1	0.81 (0.24)	0.75 (0.31)	0.79 (0.23)
	MT	0.9 (0.13)	1	0.85 (0.19)	0.94 (0.09)
	GDT-TS*	0.79 (0.24)	0.78 (0.24)	1	0.82 (0.2)
	Q*	0.78 (0.22)	0.88 (0.14)	0.8 (0.23)	1
CASP10-stage2	RMSD	1	0.76 (0.22)	0.71 (0.3)	0.63 (0.29)
	MT	0.83 (0.18)	1	0.73 (0.24)	0.83 (0.19)
	GDT-TS*	0.73 (0.26)	0.65 (0.24)	1	0.59 (0.29)
	Q*	0.62 (0.29)	0.82 (0.18)	0.62 (0.23)	1
TSA	RMSD	1	0.9 (0.11)	0.84 (0.19)	0.81 (0.18)
	MT	0.94 (0.07)	1	0.88 (0.14)	0.92 (0.09)
TM-score> 0.5	GDT-TS*	0.85 (0.16)	0.79 (0.21)	1	0.73 (0.24)
	Q*	0.79 (0.18)	0.89 (0.11)	0.81 (0.16)	1
TSA	RMSD	1	0.83 (0.19)	0.73 (0.27)	0.71 (0.27)
	MT	0.87 (0.14)	1	0.74 (0.27)	0.88 (0.14)
TM-score <0.5	GDT-TS*	0.74 (0.27)	0.7 (0.27)	1	0.67 (0.27)
	Q*	0.68 (0.27)	0.85 (0.16)	0.68 (0.27)	1

[a]R-score $R(d_1,d_2)$ between the two distance measures d_1 and d_1. We provide both the average value and the mean absolute deviation (in parenthesis) over the data set considered.

differences between their ranking of the decoys included in CASP10 Stage_2 test sets.

Training knowledge-based potentials with different distance measures

We have derived two new smooth knowledge-based residue pair potentials, PPD and PPE. Both potentials are based on distances between the $C\alpha$ atoms of the protein structure of interest. For each of the 210 types of amino acid pairs, the two potentials are written as a weighted sum of smooth spline functions, whose weights are optimized so that the total energy of a protein model resembles the distance between the model and a reference structure (usually taken to be the native structure), as described by equation 1. The two potentials differ however on which pairs of residues are taken into account. While PPD includes all pairs of residues from the protein structure P considered, PPE only include those pairs whose inter $C\alpha$ distance is consistently below a cutoff value in an ensemble of protein models similar to P. The idea behind PPE, derived from Eickholt et al. [36], is that the various models in the ensemble contain complementary information which can be pooled together to build a contact map of consistent residue-residue contacts that are more likely to be informative. Our interest here is to assess the influence of the distance measure used to train the two potentials. We have trained PPD and PPE on the Titan-HRD* training set with the four distance measures

introduced above separately, and tested the corresponding four versions of the potentials against the Titan-HRD, CASP-HRD, and TSA test sets in their abilities to mimic any of the four distance measures. All parameters describing the amino acid pair spline potentials are listed in the file Force Field S1. The encoding used and the spline basis used is described in Readme Force Field S1. Both files are in the supporting information.

Figure 1 shows some examples of the b-spline expanded pair potentials. As expected, the pair potentials are repulsive for short inter-residue distances and have a first minimum between 4 Å and 6 Å and this preferred distance relatively independent of the training metric. For longer pair distances it is seen that most PPD pair potentials have a local minimum around 10 Å whereas the PPE pair potentials tend to have a local maximum at this distance. One plausible explanation is that as PPE does not identify new contacts for these large distances; it may then set higher energy values for remote decoys. The exact placement of the minimum as well as the depth of the potential differs for the different pair potentials. While these differences may seem small, they add up when we sum over all the interactions.

We computed both the correlations between energy and the distance measure, and the R scores that compare the best decoys picked based on energy with the decoys closest to their corresponding native structures. Results are given in Table 4 for the correlation coefficients, Table 5 for the R scores, and in

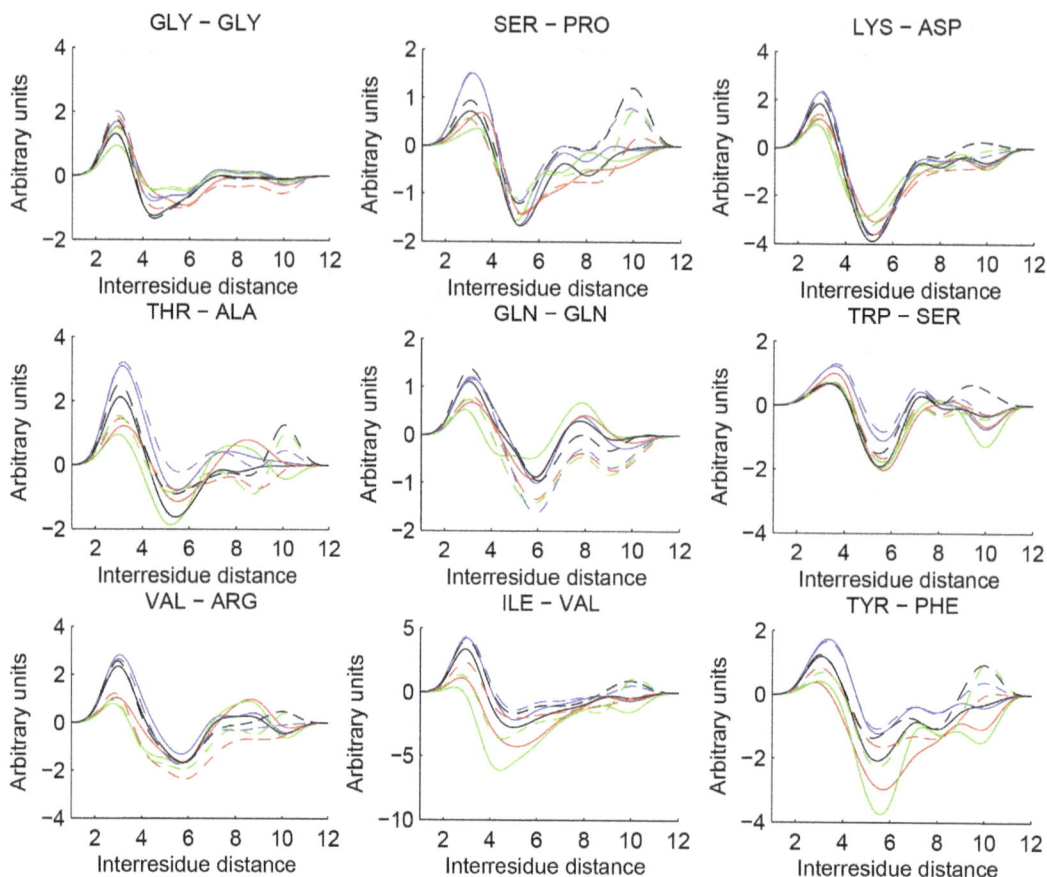

Figure 1. Showing nine different types of residue pair interactions for our single model method PPD (continuous lines) and our consensus method PPE (dotted lines) when trained on RMSD (blue), MT(red), GDT-TS(green) and Q(black).

Figures 2 and 3 for a comparison of these scores. We draw from these tables and figures the four main conclusions described below.

First, we find that both potentials PPD and PPE perform very well on the Titan-HRD test set, for all distance measures used for training and testing the potential. The corresponding mean correlation coefficients (averaged over all decoys sets in Titan-HRD) are usually above 0.8, indicating that the energy functions order the decoys in the same manner as the distance measures. In parallel, the R scores are also high, with most values well above 0.65, indicating that the decoys with the lowest energies are usually among the decoys that are close to the corresponding native structures. We should note however that PPD and PPE were trained on Titan-HRD*. While Titan-HRD and Titan-HRD* are different (see Methods), they both contain decoys that were generated with the same principles, with the significant constraint that they maintain the hydrophobic cores of the corresponding native structures. The exceptional performance of PPD and PPE may therefore not be surprising in light of this comment. Indeed, as we test these potentials on different decoy sets with more diverse populations of decoys, we observe a decrease in performance that follows the increase in diversity (in the order Titan-HRD - TSA (TM >0.5) - CASP-HRD - TSA (TM <0.5). This decrease in performance is illustrated in Figure 2.

Second, the ensemble potential PPE performs better than the single structure potential PPD, again for all the distance measures used to train and test the potentials. The differences between the two potentials are large for the high resolution decoys sets in Titan-HRD and TSA (TM>0.5), but become statistically insig-

nificant for very diverse decoy sets such as those in TSA (TM < 0.5). We believe that these differences illustrate the power of generating consensus information from an ensemble. In PPE, we only consider those contacts there are consistently below a given distance cutoff in the whole decoy set to which the protein of interest belongs. This initial filtering is clearly an advantage for Titan-HRD, as it will select the contacts in the hydrophobic cores which are native, and will ignore the contacts that fluctuate significantly due to the sampling procedure used to generate the decoys. It remains an advantage for high quality decoy but becomes less pertinent for highly diverse decoys.

Third, the performances of the two potentials PPD and PPE depend on the choice of the distance used in the training step. For example, the correlations between PPE and any of the four distance measures increase on average by 0.09 when it is trained on MT instead of RMSD (Table 4). Similar differences are observed for the R scores between PPE and the four distance measures (Table 5). More generally, it is best to train the potentials on a distance measure that is directly based on intrinsic inter-residue distances, such as MT that follows the elastic network of the protein of interest, or Q* that counts the number of contacts that fall below a given distance cutoff, than on a distance measure based on extrinsic Euclidean distances, such as RMSD. Interestingly, we find that GDT-TS* behaves more like the intrinsic distance measures MT and Q* than RMSD, even though it is also based on extrinsic distances. The reason for this discrepancy is unclear.

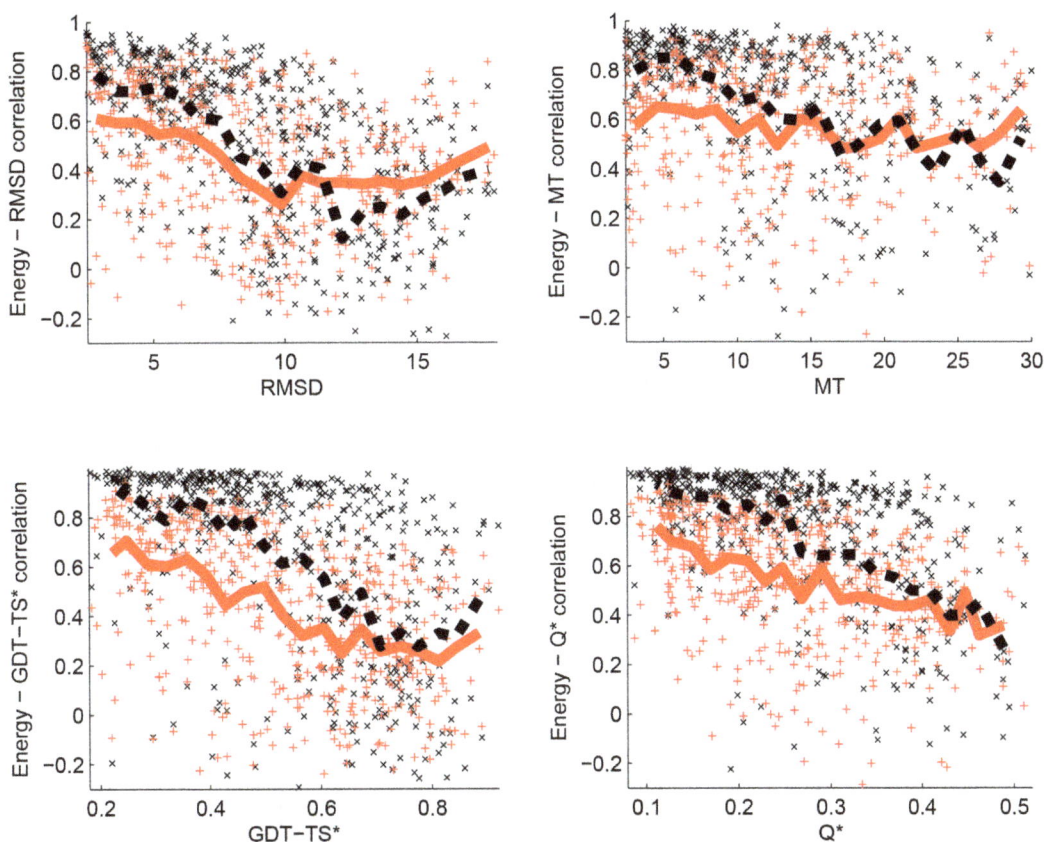

Figure 2. Energy-distance correlations as a function of the quality of the decoy set. For each decoy set in Titan-HRD, CASP-HRD, and TSA (a total of 797 sets), we plot the correlation Corr(E, d_1) as a function of the mean value of d_1 over the decoy set, where E is either the PPD energy (red, plus sign +) or the PPE energy (black, cross sign x) trained on the set Titan-HRD with the distance measure d_1, and d_1 is one of the fourth distance measures considered, namely RMSD (panel A), MT (panel B), GDT-TS* (panel C), and Q* (panel D). The corresponding running means computed over 20 equidistant intervals for PPD (red, solid line) and PPE (black, dashed line) are shown. Clearly, the quality of the correlation energy-distance decreases as the diversity of the decoy set increases.

Finally, we observe that the ability of an energy function to pick a "good" decoy (i.e. with native-like characteristics) is contingent to how well this energy function correlates with a distance measure between decoys and native structure. This is illustrated in Figure 2. This observation validates the approach of sculpting (training) a potential to mimic a distance measure.

Comparison with other energy functions

We have compared the two energy functions PPD and PPE with two well established all-atom statistical potentials RAPDF [49] and GOAP [7] and with a semi-empirical physical potential, AMBER99SB-ILDN [50], for all decoy sets in Titan-HRD, CASP-HRD, and TSA. Results for correlations between energy and distance measures and for R scores are given in Tables 4 and 5, respectively.

As intuitively expected, the performances of AMBER99SB-ILDN are very poor. This is most likely an artifact due to the presence of a few steric clashes in the decoys, and not a reflection of the quality of this potential. While it would be possible to improve this performance by applying an initial energy minimization on all decoys, this result by itself highlights that such a physical potential cannot be used directly to order a set of decoys, unless some pre-processing is applied.

RAPDF is a knowledge-based statistical potential that is based on a direct conversion of the distributions of inter-atomic distances

observed in native protein structures into energy values that are then used to assess how native-like a model is [49]. It is not based on any information from existing decoy sets, and it is not trained to mimic some differences between decoys and native structures. It is therefore not surprising that it does not perform as well as PPD and PPE, especially on the Titan-HRD as both PPD and PPE were trained on decoys resembling those included in this data set.

GOAP is an all-atom orientation-dependent knowledge-based statistical potential that includes a distance-based term and an angle-dependent contribution [7]. The distance-based term is an all-atom statistical potential that is based on the reference state that was introduced with the DFIRE potential [51]. The angle dependent component of GOAP is based on the geometric orientation of local planes. GOAP is found to perform significantly better than RAPDF on all datasets tested in this study. This is not a surprise, as GOAP includes much more information than RAPDF due to its angle term. We find however that GOAP performs only marginally better than PPD and worse than PPE. This illustrates the benefit of training a potential on a decoy set. PPD and PPE are only Ca based potentials; they have been trained however to mimic distances between non-native models and native structures of proteins.

The performances of RAPDF and GOAP depend on the distance measure used for testing. We observe that they are particularly good when the statistical potentials are tested on

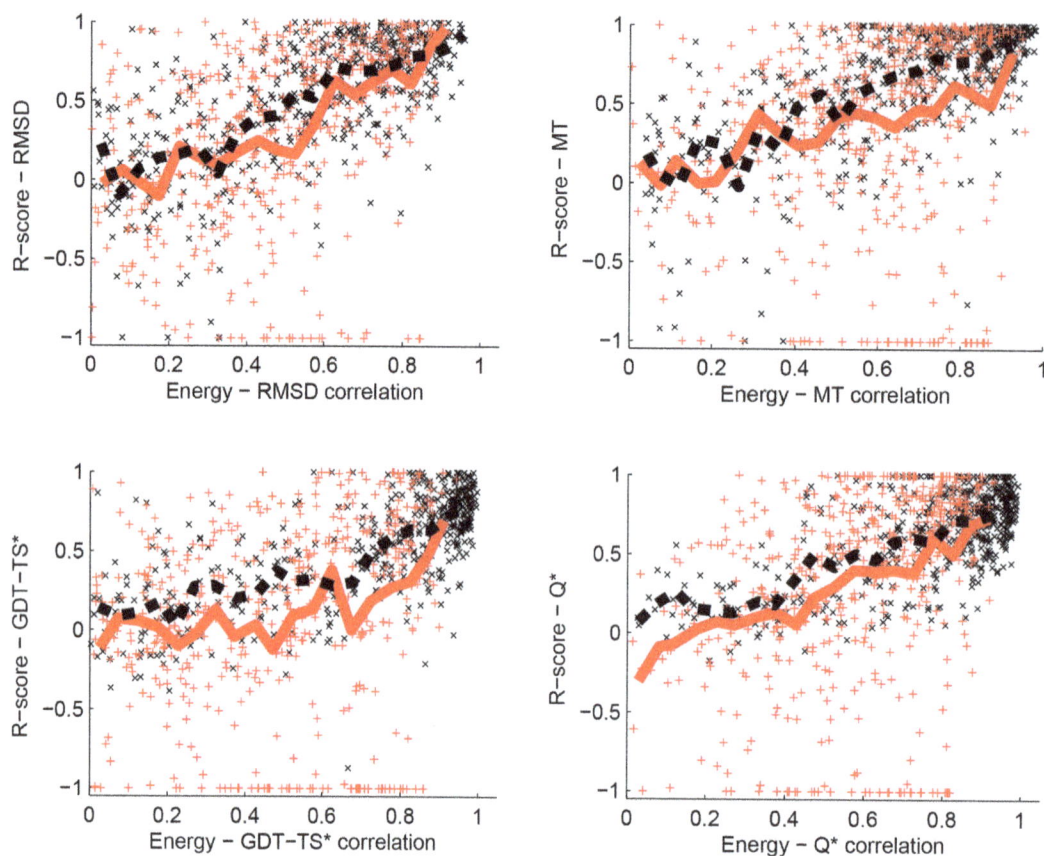

Figure 3. R scores versus Energy-distance correlations. For each decoy set in Titan-HRD, CASP-HRD, and TSA, we plot the R score $R(d_1, E)$ as a function of the correlation coefficient $Corr(d_1, E)$, where E is either the PPD energy (red, plus sign +) or the PPE energy (black, cross sign x) trained on the set Titan-HRD with the distance measure d_1, and d_1 is one of the fourth distance measures considered, namely RMSD (panel A), MT (panel B), GDT-TS* (panel C), and Q* (panel D). The corresponding running means computed over 20 equidistant intervals for PPD (red, solid line) and PPE (black, dashed line) are shown. Note that $R(d_1, E)$ compares the best decoy picked based on the energy value E with the decoy closest to the native structure according to the distance measure d_1. There is a clear correlation between these two values for all four distance measures.

GDT-TS*, reflecting the differences between these distance measures (see Table 2 and 3).

Performance in the CASP 10 quality assessment category

As part of the CASP experiment, state-of-the-art methods for protein structure assessment are judged on their ability to evaluate the quality of the predictions submitted as models for the targets considered in that specific experiment: this is the quality assessment category (QA). In 2012 as part of CASP10, 37 groups participated [48]. They were asked to evaluate the quality of sets of predictions (decoys) in two rounds designated as Stage_1 (20 decoys with a large variation in quality as measured by GDT-TS) and Stage_2 (150 decoys with homogeneous quality as measured by GDT-TS). The main reason for providing a small number of decoys in Stage_1 was to allow for judging assessment methods that rely on a single model independently from methods that rely on an ensemble of decoys (consensus methods), that would be tested extensively with the Stage_2 decoy sets. The three main conclusions drawn from these experiments were [48]: 1) The performances of the single model methods are usually worse than the the performances of consensus methods, 2) The Stage_2 sets are usually more difficult to rank than the Stage_1 sets, and 3) No methods were able to consistently pick the best decoy in an ensemble. The results for the participating groups can be seen in Figure 2 (average correlation) and Figure 3 (ability to pick the best

decoy) in [48]. We note that the single model method GOAP used in this study differs from the quasi-single model method GOAPQA used in CASP10QA. For the latter, the TM-score [46] to the top 5 ranked models is used as a measure of model quality.

The CASP 10 datasets have average native-decoy RMSDs of 11–13 Å. These differences are significantly larger than the 2.4 Å RMSD found in our training sets (see Table 1). Our analyses of the performances of PPD (single model) and PPE (ensemble of decoys) on the other datasets considered in this study have shown that for decoys that are far from their native counterparts, the two methods perform similarly, and in fact poorly (see top left panel of Figure 2 and Table Table 4). We observe the same behavior when PPD and PPE are applied on the CASP10 datasets (Tables 4 and 5). Similarly we expect and indeed find that the ensemble method PPE is ineffective in ranking the decoys of the CASP10 datasets when its performance is measured against the MT distance measure, and shows some prospects when its performance is measured against the GDT-TS* and Q* distance measures. The energy-GDT-TS correlations of 0.51(0.63) and 0.29(0.44) for PPD(resp. PPE) on Stage_1 and Stage_2 respectively are amongst the lowest reported for single model(resp. ensemble) methods in CASP10QA [48]. The low energy-distance correlations reported usually leads to a bad pick for the best decoy, see Figure 3. It is therefore surprising that the average ΔGDT-TS* of 0.07 between the GDT-TS*-closest decoy and the lowest energy decoy picked

Table 4. Energy-distance correlations.

Decoy set	Test Distance d_2[d]	PPD Training distance d_1[d]				PPE Training distance d_1[d]				RAPDF[a]	GOAP[b]	AMBER[c]
		RMSD	MT	GDT-TS*	Q*	RMSD	MT	GDT-TS*	Q*			
Titan-HRD	RMSD	0.77 (0.05)[e]	0.82 (0.04)	0.81 (0.05)	0.79 (0.04)	0.81 (0.05)	0.88 (0.03)	0.85 (0.03)	0.82 (0.04)	0.5 (0.14)	0.64(0.11)	0.01 (0.02)
	MT	0.82 (0.04)	0.89 (0.03)	0.87 (0.03)	0.87 (0.03)	0.86 (0.04)	0.95 (0.01)	0.92 (0.02)	0.89 (0.02)	0.47 (0.16)	0.63(0.11)	0.01 (0.02)
	GDT-TS*	0.83 (0.06)	0.91 (0.02)	0.91 (0.02)	0.9 (0.03)	0.86 (0.04)	0.93 (0.02)	0.95 (0.01)	0.92 (0.02)	0.43 (0.18)	0.63(0.12)	0.001 (0.02)
	Q*	0.82 (0.05)	0.92 (0.2)	0.92 (0.02)	0.93 (0.02)	0.87 (0.03)	0.95 (0.01)	0.97 (0.01)	0.95 (0.01)	0.37 (0.22)	0.57(0.13)	0.002 (0.02)
CASP-HRD	RMSD	0.32 (0.16)	0.31 (0.16)	0.31 (0.16)	0.26 (0.16)	0.42 (0.16)	0.51 (0.17)	0.51 (0.18)	0.45 (0.17)	0.31(0.15)	0.3 (0.15)	0.01 (0)
	MT	0.45 (0.11)	0.49 (0.11)	0.49 (0.12)	0.43 (0.12)	0.55 (0.13)	0.69 (0.11)	0.68 (0.13)	0.61 (0.13)	0.37 (0.14)	0.41(0.13)	0.02 (0)
	GDT-TS*	0.38 (0.14)	0.39 (0.13)	0.39 (0.13)	0.33 (0.14)	0.51 (0.15)	0.6 (0.14)	0.65 (0.14)	0.58 (0.17)	0.39 (0.14)	0.43(0.11)	0.02 (0)
	Q*	0.46 (0.12)	0.6(0.11)	0.64 (0.1)	0.57 (0.1)	0.54 (0.12)	0.69 (0.12)	0.75 (0.12)	0.71 (0.1)	0.32 (0.15)	0.42(0.12)	0.02 (0)
CASP10-stage1	RMSD	0.44 (0.22)	0.53 (0.18)	0.53 (0.18)	0.48 (0.2)	0.5 (0.22)	0.54 (0.22)	0.52 (0.21)	0.5 (0.21)	0.18(0.24)	0.32 (0.26)	−0.03 (0)
	MT	0.47 (0.19)	0.61(0.17)	0.62 (0.17)	0.55 (0.18)	0.56 (0.16)	0.63 (0.13)	0.61 (0.13)	0.57 (0.13)	0.13 (0.22	0.34(0.24)	−0.06 (0)
	GDT-TS*	0.4 (0.21)	0.49 (0.23)	0.51 (0.2)	0.43 (0.21)	0.57 (0.21)	0.63 (0.16)	0.63 (0.16)	0.59 (0.2)	0.22 (0.28)	0.4(0.2)	−0.05 (0)
	Q*	0.51 (0.22)	0.63(0.16)	0.63 (0.16)	0.56 (0.18)	0.68 (0.12)	0.75 (0.06)	0.75 (0.07)	0.72 (0.1)	0.24 (0.3)	0.41(0.2)	−0.05 (0)
CASP10-stage2	RMSD	0.33 (0.18)	0.34 (0.2)	0.34 (0.2)	0.31 (0.2)	0.31 (0.18)	0.37 (0.19)	0.34 (0.16)	0.3 (0.21)	0.15(0.13)	0.2 (0.14)	−0.005 (0)
	MT	0.42 (0.16)	0.49 (0.16)	0.49 (0.15)	0.45 (0.14)	0.4 (0.16)	0.5 (0.19)	0.48 (0.15)	0.42 (0.16)	0.19 (0.14)	0.29(0.14)	0.003 (0)
	GDT-TS*	0.31 (0.15)	0.29 (0.14)	0.29 (0.13)	0.25 (0.14)	0.37 (0.16)	0.42 (0.19)	0.44 (0.18)	0.38 (0.18)	0.29 (0.14)	0.37(0.17)	0.007 (0)
	Q*	0.45 (0.22)	0.56(0.14)	0.58 (0.12)	0.52 (0.15)	0.51 (0.18)	0.62 (0.17)	0.66 (0.14)	0.62(0.15)	0.28(0.13)	0.41(0.13)	−0.006 (0)
TSA	RMSD	0.62 (0.12)	0.62 (0.13)	0.63 (0.13)	0.59 (0.14)	0.74 (0.09)	0.8 (0.07)	0.78 (0.08)	0.73 (0.09)	0.5 (0.14)	0.58(0.13)	0.02 (0.01)
	MT	0.65 (0.11)	0.69 (0.1)	0.7 (0.1)	0.65 (0.12)	0.75 (0.08)	0.83 (0.06)	0.8 (0.06)	0.74 (0.07)	0.5 (0.16)	0.58(0.12)	0.03 (0.01)
TM-score > 0.5	GDT-TS*	0.6 (0.15)	0.59 (0.14)	0.6 (0.13)	0.54 (0.16)	0.78 (0.06)	0.85 (0.04)	0.84 (0.04)	0.79 (0.06)	0.61 (0.11)	0.7(0.1)	0.03 (0.01)
	Q*	0.69 (0.11)	0.71 (0.09)	0.72 (0.1)	0.68 (0.1)	0.87 (0.04)	0.94 (0.02)	0.93 (0.02)	0.9 (0.03)	0.57 (0.13)	0.67(0.11)	0.03 (0.01)
TSA	RMSD	0.3 (0.16)	0.34 (0.18)	0.34 (0.18)	0.32 (0.18)	0.29 (0.23)	0.36 (0.27)	0.34 (0.27)	0.29 (0.23)	0.16 (0.13)	0.25(0.15)	0 (0.01)
	MT	0.38 (0.14)	0.47 (0.15)	0.47 (0.15)	0.45 (0.17)	0.34 (0.23)	0.45 (0.24)	0.41 (0.23)	0.35 (0.22)	0.19 (0.14)	0.29(0.13)	−0.003(0.01)
TM-score < 0.5	GDT-TS*	0.27 (0.19)	0.27 (0.2)	0.28 (0.19)	0.24 (0.18)	0.36 (0.29)	0.44 (0.33)	0.42 (0.32)	0.36 (0.29)	0.26 (0.16)	0.33(0.19)	0.004 (0.02)
	Q*	0.41 (0.17)	0.47 (0.16)	0.46 (0.17)	0.45 (0.15)	0.53 (0.19)	0.63 (0.18)	0.61 (0.18)	0.57 (0.19)	0.23 (0.17)	0.3(0.19)	−0.004 (0.02)

[a]All-atom statistical distance-based potential [49].

[b]All-atom orientation-dependent statistical potential [7].

[c]The semi-empirical physical potential AMBER99SB-ILDN[50].

[d]PPD and PPE have been trained on the distance measure d_1 and tested against the distance measure d_2.

[e]Average value, and mean absolute deviation (in parenthesis) over the data set.

Table 5. Energy-distance Rvalues.

Decoy set	Test Distance d_2 [d]	PPD — Training distance d_1 [d]				PPE — Training distance d_1 [d]				RAPDF [a]	GOAP [b]	AMBER [c]
		RMSD	MT	GDT-TS*	Q*	RMSD	MT	GDT-TS*	Q*			
Titan-HRD	RMSD	0.57 (0.17)[e]	0.52 (0.17)	0.51 (0.19)	0.48 (0.2)	0.61 (0.14)	0.63 (0.17)	0.62 (0.14)	0.61 (0.15)	0.43 (0.18)	0.56 (0.15)	0.26 (0.47)
	MT	0.73 (0.11)	0.71 (0.11)	0.69 (0.13)	0.69 (0.15)	0.76 (0.11)	0.8 (0.11)	0.79 (0.12)	0.8 (0.1)	0.5 (0.18)	0.6 (0.17)	0.23 (0.53)
	GDT-TS*	0.78 (0.08)	0.74 (0.09)	0.74 (0.08)	0.73 (0.09)	0.82 (0.07)	0.86 (0.06)	0.86 (0.07)	0.86 (0.07)	0.52 (0.22)	0.67 (0.12)	0.18 (0.58)
	Q*	0.73 (0.11)	0.76 (0.12)	0.73 (0.1)	0.77 (0.1)	0.77 (0.09)	0.84 (0.09)	0.85 (0.09)	0.86 (0.08)	0.37 (0.19)	0.53 (0.17)	0.16 (0.51)
CASP-HRD	RMSD	0.19 (0.31)	0.00 (0.42)	-0.04 (0.48)	0.03 (0.4)	0.27 (0.3)	0.33 (0.31)	0.34 (0.27)	0.3 (0.31)	0.14 (0.37)	0.22 (0.37)	-0.11 (0.34)
	MT	0.31 (0.26)	0.24 (0.3)	0.14 (0.31)	0.16 (0.36)	0.38 (0.24)	0.43 (0.22)	0.43 (0.21)	0.38 (0.25)	0.24 (0.4)	0.46 (0.32)	-0.09 (0.47)
	GDT-TS*	0.14 (0.3)	-0.09 (0.4)	-0.08 (0.42)	-0.06 (0.43)	0.28 (0.23)	0.31 (0.25)	0.32 (0.24)	0.28 (0.24)	0.12 (0.44)	0.34 (0.33)	-0.22 (0.38)
	Q*	0.27 (0.25)	0.35 (0.33)	0.34 (0.32)	0.33 (0.31)	0.28 (0.2)	0.39 (0.24)	0.4 (0.25)	0.41 (0.26)	0.13 (0.4)	0.43 (0.26)	-0.17 (0.42)
CASP10-stage1	RMSD	0.55 (0.23)	0.53 (0.3)	0.57 (0.26)	0.48 (0.39)	0.5 (0.29)	0.52 (0.26)	0.53 (0.26)	0.51 (0.27)	0.32 (0.34)	0.44 (0.26)	0.13 (0.6)
	MT	0.69 (0.1)	0.7 (0.12)	0.72 (0.12)	0.64 (0.14)	0.58 (0.15)	0.62 (0.12)	0.63 (0.13)	0.6 (0.15)	0.37 (0.28)	0.52 (0.2)	0.16 (0.6)
	GDT-TS*	0.52 (0.32)	0.47 (0.39)	0.52 (0.37)	0.42 (0.43)	0.43 (0.4)	0.46 (0.37)	0.51 (0.33)	0.46 (0.36)	0.27 (0.44)	0.53 (0.22)	0.15 (0.37)
	Q*	0.6 (0.24)	0.64 (0.15)	0.67 (0.14)	0.57 (0.19)	0.57 (0.22)	0.6 (0.18)	0.63 (0.17)	0.61 (0.18)	0.32 (0.38)	0.47 (0.31)	0.19 (0.5)
CASP10-stage2	RMSD	0.38 (0.29)	0.23 (0.36)	0.29 (0.3)	0.26 (0.35)	0.36 (0.31)	0.35 (0.32)	0.32 (0.32)	0.35 (0.34)	0.29 (0.28)	0.39 (0.29)	0.11 (0.42)
	MT	0.55 (0.23)	0.46 (0.31)	0.5 (0.29)	0.49 (0.33)	0.45 (0.32)	0.47 (0.32)	0.47 (0.32)	0.48 (0.3)	0.44 (0.32)	0.52 (0.24)	0.17 (0.45)
	GDT-TS*	0.23 (0.35)	0.11 (0.33)	0.14 (0.36)	0.14 (0.34)	0.25 (0.32)	0.23 (0.28)	0.29 (0.32)	0.25 (0.32)	0.23 (0.3)	0.39 (0.3)	0.01 (0.27)
	Q*	0.45 (0.32)	0.46 (0.31)	0.51 (0.29)	0.48 (0.3)	0.44 (0.3)	0.47 (0.24)	0.51 (0.27)	0.53 (0.27)	0.33 (0.28)	0.41 (0.27)	0.13 (0.43)
TSA, TM-score> 0.5	RMSD	0.47 (0.24)	0.22 (0.41)	0.21 (0.41)	0.22 (0.41)	0.69 (0.14)	0.69 (0.13)	0.69 (0.13)	0.68 (0.14)	0.44 (0.25)	0.59 (0.19)	0.24 (0.38)
	MT	0.62 (0.14)	0.4 (0.24)	0.39 (0.27)	0.41 (0.24)	0.8 (0.08)	0.81 (0.07)	0.82 (0.08)	0.8 (0.08)	0.54 (0.18)	0.74 (0.1)	0.32 (0.33)
	GDT-TS*	0.29 (0.28)	0.09 (0.43)	0.08 (0.47)	0.09 (0.45)	0.58 (0.16)	0.6 (0.16)	0.61 (0.16)	0.57 (0.18)	0.37 (0.31)	0.56 (0.41)	0.09 (0.26)
	Q*	0.49 (0.19)	0.32 (0.3)	0.3 (0.33)	0.33 (0.28)	0.68 (0.14)	0.72 (0.14)	0.74 (0.13)	0.74 (0.13)	0.38 (0.29)	0.59 (0.43)	0.14 (0.2)
TSA, TM-score <0.5	RMSD	0.16 (0.35)	0.19 (0.3)	0.19 (0.34)	0.16 (0.32)	0.26 (0.35)	0.33 (0.37)	0.32 (0.36)	0.29 (0.35)	0.19 (0.35)	0.27 (0.4)	0.04 (0.41)
	MT	0.27 (0.34)	0.38 (0.3)	0.39 (0.33)	0.37 (0.29)	0.36 (0.33)	0.44 (0.31)	0.42 (0.32)	0.41 (0.31)	0.28 (0.34)	0.4 (0.33)	0.04 (0.46)
	GDT-TS*	0.07 (0.3)	0.07 (0.28)	0.1 (0.3)	0.06 (0.29)	0.26 (0.3)	0.32 (0.3)	0.32 (0.3)	0.29 (0.3)	0.19 (0.3)	0.28 (0.36)	0.05 (0.3)
	Q*	0.18 (0.35)	0.28 (0.34)	0.29 (0.35)	0.3 (0.31)	0.42 (0.27)	0.5 (0.27)	0.49 (0.29)	0.49 (0.27)	0.19 (0.29)	0.31 (0.32)	0.01 (0.31)

[a] All-atom statistical distance-based potential [49].
[b] All-atom orientation-dependent statistical potential [7].
[c] The semi-empirical physical potential AMBER99SB-ILDN [50].
[d] PPD and PPE have been trained on the distance measure d_1, and tested against the distance measure d_2.
[e] Average value, and mean absolute deviation (in parenthesis) over the data set.

Table 6. Assessing the best decoys selected by energy functions on different decoy datasets.

		Best	PPD	PPE	RAPDF [a]	AMBER [b]	GOAP [c]
Titan-HRD	RMSD	1.1(0.21)[d]	1.7(0.29)	1.6(0.27)	1.9(0.4)	2.1(0.55)	1.7(0.3)
	MT	0.75(0.22)	1.4(0.38)	1.2(0.38)	1.8(0.62)	2.3(0.89)	1.6(0.54)
	GDT-TS	0.94(0.02)	0.89(0.03)	0.92(0.03)	0.85(0.05)	0.8(0.09)	0.88(0.03)
	Q	0.94(0.01)	0.92(0.02)	0.93(0.02)	0.88(0.03)	0.86(0.04)	0.89(0.03)
4-state	RMSD	1.1(0.1)	3.8(0.44)	2.2(0.21)	2.1(0.22)	3.6(1.5)	1.6(0.24)
	MT	0.9(0.33)	5.8(2.1)	1.2(0.31)	2.6(0.52)	6.2(3.4)	1.5(0.38)
	GDT-TS	0.91(0.03)	0.55(0.06)	0.86(0.08)	0.8(0.04)	0.67(0.1)	0.86(0.04)
	Q	0.94(0.02)	0.75(0.03)	0.92(0.02)	0.87(0.04)	0.79(0.1)	0.9(0.02)
fisa	RMSD	3.7(0.76)	5.7(0.78)	6.5(1.4)	4.4(0.72)	8.5(1.5)	4.5(0.45)
	MT	3.8(1.5)	7.9(3.7)	5.5(2)	5.4(2.5)	10(4.2)	4.9(1.9)
	GDT-TS	0.65(0.07)	0.51(0.14)	0.54(0.08)	0.6(0.06)	0.46(0.08)	0.59(0.06)
	Q	0.82(0.02)	0.79(0.03)	0.78(0.03)	0.78(0.02)	0.73(0.02)	0.79(0.03)
fisa CASP3	RMSD	6(2)	12(1.6)	12(2.4)	12(4)	12(1)	11(1.6)
	MT	8.7(4.2)	21(7.3)	17(8.2)	23(4.5)	19(2.5)	18(7.7)
	GDT-TS	0.47(0.12)	0.32(0.01)	0.34(0.02)	0.32(0.02)	0.29(0.01)	0.33(0.04)
	Q	0.76(0.06)	0.72(0.04)	0.72(0.04)	0.68(0.04)	0.67(0.07)	0.69(0.06)
hg Structal	RMSD	1.9(0.5)	2.6(1)	2.5(0.56)	2.2(0.5)	3.3(0.71)	2.4(0.6)
	MT	1.8(0.3)	2.5(0.61)	3(0.28)	2.4(0.35)	3.7(0.8)	2.7(0.28)
	GDT-TS	0.86(0.06)	0.82(0.14)	0.85(0.07)	0.84(0.07)	0.77(0.08)	0.84(0.08)
	Q	0.93(0.03)	0.92(0.03)	0.92(0.03)	0.92(0.04)	0.89(0.03)	0.92(0.04)
lmds	RMSD	5.7(0.33)	9.9(0.72)	9.8(0.89)	9.8(0.92)	10(0.61)	10(0.65)
	MT	8(0.78)	14(3.7)	17(5.5)	16(2.5)	19(1.6)	19(4.5)
	GDT-TS	0.45(0.04)	0.29(0.04)	0.32(0.05)	0.31(0.05)	0.28(0.03)	0.3(0.03)
	Q	0.74(0.02)	0.67(0.05)	0.67(0.04)	0.65(0.04)	0.63(0.03)	0.63(0.05)
lattice ssfit	RMSD	3.8(0.46)	7.6(1.3)	7.4(1.6)	7.7(1.9)	8(2.6)	8.5(1.2)
	MT	5.2(2.2)	9.8(4.5)	10(5.1)	11(4.8)	12(6.4)	12(5)
	GDT-TS	0.62(0.06)	0.45(0.07)	0.48(0.07)	0.49(0.12)	0.45(0.07)	0.44(0.04)
	Q	0.8(0.06)	0.74(0.07)	0.75(0.04)	0.74(0.05)	0.72(0.07)	0.72(0.06)
CASP5	RMSD	6.7(2.9)	13(6.2)	11(6)	10(5.2)	11(6.1)	10(5.2)
	MT	8.5(4.2)	20(11)	18(7.7)	20(12)	22(11)	20(9.9)
	GDT-TS	0.58(0.19)	0.36(0.17)	0.48(0.24)	0.44(0.19)	0.46(0.21)	0.5(0.23)
	Q	0.82(0.09)	0.72(0.13)	0.77(0.09)	0.7(0.1)2	0.72(0.13)	0.75(0.1)
CASP6	RMSD	4.8(1.5)	10(5.1)	11(4.5)	9.7(5.1)	12(5.9)	8(3.1)
	MT	5.4(1.9)	23(11)	18(4.3)	19(5.7)	24(15)	12(4)
	GDT-TS	0.64(0.14)	0.33(0.14)	0.52(0.18)	0.49(0.27)	0.38(0.17)	0.54(0.17)

Table 6. Cont.

		Best	PPD	PPE	RAPDF[a]	AMBER[b]	GOAP[c]
CASP7	Q	0.85(0.06)	0.7(0.07)	0.79(0.08)	0.75(0.11)	0.68(0.11)	0.79(0.09)
	RMSD	4.5(1.8)	8.8(4.9)	7.1(3.1)	7.9(3.9)	11(5.1)	7.8(3.4)
	MT	3.8(1.6)	9.5(3.8)	6.6(2.8)	10(4.3)	18(9.1)	8.3(3.5)
	GDT-TS	0.66(0.13)	0.49(0.21)	0.56(0.14)	0.56(0.17)	0.43(0.21)	0.58(0.13)
CASP8	Q	0.88(0.04)	0.81(0.08)	0.85(0.06)	0.81(0.06)	0.74(0.08)	0.82(0.06)
	RMSD	4.1(1.3)	7.4(2.7)	6.4(1.8)	9.8(5.5)	9.7(5.2)	7.5(3.1)
	MT	3.2(1.3)	10(4.1)	6.1(2.2)	14(6.4)	15(8)	8.7(2.7)
	GDT-TS	0.7(0.1)	0.53(0.17)	0.63(0.13)	0.51(0.22)	0.47(0.19)	0.61(0.16)
CASP9	Q	0.89(0.04)	0.81(0.07)	0.85(0.06)	0.79(0.09)	0.75(0.1)	0.83(0.07)
	RMSD	4.8(1.4)	9.7(4.9)	7.5(2.6)	9.4(4.7)	9.8(4.9)	8.2(3.3)
	MT	3.7(1.3)	14(7.8)	7.3(2.6)	12(3.9)	13(5.2)	8.5(2.6)
	GDT-TS	0.68(0.1)	0.4(0.15)	0.6(0.12)	0.52(0.19)	0.51(0.21)	0.57(0.15)
TASSER Set II	Q	0.88(0.04)	0.75(0.12)	0.85(0.04)	0.8(0.09)	0.79(0.09)	0.83(0.07)
	RMSD	3.2(1)	5.6(1.9)	5.2(1.4)	5.2(1.6)	6.4(2.1)	5.4(1.9)
	MT	3.7(1.2)	7.1(2.2)	6.6(2.5)	6.7(2.2)	11(5.4)	6.8(2.2)
	GDT-TS	0.69(0.09)	0.57(0.12)	0.59(0.12)	0.59(0.1)	0.52(0.13)	0.59(0.12)
Rosetta-All	Q	0.85(0.05)	0.81(0.06)	0.82(0.06)	0.8(0.05)	0.75(0.08)	0.79(0.05)
	RMSD	6.4(1.3)	11(2.1)	11(2.2)	12(3.2)	16(4.7)	11(2.6)
	MT	12(4.7)	22(8.9)	26(7.2)	25(9.9)	47(14)	24(7.5)
	GDT-TS	0.41(0.06)	0.29(0.04)	0.29(0.04)	0.28(0.04)	0.25(0.04)	0.28(0.04)
Rosetta-Baker	Q	0.72(0.06)	0.64(0.07)	0.64(0.07)	0.62(0.07)	0.59(0.07)	0.62(0.07)
	RMSD	4.7(2.2)	7.5(3.4)	8.4(4.3)	7.6(4.1)	8.2(2.7)	6.9(3.6)
	MT	6.9(3)	13(7.5)	15(9.2)	13(7.8)	13(6.8)	11(5.4)
	GDT-TS	0.6(0.21)	0.47(0.15)	0.46(0.13)	0.48(0.15)	0.46(0.13)	0.5(0.18)
Rosetta-Tsai	Q	0.84(0.08)	0.77(0.1)	0.77(0.11)	0.77(0.01)	0.76(0.07)	0.79(0.09)
	RMSD	2.8(0.8)	6.9(2.8)	5(1.4)	7.3(1.8)	5.7(2)	6.2(2.1)
	MT	3(1.1)	9.3(4.9)	5.5(2.2)	10(5.8)	8.3(3)	7.1(2.3)
	GDT-TS	0.72(0.08)	0.47(0.08)	0.59(0.09)	0.45(0.06)	0.54(0.09)	0.52(0.12)
CASP-HRD	Q	0.86(0.05)	0.77(0.08)	0.81(0.08)	0.74(0.07)	0.77(0.07)	0.77(0.06)
	RMSD	2(0.56)	2.6(0.73)	2.6(0.71)	2.6(0.76)	2.8(0.7)	2.6(0.74)
	MT	1.1(0.5)	2(0.8)	1.7(0.61)	2(0.83)	2.3(0.11)	1.7(0.79)
	GDT-TS	0.83(0.07)	0.75(0.06)	0.78(0.07)	0.77(0.07)	0.75(0.06)	0.78(0.07)
CASP10-stage1[e]	Q	0.93(0.03)	0.9(0.03)	0.91(0.03)	0.89(0.04)	0.88(0.04)	0.91(0.03)
	RMSD	4.7(1.2)	6.6(2.3)	7.1(2.4)	8.2(3.2)	12(4.7)	7.3(2.4)
	MT	4(2)	6.6(2)	8(1.6)	10(3.5)	20(3.4)	7.9(2.2)

Table 6. Cont.

		Best	PPD	PPE	RAPDF [a]	AMBER [b]	GOAP [c]
	GDT-TS	0.71(0.1)	0.62(0.13)	0.63(0.15)	0.57(0.16)	0.55(0.19)	0.63(0.14)
	Q	0.88(0.04)	0.84(0.04)	0.85(0.04)	0.82(0.06)	0.8(0.06)	0.84(0.06)
CASP10-stage2 [e]	RMSD	4(1.1)	5.9(1.7)	5.9(1.6)	6.7(1.9)	9(2.2)	6.2(1.6)
	MT	3.1(1)	5.4(1.6)	5.8(1.6)	6.6(1.8)	14(1.9)	5.4(1.5)
	GDT-TS	0.73(0.09)	0.64(0.14)	0.66(0.11)	0.65(0.12)	0.65(0.11)	0.67(0.11)
	Q	0.89(0.04)	0.86(0.03)	0.87(0.04)	0.86(0.05)	0.85(0.04)	0.86(0.04)

[a] All-atom statistical distance-based potential [49].
[b] The semi-empirical physical potential AMBER99SB-ILDN [50].
[c] All-atom orientation-dependent statistical potential [7].
[d] Average value, and mean absolute deviation (in parenthesis) over the data set.
[e] Only ensembles who contains a decoy with a GDT-TS> 0.4 are included. Compare with Figure 2 in [48].

by PPE on the CASP10 Stage_2 data sets places PPE in the middle of the CASP10 participating methods (see [48] Figure 2(A)).

The results for PPD, PPE, AMBER99SB-ILDN, RAPDF and GOAP on CASP 10 stages 1 and 2 are given in Tables 4 - 6 where PPD and PPE were trained and tested on the same distance measure. Clearly, GOAP has a better performance than PPD when GDT-TS* is chosen as a measure of distance. It is however noteworthy that PPD performs better than GOAP when measured by RMSD and MT instead. It is encouraging that the distance dependent C-alpha potential, PPD, as a single model method has a performance that is comparable to the state-of-the-art orientation-dependent all-atom potential, GOAP. We find that PPD is good at selecting a decoy that is close to the native structure (Table 6).

Concluding Remarks

The recent literature on generating knowledge-based potentials for protein structure modeling makes no secrets of their limitations and problems. Knowledge-based potentials are energy functions derived primarily from databases of protein structures and sequences. They can be divided into two classes. Potentials from the first class are based on a direct conversion of the distributions of some geometric properties observed in native protein structures into energy values, while potentials from the second class are trained to mimic quantitatively the geometric differences between incorrectly folded models (also called decoys) and native structures. Both potentials are designed to assess how native-like a model structure is. There is no consensus however on which geometric property should be considered, on how to convert a statistical distribution into an energy for the first class, and on how energy and geometry should be related in the second class.

In this paper, we focused on the relationship between energy and geometry when training knowledge-based potentials from the second class. We assumed that the difference between the energy of a decoy and the energy of its corresponding native structure must be linearly related to the distance between the decoy and the native structure. We trained two distance-based Cα potentials accordingly, one based on all inter-residue distances (PPD), while the other had the set of all these distances filtered to reflect consistency in an ensemble of decoys (PPE). Compared to other methods that follow the same approach however, we did not assume that the distance between a decoy and the native structure is the traditional RMSD. Instead, we tested four different distance measures, two based on extrinsic geometry (RMSD and GTD-TS*), and two based on intrinsic geometry (Q* and MT). We found that it is usually better to train the potentials using the latter type of distances.

We have found that both PPD and PPE perform extremely well on the high resolution decoy set Titan-HRD, with correlation coefficients between energy and distance usually well above 0.8. PPE always performs better than PPD on this set, emphasizing the benefits of capturing consistent information in an ensemble. While we trust the general trends highlighted by these results, we tone down the importance of In extensive testing on available decoy sets and models from the Critical Assessment of protheir exceptional character as they may only reflect the specificity of the Titan-HRD data set. tein Structure Prediction (CASP) experiments we find that PPD yields better energy-distance correlations than one of the state of the art single model potentials, GOAP [7]. We note however that the sophisticated distance-based and orientation-based statistical potential GOAP is better at picking the best decoys and has a better though comparable performance for fixed energy-distance correlation. It should be noted that PPD and PPE are Cα-based, while GOAP is an all-atom potential. We believe that this demonstrates that a very efficient training of a simple distance-based pair potential can generate a very effective measure for assessing protein structure models.

There is still room for improvement in training knowledge-based potentials. We limited our study to pairwise potentials; we will test different geometric properties of protein structures in future studies. We plan to include the potentials described here into a structure minimization package, to assess their performances in improving non-native protein structure models.

References

1. Zhang Y (2009) Protein structure prediction: when is it useful? Curr Opin Struct Biol 19: 145–155.
2. Moult J, Fidelis K, Kryshtafovych A, Tramontano A (2011) Critical assessment of methods of protein structure prediction (CASP)-round IX. Proteins: Struct Func Bioinfo 79: 1–5.
3. Cozzetto D, Kryshtafovych A, Tramontano A (2009) Evaluation of CASP8 model quality predictions. Proteins: Struct Func Bioinfo 77: 157–166.
4. Kryshtafovych A, Fidelis K, Tramontano A (2011) Evaluation of model quality predictions in CASP9. Proteins: Struct Func Bioinfo 79: 91–106.
5. Anfinsen C (1973) Principles that govern the folding of protein chains. Science 181: 223–230.
6. Lazaridis T, Karplus M (2000) Effective energy functions for protein structure prediction. Curr Opin Struct Biol 10: 139–145.
7. Zhou H, Skolnick J (2011) GOAP: a generalized orientation-dependent, all-atom statistical potential for protein structure prediction. Biophys J 101: 2043–2052.
8. Skolnick J (2006) In quest of an empirical potential for protein structure prediction. Curr Opin Struct Biol 16: 166–171.
9. Summa C, Levitt M (2007) Near-native structure refinement using *in vacuo* energy minimization. Proc Natl Acad Sci (USA) 104: 3177–3182.
10. Zhu J, Fan H, Peiole X, Honig B, Mark A (2008) Refining homology models by combining replica-exchange molecular dynamics and statistical potentials. Proteins: Struct Func Bioinfo 72: 1171–1188.
11. Chopra G, Kalisman N, Levitt M (2010) Consistent refinement of submitted models at CASP using a knowledge-based potential. Proteins: Struct Func Bioinfo 78: 2668–2678.
12. Amautova Y, Scheraga H (2008) Use of decoys to optimize an all-atom forcefield including hydration. Biophys J 95: 2434–2449.
13. Bhattachary D, Cheng J (2013) 3Drefine: consistent protein structure refinement by optimizing hydrogen bonding network and atomic level refinement. Proteins: Struct Func Bioinfo 81: 119–131.
14. Rohl C, Strauss C, Misura K, Baker D (2004) Protein structure prediction using Rosetta. Methods Enzymol 383: 66–93.
15. Zhang Y, Kolinski A, Skolnick J (2003) Touchstone II: A new approach to ab initio protein structure prediction. Biophys J 85: 1145–1164.
16. Benkert P, Tosatto S, Schomburg D (2008) QMEAN: A comprehensive scoring function for model quality assessment. Proteins: Struct Func Bioinfo 71: 261–277.
17. Zhang Y, Skolnick J (2004) Automated structure prediction of weakly homologous proteins on a genomic scale. Proc Natl Acad Sci (USA) 101: 7594–7599.
18. Zemla A (2003) LGA: a method for finding 3D similarities in protein structures. Nucl Acids Res 31: 3370–3374.
19. Perez A, Yang Z, Bahar I, Dill K, MacCallum J (2012) FlexE: using elastic network models to compare models of protein structure. J Chem Theory Computat 8: 3985–3991.
20. Rajgaria R, McAllister S, Floudas C (2006) A novel high resolution Cα–Cα distance dependent force field based on a high quality decoy set. Proteins: Struct Func Bioinfo 65: 726–741.
21. Samudrala R, Levitt M (2008) Decoys 'R'Us: A database of incorrect conformations to improve protein structure prediction. Protein Science 9: 1399–1401.
22. Tsai J, Bonneau R, Morozov A, Kuhlman B, Rohl C, et al. (2003) An improved protein decoy set for testing energy functions for protein structure prediction. Proteins: Struct Func Bioinfo 53: 76–87.
23. McLachlan A (1979) Gene duplications in the structural evolution of chymotrypsin. J Mol Biol 128: 49–80.
24. Horn B (1987) Closed form solution of absolute orientation using unit quaternions. J Opt Soc Am 4: 629–642.
25. Coutsias E, Seok C, Dill K (2004) Using quaternions to calculate RMSD. J Comp Chem 25: 1849–1857.
26. Kaindl K, Steipe B (1997) Metric properties of the root-mean square deviation of vector sets. Acta Cryst A 53: 809.
27. Tirion M (1996) Large amplitude elastic motions in proteins from a single-parameter, atomic analysis. Phys Rev Lett 77: 1905–1908.
28. Tama F, Sanejouand Y (2001) Conformational change of proteins arising from normal mode calculations. Protein Eng 14: 1–6.
29. Bohr J, Bohr H, Brunak S, Cotterill R, Fredholm H, et al. (1993) Protein structures from distance inequalities. J Mol Biol 231: 861–869.
30. Summa C, Levitt M (2007) Near-native structure refinement using in vacuo energy minimization. Proc Natl Acad Sci (USA) 104: 3177–3182.
31. Bahar I, Atilgan A, Erman B (1997) Direct evaluation of thermal fluctuations in proteins using a single-parameter harmonic potential. Folding and Design 2: 173–181.
32. Atilgan A, Durell S, Jernigan R, Demirel M, Keskin O, et al. (2001) Anisotropy of fluctuation dynamics of proteins with an elastic network model. Biophys J 80: 505–515.
33. Toda M (1967) Vibration of a chain with nonlinear interaction. J Phys Soc Japan 22: 431–436.
34. Røgen P, Koehl P (2013) Extracting knowledge from protein structure geometry. Proteins: Struct Func Bioinfo 81: 841–851.
35. de Boor C (1978) A practical guide to splines. New York: Springer-verlag.
36. Eickholt J, Wang Z, Cheng J (2011) A conformation ensemble approach to protein residue-residue contact. BMC structural biology 11: 38.
37. Handl J, Knowles J, Lovell S (2009) Artefacts and biases affecting the evaluation of scoring functions on decoy sets for protein structure prediction. Bioinformatics 25: 1271–1279.
38. Güntert P, Mumenthaler C, Wüthrich K (1997) Torsion angle dynamics for NMR structure calculation with the new program DYANA. J Mol Biol 273: 283–298.
39. Park B, Levitt M (1996) Energy functions that discriminate x-ray and near-native folds from well-constructed decoys. J Mol Biol 258: 367–392.
40. Simons K, Kooperberg C, Huang E, Baker D (1997) Assembly of protein tertiary structures from fragments with similar local sequences using simulated annealing and bayesian scoring functions. J Mol Biol 268: 209–225.
41. Keasar C, Levitt M (2003) A novel approach to decoy set generation: designing a physical energy function having local minima with native structure characteristics. J Mol Biol 329: 159–174.
42. Huang E (1999) A combined approach for ab initio construction of low resolution protein tertiary structures from sequence. In: Pacific Symposium on Biocomputing. volume 4, pp. 505–516.
43. Xia Y, Huang E, Levitt M, Samudrala R (2000) Ab initio construction of protein tertiary structures using a hierarchical approach. J Mol Biol 300: 171–185.
44. Simons K, Ruczinski I, Kooperberg C, Fox B, Bystroff C, et al. (1999) Improved recognition of native-like protein structures using a combination of sequence-dependent and sequence-independent features of proteins. Proteins: Struct Func Bioinfo 34: 82–95.
45. Zhang J, Zhang Y (2010) A novel side-chain orientation dependent potential derived from random-walk reference state for protein fold selection and structure prediction. PloS One 5: e15386.
46. Zhang Y, Skolnick J (2004) Scoring function for automated assessment of protein structure template quality. Proteins: Struct Func Bioinfo 57: 702–710.
47. Xu J, Zhang Y (2010) How significant is a protein structure similarity with TM-score = 0.5? Bioinformatics 26: 889–895.

Acknowledgments

The authors want to thank the anonymous reviewers for constructive criticism and careful reading of the first version of this manuscript.

Author Contributions

Conceived and designed the experiments: MC PK PR. Performed the experiments: MC. Analyzed the data: MC PK PR. Contributed reagents/materials/analysis tools: MC. Wrote the paper: MC PK PR.

48. Kryshtafovych A, Barbato A, Fidelis K, Monastyrskyy B, Schwede T, et al. (2014) Assessment of the assessment: evaluation of the model quality estimates in CASP10. Proteins: Struct Func Bioinfo 82: 112–126.

49. Samudrala R, Moult J (1998) An all-atom distance-dependent conditional probability discriminatory function for protein structure prediction. J Mol Biol 275: 895–916.

50. Lindorff-Larsen K, Piana S, Palmo K, Maragakis P, Klepeis JL, et al. (2010) Improved side-chain torsion potentials for the Amber ff99SB protein force field. Proteins: Struct Func Bioinfo 78: 1950–1958.

51. Zhou H, Zhou Y (2002) Distance-scaled, finite ideal-gas reference state improves structure-derived potentials of mean force for structure selection and stability prediction. Protein Sci 11: 2714–2726.

12

How Does Domain Replacement Affect Fibril Formation of the Rabbit/Human Prion Proteins

Xu Yan, Jun-Jie Huang, Zheng Zhou, Jie Chen, Yi Liang*

State Key Laboratory of Virology, College of Life Sciences, Wuhan University, Wuhan, China

Abstract

Background: It is known that *in vivo* human prion protein (PrP) have the tendency to form fibril deposits and are associated with infectious fatal prion diseases, while the rabbit PrP does not readily form fibrils and is unlikely to cause prion diseases. Although we have previously demonstrated that amyloid fibrils formed by the rabbit PrP and the human PrP have different secondary structures and macromolecular crowding has different effects on fibril formation of the rabbit/human PrPs, we do not know which domains of PrPs cause such differences. In this study, we have constructed two PrP chimeras, rabbit chimera and human chimera, and investigated how domain replacement affects fibril formation of the rabbit/human PrPs.

Methodology/Principal Findings: As revealed by thioflavin T binding assays and Sarkosyl-soluble SDS-PAGE, the presence of a strong crowding agent dramatically promotes fibril formation of both chimeras. As evidenced by circular dichroism, Fourier transform infrared spectroscopy, and proteinase K digestion assays, amyloid fibrils formed by human chimera have secondary structures and proteinase K-resistant features similar to those formed by the human PrP. However, amyloid fibrils formed by rabbit chimera have proteinase K-resistant features and secondary structures in crowded physiological environments different from those formed by the rabbit PrP, and secondary structures in dilute solutions similar to the rabbit PrP. The results from transmission electron microscopy show that macromolecular crowding caused human chimera but not rabbit chimera to form short fibrils and non-fibrillar particles.

Conclusions/Significance: We demonstrate for the first time that the domains beyond PrP-H2H3 (β-strand 1, α-helix 1, and β-strand 2) have a remarkable effect on fibrillization of the rabbit PrP but almost no effect on the human PrP. Our findings can help to explain why amyloid fibrils formed by the rabbit PrP and the human PrP have different secondary structures and why macromolecular crowding has different effects on fibrillization of PrPs from different species.

Editor: Human Rezaei, INRA, France

Funding: This study was supported by National Key Basic Research Foundation of China (http://www.most.gov.cn/, Grant nos. 2013CB910702 and 2012CB911003, YL), National Natural Science Foundation of China (http://www.nsfc.gov.cn/, Grant nos. 31370774 and 31170744, YL), and Fundamental Research Funds for the Central Universities of China (http://www.moe.edu.cn/, Grant no. 1104006, YL). The funders had no role in study design, data collection and analysis, decision to publish, or preparation of the manuscript.

Competing Interests: The authors have declared that no competing interests exist.

* Email: liangyi@whu.edu.cn

Introduction

Transmissible spongiform encephalopathies, also known as prion diseases, are infectious fatal neurodegenerative diseases that affect the nervous system in humans and animals [1]. The key procedure of prion diseases is believed to be the conversion of the normal protease-sensitive cellular prion protein (PrP^{C}) in such mammals into the aberrant protease-resistant pathogenic prion protein (PrP^{Sc}) [1–4]. Rabbits are among the few animal species that have resistance to prions from other animal species [5]. However, the three-dimensional structure of the recombinant protein rabbit PrP^{C} is composed of an unstructured flexible N-terminal region and a C-terminal globular domain which comprises two short anti-parallel β-strands and three α-helices similar to the structures of other mammalian PrP^{C} [6–8]. There are few difference among the C-terminal domains, and some

research believes that the unique primary structure of rabbit PrP^{C} inhibits formation of its abnormal isoform [6,9].

A cell-free conversion system for reconstitution of the infectious PrP^{Sc} from recombinant PrP *in vitro* has been developed [10–12]. The Baskakov lab has provided the first demonstration that recombinant full-length prion protein (PrP) with an intact S-S bond can be folded into amyloid conformation *in vitro* [13]. Using a serial protein misfolding cyclic amplification (PMCA) protocol [14], a prion has been generated with bacterially expressed recombinant PrP in the presence of synthetic anionic phospholipid and RNA [15]. Furthermore, Kim and co-workers have demonstrated that mammalian prions can also be generated from bacterially expressed PrP in the absence of any mammalian cofactors [16]. De novo rabbit prions have been produced by rabbit brain homogenate *in vitro* from unseeded material [17]. All of these works suggest that amyloid fibrils generated *in vitro* are infectious something like PrP^{Sc} and there could be some similarities between them. The Baskakov lab has also provided compelling evidence that noninfectious amyloids with a structure different from that of PrP^{Sc} could lead to transmissible prion diseases and raised a new mechanism responsible for prion diseases [18], which is different from the PrP^{Sc}-templated mechanism [19] or spontaneous conversion of PrP^C into PrP^{Sc} [20].

Amyloidogenic proteins form fibril deposits in crowded physiological environments [21–28]. It is known that *in vivo* human PrP have the tendency to form fibril deposits and are associated with infectious fatal prion diseases [1,4], while the rabbit PrP does not readily form fibrils and is unlikely to cause prion diseases [9,27,28]. Although we have previously demonstrated that amyloid fibrils formed by the rabbit PrP and the human PrP have different secondary structures and macromolecular crowding has different effects on fibril formation of the rabbit/human PrPs [27,28], the molecular mechanism of this phenomenon is not clear, and we do not know which domains of PrPs cause such differences. The Rezaei lab has found that the α-helical 2 and α-helix 3 (H2H3) domain of the mouse PrP forms amyloid fibrils morphologically similar to those obtained for the full-length PrP and generates insoluble proteinase K (PK)-resistant aggregates [29–31]. We want to know the role of the H2H3 domain in fibrillization of the rabbit/human PrPs in crowded physiological environments.

In this study, we constructed two PrP chimeras, in one of which the H2H3 domain of the human PrP was replaced with that of the rabbit PrP (termed rabbit chimera) and in another of which the H2H3 domain of the rabbit PrP is replaced with that of the human PrP (termed human chimera), and used site-directed mutation to analysis how domain replacement affects fibril formation of the rabbit/human PrPs. As revealed by thioflavin T (ThT) binding assays and Sarkosyl-soluble SDS-PAGE, the addition of high concentrations of crowding agents (Ficoll 70 or Ficoll 400) significantly enhanced fibril formation of both PrP chimeras. As evidenced by circular dichroism (CD), Fourier transform infrared (FTIR) spectroscopy, and PK digestion assays, the overall secondary structure and PK resistance of the fibrils formed by human chimera are similar to those formed by the human PrP. However, amyloid fibrils formed by rabbit chimera had PK-resistant features and secondary structures in crowded physiological environments different from those formed by the rabbit PrP, and secondary structures in dilute solutions similar to the rabbit PrP. We also demonstrated that the amino acids beyond PrP-H2H3 had almost no effect on fibrillization of the human PrP but a remarkable effect on the rabbit PrP.

Materials and Methods

Ethics statement

All research involving original human work was approved by the Institutional Review Board of the College of Life Sciences, Wuhan University (Wuhan, China), leaded by Dr. Bao-Liang Song, the Dean of the college, in accordance with the guidelines for the protection of human subjects. Written informed consent for the original human work that produced the plasmid samples was obtained. This study was approved by the ethics committee of the College of Life Sciences, Wuhan University, leaded also by Dr. Bao-Liang Song. The vector pET-30a (+) contain mature human prion protein(23–231) (pET-30-hPrP) and the vector pET-30a (+) contain mature rabbit prion protein(23–228) (pET-30-rPrP) were kindly donated by Prof. Geng-Fu Xiao (Wuhan Institute of Virology, Chinese Academy of Sciences). Written informed consent for using and modifying these plasmids was obtained.

The plasmids of pET-28a-hH2H3 and pET-28a-rH2H3 were constructed from pET-30-hPrP/pET-30-rPrP and pET-28a (+) by ourselves. All the proteins (human PrP, rabbit PrP, human chimera, rabbit chimera, chimera H, chimera R, human PrP-H2H3, and rabbit PrP-H2H3) involved in this work were produced by *E. coli* BL21 (DE3) expression system. There is not any animal work or *in vivo* experiments performed in this study.

Materials

The crowding agents, Ficoll 70 and Ficoll 400, were purchased from Sigma-Aldrich (St. Louis, MO). ThT was also obtained from Sigma-Aldrich. Guanidine hydrochloride (GdnHCl) was obtained from Promega (Madison, WI). Proteinase K and Triton X-100 were purchased from Ameresco (Solon, OH). All other chemicals used were made in China and were of analytical grade.

Plasmid construction and prion protein purification

Standard cloning procedures were used. Using KOD-Plus-Mutagenesis Kit, the H2H3 of the human PrP was replaced by the H2H3 of the rabbit PrP, termed rabbit chimera, which just mutated the sites 174N, 184I, 203V, 205M, 219E, 220R, 225Y, 229G and 230S of human PrP to 174S, 184V, 203I, 205I, 219Q, 220Q, 225A, 229A and 230Stop, using pET-30-hPrP as template. The primers used can be seen in Table S1. The H2H3 of the rabbit PrP was replaced by the H2H3 of the human PrP, termed human chimera, in the same way with primers shown in Table S2, using pET-30-rPrP as template.

The primers used to construct chimera H in which the human PrP-B1H1B2 (β-strand 1, α-helix 1, and β-strand 2) was replaced by the rabbit PrP-B1H1B2 is shown in Table S3, and the primers used to construct chimera R in which the rabbit PrP-B1H1B2 was replaced by the human PrP-B1H1B2 is shown in Table S4.

Human PrP-H2H3 (169–231) was amplified by polymerase chain reaction (PCR) using pET-30-hPrP as template with primers:

SH-H2H3 5′GGGTAATCCATATGTACAGCAACCAGA-AC

AH-H2H3 5′ ACAGAATTCTCATCACGATCCTCTCTG-GT

The PCR product was cloned between restriction sites Nde I and EcoR I of the vector pET-28a (+), termed pET-28a-hH2H3.

Rabbit PrP-H2H3 (168–228) was amplified by PCR using pET-30-rPrP as template with primers:

SR-H2H3 5′ GGGTGCTCATATGTACAGCAACCAGAA-CAG

AR-H2H3 5′ ACAGAATTCTTATGCCGCCCTCTGGTA-GGC

The PCR product was cloned between restriction sites Nde I and EcoR I of the vector pET-28a (+), termed pET-28a-rH2H3.

Recombinant full-length prion proteins, human chimera, rabbit chimera, chimeras H and R, and PrP-H2H3 were expressed in *E. coli* BL21 (DE3) (Novagen), and purified by HPLC on a C4 reversed-phase column (Shimadzu, Kyoto, Japan) as described by Bocharova and co-workers [32].

Thioflavin T binding assays

The methods for fibrillization of full-length prion proteins, human chimera, and rabbit chimera were the same as the methods described by the Liang lab [28], and 10 μM PrP was incubated at 37°C in PBS buffer (pH 7.0) containing 2 M GdnHCl in the absence and presence of crowding agents with continuous shaking at 220 rpm, and samples (50 μl) were diluted into PBS buffer containing 12.5 μM ThT, giving a final volume of 2.5 ml.

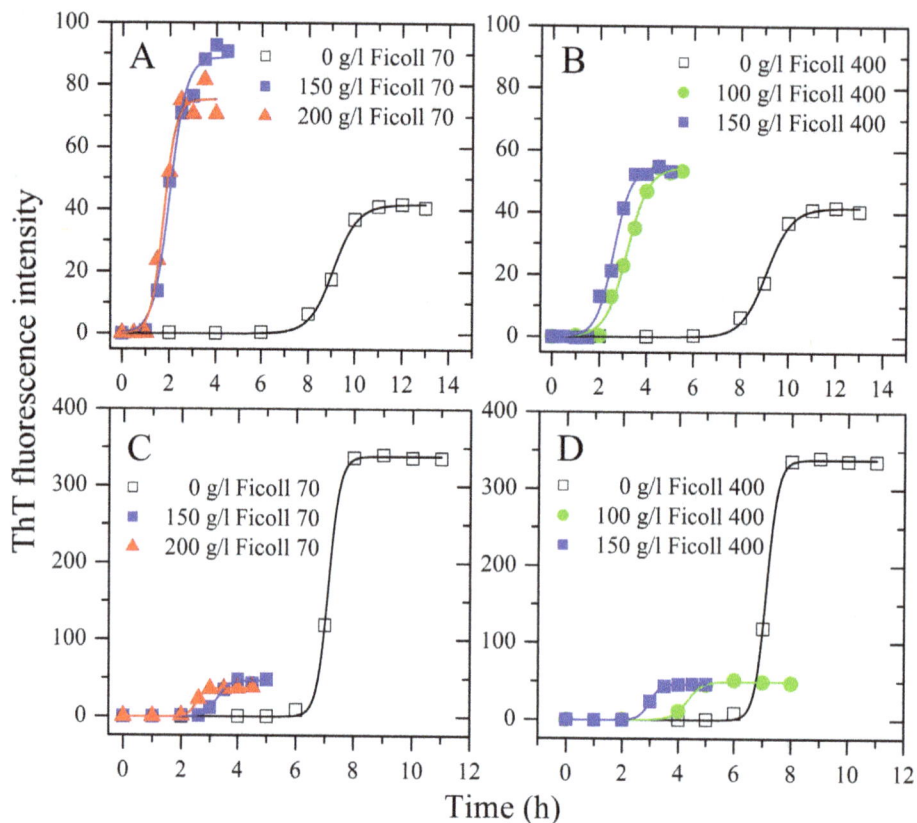

Figure 1. Effects of macromolecular crowding on amyloid formation of human chimera and rabbit chimera of prion proteins. Human chimera (A and B) and rabbit chimera (C and D) in the absence and presence of Ficoll 70 (A and C) or Ficoll 400 (B and D), monitored by ThT fluorescence. All experiments were repeated at least three times. The final concentrations of all prion proteins were 10 µM. The crowding agent concentrations were 0 (open square), 100 g/l (solid circle), 150 g/l (solid square), and 200 g/l (solid triangle), respectively. A sigmoidal equation was fitted to the data and the solid lines represented the best fit. The corresponding kinetic parameters from A–D are summarized in Table 1.

The method for fibrillization of PrP-H2H3 was similar to the method described by the Rezaei lab [30]. Because urea is not stable at pH 5.0, we modified the method as follow: 10 µM PrP-H2H3 was incubated at 37°C in 20 mM NaAc buffer (pH 5.0) containing 200 mM NaCl and 1 M GdnHCl in the absence and presence of crowding agents with continuous shaking at 220 rpm, and samples (50 µl) were diluted into NaAc buffer buffer containing 12.5 µM ThT, giving a final volume of 2.5 ml.

Table 1. Kinetic parameters of amyloid formation of human chimera or rabbit chimera in the absence and presence of Ficoll 70 or Ficoll 400 by ThT binding assays at 37°C.

PrP chimera	Crowding agent	A	Lag time (h)	k (h^{-1})
Human chimera	0	41.7±0.7	8.05±0.15	1.92±0.18
	150 g/l Ficoll 70	88.8±2.7	1.31±0.17	2.89±0.45
	200 g/l Ficoll 70	75.3±2.5	1.23±0.16	3.83±0.72
	100 g/l Ficoll 400	54.7±1.5	2.26±0.16	2.19±0.23
	150 g/l Ficoll 400	54.5±1.4	1.86±0.14	2.74±0.32
Rabbit chimera	0	338±2	6.72±0.09	5.01±0.88
	150 g/l Ficoll 70	46.7±1.3	2.82±0.10	4.66±0.63
	200 g/l Ficoll 70	37.1±0.4	2.24±0.06	6.88±1.09
	100 g/l Ficoll 400	49.8±0.8	3.79±0.12	3.70±0.47
	150 g/l Ficoll 400	46.5±0.1	2.61±0.01	5.31±0.15

Best-fit values of these kinetic parameters were derived from non-linear least squares modeling of a sigmoidal equation as described in the "Materials and Methods" to the data plotted in Fig. 1. Errors shown are ± S.E.

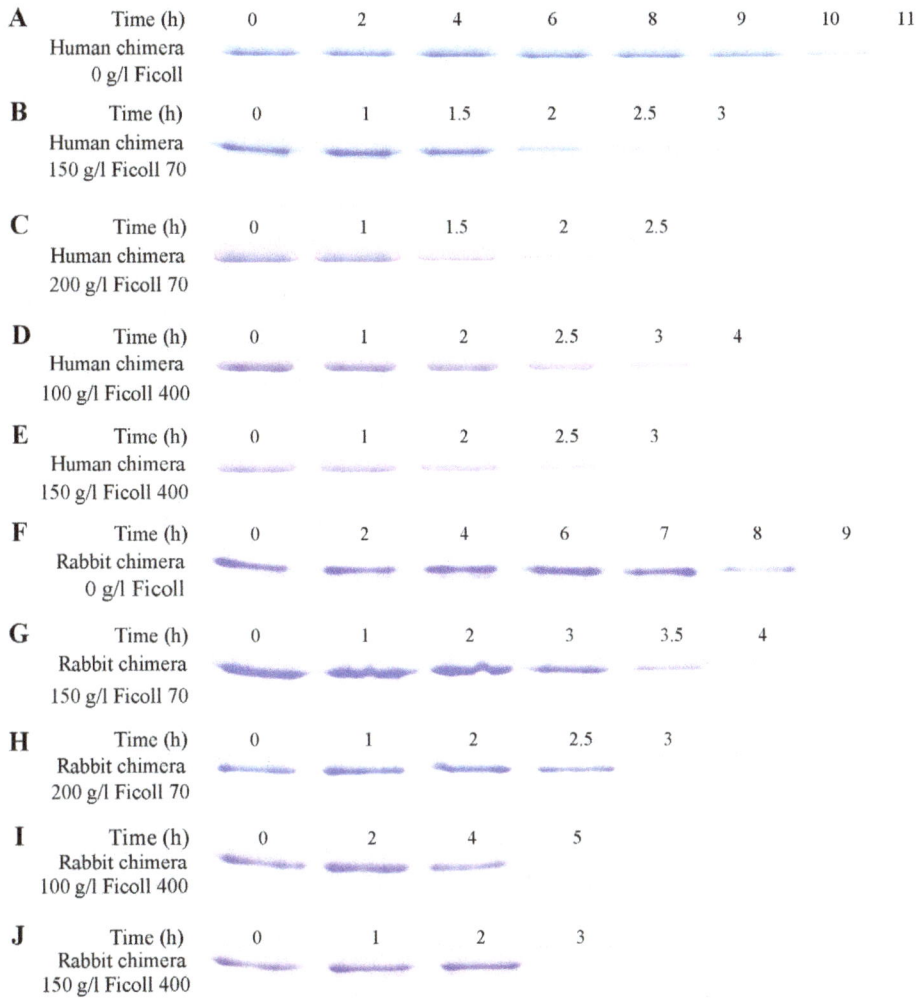

Figure 2. Time-dependent SDS-PAGE analysis of Sarkosyl-soluble human chimera (A–E) and rabbit chimera (F–J) of prion proteins incubated in 0 g/l Ficoll (A and F), 150 g/l (B and G) and 200 g/l (C and H) Ficoll 70, and 100 g/l (D and I) and 150 g/l (E and J) Ficoll 400. We took and dialyzed the samples against 10 mM Tris-HCl (pH 7.0), and incubated them with 10 mM Tris-HCl buffer containing 2% Sarkosyl for 30 min. Then we centrifugated the samples at 17,000 g for 30 min and mixed the supernatants with 2× loading buffer and separated them by 15% SDS-PAGE. The human/rabbit PrPs were denatured in PBS buffer (pH 7.0) containing 2 M GdnHCl.

As described in detail previously [28,33–35], the fluorescence of ThT was excited at 450 nm (slit-width, 5 nm) and the emission was measured at 480 nm (slit-width, 5 nm) for PrPs on an LS-55 luminescence spectrometer (PerkinElmer Life Sciences, Shelton, CT). The fluorescence intensity at 480 nm was averaged over 60 s to increase the signal-to-noise ratio of the measurements. Control experiments were performed to ensure that the crowding agents had no influence on the ThT binding assays for PrPs. Kinetic parameters were determined by fitting ThT fluorescence intensity *versus* time to a sigmoidal equation, as described in detail previously [25,27,33,36,37]:

$$F = F_0 + (A + ct)/\{1 + \exp[k(t_m - t)]\} \qquad (1)$$

where F is the fluorescence intensity, k the rate constant for the growth of fibrils, and t_m the time to 50% of maximal fluorescence. F_0 describes the initial baseline during the lag time. $A+ct$ describes the final baseline after the growth phase has ended. The lag time is determined to be $t_m - 2/k$.

Sarkosyl-soluble SDS-PAGE

The protocol for Sarkosyl-soluble SDS-PAGE was described in detail previously [28,33]. Briefly, amyloid formation of 10 μM PrP was carried out as state above, during the incubation time, 20 μl samples were taken out and added with 2.5 μl of 100 mM Tris-HCl (pH 7.0) and 2.5 μl of 20% Sarkosyl. The mixture were left at room temperature for 30 min, then mixed with 2× loading buffer (without SDS and no heating) and separated by 15% SDS-PAGE. Gels were stained by Coomassie Blue.

Transmission electron microscopy

The formation of fibrils by PrPs was confirmed by electron microscopy of negatively stained samples. The preparation for negatively stained samples was described in detail previously [28,33–35]. Briefly, the incubation time was chosen within a time range of the plateau of each kinetic curve of ThT fluorescence shown in Fig. 1. Sample aliquots of 10 μl were placed on copper grids and left at room temperature for 1–2 min, rinsed twice with H_2O, and then stained with 2% (w/v) uranyl acetate for another 1–2 min. The stained samples were examined using an FEI

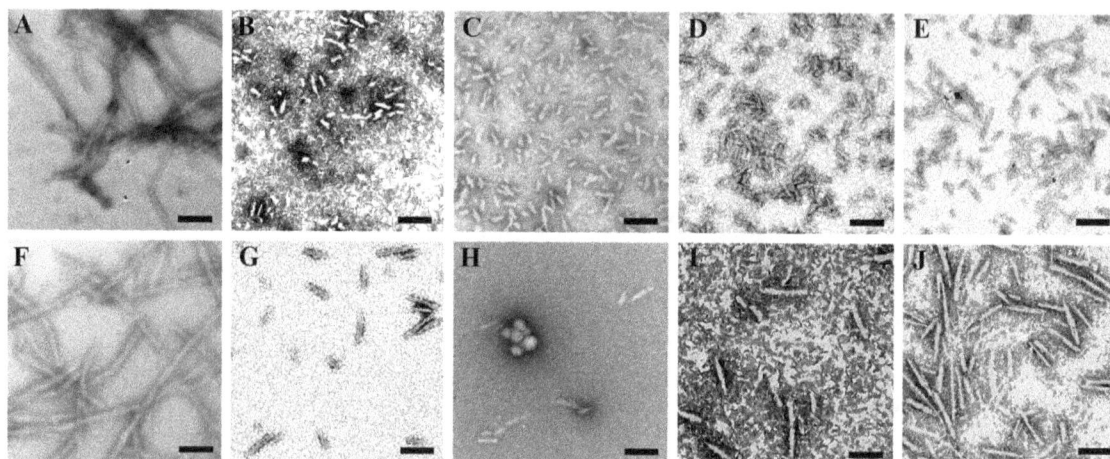

Figure 3. Transmission electron micrographs of human/rabbit chimera PrP samples at physiological pH after incubation under different conditions. Human (A–E) and rabbit (F–J) chimera PrP samples were incubated for 10 h (A) or 8 h (F) or 4 h (D, G, and I) or 3 h (B, E, H, and J) or 2 h (C) in the absence of a crowding agent (A and F) and in the presence of 150 g/l Ficoll 70 (B and G) or 200 g/l Ficoll 70 (C and H) or 100 g/l Ficoll 400 (D and I) or 150 g/l Ficoll 400 (E and J), respectively. We used a 2% (w/v) uranyl acetate solution to stain the fibrils negatively. The scale bars represent 200 nm.

Tecnai G2 20 transmission electron microscope (Hillsboro, OR) operating at 200 kV or an H-8100 transmission electron microscope (Hitachi, Tokyo, Japan) operating at 100 kV.

CD measurements

Circular dichroism spectra were obtained by using a Jasco J-810 spectropolarimeter (Jasco Corp., Tokyo, Japan) with a thermostated cell holder, as described in detail previously [34,35]. Briefly, quartz cell with a 1 mm light-path was used for measurements in the far-UV region. Spectra were recorded from 195 to 250 nm for far-UV CD. PrP fibril samples were subjected to extensive dialysis against NaAc buffer (pH 5.0) to remove guanidine hydrochloride. The final concentration of PrP was kept at 10 μM. The averaged spectra of several scans were corrected relative to the buffer blank or the buffer containing crowding agents. Measurements were made at 25°C. All CD experiments were repeated three times. The experiments were pretty reproducible.

Fourier transform infrared spectroscopy

Attenuated total reflection FTIR spectra of PrP fibril samples were recorded using a Nicolet 5700 FTIR spectrophotometer (Thermo Electron, Madison, WI), as described in detail previously [27]. Briefly, 3 ml PrP fibril samples were harvested (~150000 g) within a time range of the plateau of each kinetic curve of ThT fluorescence, and then washed by H$_2$O and dried by vacuum drying. The dried samples were prepared in D$_2$O and FTIR spectra were recorded in the range from 400 to 4000 cm^{-1} at 4 cm^{-1} resolution. The sample was scanned 32 times in each FTIR measurement, and the spectrum acquired is the average of all these scans. Spectra were corrected for the D$_2$O and water vapors. Measurements were made at 25°C.

PK digestion assays

PrP fibril samples were prepared in 100 mM Tris-HCl buffer (pH 7.5) and incubated with PK at a PK: PrP molar ratio of 1:100 to 1:50 for 1 h at 37°C. Digestion was stopped by the addition of 2 mM phenylmethylsulfonyl fluoride (PMSF), and samples were analyzed in 15% SDS-PAGE and detected by silver staining.

Results

Amyloid fibrils formed by human chimera had secondary structures and PK-resistant features similar to those formed by the human PrP

Our previous studies [27,28] have shown that macromolecular crowding significantly accelerated fibril formation of the human PrP. In this study, the effects of nominally inert polymeric additives, Ficoll 70 and Ficoll 400, on kinetics of amyloid fibril formation of the recombinant human PrP chimera were examined by ThT binding assays (Fig. 1, A and B), as a function of crowder concentration.

As shown in Fig. 1A, the presence of Ficoll 70 at concentrations of 150 g/l and 200 g/l in PBS buffer (pH 7.0) containing 2 M GdnHCl significantly accelerated amyloid fibril formation of human chimera on the investigated time scale, accompanied by a remarkable increase in the maximum ThT intensity, which is very similar to that of the human PrP under the same conditions [28]. As shown in Fig. 1B, the presence of Ficoll 400 at concentrations of 100 g/l and 150 g/l also strongly enhanced fibril formation of human chimera on the investigated time scale, accompanied by an increase in the maximum ThT intensity. To elucidate the detailed effects of macromolecular crowding on amyloid fibril formation of human chimera, a sigmoidal equation was used to fit the kinetic data, yielding three kinetic parameters which are summarized in Table 1. As shown in Table 1, the addition of 150 g/l Ficoll 70 dramatically accelerated both nucleation and elongation steps of human chimera fibrillization, resulting in a lag time of 1.31 h in the presence of Ficoll 70, which is 6.1-fold decreased compared with that in the absence of a crowding agent (8.05 h), and a rate constant for the growth of fibrils of 2.89 h^{-1}, which is 1.5-fold larger than that in the absence of a crowding agent (1.92 h^{-1}). Similarly, the addition of 150 g/l Ficoll 400 also markedly accelerated both steps of human chimera fibrillization (Table 1).

After centrifugation assays, Sarkosyl-soluble SDS-PAGE experiments were carried out to semi-quantify the decrease of PrP monomers in the presence of crowders. As shown in Fig. 2, A–E, when human chimera was incubated in the absence of a crowding agent for 10 h, a clear band corresponding to Sarkosyl-soluble human chimera monomers was observed, while when human

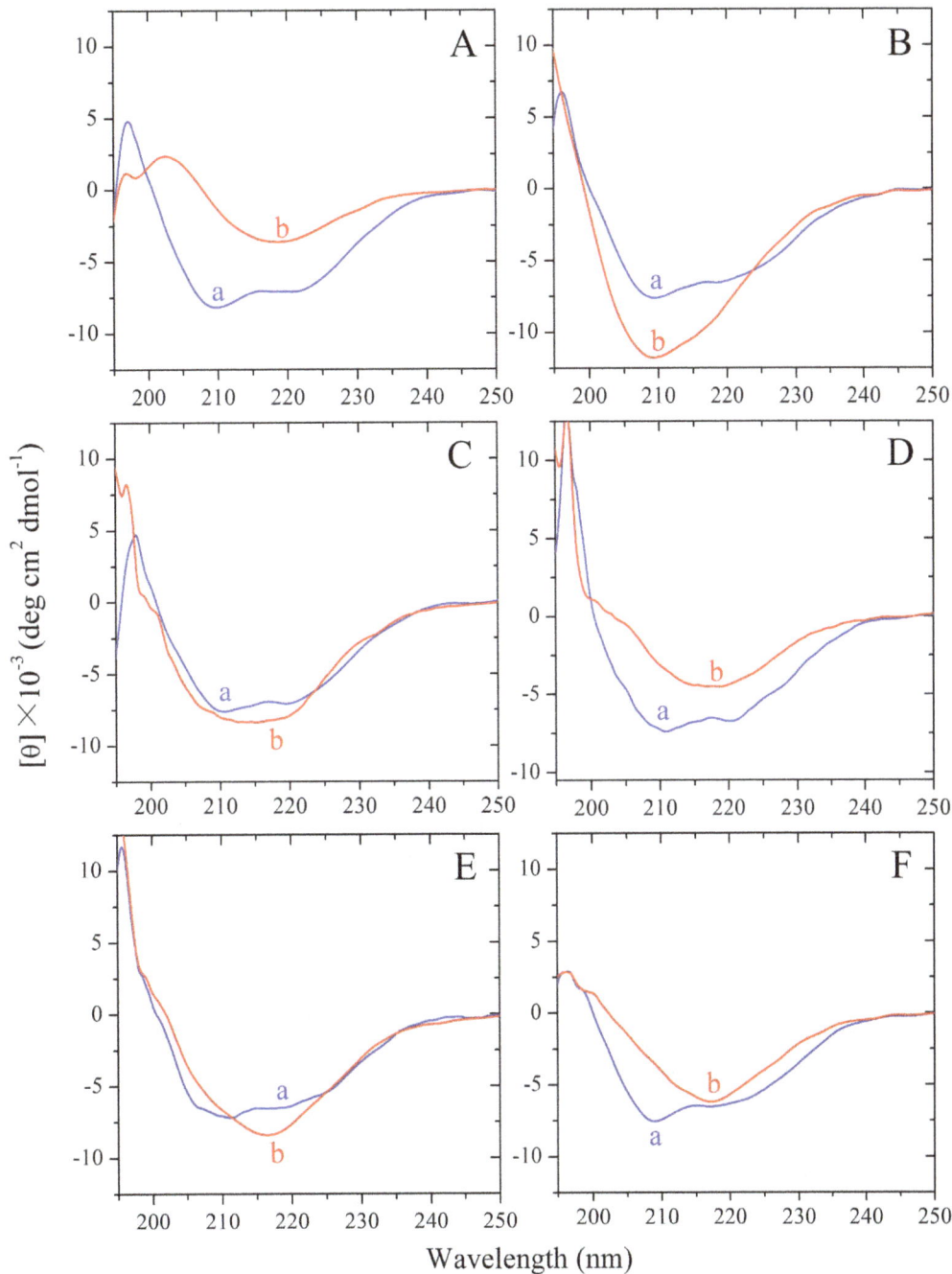

Figure 4. Secondary structural changes of human (A, C, and E) and rabbit (B, D, and F) chimera PrP isoforms monitored by far-UV CD. Curve a: native human/rabbit chimera prion proteins, and Curve b: amyloid fibrils produced from human chimera and rabbit chimera, in the absence of a crowding agent (A and B) and in the presence of 150 g/l Ficoll 70 (C and D) or 100 g/l Ficoll 400 (E and F), respectively. The incubation time was chosen within a time range of the plateau of each kinetic curve of ThT fluorescence shown in Fig. 1.

chimera was incubated with 150 g/l Ficoll 70/Ficoll 400 for 2.5–3 h, a much shorter time than 10 h, the human chimera monomer band was observed. The above results indicate that crowding agents dramatically promoted fibril formation of human chimera, similar to that of the human PrP [28].

TEM was then employed to characterize the morphology of human chimera aggregates formed in the absence and in the presence of crowding agents. Fig. 3, A–E, shows TEM images of

human chimera samples incubated in the solution of a crowding agent (Ficoll 70 or Ficoll 400), compared with those in dilute solutions. In absence of a crowding agent, human chimera formed fibrils with a long, twisted, and branched structure after incubation for 10 h (Fig. 3A), which is similar to those produced from the human PrP or the bovine PrP in the absence of a crowding agent [25,27]. In the presence of 150 g/l, 200 g/l Ficoll 70 or 100 g/l, 150 g/l Ficoll 400, however, abundant short fibrillar fragments

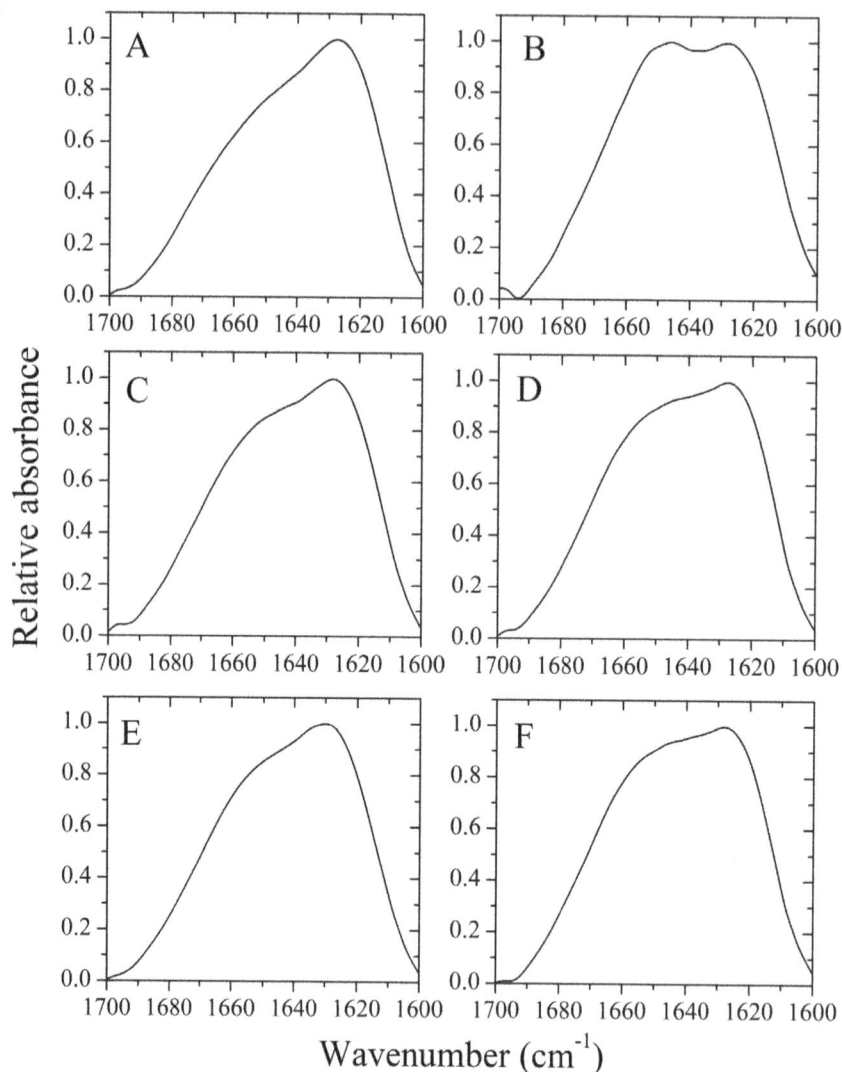

Figure 5. Secondary structural changes of human (A, C, and E) and rabbit (B, D, and F) chimera PrP isoforms monitored by FTIR. The FTIR spectra of amyloid fibrils produced from human chimera and rabbit chimera, in the absence of a crowding agent (A and B) and in the presence of 150 g/l Ficoll 70 (C and D) or 100 g/l Ficoll 400 (E and F), respectively. The incubation time was chosen within a time range of the plateau of each kinetic curve of ThT fluorescence shown in Fig. 1.

and spherical or ellipsoidal particles were observed when human chimera samples were incubated for 2–4 h (Fig. 3, B–E), which is similar to those produced from the human PrP or the bovine PrP in the presence of 150 g/l Ficoll 70 [25,27]. The above results indicated that macromolecular crowding also caused human chimera to form short fibrils and non-fibrillar particles, which is very similar to that of the human PrP or the bovine PrP [25,27].

CD spectroscopy was used to confirm the formation of amyloid fibrils by human chimera. Fig. 4, A, C, and E, shows the CD spectra of native human chimera and human chimera fibrils formed in the absence and presence of 150 g/l Ficoll 70 or 100 g/l Ficoll 400. Under all conditions, human chimera formed amyloid fibrils with β-sheet-rich conformation (a single minimum around 218 nm was observed) from the native state which has predominant α-helix conformation with double minima at 208 and 222 nm.

CD is not the best approach to investigate the secondary structure of hyper-large protein assemblies as at this scale anisotropic light scattering could take place. The shape and position of amide I′ (1600–1700 cm^{-1}) of FTIR bands provide detailed information on the secondary structure of proteins, and the amide I′ band at 1630 cm^{-1} is characteristic for β-sheet formed by amyloid fibrils [27]. Considering this, we performed FTIR experiments to investigate the secondary structure of amyloid fibrils formed by human chimera. Fig. 5, A, C, and E, shows the FTIR spectra in the amide I′ region of human chimera fibrils formed in the absence and presence of 150 g/l Ficoll 70 or 100 g/l Ficoll 400. Under all conditions, human chimera formed amyloid fibrils with β-sheet-rich conformation (a single amide I′ band around 1630 cm^{-1} was observed).

PK resistance activity has been widely used to distinguish PrP^{C} from PrP^{Sc} since the pioneering studies of Prusiner and co-workers [38]. As shown in Fig. 6A, amyloid fibrils produced from human chimera generated PK-resistant fragments of 15–16-kDa after PK digestion for 1 h, which are similar to those of the human PrP [25,27]. Human chimera fibrils also generated three short fragments (12-, 10-, and 8-kDa bands), which are similar to those of the human PrP reported previously [25,27]. Taken together,

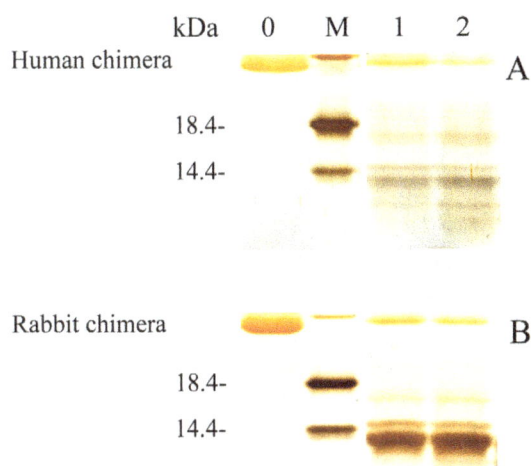

Figure 6. Concentration-dependent proteinase K-digestion assays of human (A) and rabbit (B) chimera PrP fibrils. Samples were treated with PK for 1 h at 37°C at PK: PrP molar ratios as follows: 1:100 (lane 1) and 1:50 (lane 2). PK concentration: 0.4 μg/ml (lane 1) and 0.8 μg/ml (lane 2). The controls with zero protease in the absence of a crowding agent were loaded in lane 0. Protein molecular weight markers were loaded on lane M: restriction endonuclease Bsp98 I (25.0 kDa), β-lactoglobulin (18.4 kDa), and lysozyme (14.4 kDa). Amyloid fibrils were produced from human chimera and rabbit chimera in the absence of a crowding agent. Protein fragments were separated by SDS-PAGE and detected by silver staining.

our CD, FTIR and PK digestion data demonstrate that amyloid fibrils formed by human chimera have secondary structures and PK-resistant features similar to those formed by the human PrP.

Amyloid fibrils formed by rabbit chimera had PK-resistant features and secondary structures in crowded physiological environments different from those formed by the rabbit PrP

We have demonstrated previously that macromolecular crowding remarkably inhibited fibril formation of the rabbit PrP [27,28]. In this study, the effects of two macromolecular crowding agents, Ficoll 70 and Ficoll 400, on kinetics of amyloid fibril formation of the recombinant rabbit PrP chimera were examined by ThT binding assays (Fig. 1, C and D), as a function of crowder concentration.

To our surprise, the presence of Ficoll 70 at concentrations of 150 g/l and 200 g/l significantly accelerated the nucleation step of fibril formation of rabbit chimera (a significant decrease in the lag time) (Fig. 1C), which is different from that of the rabbit PrP under the same conditions [27,28], but accompanied by a decrease of one order of magnitude in the maximum ThT intensity (Fig. 1C), which is similar to that of the rabbit PrP [27,28]. As shown in Fig. 1D, the presence of Ficoll 400 at concentrations of 100 g/l and 150 g/l also strongly enhanced the nucleation step of fibril formation of rabbit chimera but accompanied by a decrease of one order of magnitude in the maximum ThT intensity. Clearly, the impact of crowding agents on fibril formation of rabbit chimera is more dramatic than that of human chimera. To elucidate the detailed effects of macromolecular crowding on amyloid fibril formation of rabbit chimera, a sigmoidal equation was used to fit the kinetic data, yielding three kinetic parameters which are also summarized in Table 1. As shown in Table 1, the addition of 150 g/l Ficoll 70 remarkably accelerated the nucleation step of rabbit chimera fibrillization, resulting in a lag time of 2.82 h in the

presence of Ficoll 70, which is 2.4-fold decreased compared with that in the absence of a crowding agent (6.72 h), and the addition of 150 g/l Ficoll 400 also markedly accelerated the nucleation step of rabbit chimera fibrillization. However, the addition of 150 g/l Ficoll 70 or 150 g/l Ficoll 400 significantly decreased the amount of rabbit chimera fibrils represented by A (Table 1).

In order to confirm the enhancing effect of macromolecular crowding on the nucleation step of fibril formation of rabbit chimera, we carried out Sarkosyl-soluble SDS-PAGE experiments. As shown in Fig. 2, F–J, when rabbit chimera was incubated in the absence of a crowding agent for 8 h, a clear band corresponding to Sarkosyl-soluble rabbit chimera monomers was observed, while when rabbit chimera was incubated with 150 g/l Ficoll 70/Ficoll 400 the band disappeared within 4–5 h. The above results indicate that crowding agents remarkably promoted the nucleation step of fibril formation of rabbit chimera, different from that of the rabbit PrP [27,28].

The morphology of rabbit chimera aggregates formed in the absence and in the presence of crowding agents was characterized by using TEM. Fig. 3, F–J, shows TEM images of rabbit chimera samples incubated in the solution of a crowding agent (Ficoll 70 or Ficoll 400), compared with those in dilute solutions. In dilute solutions, rabbit chimera formed long and branched fibrils after incubation for 8 h (Fig. 3F). However, rabbit chimera formed some short amyloid fibrils and spherical particles when incubated with 150–200 g/l Ficoll 70 or 100–150 g/l Ficoll 400 for 3–4 h (Fig. 3, G–J). The amount of fibrils formed by rabbit chimera incubated with crowding agents (Fig. 3, G–J) was remarkably less than that in dilute solutions (Fig. 3F) on the same time scale, consistent with the conclusion from ThT binding assays that macromolecular crowding significantly decreased the amount of rabbit chimera fibrils.

CD spectroscopy was used to confirm the significant differences in secondary structures between rabbit chimera fibrils formed in crowded physiological environments and those formed in dilute solutions. Fig. 4, B, D, and F, shows the CD spectra of native rabbit chimera and rabbit chimera fibrils formed in the absence and presence of 150 g/l Ficoll 70 or 100 g/l Ficoll 400. As shown in Fig. 4, B, D, and F, in the absence and presence of a crowding agent and before incubation, the CD spectra measured for rabbit chimera sample had double minima at 208 and 222 nm, indicative of predominant α-helical structure. After incubation for 4 h, a single minimum around 218 nm was observed for rabbit chimera fibril samples in crowded physiological environments (Fig. 4, D and F), which is typical of predominant β-sheet structure and a characteristic for amyloid formation. After incubation for 8 h, however, a single minimum around 210 nm (but not 218 nm) was observed for rabbit chimera fibril samples in dilute solutions (Fig. 4B), which is similar to that of the rabbit PrP in dilute solutions [27]. Then we performed FTIR experiments to investigate the secondary structure of amyloid fibrils formed by rabbit chimera. Fig. 5, B, D, and F, shows the FTIR spectra in the amide I' region of rabbit chimera fibrils formed in the absence and presence of 150 g/l Ficoll 70 or 100 g/l Ficoll 400. After incubation for 4 h, a single amide I' band around 1630 cm^{-1} was observed for rabbit chimera fibril samples in crowded physiological environments (Fig. 5, D and F), which is typical of predominant β-sheet structure and a characteristic for amyloid formation. After incubation for 8 h, however, two amide I' bands around 1650 cm^{-1} and 1630 cm^{-1} were observed for rabbit chimera fibril samples in dilute solutions (Fig. 5B), which is similar to that of the rabbit PrP in dilute solutions [27]. The above data demonstrate that some PrP chimeras could form amyloid fibrils with different structural features in absence and presence of

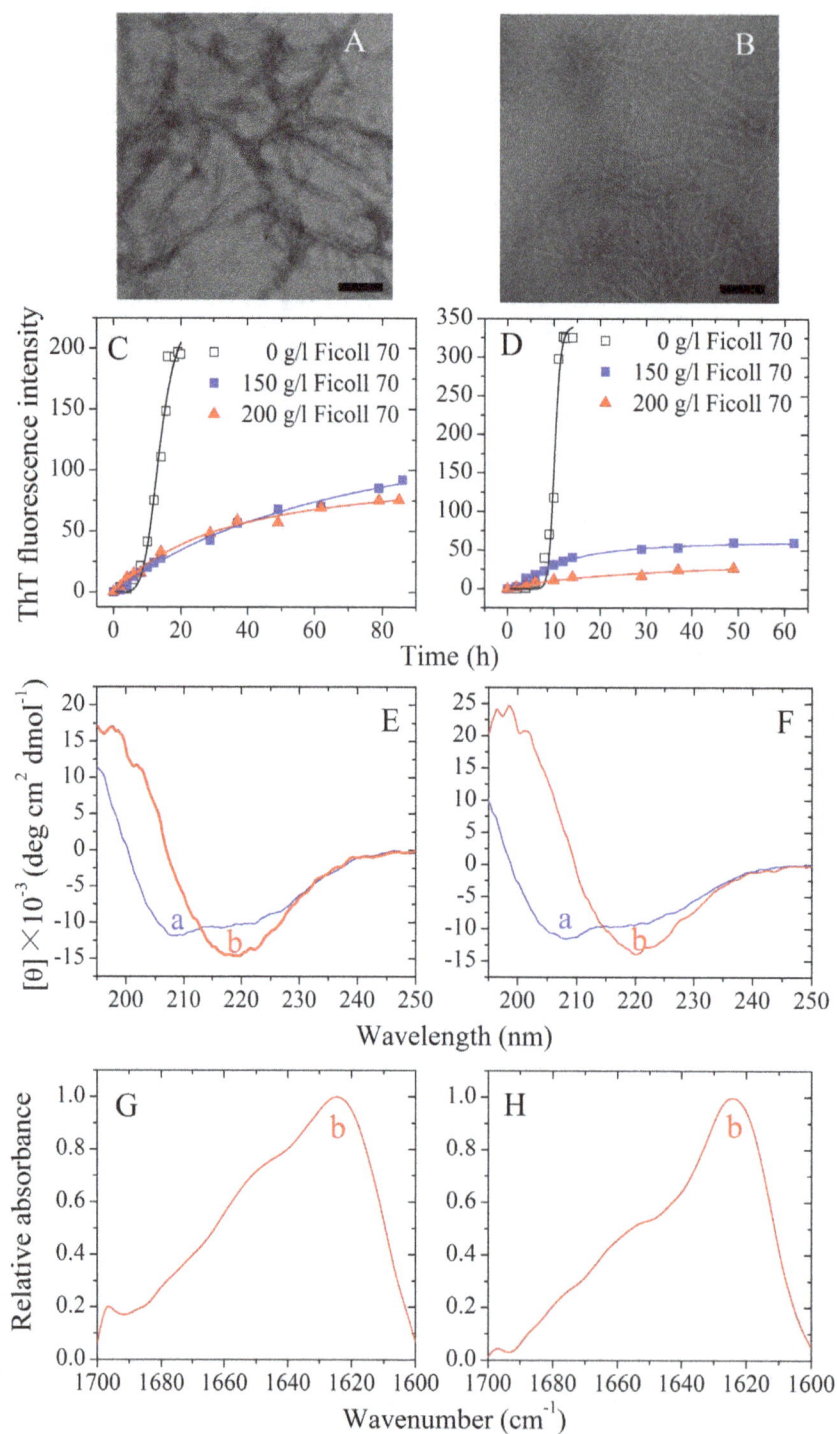

Figure 7. Amyloid formation of human PrP-H2H3 and rabbit PrP-H2H3. Transmission electron micrographs of human PrP-H2H3 (A) and rabbit PrP-H2H3 (B) were made after incubation under different conditions. PrP-H2H3 samples were incubated in the absence of a crowding agent for 14 h (A) and 10 h (B), respectively. Fibril formation of human PrP-H2H3 (C) and rabbit PrP-H2H3 (D) in the absence and in the presence of Ficoll 70, monitored by ThT fluorescence. All experiments were repeated at least three times. The final concentrations of all prion proteins were 10 μM. The crowding agent concentrations were 0 (open square), 150 g/l (solid square), and 200 g/l (solid triangle), respectively. The empirical Hill equation was fitted to the data and the solid lines represented the best fit. Secondary structural changes of human PrP-H2H3 (E and G) and rabbit PrP-H2H3 (F and H) isoforms monitored by far-UV CD and FTIR, respectively. Curve a: native human/rabbit PrP-H2H3. Curve b: amyloid fibrils produced from human PrP-H2H3 and rabbit PrP-H2H3 incubated for 20 h in the absence of a crowding agent.

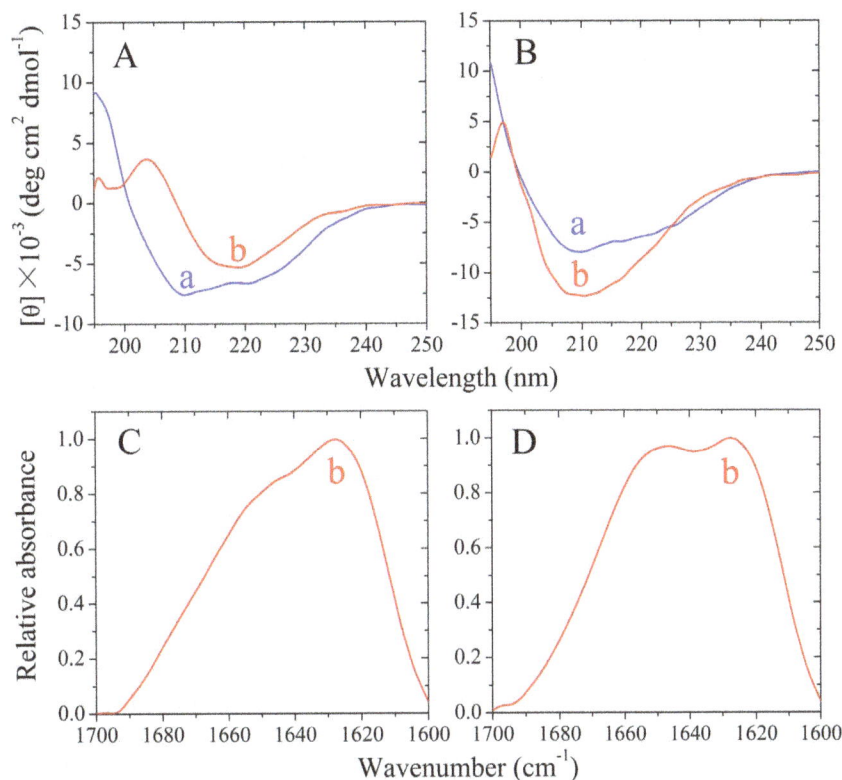

Figure 8. Secondary structural changes of chimera H (A and C) and chimera R (B and D) isoforms monitored by far-UV CD and FTIR, respectively. Curve a: native chimera prion proteins. Curve b: amyloid fibrils produced from chimera prion proteins in the absence of a crowding agent. The incubation time was chosen within a time range of the plateau of each kinetic curve of ThT fluorescence shown in Fig. S1.

crowding agents, revealing the importance of macromolecular crowding on protein misfolding.

PK digestion assays was used to confirm the significant differences in PK resistance activities between rabbit chimera fibrils and rabbit PrP fibrils. As shown in Fig. 6B, amyloid fibrils produced from rabbit chimera did generate PK-resistant fragments of 15–16-kDa after PK digestion for 1 h, which are different from those of the rabbit PrP but similar to those of the human PrP [25,27]. Taken together, our CD, FTIR and PK digestion data demonstrate that amyloid fibrils formed by rabbit chimera had PK-resistant features and secondary structures in crowded physiological environments different from those formed by the rabbit PrP.

Amyloid formation of human PrP-H2H3 and rabbit PrP-H2H3

In order to know the role of the H2H3 domain in amyloid fibril formation of the rabbit/human PrPs, we investigated amyloid formation of human PrP-H2H3 and rabbit PrP-H2H3, by using ThT binding assays, CD, FTIR, and TEM methods. Fig. 7, A and B, shows TEM images of human PrP-H2H3 and rabbit PrP-H2H3 samples. In absence of a crowding agent, both human PrP-H2H3 and rabbit PrP-H2H3 formed fibrils with a long, twisted, and branched structure after incubation for 14 and 10 h, respectively. Fig. 7, C and D, shows the effects of macromolecular crowding on fibril formation of human/rabbit PrP-H2H3 examined by ThT binding assays. To our surprise, the kinetics of fibril formation of both human PrP-H2H3 and rabbit PrP-H2H3 are similar with a similar lag time of 8 h, and the effects of macromolecular crowding on fibril formation of human/rabbit

PrP-H2H3 are also similar (Fig. 7, C and D). As shown in Fig. 7, C and D, the presence of Ficoll 70 at concentrations of 150 g/l and 200 g/l significantly accelerated the nucleation step of fibril formation of both human PrP-H2H3 and rabbit PrP-H2H3 (a significant decrease in the lag time) but accompanied by a remarkable decline of the maximum ThT intensity. Fig. 7, E and F, shows the CD spectra of native human/rabbit PrP-H2H3 and human/rabbit PrP-H2H3 fibrils formed in the absence of a crowding agent. Under such conditions, both human PrP-H2H3 and rabbit PrP-H2H3 formed amyloid fibrils with β-sheet-rich conformation (a single minimum around 218 nm was observed) from the native state which has predominant α-helix conformation with double minima at 208 and 222 nm. Fig. 7, G and H, shows the FTIR spectra in the amide I′ region of human/rabbit PrP-H2H3 fibrils formed in the absence of a crowding agent. Under such conditions, both human PrP-H2H3 and rabbit PrP-H2H3 formed amyloid fibrils with β-sheet-rich conformation (a single amide I′ band around 1630 cm^{-1} was observed). Clearly, amyloid formation of human PrP-H2H3 is similar to that of rabbit PrP-H2H3, and the amino acids beyond PrP-H2H3 need to be explored in order to explain why amyloid fibrils formed by the rabbit PrP and the human PrP have different secondary structures and why macromolecular crowding has different effects on fibril formation of the rabbit/human PrPs.

The amino acids beyond PrP-H2H3 had a remarkable effect on fibrillization of the rabbit PrP

Then we want to know whether the amino acids beyond PrP-H2H3 can influence fibrillization of the rabbit/human PrPs. We thus constructed chimera H, in which the human PrP-B1H1B2 (β-

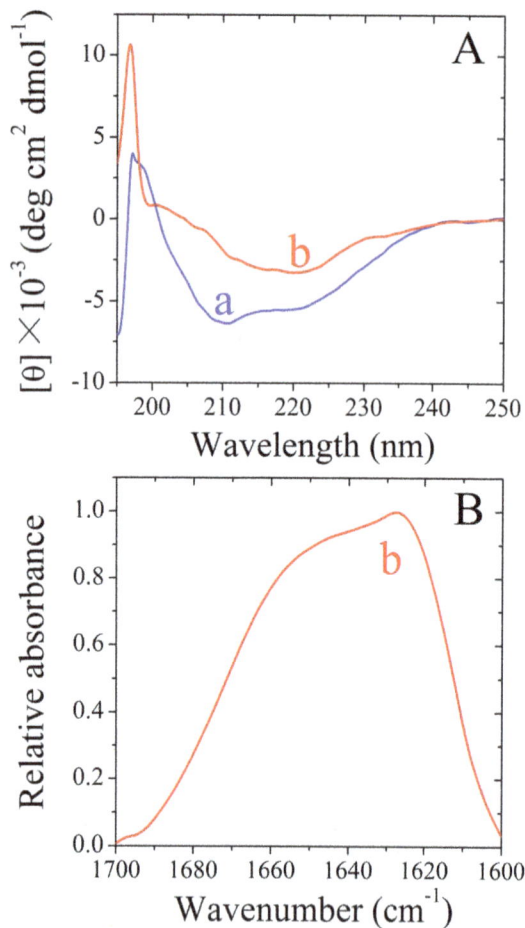

Figure 9. Secondary structural changes of chimera R isoforms in the presence of Ficoll 70 monitored by far-UV CD (A) and FTIR (B). Curve a: native chimera R. Curve b: amyloid fibrils produced from chimera R in the presence of 150 g/l Ficoll 70. The incubation time was chosen within a time range of the plateau of each kinetic curve of ThT fluorescence shown in Fig. S1.

strand 1, α-helix 1, and β-strand 2) domain was replaced by the rabbit PrP-B1H1B2 domain, and chimera R, in which the rabbit PrP-B1H1B2 domain was replaced by the human PrP-B1H1B2 domain, and studied amyloid formation of chimeras H and R, compared with that of the human PrP and the rabbit PrP. As shown in Fig. S1A, the presence of 200 g/l Ficoll 70 significantly accelerated both nucleation and elongation steps of fibril formation of chimera H, which is not only similar to that of the human PrP [28] but also similar to that of human chimera (Fig. 1A). Fig. 8A shows the CD spectra of native chimera H and chimera H fibrils formed in dilute solutions, and Fig. 8C shows the FTIR spectra in the amide I′ region of chimera H fibrils formed in dilute solutions. Similar to human chimera and the human PrP, chimera H also formed amyloid fibrils with β-sheet-rich conformation from the native state which has predominant α-helix conformation. As shown in Fig. S1B, the presence of 200 g/l Ficoll 70 remarkably enhanced the nucleation step of fibril formation of chimera R but accompanied by a very strong decline the maximum ThT intensity, which is similar to that of rabbit chimera (Fig. 1C). CD spectroscopy and FTIR spectroscopy were also employed to confirm the significant differences in secondary structures between chimera R fibrils formed in crowded physiological environments and those formed in dilute solutions. Figs. 8B and 9A show the CD spectra of native chimera R and chimera R fibrils formed in the absence and presence of 150 g/l Ficoll 70, respectively. Figs. 8D and 9B show the FTIR spectra in the amide I′ region of chimera R fibrils formed in the absence and presence of 150 g/l Ficoll 70, respectively. As shown in Figs. 8B and 9A, in the absence and presence of a crowding agent and before incubation, the CD spectra measured for chimera R sample had double minima at 208 and 222 nm, indicative of predominant α-helical structure. After incubation for 3 h, a single minimum around 218 nm or a single amide I′ band around 1630 cm^{-1} was observed for chimera R fibril samples in crowded physiological environments (Fig. 9, A and B), which is typical of predominant β-sheet structure and different from that of the rabbit PrP in crowded physiological environments [27]. After incubation for 5 h, however, a single minimum around 210 nm or two amide I′ bands around 1650 cm^{-1} and 1630 cm^{-1} were observed for chimera R fibril samples in dilute solutions (Fig. 8, B and D), which is similar to not only that of rabbit chimera in dilute

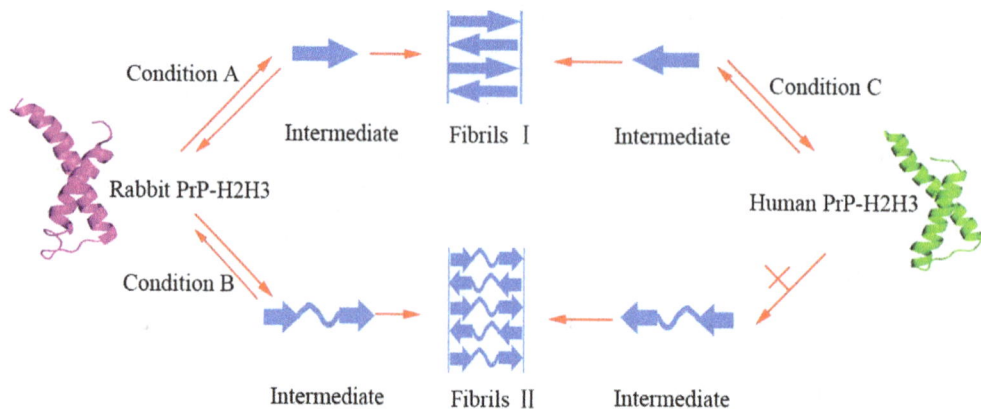

Figure 10. A hypothetical model to explain why amyloid fibrils formed by the rabbit PrP and the human PrP have different structural features. Some chimeras containing rabbit PrP-H2H3 domain could form fibrils I and fibrils II with different structures under different conditions (Conditions A and B), but PrPs containing human PrP-H2H3 domain could only form fibrils I under all conditions (Condition C). A single minimum around 218 nm is observed for fibrils I with β-sheet-rich conformation, but a single minimum around 210 nm is observed for fibrils II with random coils and less β-sheet conformation.

solutions (Figs. 4B and 5B) but also that of the rabbit PrP in dilute solutions [27]. The above data once again demonstrate that some PrP chimeras could form amyloid fibrils with different structural features in absence and presence of crowding agents.

Discussion

Prion protein, a unique infectious amyloid disease-associated protein, causes many lethal human and animal prion diseases [1]. Rabbits are only sensitive to artificial rabbit prion [17] and resistant to prions from other animal species [5]. It has been demonstrated that the unique rabbit PrP sequence, $\beta2$-$\alpha2$ helix-cap, and the residues surrounding the glycosylphosphatidylinositol anchor attachment site could contribute to its resistance to prion diseases [9,39–41]. Furthermore, the presence of either serine (rabbit) or asparagine (hamster) residues in positions 170 and 174 of PrP not only affect the secondary structure of the $\beta2$-$\alpha2$ loop but also the propensity with which the PrP misfolds into β-state-rich octamers [42]. A recent molecular dynamics study has indicated that the sites I214 and S173 of the rabbit PrP can influence the stability of rabbit PrP native state [43]. We have demonstrated previously that macromolecular crowding remarkably inhibits fibril formation of the rabbit PrP but significantly accelerates fibril formation of the human/bovine PrPs [27,28]. In this study, we want to know which domains of PrPs cause such differences. To align the rabbit PrP23-228 with the human PrP23-231, we found that there are 88% identities between their sequences. Hydrogen/deuterium exchange and solid-state NMR results have demonstrated that PrP fibrils contain in-register parallel β-sheets and that the structurally ordered fibril core includes the C-terminal segment, approximately residues 175–225, which includes the H2H3 of monomeric PrP [44,45]. In this study the effects of replacement certain domains in rabbit and human PrPs on fibril formation under natural crowding conditions were investigated by a series of biochemical experiments using PrP chimeras containing the H2H3 domain. We demonstrated that two rabbit PrP chimeras we designed (rabbit chimera and chimera R) did form amyloid fibrils with different structural features in absence and presence of crowding agents. By contrast, the rabbit PrP forms amyloid fibrils with same structural features in absence and presence of crowding agents [27]. We also demonstrated that two human PrP chimeras we designed (human chimera and chimera H) did form amyloid fibrils with same structural features in absence and presence of crowding agents, which is in agreement with those observed in human PrP fibrils [25,27]. Furthermore, we found that PK resistance of the fibrils from the rabbit chimera is different from that observed in rabbit PrP fibrils [27] but PK resistance of the fibrils from the human chimera is similar to that observed in human PrP fibrils [25,27]. We then investigated amyloid formation of rabbit PrP-H2H3 and human PrP-H2H3. To our surprise, macromolecular crowding accelerated the nucleation step of fibril formation of both rabbit PrP-H2H3 and human PrP-H2H3, and both rabbit PrP-H2H3 and human PrP-H2H3 formed amyloid fibrils with β-sheet-rich conformation from the native state which has predominant α-helix conformation, which are similar to those of the human PrP [27,28]. Our data indicate that the structure of amyloid fibrils formed by rabbit PrP-H2H3 is different from that formed by the rabbit PrP [27] while the structure of amyloid fibrils formed by human PrP-H2H3 is similar to that formed by human PrP [25,27]. All the results above suggest that the amino acids beyond PrP-H2H3 (herein PrP-B1H1B2 domain) have a remarkable effect on fibrillization of the rabbit PrP but almost no effect on the human PrP. In other words, not only the H2H3 domain could play an important role, but also

the B1H1B2 domain could take part in fibrillization of the rabbit PrP. Our conclusion that the H2H3 domain plays an important role in PrP fibrillization is in agreement with previously published work [29–31,44,45]. Rabbit/human chimera formed amyloid fibrils with the same structural features as those of chimera R/H, and the difference between Rabbit/human chimera and chimera R/H is the N-terminal flexibly disordered tail, indicating that an N-terminal flexibly disordered tail of PrP is not important for fibrillization of the rabbit/human PrPs.

First, this paper demonstrated that the B1H1B2 domain causes the difference between fibrillization of the rabbit PrP and the human PrP. Therefore, the present study provides an explanation of how domain replacement affects fibril formation of prion proteins from different species, much better than our previous studies [27,28]. Second, this paper employed ThT binding assays, Sarkosyl-soluble SDS-PAGE, CD, FTIR, and TEM, similar to our previous studies [28,33–35], but designed four novel PrP chimeras. Therefore, the present study presents a new finding concerning the influence of the amino acids among and beyond PrP-H2H3 on fibrillization of prion proteins from different species, as compared with our previous studies [27,28].

Based on our data and the reported results [27,28], we propose a valuable hypothetical model to explain why amyloid fibrils formed by the rabbit PrP and the human PrP have different structural features (Fig. 10). Some chimeras containing rabbit PrP-H2H3 domain (rabbit chimera and chimera R) could form fibrils I and fibrils II with different structures under different conditions, but PrPs with human PrP-H2H3 domain (such as wild-type human PrP, human chimera, and chimera H) could only form fibrils I under all conditions and wild-type rabbit PrP could only form fibrils II under all conditions. A single minimum around 218 nm or a single amide I' band around 1630 cm^{-1} is observed for fibrils I with β-sheet-rich conformation, but a single minimum around 210 nm or two amide I' bands around 1650 cm^{-1} and 1630 cm^{-1} are observed for fibrils II with random coils and less β-sheet conformation.

In conclusion we have shown that: (i) the presence of a strong crowding agent dramatically promoted fibril formation of four PrP chimeras we designed, and amyloid fibrils from human chimera had biochemical features similar to those from the human PrP; (ii) amyloid fibrils from rabbit chimera had biochemical features in crowded physiological environments different from those from the rabbit PrP. This work is novel as demonstrates that certain protein domains have a significant effect on fibril formation of PrP from rabbit but almost no effect on human PrP, and has the following biochemical and biomedical implications. First, our findings can help to explain why amyloid fibrils formed by the rabbit PrP and the human PrP have different secondary structures and why macromolecular crowding has different effects on fibrillization of PrPs from different species. Second, our findings may help to better understand why human PrP *in vivo* tends to form fibril deposits associated with serious infectious diseases while the rabbit PrP does not and thus is unlikely that will cause prion diseases.

Supporting Information

Figure S1 Effects of macromolecular crowding on amyloid formation of chimeras H and R. Chimera H (A) and chimera R (B) in the absence and presence of Ficoll 70, monitored by ThT fluorescence. All experiments were repeated at least three times. The final concentrations of all prion proteins were 10 μM. The crowding agent concentrations were 0 (open square) and 200 g/l (solid triangle), respectively. A sigmoidal equation was fitted to the data and the solid lines represented the best fit. The insets in B show the kinetics of amyloid formation of chimera R in the presence of Ficoll 70 more clearly.

Table S1 The primers used to replace the human PrP-H2H3 by the rabbit PrP-H2H3.

Table S2 The primers used to replace the rabbit PrP-H2H3 by the human PrP-H2H3.

Table S3 The primers used to construct chimera H in which the human PrP-B1H1B2 (β-strand 1, α-helix 1, and β-strand 2) was replaced by the rabbit PrP-B1H1B2.

Table S4 The primers used to construct chimera R in which the rabbit PrP-B1H1B2 (β-strand 1, α-helix 1, and β-strand 2) was replaced by the human PrP-B1H1B2.

Acknowledgments

We sincerely thank Prof. Geng-Fu Xiao (Wuhan Institute of Virology, Chinese Academy of Sciences) for his kind gift of the human/rabbit PrP^C plasmids. We thank Dr. Li Li in this college and Dr. Zhi-Ping Zhang (Wuhan Institute of Virology, Chinese Academy of Sciences) for their technical assistances on TEM.

Author Contributions

Conceived and designed the experiments: YL. Performed the experiments: XY J-JH ZZ. Analyzed the data: XY YL. Contributed reagents/materials/analysis tools: JC. Wrote the paper: XY YL.

References

1. Prusiner SB (1998) Prions. Proc Natl Acad Sci USA 95: 13363–13383.
2. Stohr J, Weinmann N, Wille H, Kaimann T, Nagel-Steger L, et al. (2008) Mechanisms of prion protein assembly into amyloid. Proc Natl Acad Sci USA 105: 2409–2414.
3. Aguzzi A, Sigurdson C, Heikenwaelder M (2008) Molecular mechanisms of prion pathogenesis. Annu Rev Pathol 3: 11–40.
4. Soto C (2011) Prion hypothesis: the end of the controversy? Trends Biochem Sci 36: 151–158.
5. Barlow RM, Rennie JC (1976) The fate of ME7 scrapie infection in rats, guinea-pigs and rabbits. Res Vet Sci 21: 110–111.
6. Wen Y, Li J, Yao W, Xiong M, Hong J, et al. (2010) Unique structural characteristics of the rabbit prion protein. J Biol Chem 285: 31682–31693.
7. Zahn R, Liu A, Luhrs T, Riek R, von Schroetter C, et al. (2000) NMR solution structure of the human prion protein. Proc Natl Acad Sci USA 97: 145–150.
8. Lopez Garcia F, Zahn R, Riek R, Wuthrich K (2000) NMR structure of the bovine prion protein. Proc Natl Acad Sci USA 97: 8334–8339.
9. Vorberg I, Groschup MH, Pfaff E, Priola SA (2003) Multiple amino acid residues within the rabbit prion protein inhibit formation of its abnormal isoform. J Virol 77: 2003–2009.
10. Hill AF, Antoniou M, Collinge J (1999) Protease-resistant prion protein produced in vitro lacks detectable infectivity. J Gen Virol 80 (Pt 1): 11–14.
11. Kirby L, Birkett CR, Rudyk H, Gilbert IH, Hope J (2003) In vitro cell-free conversion of bacterial recombinant PrP to PrPres as a model for conversion. J Gen Virol 84: 1013–1020.
12. Swietnicki W, Morillas M, Chen SG, Gambetti P, Surewicz WK (2000) Aggregation and fibrillization of the recombinant human prion protein huPrP90-231. Biochemistry 39: 424–431.
13. Bocharova OV, Breydo L, Parfenov AS, Salnikov VV, Baskakov IV (2005) In vitro conversion of full-length mammalian prion protein produces amyloid form with physical properties of PrP^{Sc}. J Mol Biol 346: 645–659.
14. Morales R, Duran-Aniotz C, Diaz-Espinoza R, Camacho MV, Soto C (2012) Protein misfolding cyclic amplification of infectious prions. Nat Protoc 7: 1397–1409.
15. Wang F, Wang X, Yuan CG, Ma J (2010) Generating a prion with bacterially expressed recombinant prion protein. Science 327: 1132–1135.
16. Kim JI, Cali I, Surewicz K, Kong Q, Raymond GJ, et al. (2010) Mammalian prions generated from bacterially expressed prion protein in the absence of any mammalian cofactors. J Biol Chem 285: 14083–14087.
17. Chianini F, Fernandez-Borges N, Vidal E, Gibbard L, Pintado B, et al. (2012) Rabbits are not resistant to prion infection. Proc Natl Acad Sci USA 109: 5080–5085.
18. Makarava N, Kovacs GG, Savtchenko R, Alexeeva I, Ostapchenko VG, et al. (2012) A new mechanism for transmissible prion diseases. J Neurosci 32: 7345–7355.
19. Marijanovic Z, Caputo A, Campana V, Zurzolo C (2009) Identification of an intracellular site of prion conversion. PLoS Pathog 5: e1000426.
20. Baskakov IV, Legname G, Gryczynski Z, Prusiner SB (2004) The peculiar nature of unfolding of the human prion protein. Protein Sci 13: 586–595.
21. Bellotti V, Chiti F (2008) Amyloidogenesis in its biological environment: challenging a fundamental issue in protein misfolding diseases. Curr Opin Struc Biol 18: 771–779.
22. Ellis RJ (2001) Macromolecular crowding: an important but neglected aspect of the intracellular environment. Curr Opin Struc Biol 11: 114–119.
23. White DA, Buell AK, Knowles TP, Welland ME, Dobson CM (2010) Protein aggregation in crowded environments. J Am Chem Soc 132: 5170–5175.
24. Jiao M, Li HT, Chen J, Minton AP, Liang Y (2010) Attractive protein-polymer interactions markedly alter the effect of macromolecular crowding on protein association equilibria. Biophys J 99: 914–923.
25. Zhou Z, Fan JB, Zhu HL, Shewmaker F, Yan X, et al. (2009) Crowded cell-like environment accelerates the nucleation step of amyloidogenic protein misfolding. J Biol Chem 284: 30148–30158.
26. Zhou HX, Rivas G, Minton AP (2008) Macromolecular crowding and confinement: biochemical, biophysical, and potential physiological consequences. Annu Rev Biophys 37: 375–397.
27. Zhou Z, Yan X, Pan K, Chen J, Xie ZS, et al. (2011) Fibril formation of the rabbit/human/bovine prion proteins. Biophys J 101: 1483–1492.
28. Ma Q, Fan JB, Zhou Z, Zhou BR, Meng SR, et al. (2012) The contrasting effect of macromolecular crowding on amyloid fibril formation. PLoS One 7: e36288.
29. Xu Z, Prigent S, Deslys JP, Rezaei H (2011) Dual conformation of H2H3 domain of prion protein in mammalian cells. J Biol Chem 286: 40060–40068.
30. Adrover M, Pauwels K, Prigent S, de Chiara C, Xu Z, et al. (2010) Prion fibrillization is mediated by a native structural element that comprises helices H2 and H3. J Biol Chem 285: 21004–21012.
31. Chakroun N, Prigent S, Dreiss CA, Noinville S, Chapuis C, et al. (2010) The oligomerization properties of prion protein are restricted to the H2H3 domain. FASEB J 24: 3222–3231.
32. Makarava N, Baskakov IV (2008) Expression and purification of full-length recombinant PrP of high purity. Methods Mol Biol 459: 131–143.
33. Xu LR, Liu XL, Chen J, Liang Y (2013) Protein disulfide isomerase interacts with Tau protein and inhibits its fibrillization. PLoS One 8: e76657.
34. Meng SR, Zhu YZ, Guo T, Liu XL, Chen J, et al. (2012) Fibril-forming motifs are essential and sufficient for the fibrillization of human Tau. PLoS One 7: e38903.
35. Yang F, Jr., Zhang M, Zhou BR, Chen J, Liang Y (2006) Oleic acid inhibits amyloid formation of the intermediate of α-lactalbumin at moderately acidic pH. J Mol Biol 362: 821–834.
36. Chattopadhyay M, Durazo A, Sohn SH, Strong CD, Gralla EB, et al. (2008) Initiation and elongation in fibrillation of ALS-linked superoxide dismutase. Proc Natl Acad Sci USA 105: 18663–18668.
37. Mo ZY, Zhu YZ, Zhu HL, Fan JB, Chen J, et al. (2009) Low micromolar zinc accelerates the fibrillization of human Tau via bridging of Cys-291 and Cys-322. J Biol Chem 284: 34648–34657.
38. McKinley MP, Bolton DC and Prusiner SB (1983) A protease-resistant protein is a structural component of the scrapie prion. Cell 35: 57–62.
39. Khan MQ, Sweeting B, Mulligan VK, Arslan PE, Cashman NR, et al. (2010) Prion disease susceptibility is affected by β-structure folding propensity and local side-chain interactions in PrP. Proc Natl Acad Sci USA 107: 19808–19813.
40. Nisbet RM, Harrison CF, Lawson VA, Masters CL, Cappai R, et al. (2010) Residues surrounding the glycosylphosphatidylinositol anchor attachment site of PrP modulate prion infection: insight from the resistance of rabbits to prion disease. J Virol 84: 6678–6686.
41. Fernandez-Funez P, Zhang Y, Casas-Tinto S, Xiao X, Zou WQ, et al. (2010) Sequence-dependent prion protein misfolding and neurotoxicity. J Biol Chem 285: 36897–36908.
42. Sweeting B, Brown E, Khan MQ, Chakrabartty A, Pai EF (2013) N-terminal helix-cap in α-helix 2 modulates beta-state misfolding in rabbit and hamster prion proteins. PLoS One 8: e63047.
43. Zhang J, Zhang Y (2014) Molecular dynamics studies on the NMR and X-ray structures of rabbit prion proteins. J Theor Biol 342: 70–82.
44. Smirnovas V, Kim JI, Lu X, Atarashi R, Caughey B, et al. (2009) Distinct structures of scrapie prion protein (PrP^{Sc})-seeded versus spontaneous recombinant prion protein fibrils revealed by hydrogen/deuterium exchange. J Biol Chem 284: 24233–24241.
45. Tycko R, Savtchenko R, Ostapchenko VG, Makarava N, Baskakov IV (2010) The α-helical C-terminal domain of full-length recombinant PrP converts to an in-register parallel β-sheet structure in PrP fibrils: evidence from solid state nuclear magnetic resonance. Biochemistry 49: 9488–9497.

13

Structural and Biophysical Characterization of *Bacillus thuringiensis* Insecticidal Proteins Cry34Ab1 and Cry35Ab1

Matthew S. Kelker[1]*, Colin Berry[2], Steven L. Evans[1], Reetal Pai[1], David G. McCaskill[1], Nick X. Wang[1], Joshua C. Russell[1¤a], Matthew D. Baker[2], Cheng Yang[3], J. W. Pflugrath[3], Matthew Wade[4], Tim J. Wess[4¤b], Kenneth E. Narva[1]

1 Dow AgroSciences, LLC, Indianapolis, Indiana, United States of America, 2 Cardiff School of Biosciences, Cardiff University, Cardiff, Wales, United Kingdom, 3 Rigaku Americas Corporation, The Woodlands, Texas, United States of America, 4 School of Optometry & Vision Sciences, Cardiff University, Cardiff, Wales, United Kingdom

Abstract

Bacillus thuringiensis strains are well known for the production of insecticidal proteins upon sporulation and these proteins are deposited in parasporal crystalline inclusions. The majority of these insect-specific toxins exhibit three domains in the mature toxin sequence. However, other Cry toxins are structurally and evolutionarily unrelated to this three-domain family and little is known of their three dimensional structures, limiting our understanding of their mechanisms of action and our ability to engineer the proteins to enhance their function. Among the non-three domain Cry toxins, the Cry34Ab1 and Cry35Ab1 proteins from *B. thuringiensis* strain PS149B1 are required to act together to produce toxicity to the western corn rootworm (WCR) *Diabrotica virgifera virgifera* Le Conte via a pore forming mechanism of action. Cry34Ab1 is a protein of ~14 kDa with features of the aegerolysin family (Pfam06355) of proteins that have known membrane disrupting activity, while Cry35Ab1 is a ~44 kDa member of the toxin_10 family (Pfam05431) that includes other insecticidal proteins such as the binary toxin BinA/BinB. The Cry34Ab1/Cry35Ab1 proteins represent an important seed trait technology having been developed as insect resistance traits in commercialized corn hybrids for control of WCR. The structures of Cry34Ab1 and Cry35Ab1 have been elucidated to 2.15 Å and 1.80 Å resolution, respectively. The solution structures of the toxins were further studied by small angle X-ray scattering and native electrospray ion mobility mass spectrometry. We present here the first published structure from the aegerolysin protein domain family and the structural comparisons of Cry34Ab1 and Cry35Ab1 with other pore forming toxins.

Editor: Juan Luis Jurat-Fuentes, University of Tennessee, United States of America

Funding: Portions of this work were performed at the DuPont-Northwestern-Dow Collaborative Access Team (DND-CAT) located at Sector 5 of the Advanced Photon Source (APS). DND-CAT is supported by E.I. DuPont de Nemours & Co., The Dow Chemical Company and Northwestern University. Use of the APS, an Office of Science User Facility operated for the U.S. Department of Energy (DOE) Office of Science by Argonne National Laboratory, was supported by the U.S. DOE under Contract No. DE-AC02-06CH11357. Molecular graphics and analyses were performed with the UCSF Chimera package. Chimera was developed by the Resource for Biocomputing, Visualization, and Informatics at the University of California, San Francisco (supported by NIGMS P41-GM103311). The funders had no role in study design, data collection and analysis, decision to publish, or preparation of the manuscript.

Competing Interests: The authors have read the journal's policy and have the following competing interests: MSK, SLE, RP, DM, NXW and KEN are current employees of Dow AgroSciences. JCR was formerly employed by Dow AgroSciences. CY and JWP are employed by Rigaku Americas Corporation. A Material Transfer Agreement was in place between Dow AgroSciences and Cardiff University. Cry34Ab1 and Cry35Ab1 are protected under US patents 7,309,785 and family members and 7,524,810 and family members and are part of Dow AgroSciences' product, Herculex RW.

* Email: mskelker@dow.com

¤a Current address: Department of Biochemistry, University of Washington, Seattle, Washington, United States of America
¤b Current address: Office of the Dean of Science, Charles Sturt University, New South Wales, Victoria, Australia

Introduction

Bacillus thuringiensis strains are well-known for the production of insecticidal toxins on sporulation and these proteins are deposited in parasporal crystalline inclusions, closely associated with the spore. To date, many crystal toxins (Cry) have been discovered and these are currently divided into 73 major classes (see http://www.lifesci.susx.ac.uk/home/Neil_Crickmore/Bt/for an updated list). The great majority of these toxins belong to a single structural class of proteins, exhibiting 3 domains in the mature toxin sequence. However, an increasing number of other Cry toxins are structurally and evolutionarily unrelated to this three-domain family. Unfortunately, little is known of their three dimensional structures, limiting our understanding of their mechanisms of action and our ability to engineer the proteins to enhance their function. Amongst the non-three domain Cry toxins, the Cry34Ab1 and Cry35Ab1 proteins from *B. thuringiensis* strain PS149B1 are required to act together to produce toxicity to the western corn rootworm (WCR) *Diabrotica virgifera virgifera* via a pore forming mechanism of action [1,2,3]. Very few

Cry proteins have been described with activity against WCR [4] and among them the binary mode of action of Cry34Ab1/Cry35Ab1 and related Cry family members is unique. The Cry34Ab1/Cry35Ab1 proteins are important for WCR resistance trait technology, having been introduced to corn hybrids through genetic transformation event DAS-59122-7 to provide protection from WCR feeding in commercialized corn hybrids [5].

Cry34Ab1 is a protein of ~14 kDa with features of the aegerolysin family (Pfam06355) of proteins that have known ability to interact with membranes to form pores [6] while Cry35Ab1 appears to contain β-trefoil sequences reminiscent of the carbohydrate-binding domain of ricin B subunit (Pfam00652) and is a member of the toxin_10 family (Pfam05431) that includes the binary toxin BinA/BinB, the Cry49Aa1 component of a second binary toxin (Cry48Aa1/Cry49Aa1) from *Lysinibacillus sphaericus*, the Cry36Aa1 protein of *B. thuringiensis* and a hypothetical protein from *Chlorobium phaeobacteroides* [7,8,9]. In this study we have elucidated the structures of both Cry34Ab1 and Cry35Ab1 and further probed the consistency of the crystal structure data with their structures in solution using small angle X-ray scattering (SAXS) and native electrospray ion mobility mass spectrometry. We present here the first published structure from the aegerolysin protein family and compare the Cry34Ab1 and Cry35Ab1 protein structures with other pore forming toxins.

Materials and Methods

Expression and purification of full length Cry34Ab1 and Cry35Ab1

Full length Cry35Ab1 and Cry34Ab1 toxins were over-expressed in the inclusion body fraction of recombinant *Pseudomonas fluorescens* (*Pf*) [10] and were purified as follows. Whole broth, including *Pf* cells, was frozen at −20°C. To isolate and wash the inclusion bodies, the broth was thawed at 37°C and 200 mL lysis buffer (50 mM Tris-HCl, pH 7.5, 0.2 M NaCl, 5% glycerol, 1 mM DTT 20 mM EDTA and 0.5% Triton X-100) was added for every 60 grams of frozen broth, mixed and centrifuged at 10,000 g for 20 minutes at 4°C. The cell pellet was resuspended at 200 mg cell pellet/mL cold lysis buffer and the cells disrupted by micro-fluidization using a pressure difference of 16,000 psi. Lysozyme was then added to 0.6 mg/mL and the mixture briefly incubated at 37°C, then placed on ice for one hour with stirring. Cell lysis was confirmed by microscopy. Magnesium sulfate was added to 60 mM and DNase I added to 0.25 mg/mL and the mixture incubated overnight at 4°C with stirring. The mixture was gently homogenized using a hand held homogenizer to shear any undigested genomic DNA and centrifuged at 10,000 g for 20 minutes at 4°C. The pellet was washed in lysis buffer and centrifuged three more times.

Purification and crystallization of Cry35Ab1

Freshly prepared Cry35Ab1 inclusions (100 mg) were solubilized in 100 mL of 50 mM sodium citrate, pH 3.5, precipitated by the addition of 80% ammonium sulfate then isolated by centrifugation at 15,000 g for 15 minutes at 4°C. The pellet was resuspended in 4.0 mL of 15 mM sodium citrate, pH 3.5 and dialyzed against 6 L of the same buffer overnight using a 10,000 MWCO dialysis membrane. The pellet was completely dissolved. This method reliably yielded 60 mg of highly pure Cry35Ab1. The final concentration of Cry35Ab1 used in crystallization experiments was 10–15 mg/mL in 15 mM sodium citrate, pH 3.5. The results from SDS-PAGE analysis and dynamic light scattering scan indicated Cry35Ab1 had reached>90% purity and was monodisperse in solution.

Microbatch crystallization experiments using the Hampton Index screen were set up with the Cry35Ab1 protein. Small crystals appeared within several conditions after two days. After optimization, the best crystals were produced in 0.72 M NaH_2PO_4, pH 4.5, 80 mM K_2HPO_4 and 0.2 M NaCl with ~15 mg/mL Cry35Ab1 protein at 16°C.

Purification and crystallization of Cry34Ab1

In order to achieve a highly concentrated protein solution for crystallization, washed Cry34Ab1 inclusions were dissolved in 7 M urea then refolded using a specially designed dialyzer. This dialyzer has an open-end tube with a 10 K cutoff membrane at the bottom end, which was hung in the center of a micro-centrifuge tube. The membrane allows the solution outside of the dialysis tube to equilibrate slowly into the tube. The small membrane surface limits the rate of diffusion and may also create gradients of components of dialysate in the tube.

About 8.9 mg of Cry34Ab1 powder was initially dissolved in 0.5 mL of 7 M urea, 20 mM potassium phosphate buffer pH 7.8. A 40 μL sample of this protein solution was transferred into the microdialysis apparatus described above and dialyzed against a 3.0 mL solution of 25% (v/v) PEG 400 and 50 mM sodium acetate, pH 4.4, at 16°C. The results from SDS-PAGE analysis and dynamic light scattering scan indicated Cry34Ab1 had reached>90% purity and were monodisperse in solution. After two days, small hexagonally-shaped crystals ($0.15 \times 0.15 \times 0.03$ mm^3) were observed on the membrane.

Data collection and phasing of Cry34Ab1

Data collection on both Cry34Ab1 (Table 1) and full length Cry35Ab1 (Table 2) crystals were carried out at −180°C on a home X-ray system (Rigaku MicroMax-007 X-ray generator and R-AXIS IV^{++} detector). Two different wavelengths of X-ray radiation (Cu Kα, 1.54 Å and Cr Kα, 2.29 Å) were used to collect data sets for structure refinement and enhanced anomalous signal to phase the protein diffraction data. All data sets were processed using the d*TREK data processing package [11].

As a novel structure, the multiple isomorphous replacement (MIR) method was employed to phase Cry34Ab1 diffraction data. After numerous trials, a Pb-derivatized crystal was prepared by soaking a native Cry34Ab1 crystal in a crystallization solution containing 10 mM lead acetylacetonate for 24 hours. The diffraction data were collected at −180°C (Table 1). One major lead site was found on three Harker sections of an isomorphous difference Patterson map. Two more minor lead sites were found in the isomorphous difference Fourier map of $F_{PH} - F_P$ using the phases calculated from the major lead site. Three lead sites were initially used in phase calculation up to 3.0 Å resolution with the program MLPHARE in CCP4 [12]. The figure of merit (FOM) of the single isomorphous replacement with anomalous scattering (SIRAS) phase set was 0.41, while an electron density map calculated with these phases showed clear protein-solvent boundaries but with many broken regions. The anomalous differences of the native data with the initial SIRAS phases were used to generate an anomalous difference Fourier map. A peak with a height of ~5 sigma above the average value was found in this anomalous Fourier map. It was considered as a sulfur site arising from one of the two methionine residues (the other methionine residue at the N-terminus is disordered and not visible in these maps). The sulfur position and the anomalous difference in the native data were used in further phase calculation. The FOM of the MIR phase set was improved to 0.45. Furthermore, since the anomalous differences of the native data were included as a new independent data set in the phase calculation, it greatly enhanced

Table 1. Cry34Ab1 (4JOX) data processing, model and refinement statistics.

Data processing statistics		
	Native	**Pb derivative**
Radiation Wavelength	1.54	1.54
Resolution (Å)	22.37–2.15 (2.23–2.15)	26.15–3.00 (3.11–3.00)
Space group	I422	I422
Unit cell length (Å)		
a	100.6	100.3
b	100.6	100.3
c	56.2	56.3
Number of reflections (>1σ)	299,280 (29284)	84,864 (8538)
Unique reflections	14,955 (1500)	5,511 (559)
Average I/s (I)	17.7(3.5)	17.5 (7.0)
R$_{merge}$[a] (%)	9.1 (29.1)	10.0 (31.7)
Completeness (%)	99.8 (99.6)	100.0 (100.0)
FOM after MIR	0.45	
Model and refinement statistics		
Resolution range (Å)	12.0–2.15	
Reflections (total)	7271	
Reflections (test)	729	
R$_{cryst}$[b] (%)	22.4	
R$_{free}$[b] (%)	27.6	
RMSD Bond length (Å)	0.01	
RMSD Bond Angle (°)	1.70	
RMSD Dihedrals (°)	0.10	
Number of protein atoms	915	
Number of water atoms	67	
<B> All protein atoms (Å^2)	26.7	
<B> Side chain atoms (Å^2)	27.9	
<B> Main chain atoms (Å^2)	25.6	
<B> Water molecules (Å^2)	40.2	
Ramachandran plot[c]	92.1/7.9	

[a] $R_{merge} = 100\Sigma(h)\Sigma(i)|I(i)-<I>|/\Sigma(h)\Sigma(i)I(i)$ where I(i) is the ith intensity measurement of reflection h, and <I> is the average intensity from multiple observations.
[b] $R_{cryst} = \Sigma||F_{obs}|-|F_{calc}||/\Sigma|F_{obs}|$. Where F_{obs} and F_{calc} are the structure factor amplitudes from the data and the model, respectively. R_{free} is R_{cryst} with 10% of the structure factors.
[c] Number of residues in favored/additionally favored outlier region. Calculated using PROCHECK [14].

the power to resolve the phase ambiguity of the initial SIRAS phases from the Pb derivative. The electron density map calculated with these new phases was improved and revealed clear and recognizable regions, such as several β-strands and some large side chain electron density. A solvent-flattening procedure was employed to improve the quality of this electron density map by using program DM [12].

Data collection and phasing of Cry35Ab1

A platinum derivative of Cry35Ab1 was prepared by an overnight soaking of a native crystal in 20 mM platinum diammine dichloride and the crystallization condition. Crystals were cryoprotected by addition of a final concentration of 20% (v/v) of glycerol to the well condition. Diffraction data sets from both native and heavy-atom derivatives were collected at 100 K with home source X-ray equipment. All the data sets were processed

using the d*TREK data processing package [11]. The statistics of data collection are listed in Table 2.

Structure determination and refinement of Cry34Ab1

The electron density map was used to build an initial model with the program O [13]. The value of the Matthews number indicated that Cry34Ab1 crystals have one molecule per asymmetric unit. The chain tracing and sequence match started at the position of Met54 that was recognized in the anomalous difference Fourier map of the native data set and were extended from there in both directions. Initially, 78 amino acids out of 123 residues were fitted into their densities. The model was refined using the program REFMAC5 [12] and improved by rebuilding after recalculation of the electron density map using weighted combinations of model and MIR phases. Some regions missing in the MIR electron density map gradually appeared in the partial model combined electron density maps. The rigid-body, overall

Table 2. Cry35Ab1 (4JP0) data processing, model and refinement statistics.

Data processing statistics

	Native		Pt derivative	
Radiation Wavelength	1.54	2.29	1.54	2.29
Resolution (Å)	26.37–1.80 (1.88–1.80)	25.37–2.7 (2.78–2.7)	24.51–2.5 (2.57–2.5)	21.98–2.85 (2.91–2.85)
Space group	$P2_12_12_1$	$P2_12_12_1$	$P2_12_12_1$	$P2_12_12_1$
Unit cell length				
a	48.7	48.6	48.5	48.5
b	65.1	64.3	65.0	64.4
c	117.4	117.2	117.2	117.1
Number of reflections (>1σ)	403,522 (39,620)	122,443 (11937)	36,245 (3593)	107,428 (11890)
Unique reflections	35,100 (4,862)	10,791 (1203)	14,538 (1498)	8,968 (9133)
Average I/s (I)	21.3 (2.5)	15.3 (5.6)	12.5(3.0)	15.9 (6.5)
R_{merge}[a] (%)	5.6 (29.2)	8.4 (29.6)	7.6 (27.9)	11.3 (32.8)
Completeness (%)	99.3 (95.3)	94.4 68.4)	96.5 (96.5)	99.7 (94.0)

Model and refinement statistics

Resolution range (Å)	15.0–1.80
Reflections (total)	31,479
Reflections (test)	3,504
R_{cryst}[b] (%)	18.1
R_{free}[b] (%)	23.4
RMSD Bond length (Å)	0.01
RMSD Bond Angle (°)	1.53
RMSD Dihedrals (°)	0.113
Number of protein atoms	3053
Number of water atoms	295
<B> All protein atoms (Å²)	25.1
<B> Side chain atoms (Å²)	26.7
<B> Main chain atoms (Å²)	23.5
<B> Water molecules (Å²)	33.3
Ramachandran plot[c]	89.2/10.8

[a] $R_{merge} = 100\Sigma(h)\Sigma(i)|I(i)-<I>|/\Sigma(h)\Sigma(i)I(i)$ where I(i) is the ith intensity measurement of reflection h, and $<I>$ is the average intensity from multiple observations.
[b] $R_{cryst} = \Sigma||F_{obs}|-|F_{calc}||/\Sigma|F_{obs}|$. Where F_{obs} and F_{calc} are the structure factor amplitudes from the data and the model, respectively. R_{free} is R_{cryst} with 10% of the structure factors.
[c] Number of residues in favored/additionally favored outlier region. Calculated using PROCHECK [14].

B-factor, individual B-factor and TLS refinement procedure were iterated a number of times to refine the model. The final model was refined to 2.15 Å and contains 117 (from Ala3 to Tyr119) out of 123 amino acids and 67 water molecules. R_{cryst} and R_{free} factors for the final model were 22.4% and 27.6%, respectively. Analysis of the model by the program PROCHECK [14] indicated that 92.1% of the residues fell into the most favored regions of a Ramachandran plot while the remaining 7.9% occurred in additionally allowed regions. The refinement statistics and structure analysis are listed in Table 1. Coordinates and reflection files were assigned the PDB accession code, 4JOX.

Structure determination and refinement of Cry35Ab1

The initial model was built using the program O [13]. One Cry35Ab1 molecule was determined to be in an asymmetric unit based on calculated Matthews number. The chain tracing and sequence match of Cry35Ab1 was started simultaneously at the position of Cys183 and Met185. They were recognized through their unique densities in the anomalous difference Fourier map of Cr Kα derived native data set. The sequence match was further confirmed by the unique motif of electron densities of Met176, Gly177 and Trp178 and anomalous peak of the sulfur of Met176. About 200 amino acids were fitted into their densities in the first round of map fitting. The model was refined using the program REFMAC5 [12] which includes the procedures of idealization, rigid-body, overall B-factor, TLS and individual B-factor. The electron density map was recalculated using weighted combinations of model and MIR phases. During the refinement process, the electron density was improved in each new map, especially in some uninterpretable regions. The final model contains 378 (from Leu2 to His381) out of 385 amino acids and 295 water molecules. Pro163 and Thr164 were excluded from final model. Some of their electron density was observed in the maps of later cycles, but these two residues cannot be refined into a conformation with both good geometry and density coverage. It may result from their structural location at a loop region with high thermomobility. The

Table 3. Measured and theoretical values for CCS for Cry34Ab1 and trCry35Ab1.

Protein	component	charge (z)	DT (ms, obs.)	CCS Ω (Å²)	Theoretical PA (Å²)	EHHS (Å²)	TM (Å²)
Cry34Ab1	1	7	27.57	1404.7	1368.6	1720.2	1706.6
	2	7	29.86	1477.5			
	3	8	27.39	1633.9			
	4	9	25.06	1675.9			
	5	9	27.22	1819.1			
	6	9	31.04	2092.9			
	7	9	36.6	2497.0			
	8	10	25.32	1841.1			
	9	10	27.34	2047.1			
	10	10	32.6	2415.3			
	11	10	34.84	2597.5			
	12	10	36.79	2771.9			
	13	11	23.91	1922.0			
	14	11	26.55	2253.2			
	15	11	30.35	2406.4			
	16	11	32.52	2837.2			
	17	11	34.7	2872.5			
trCry35Ab1	1	14	32.66	3466.4	3201.0	4146.0	4211.5
	2	14	34.34	3487.8			
	3	15	32.47	3673.3			
	4	15	32.56	3717.9			
	5	16	31.09	3703.6			
	6	17	30.03	3762.0			
	7	17	36.76	4808.8			
	8	18	29.14	3819.5			
	9	18	35.63	5143.1			

Theoretical values were calculated using the projection approximation (PA), exact hard sphere scattering (EHSS) and trajectory method (TM) with helium as the collision gas. Experimentally measured CCS used nitrogen as the collision gas.

final model was refined to 1.80 Å. R_{cryst} and R_{free} factors for the final model were 18.1% and 23.4%, respectively. The analysis of the model by the program PROCHECK [14] indicated 89.2% of the residues fell into the most favored regions of a Ramachandran plot while the remaining 10.8% occurred in additionally allowed regions. The refinement statistics and structure analysis are listed in Table 2. Coordinates and reflection files were assigned the PDB accession code, 4JP0.

Expression and purification of soluble, truncated Cry35Ab1

A transgenic corn line encoding full length versions of both Cry34Ab1and Cry35Ab1 was jointly developed by Dow AgroSciences and Pioneer Hi-Bred International [15,16] and sold under the brand name HERCULEX RW. Full length Cry35Ab1 is 44 kDa; however, during characterization of the proteins expressed in transgenic corn, a 40 kDa C-terminal truncation of Cry35Ab1 (trCry35Ab1) was isolated. Interestingly, this 40 kDa form retains both the insecticidal activity and immunoreactivity of the full length Cry35Ab1 [17]. In this study we wished to use this construct to examine the solution state of the truncated molecule in comparison to the crystal structure. In addition, trCry35Ab1 is highly soluble and stable over the time course of experimentation. A plasmid encoding residues 1–354 of Cry35Ab1 (trCry35Ab1; lacking 31 residues at the C-terminus) was transformed into a Dow AgroSciences *P. fluorescens* expression strain [10]. Seed cultures were grown overnight. Production cultures were inoculated with 2% volume of the overnight culture and grown in production media with trace elements and fermented in 2 L controlled bioreactors. Twenty-four hours post inoculation, the cultures were induced with 0.3 mM IPTG. The cells were harvested at 48 hours post-induction by centrifugation. The pellets were stored at − 80°C until purification. Routine expression levels are ~30 grams of soluble trCry35Ab1 per liter of cell culture.

The trCry35Ab1 is expressed in the soluble fraction of the cell lysate. Frozen cell pellets were resuspended in 0.1 M Na acetate, pH 3.3, 1 mM EDTA and 1 mM TCEP (tris(2-carboxyethyl)phosphine). The suspension was sonicated for 30 seconds, followed by a 1 minute rest on ice, three times. After lysis, the lysate was centrifuged at 19,000 rpm for 20 minutes at 4 °C. The supernatant was filtered through a 0.22 μm filter. Purified protein was obtained by using cation exchange chromatography with a Source 30S 16/20 column pre-equilibrated in 0.1 M sodium acetate pH 3.3, and gradient elution with 0.1 M sodium acetate pH 3.3 and 1 M NaCl. The fractions containing trCry35Ab1 were concentrated with 10,000 MWCO 15 mL, Amicon concentrators, centrifuged at 5000 g for 10 minutes. Final samples were filtered through a 0.22 μm filter and applied to a Superdex 75 26/90 column pre-equilibrated in 20 mM sodium citrate, pH 3.3.

Small angle X-ray scattering

Full length Cry34Ab1 and trCry35Ab1 samples, prepared as described above, were diluted to various concentrations between 2.07 to 6.86 mg/mL in 20 mM sodium citrate pH 3.3. Immediately prior to the data collection, both samples were centrifuged at 14,000 rpm in a tabletop centrifuge for one hour, then filtered through a 0.22 μm syringe filter. Synchrotron scattering data were collected and processed at beamline 5-ID-D at the Advanced Photon Source at Argonne National Laboratories, Illinois, USA.

Data were analyzed using the ATSAS package [18]. Buffer scattering intensities were subtracted from the sample image to remove background scattering using PRIMUS [19]. For the SAXS data, the radius of gyration and the particle distance distribution function, p(r) were evaluated with the GNOM program [20].

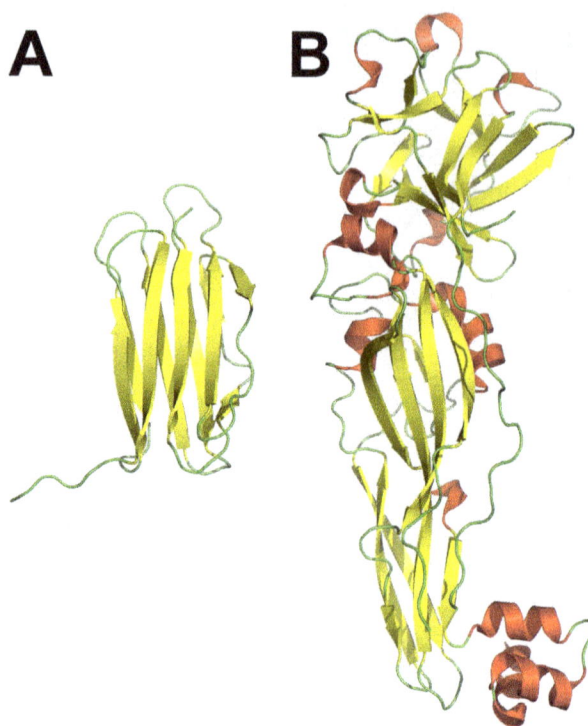

Figure 1. Crystal structures of Cry34Ab1 and Cry35Ab1. (A) The structure of Cry34Ab1 is a β-sandwich of 10 strands. (B) Cry35Ab1 contains two domains. The N-terminal trefoil domain contains α-helices and three β-sheets. The C-terminal domain is terminated with a three helix fold which is not required for activity [17]. This figure, and all subsequent structure representations, were made with PyMOL [66].

Particle shapes were generated using the *ab-initio* software program DAMMIN [21]. Multiple DAMMIN runs were performed (~25) to check the 'uniqueness' of the solution and to generate 25 similar shapes that were combined and filtered to produce an averaged model using the DAMAVER and DAMFILT programs [22].

Cry34Ab1 and Cry35Ab1 crystal structures were docked with the SAXS calculated envelopes using the Chimera program [23]. The C-terminal residues 355–381 of the full length Cry35Ab1 crystal structure were removed for SAXS docking purposes.

Native electrospray ion mobility mass spectrometry

The behavior of Cry34Ab1 and trCry35Ab1 in solution were probed using native electrospray ion mobility mass spectrometry. Stock solutions of Cry34Ab1 (3.5 mg/mL stored in 20 mM sodium citrate buffer, pH 3.5) and a trCry35Ab1 (3.2 mg/mL stored in 20 mM sodium citrate buffer, pH 3.5) were used for direct infusion under non-denaturing nano-electrospray conditions.

In the case of the Cry34Ab1 sample, the stock solution was diluted 4 – fold with 0.1% formic acid and buffer exchanged into 0.1% formic acid (pH 3.0) using a Zeba spin desalting column (ThermoFisher Scientific) pre-equilibrated with 0.1% formic acid. The buffer exchanged sample was subsequently diluted 5 – fold to give a stock solution of approximately 12.9 μM.

The trCry35Ab1 sample was buffer exchanged without initial dilution into 0.1% formic acid using a Zeba spin desalting column pre-equilibrated with 0.1% formic acid. The resulting buffer

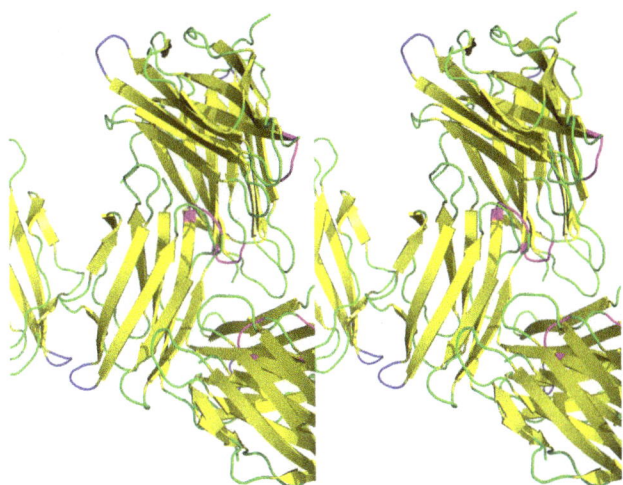

Figure 2. Stereogram of Cry34Ab1 crystal packing. Symmetry related molecules of Cry34Ab1 are shown with loop regions Asn66 to Gln68 colored blue and Gly103 to Gln105 colored magenta. Neither loop region is stabilized by crystal contacts or intramolecular interactions which results in elevated temperature factors and diminished electron density quality.

exchanged material was diluted 5 – fold to give a stock solution of approximately 15.9 μM.

Electrospray mass spectrometry of the Cry34Ab1 and trCry35Ab1 stock solutions were carried out by directly infusing the proteins with a syringe pump at 500 nL/min with an unheated nanospray inlet. Detection and ion mobility measurements of the proteins was carried out using a prototype ion mobility quadrupole time-of-flight (model 6560 IM-QTOF) mass spectrometer at Agilent Technologies (Santa Clara, CA). This instrument utilizes a drift tube configuration with nitrogen collision gas for ion mobility measurements. The drift tube was operated at 27°C, with 4 Torr of nitrogen collision gas.

For calculation of the measured collisional cross sectional areas (CCS) of the analyzed proteins, the drift tube was calibrated according to the manufacturer's directions using infusion of a colchicine standard (400 m/z, literature value for CCS = 196.2 Å), and the calibration was confirmed using infusion of a standard of ondansetrone (m/z 294, measured CCS value = 172.5 Å, literature value = 172.7 Å) [24]. Measured CCS values for Cry34Ab1 and trCry35Ab1 were calculated using these calibration values with software provided by Agilent.

Determining collision cross sectional areas by MOBCAL

Theoretical collision cross sectional areas (CCS) of Cry34Ab1 and trCry35Ab1 were calculated using the open source software program MOBCAL [25,26]. MOBCAL source code was downloaded from the website of Professor M.F. Jarrold's group at Indiana University (http://www.indiana.edu/~nano/software. html) and compiled with Fortran 95 in an in-house Linux work station. The MOBCAL program was further modified to process protein systems up to 15,000 atoms. PDB files of Cry34Ab1 and Cry35Ab1 were used as input files. The calculations were carried out with a uniform charge distribution. A scaling factor of 1.0 was applied throughout the calculations. MOBCAL implements three different types of calculations to derive the CCS area between a protein and helium buffer gas: the projection approximation (PA), the exact hard sphere scattering (EHSS) and the trajectory method (TM). In this study, the PA values are consistently in better

agreement with experimental IM-MS measurements. All calculated CCS values from three the methods were included in Table 3.

Protein structure alignment by combinatorial extension

Structures of Cry34Ab1 and Cry35Ab1 were aligned against all the 3D structures in the Protein Data Bank (http://www.rcsb.org/pdb/home/home.do). These alignments were performed using the Combinatorial Extension (CE) algorithm [27]. This method is a fast and accurate way to perform structural alignment against large protein databases. It identifies the optimal alignment between any two structures by defining an alignment path between aligned fragment pairs in the two structures. Similarity between fragment pairs is calculated on the basis of inter-residue distances between the fragments after the superposition. Other structural features like secondary structure, solvent exposure, dihedral angles, etc. are also included to increase the accuracy of the alignment between the fragment pairs. The algorithm provides the sequence identity, r.m.s. of superposition and a Z-score for each alignment.

Modeling of related proteins

Using the coordinates of Cry35Ab1 and BinB as a template for Cry49Aa1 and Cry34Ab1 as a template for Pam, the possible structures of the related proteins were modeled using Modeller 9.11 [28,29]. Briefly, for Cry49Aa1 modelling, structure-based sequence alignments were performed using the amino acid sequences of Cry49Aa1, BinA and BinB as template sequences along with the structure of Cry35Ab1 (PDB ID 4JPO) and BinB (PDB ID 3WA1), followed by automated model building and minimization. Manual inspection of clashes and rebuilding of surface loops was performed using Chimera [23]. Final model selection was based on the GA341 score of Modeller [30] and Ramachandran plots.

Results and Discussion

Cry34Ab1 and Cry35Ab1 crystal structures

The crystal structure of the *Bacillus thuringiensis* Cry34Ab1 protein was refined to 2.15 Å resolution (Table 1). The Cry34Ab1 structure has one distinct structural domain containing 117 amino acids (Figure 1A). The protein folds in a typical β-sandwich conformation, which has two β-sheets packed against each other. β-sheet I containing the N- and C-termini is composed of four β-strands while β-sheet II has five β-strands. All β-strands, except the adjacent N- and C-terminal strands, are antiparallel. N- and C-terminal strands are located at the center of sheet I and parallel to each other. The peptide fragment comprising residues Thr115 to Tyr119 extends beyond its β-sheet toward a symmetry-related molecule within the crystal lattice. The entire β-sandwich has a relatively flat layer-like conformation and two slightly twisted β-sheets. When the side chains are excluded, the distance between the two β-sheets is between ~7–10 Å. The molecule is ~45 Å in length and ~20 Å in width.

As expected, the Cry34Ab1 structure has a very hydrophobic core between the two β-sheets including Val6, Ile8, Val10, Leu18, Trp31, Ile61, Tyr63, Ile71, Leu73, Phe75, Ile96, Val108, Tyr110 and Ile112. The phenol groups of Tyr63 and Tyr110 hydrogen bond with the side chain of Ser106 and the carbonyl oxygen of Thr36, respectively. Residue Trp31 is located at the loop region between β-strand 2 and 3 and its indole group is inserted directly into the core.

Nearly every residue in the final model of the Cry34Ab1 structure has well-defined electron density except for residues in

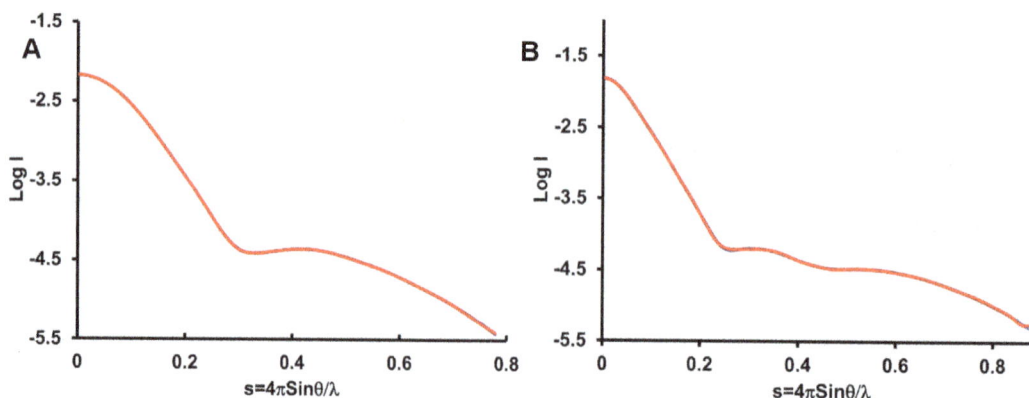

Figure 3. Scattering curve of the SAXS data and experimental fit. (A) Scattering curves for Cry34Ab1 (blue line) from the SAXS experiment and the fit made by the GNOM program [20] (red line) to scattering curve. (B) Scattering curves for trCry35Ab1. The X axis is s in arbitrary units where $s = 4\pi\sin\theta/\lambda$ and the Y axis is the log of intensity in arbitrary units.

two short loop regions (Asn66 to Gln68 and Gly103 to Gln105), which have relatively poorly defined electron density. Both these regions have higher average temperature factors for their main chain and side chain atoms than that of residues calculated over the entire structure. This indicates these loops might have multiple conformations in the crystal and reside in flexible regions due to the lack of crystal contacts to stabilize them (Figure 2).

A total of 67 water molecules were included in the final model of Cry34Ab1. Among these water molecules, ~60% of them have high temperature factors ($>40\text{Å}^2$). indicating these positions were partially occupied or otherwise disordered in the entire crystal. One water molecule (HOH1) is located at a special position of a two-fold crystallography axis and participates in hydrogen bonds with the side chains of two symmetry-related His107 residues.

The crystal structure of the Cry35Ab1 protein was refined to 1.80 Å resolution (Table 2). The Cry35Ab1 structure has an elongated rectangular shape with dimensions $42 \times 45 \times 105$ Å and is composed of two distinct domains (Figure 1B). Residues Pro163 and Thr164, had no interpretable electron density. The N-terminal domain is a β-trefoil fold, which contains a very hydrophobic core including Ala19, Val30, Leu32, Trp47, Ile59 Trp70, Val72, Ile77, Val79, Trp91, Ile93, Leu123, Trp135, Tyr100, Ile102, Leu110, Ile121, and Leu137.

The next domain contains six helices and three antiparallel β-sheets. A four antiparallel strand β-sheet sits below the N-terminal domain and another two β-strands form a β-sandwich. Within this fold, a π-π stacking interaction is formed between the phenol rings of Tyr231 and Tyr341. Additionally, Tyr341 is also hydrogen-bonded with Tyr229. The hydrophobic side chains of Val219, Leu221 and Met307 cluster around Ile299.

The N-terminal domain and C-terminal domain pack tightly against each other with more than 400 Å^2 area buried at the interface. The buried region includes hydrophobic residues Ile184, Ile197, Phe50, Phe48, Pro182, Met182, Ile58, Ile52, Ile271. In addition, a hydrogen bonding network exists between the side chains of residues Tyr82 and Glu270, the side chains and main chains of Tyr202 and Thr4 and Asp53 and Thr273, respectively, and the main chains of residues Gly270 and Asp53. These interactions appear to keep domain packing very strong and the conformation of the entire molecule very rigid. The two cysteine residues (Cys67 and Cys187) are present in the interface but their sulfur atoms are 6.1 Å apart, which is too distant to form a disulfide bond. Cys187 is conserved in all the toxins within this family, except Cry36. It is interesting to note that replacement of

the Cys187-equivalent residue in BinA (Cys195) drastically reduces its activity [31] while substitution of the equivalent in BinB (Cys241) has no effect [32].

The Cry35Ab1 structure is ended with a C-terminal cluster of three α-helices. The first two helices form a typical helix-loop-helix. The third helix is perpendicular to this helix-loop-helix and the group is held together through a hydrophobic core, consisting of Leu378, Leu353, Leu356, Ala352, Leu375, Val364. Leu353 and Leu356 are the first and last residues of a distinct sequence pattern of four tandem leucines (353-LLLL-356). Due to these structural characteristics, this C-terminal domain is very stable and tightly packed.

Cry34Ab1 and Cry35Ab1 solution structures

Analysis of the individual crystal structures suggested that both toxins are monomeric in solution, with no obvious higher order associations based upon content of the asymmetric unit or symmetry related molecules. To confirm the monomeric state in solution and under native conditions, we calculated the solution structures of both Cry34Ab1 and trCry35Ab1 by small angle X-ray scattering. The molecular envelope was generated using the ATSAS software package [18] and superimposed with the crystal structures (Dataset S1 and Dataset S2). The SAXS data indicate that both proteins exist as monomers in these conditions with the predicted radii of gyration of 14.6 and 26.7 Å calculated from the crystal structures of Cry34Ab1 and Cry35Ab1 (with the C-terminal three helix domain removed, consistent with the trCry35Ab1 sequence) respectively, matching closely with those of the SAXS models (14.9 and 25.9 Å, respectively). The overlap of the SAXS envelope and crystal structures (Figure 3) correlates well for both Cry34Ab1 (Figure 4A) and trCry35Ab1 (Figure 4B). It also suggests that the structures remain stable over a range of pH values since this match is seen despite differences in the pH of the crystallization and SAXS conditions (SAXS carried out at pH 3.3 compared to Cry35Ab1 crystallization at pH 4.5 and Cry34Ab1 crystallization at pH 7.8). Removal of the C-terminal 31 amino acids of Cry35Ab1 in the trCry35Ab1 SAXS structure does not appear to alter the core structure of the toxin in solution. When taken together, these finding support a monomeric solution state.

To expand upon the calculated SAXS solution structures, Cry34Ab1 and trCry35Ab1 were further assessed by native electrospray ion mobility mass spectrometry. Experimentally measured values for the collisional cross sectional (CCS) area of

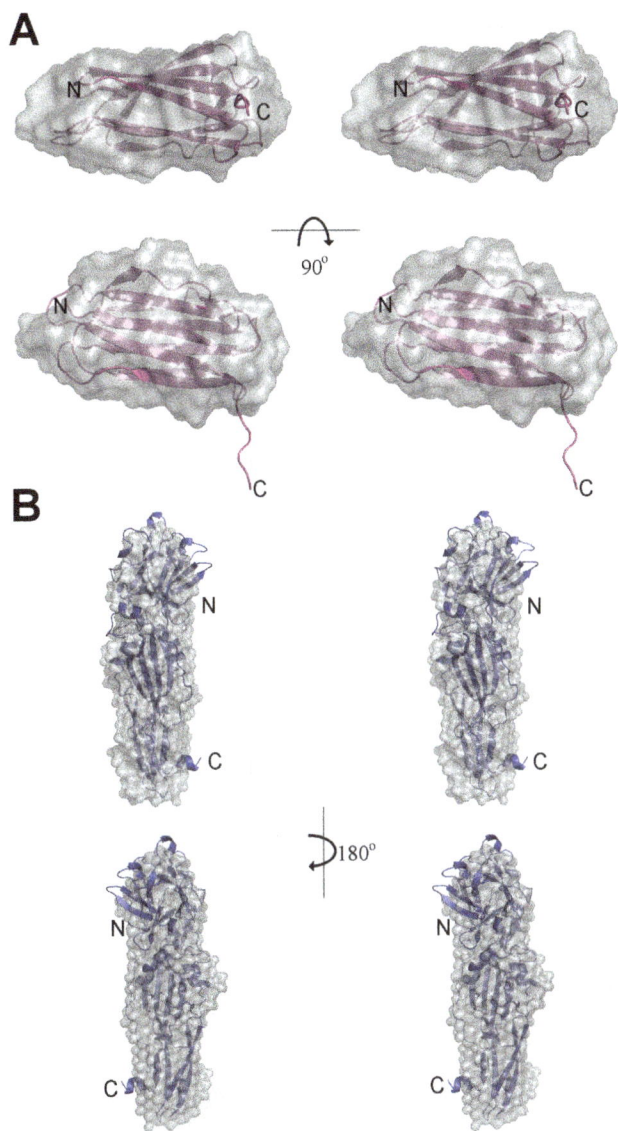

Figure 4. Stereogram of SAXS shapes of Cry34Ab1 and trCry35Ab1. (A) The SAXS calculated envelope of Cry34 matches closely to the crystal structure. The top view is related to the bottom view by 90° rotation to the bottom of the page. (B) The trCry35Ab1 SAXS structures are in good agreement with the Cry35Ab1 crystal structure. No higher order structures or oligomeric states were evident in either the Cry34Ab1 or trCry1Ab1 SAXS data. It is clear that both toxins are monomeric in solution in the absence of a receptor or binding partner.

the low charge states typically observed for electrospray of intact proteins under non-denaturing (native) conditions [33].

The measured drift traces for the five major charge states observed for Cry34Ab1 are shown in Figure 5. The lowest charge state (Figure 5A, z = 7), typically considered to correspond to the most compact conformation of the protein in solution [33,34], contains two partially resolved populations. The +8 charge state contains a single, uniform CCS. Increasing the charge on the protein from +9 to +11 leads to an increasing number of resolved populations, with the +11 charge state containing at least five partially resolved species. This observation indicates that increasing the charge on the ionized protein results in unfolding of the Cry34Ab1 protein into multiple populations of conformers.

CCS values ($\mathring{A}^2$) in nitrogen gas were calculated [34] from the measured drift values (msec) for each of the major charge states of both Cry34Ab1 and trCry35Ab1 and are reported in Table 3. For reference, the theoretical values calculated using MOBCAL for both Cry34Ab1 and trCry35Ab1 are also reported.

The lowest charge state for Cry34Ab1 has two components with CCS values of 1404 and 1477 $\mathring{A}^2$. Comparison with literature values for experimentally measured CCS using nitrogen gas indicates that these cross sections are similar to what would be expected for a protein of this size [35]. Cry34Ab1 exhibits a very large degree of conformational flexibility as the charge state of the protein increases from +7 to +11, with the CCS increasing more than 2–fold from 1404 $\mathring{A}^2$ to 2872 $\mathring{A}^2$ (Table 3). This increase in cross sectional area as the number of charges on the protein increases in the gas phase is typically considered to arise from coulombic repulsion of the positive charges as solvent is stripped away from the protein [33,34]. In the context of the crystal structure, the flexibility of Cry34Ab1 observed with the native electrospray ion mobility is consistent with an increased solvent exposed surface area to packed core ratio.

The trCry35Ab1 also shows a narrow distribution of low charge states (+14-+18) at high m/z values (>2000 m/z). As with Cry34Ab1, the lowest charge state exhibits two partially resolved components, with CCS values of 3466 and 3487 $\mathring{A}^2$. Based on calibrated values for protein standards [35], these values are also consistent with what would be expected for a protein of this size. Based on the number of different conformers observed as the charge state increases from +14 to +18 (Table 3), trCry35Ab1 appears to be more stable in the gas phase compared to Cry34Ab1 due to its increased size.

The theoretical values calculated with MOBCAL assume helium as a collision gas [25,26], and thus are not directly comparable with the experimentally measured values using nitrogen. Of the three methods used to calculate theoretical CCS values using MOBCAL, the TM method is considered to be the most reliable and accurate [34]. In addition, the theoretical value calculated for trCry35Ab1 is derived from the crystal structure data of the full length protein, whereas the experimentally measured value is from the truncated form which is missing 31 residues from the C-terminus of the protein.

In comparing the experimentally measured CCS values for both the smallest conformers of Cry34Ab1 and trCry35Ab1 with the theoretical values from MOBCAL, the experimental values are consistently smaller. This agrees with previous observations [34] and is considered to be a reflection of partial collapse of the protein structure during desolvation.

Based upon the crystallographic, SAXS and mass spectrometry data collected, it is clear that both toxins are monomeric in solution and remarkably stable. However, higher-order oligomeric organization of Cry34Ab1 and Cry35Ab1 in the presence of any putative membrane receptors cannot be ruled out.

both Cry34Ab1 and trCry35Ab1 were produced using nano electrospray of the proteins in 100% water at pH 3.0 (Dataset S3).

Proteins that are ionized under denaturing electrospray conditions typically exhibit a large number of high charge states as a result of the denaturation exposing multiple protonation sites. Ionization under non-denaturing conditions exposes a much smaller number of protonation sites on the surface of the folded protein, resulting in a narrow distribution of conformers with low charge states at high m/z values. Electrospray of both Cry34Ab1 and trCry35Ab1 under these conditions showed a narrow distribution of charge states at relatively high m/z, indicative of

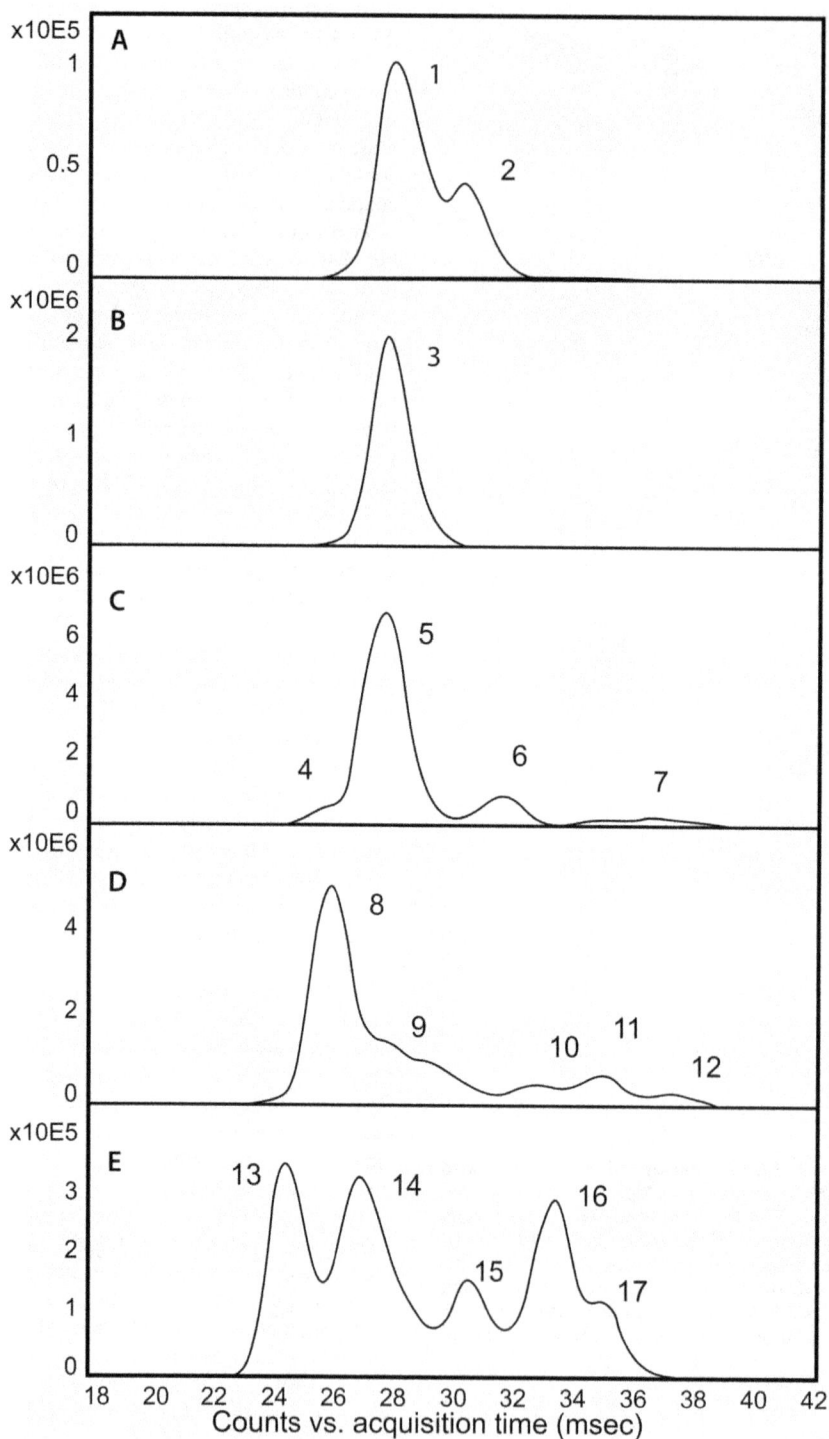

Figure 5. Collision cross sectional profiles of Cry34Ab1. Collision cross sectional profiles of Cry34Ab1. There were five major charge states observed with nanospray. They corresponded to: (A) +7 charge at m/z 1915.62–1941.80; (B) +8 charge at m/z 1680.96–1694.72; (C) +9 charge at m/z 1494.40–1508.25; (D) +10 charge at m/z 1327.13–1387.07; (E) +11 charge at m/z 1221.60–1229.82. The numbered components refer to values reported in Table 3.

Protein structure alignment by combinatorial extension

Structural alignment of protein structures in combination with sequence homology studies provides additional insight into the evolutionary relationships and modes of action of proteins. To this end, 3D structures of Cry34Ab1 and Cry35Ab1 were aligned against all the 3D structures in the Protein Data Bank (http://www.rcsb.org/pdb/home/home.do). Table 4 lists the top 10 unique structures in the pdb that have a threshold z-score of 4

Table 4. Combinatorial extension analysis of PDB submitted structures against Cry34Ab1 coordinates.

PDB ID	Name	RMS superposition	Z-score	% sequence identity	CATH Classification
1o72	Cytolysin Sticholysin II	3.47 Å	5.2	16.10%	Mainly Beta; Sandwich; Mutm (Fpg) Protein
3lim	Fragaceatoxin C	3.33 Å	5	17.70%	N/A
1iaz	Equinatoxin II	4.15 Å	5	15.00%	N/A
2qqp	Providence Virus	4.43 Å	4.6	12.00%	N/A
2qsv	Protein of unknown function	2.88 Å	4.6	10.50%	Mainly Beta; Sandwich; Immunoglobin-like; PapD-like
3a57	Thermostable direct hemolysin	2.81 Å	4.6	8.70%	N/A
1bci	Phospholipase	4.08 Å	4.6	8.10%	Mainly Beta; Sandwich; Immunoglobin-like; C2-domain Calcium/lipid binding domain
2xc8	Gene 22 product of the *Bacillus subtilis* SPP1 Phage	7.64 Å	4.6	7.10%	N/A
3l9b	Otoferlin C2A	4.52 Å	4.6	6.90%	N/A
3i6s	Plant subtilisin-like protease SBT3	2.64 Å	4.4	13.50%	N/A

Results of the top 10 unique hits ranked according to Z-score.

or higher with the Cry34Ab1 structure. It also includes the CATH description for the protein where available. Table 5 provides a similar analysis for the Cry35Ab1 structure.

The Cry34Ab1 β-sandwich architecture is most similar to a superfamily of sea anemone toxins known as actinoporins including equinatoxin [36,37], sticholysin II [38] and fragaceatoxin C [39]. Actinoporins are small, approximately 20 kDa, cytolytic proteins found in sea anemone venom. Cry34Ab1 is also structurally similar to *Vibrio parahaemolyticus* thermostable direct hemolysin [40].

Cry35Ab1 has an overall structure that is very similar to the recently described structure of BinB [41]. Both structures have an N-terminal domain with two QxW repeats and a second domain consisting of extended antiparallel beta sheets. Structural homology of BinB and proteins containing similar β-trefoil lectin-like domains are identified by high Z-scores in Table 5.

Cry34Ab1 structure comparisons

The Cry34Ab1 structure is clearly related to other membrane-interacting proteins with a beta sandwich fold, including actinoporins and hemolysin (Figure 6). Both actinoporins and hemolysin form tetrameric pores in lipid membranes [42]. The molecular mechanism of actinoporins has been extensively studied (reviewed in [43]). Actinoporins show strong specificity for sphingomyelin and form pores in membranes containing sphingomyelin. A key feature of actinoporin mechanism of action is the insertion of the N-terminal α-helical segment that precedes oligomerization and pore formation [44]. The actinoporin N-terminal α-helix is necessary for pore formation. Cry34Ab1 does not contain an analogous N-terminal helical structure, implying differences in membrane interaction mode of action when compared to actinoporins.

Further, Pfam analysis indicates that Cry34Ab1 appears to be a member of the Aegerolysin protein family [6]. The β-sandwich

Table 5. Combinatorial extension analysis of PDB submitted structures against Cry35Ab1 coordinates.

PDB ID	Name	RMS superposition	Z-score	% sequence identity	CATH Classification
3wa1	BinB	2.73 Å	7.1	17.2%	
1ups	β-galactosidase	7.10 Å	6.1	15.10%	Mainly β-sandwich; Trefoil; Jelly Rolls
2f2f	Cytolethal distending toxin	2.45 Å	6.1	12.80%	Mainly β-sandwich; Trefoil; Jelly Rolls; Alpha-Beta; 4-Layer Sandwich; DNase I
1qxm	Hemagglutinin component (HA1)	1.99 Å	6	22.70%	Mainly β-sandwich; Trefoil; Jelly Rolls
3ef2	Mushroom lectin	2.29 Å	6	19.90%	N/A
2y9g	β-trefoil lectin binding domain	2.60 Å	6	13.00%	N/A
2vsa	Mosquitocidal holotoxin	2.28 Å	5.9	19.70%	N/A
3nbd	Ricin-B like lectin domain	3.11 Å	5.9	19.40%	
1dfc	Fascin	2.91 Å	5.9	13.50%	Mainly Beta; Trefoil;
2yug	FRG1	2.50 Å	5.9	11.20%	N/A

Results of the top 10 unique hits ranked according to Z-score.

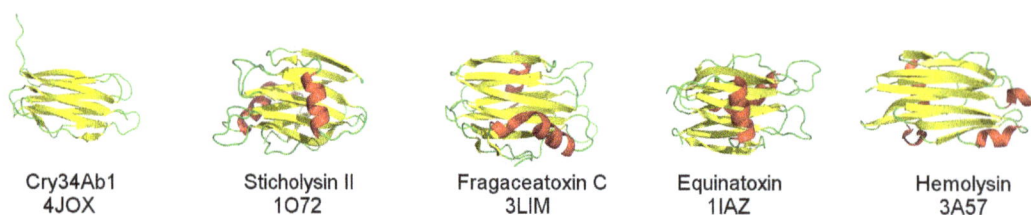

Figure 6. Comparison of proteins structurally related to Cry34Ab1. Cry34Ab1 is structurally related to a wide variety membrane-interacting proteins as assessed by combinatorial extension. All structures compared are comprised of a conserved beta-sheet core and varying loop regions.

fold exemplified in Cry34Ab1 is common among other cytolytic proteins found in nature including necrosis and ethylene-inducing peptide 1 (Nep1)-like proteins (NLPs) from microbial plant pathogens [45] and fungal fruit lectins [46]. Cry34Ab1 protein also shows some similarity to the Pam protein of *Photorhabdus asymbiotica* [47]. This similarity is sufficient to allow modeling of the Pam sequence based on the Cry34Ab1 template (Figure 7A). The model shows a similar structure but some of the β-strands appear shorter and one strand of the five-strand sheet shows a loop in the Pam model.

Cry35Ab1 structure comparisons

It is clear from primary sequence alignments that Cry35Ab1 is a member of the toxin_10 family that includes BinA, BinB, Cry36Aa1 and Cry49Aa1. The 3D structure of Cry35Ab1 gives us some insight into the structures of these related proteins as demonstrated by comparison to BinB and the theoretical model of Cry49Aa1. Building this model on both Cry35Aa1 and BinB templates resulted in a model with a probability in excess of 95% as judged by a GA341 score of 0.92133 (where a value >0.7 generally indicates a reliable model, defined as $\geq 95\%$ probability of a correct fold).

The N-terminal 40 amino acids in the Cry49Aa1 sequence appeared as a flexible region projecting beyond the extended the N-terminus of the BinB structure and could not be modelled reliably and, therefore, was removed from the structure shown in Figure 7B. Cry49Aa1 clearly shows significant similarity to Cry35Aa1, particularly in the core, β-sheet region of the C-terminal domain (Figure 7B). The high proportion of β-sheet

structure in Cry35Ab1 is consistent with CD analysis of the related BinA protein that indicated a high proportion of β-sheet and little α-helix [48,49,50].

Cry35Ab1 is also related to a 41.9 kDa toxin_10 family protein found in the genomes of a number of *Bacillus cereus* and *B. thuringiensis* strains although, to date, no toxicity has been found for this protein, which has only been tested thus far on lepidopteran targets [51]. In addition, the structure of Cry35Ab1 shows interesting similarities to several other toxins that show no significant relationship at the primary sequence level yet, like Cry35Ab1, are predominantly composed of β-sheets arranged in extended structures. These include the structure of aerolysin (PDB accession number 1PRE), parasporin 4 (PDB accession 2D42; also known as Cry45Aa1) and parasporin 2 [52,53] (PDB accession numbers 2D42 and 2ZTB; also known as Cry46Aa1) (Figure 8). Membership of the β-pore forming toxin family that includes aerolysin is consistent with Cry35Aa1 causing toxicity by participating in pore formation [1]. The mechanism of action of this family of toxins involves the oligomerisation of individual subunits followed by structural rearrangements that must occur for the penetrating β-sheet pore structures to enter the membrane. Parasporins, like Cry proteins, are produced as crystalline inclusions by *B. thuringiensis* strains but, to date, have no reported toxicity to invertebrates however, have demonstrated anti-cancer activity [54,55].

The structures of Cry34Ab1 and Cry35Ab1 also show striking similarity to the structures determined for Cry37Aa1 and Cry23Aa1 respectively [56], despite little primary sequence identity. Cry37Aa1/Cry23Aa1, like Cry34Ab1/Cry35Ab1, represents a two-component toxin active against certain coleopteran targets [57]. Cry23Aa1 appears to be a member of a family of proteins that include *B. thuringiensis* Cry38Aa1, a protein of unknown activity encoded by a gene linked with the *Cry34Aa1* and *Cry35Aa1* operon [58]. Cry33Aa1, Cry45Aa1, Cry15Aa1 and the related proteins, Bti34 and Bti36, are described as having antibacterial activity by Revina *et al.* [59]. In addition to these *B. thuringiensis* proteins, Cry23Aa is also related to the Mtx2, Mtx3 and Mtx4 proteins of *L. sphaericus* [7] and to *Aeromonas hydrophila* aerolysin [56]. Given the structural relationship of Cry23Aa1 to Cry35Aa1 and its related crystalline proteins Cry36Aa1, Cry49Aa1, BinA and BinB, nature seems to have adapted the extended antiparallel β-sheet structure extensively in the production of insecticidal toxins in the *Bacillus* and *Lysinibacillus* genera.

The Cry35Ab1 protein contains two repeats of the QxW motif found within the N-terminal, β-trefoil domain. Such features may be involved in lectin-like binding in toxins such as ricin B subunit [2]. The lectin-like domains of aerolysin and pertussis toxin have been proposed as conserved receptor binding domains [60]. Effects of sugar groups on the toxicity of the *L. sphaericus* Bin toxins have been reported previously [61,62] although these

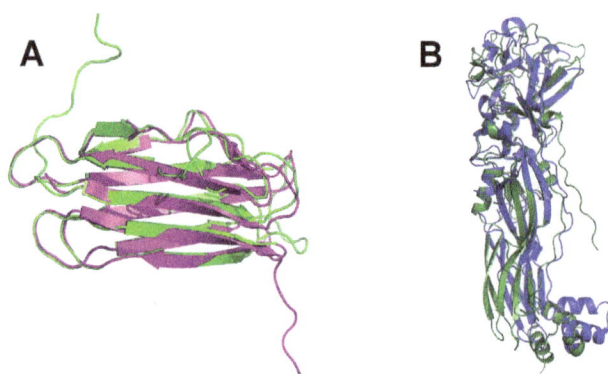

Figure 7. Structural superpositions of Cry34Ab1 and Cry35Ab1 with calculated theoretical models of Pam and Cry49. (A) Overlay of the Cry34Ab1 structure (purple) with a model of Pam (green). (B) Overlay of the Cry35Ab1 crystal structure (blue) with a model of Cry49 (green). See text for details.

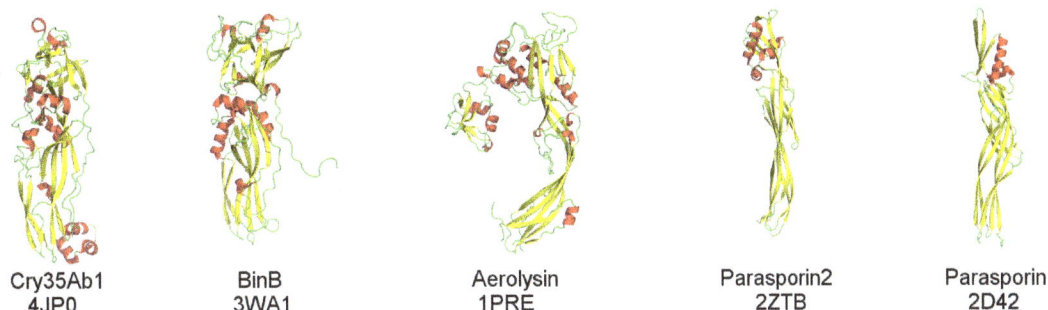

Figure 8. Comparison of proteins structurally related to Cry35Ab1. Cry35Ab1 is structurally related to a wide variety pore-forming proteins as assessed by combinatorial extension. All structures contain a conserved beta-sheet core and varying loop regions.

proteins lack the QxW motif in β-trefoil structures. BinA, BinB and Cry49Aa1 do contain an occurrence of the QxW motif in the β-sheet core of the C-terminal domain as does Cry35Ab1 although this occurrence of the motif is absent from the much more closely-related Cry35Aa1 sequence. In BinA, substitution of Trp222 in this feature, equivalent to Trp211 of Cry35Ab1, results in loss of activity although the protein is still able to permeabilize liposomes [63]. Ricin-B-like lectin repeats were also noted in the 41.9 kDa *B. thuringiensis* protein [51] although there are no QxW motifs in this protein.

The Cry35Ab1 three α-helical C-terminal motif is not required for insecticidal activity [17]. Moreover, removal of this motif to create trCry35Ab1 shifts expression entirely from the insoluble to the soluble fraction of the cell lysate of our heterologous *Pseudomonas fluorescens* expression system. This raises the possibility of a role in the formation of parasporal inclusions by functioning as an inclusion anchor or a protein: protein interaction domain.

Cry34Ab1 and Cry35Ab1 function

The structural similarities of Cry34Ab1 and Cry35Ab1 to other membrane-binding or pore forming proteins described above suggest several possibilities for a mechanism of action wherein either Cry34Ab1 and/or Cry35Ab1 might possibly initiate pore formation. However, the mechanism of interaction by which Cry34Ab1 and Cry35Ab1 function as a binary toxin is yet unknown. Both proteins have structural features allowing for conformational changes during putative pore formation event. What is currently known regarding mechanism of action is that Cry35Ab1 binds to WCR brush border membrane vesicles and Cry35Ab1 binding is dramatically enhanced by the presence of Cry34Ab1 [64] suggesting formation of a protein complex that results in pore formation. Presence of the N-terminal lectin domain on Cry35Ab1 suggests that binding to membrane glycoproteins might be involved, as has been suggested for BinB [41]. However, because iodo-radiolabelling Cry34Ab1 reduced its

biological activity, its role is more difficult to probe and therefore direct interaction of Cry34Ab1 with putative receptors cannot be ruled out. In fact, Cry34Ab1 has biological activity alone, albeit at a much reduced level compared to the binary toxin [65].

In conclusion, the Cry34Ab1 and Cry35Ab1 protein structures presented here, while sharing structural similarity with other pore forming toxins, are novel among proteins developed for corn rootworm resistance traits. This structural information provides the basis for experimentation aimed at dissecting the Cry34Ab1/Cry35Ab1 binary toxin mode of action and for protein engineering aimed at improving insecticidal properties of the proteins.

Supporting Information

Dataset S1   Processed Cry34Ab1 SAXS data.

Dataset S2   Processed trCry35Ab1 SAXS data.

Dataset S3   Processed Cry34Ab1 and trCry35Ab1 native electrospray ion mobility mass spectrometry data sets.

Acknowledgments

We wish to gratefully acknowledge the expert assistance of Steven Weigand (Advanced Photon Source) during SAXS data collection and the support of Christian Klein (Agilent Technologies) for acquisition of the native electrospray ion mobility data.

Author Contributions

Conceived and designed the experiments: MSK CY JWP JCR SLE. Performed the experiments: MSK JWP CY DGM JCR. Analyzed the data: JWP CY MSK MW MDB CB NXW DGM RP TJW. Contributed reagents/materials/analysis tools: MSK NXW DGM MDB JCR. Wrote the paper: MSK KEN CB CY RP.

References

1. Masson L, Schwab G, Mazza A, Brousseau R, Potvin L, et al. (2004) A novel *Bacillus thuringiensis* (PS149B1) containing a Cry34Ab1/Cry35Ab1 binary toxin specific for the western corn rootworm *Diabrotica virgifera virgifera LeConte* forms ion channels in lipid membranes. Biochemistry 43: 12349–12357.

2. Schnepf HE, Lee S, Dojillo J, Burmeister P, Fencl K, et al. (2005) Characterization of Cry34/Cry35 binary insecticidal proteins from diverse *Bacillus thuringiensis* strain collections. Appl Environ Microbiol 71: 1765–1774.

3. Ellis RT, Stockhoff BA, Stamp L, Schnepf HE, Schwab GE, et al. (2002) Novel *Bacillus thuringiensis* binary insecticidal crystal proteins active on western corn rootworm, *Diabrotica virgifera virgifera LeConte*. Applied and Environmental Microbiology 68: 1137–1145.

4. Narva KE, Siegfried BD, Storer NP (2013) Transgenic approaches to Western corn rootworm control. Adv Biochem Eng Biotechnol 136: 135–162.

5. (2012) U.S. Environmental Protection Agency Current & Previously Registered Section 3 PIP Registrations.

6. Berne S, Lah L, Sepcic K (2009) Aegerolysins: structure, function, and putative biological role. Protein Sci 18: 694–706.

7. Berry C (2012) The bacterium, *Lysinibacillus sphaericus*, as an insect pathogen. J Invertebr Pathol 109: 1–10.

8. Jones GW, Nielsen-Leroux C, Yang Y, Yuan Z, Dumas VF, et al. (2007) A new Cry toxin with a unique two-component dependency from *Bacillus sphaericus*. FASEB J 21: 4112–4120.

9. Rupar MJ DW, Chu C-R, Pease E, Tan Y, et al. (2003) Nucleic acids encoding coleopteran-toxic polypeptides and insect-resistant transgenic plants comprising them. Monsanto Technology LLC (St Louis, MO).

10. Charles H. Squires DMR, Lawrence C. Chew, Tom M. Ramseier, Jane C. Schneider, and Henry W. Talbot. (2004) Heterologous Protein Production in *P. fluorescens*. Bioprocess International: 54–59.

11. Pflugrath JW (1999) The finer things in X-ray diffraction data collection. Acta Crystallographica Section D-Biological Crystallography 55: 1718–1725.

12. Dodson EJ, Winn M, Ralph A (1997) Collaborative Computational Project, number 4: providing programs for protein crystallography. Methods Enzymol 277: 620–633.

13. Jones TA, Zou JY, Cowan SW, Kjeldgaard M (1991) Improved methods for building protein models in electron density maps and the location of errors in these models. Acta Crystallogr A 47 (Pt 2): 110–119.

14. Laskowski RA, MacArthur MW, Moss DS, Thornton JM (1993) PROCHECK: a program to check the stereochemical quality of protein structures. Journal of Applied Crystallography 26: 283–291.

15. Moellenbeck DJ, Peters ML, Bing JW, Rouse JR, Higgins LS, et al. (2001) Insecticidal proteins from *Bacillus thuringiensis* protect corn from corn rootworms. Nat Biotechnol 19: 668–672.

16. Narva KE, Storer NP, Meade T (2014) Discovery and Development of Insect-Resistant Crops Using Genes from *Bacillus thuringiensis*. In: Dhadialla TS, Gill SS, editors. Advances in Insect Physiology. Oxford: Academic Press.

17. Gao Y, Schafer BW, Collins RA, Herman RA, Xu XP, et al. (2004) Characterization of Cry34Ab1 and Cry35Ab1 insecticidal crystal proteins expressed in transgenic corn plants and *Pseudomonas fluorescens*. Journal of Agricultural and Food Chemistry 52: 8057–8065.

18. Petoukhov MV, Franke D, Shkumatov AV, Tria G, Kikhney AG, et al. (2012) New developments in the ATSAS program package for small-angle scattering data analysis. Journal of Applied Crystallography 45: 342–350.

19. Konarev PV, Volkov VV, Sokolova AV, Koch MHJ, Svergun DI (2003) PRIMUS: a Windows PC-based system for small-angle scattering data analysis. Journal of Applied Crystallography 36: 1277–1282.

20. Semenyuk AV, Svergun DI (1991) Gnom - a Program Package for Small-Angle Scattering Data-Processing. Journal of Applied Crystallography 24: 537–540.

21. Svergun DI (1999) Restoring low resolution structure of biological macromolecules from solution scattering using simulated annealing. Biophysical Journal 76: 2879–2886.

22. Volkov VV, Svergun DI (2003) Uniqueness of *ab initio* shape determination in small-angle scattering. Journal of Applied Crystallography 36: 860–864.

23. Pettersen EF, Goddard TD, Huang CC, Couch GS, Greenblatt DM, et al. (2004) UCSF chimera - A visualization system for exploratory research and analysis. Journal of Computational Chemistry 25: 1605–1612.

24. Campuzano I, Bush MF, Robinson CV, Beaumont C, Richardson K, et al. (2012) Structural Characterization of Drug-like Compounds by Ion Mobility Mass Spectrometry: Comparison of Theoretical and Experimentally Derived Nitrogen Collision Cross Sections. Analytical Chemistry 84: 1026–1033.

25. Mesleh MF, Hunter JM, Shvartsburg AA, Schatz GC, Jarrold MF (1996) Structural information from ion mobility measurements: Effects of the long-range potential. Journal of Physical Chemistry 100: 16082–16086.

26. Shvartsburg AA, Jarrold MF (1996) An exact hard-spheres scattering model for the mobilities of polyatomic ions. Chemical Physics Letters 261: 86–91.

27. Shindyalov IN, Bourne PE (1998) Protein structure alignment by incremental combinatorial extension (CE) of the optimal path. Protein Eng 11: 739–747.

28. N. Eswar MAM-R, B. Webb, M. S . Madhusudhan, D. Eramian, M. Shen, U. Pieper, A. Sali (2006) Comparative Protein Structure Modeling With MODELLER.: John Wiley & Sons, Inc.

29. Sali A, Blundell TL (1993) Comparative protein modelling by satisfaction of spatial restraints. J Mol Biol 234: 779–815.

30. Melo F, Sanchez R, Sali A (2002) Statistical potentials for fold assessment. Protein Sci 11: 430–448.

31. Boonyos P, Soonsanga S, Boonserm P, Promdonkoy B (2010) Role of cysteine at positions 67, 161 and 241 of a *Bacillus sphaericus* binary toxin BinB. Bmb Reports 43: 23–28.

32. Promdonkoy B, Promdonkoy P, Wongtawan B, Boonserm P, Panyim S (2008) Cys31, cys47, and cys195 in BinA are essential for toxicity of a binary toxin from *Bacillus sphaericus*. Current Microbiology 56: 334–338.

33. Konijnenberg A, Butterer A, Sobott F (2013) Native ion mobility-mass spectrometry and related methods in structural biology. Biochim Biophys Acta 1834: 1239–1256.

34. Jurneczko E, Barran PE (2011) How useful is ion mobility mass spectrometry for structural biology? The relationship between protein crystal structures and their collision cross sections in the gas phase. Analyst 136: 20–28.

35. Bush MF, Hall Z, Giles K, Hoyes J, Robinson CV, et al. (2010) Collision Cross Sections of Proteins and Their Complexes: A Calibration Framework and Database for Gas-Phase Structural Biology. Analytical Chemistry 82: 9557–9565.

36. Athanasiadis A, Anderluh G, Macek P, Turk D (2001) Crystal structure of the soluble form of equinatoxin II, a pore-forming toxin from the sea anemone Actinia equina. Structure 9: 341–346.

37. Hinds MG, Zhang W, Anderluh G, Hansen PE, Norton RS (2002) Solution structure of the eukaryotic pore-forming cytolysin equinatoxin II: Implications for pore formation. Journal of Molecular Biology 315: 1219–1229.

38. Mancheno JM, Martin-Benito J, Martinez-Ripoll M, Gavilanes JG, Hermoso JA (2003) Crystal and electron microscopy structures of sticholysin II actinoporin reveal insights into the mechanism of membrane pore formation. Structure 11: 1319–1328.

39. Mechaly AE, Bellomio A, Morante K, Gonzalez-Manas JM, Guerin DMA (2009) Crystallization and preliminary crystallographic analysis of fragaceatoxin C, a pore-forming toxin from the sea anemone *Actinia fragacea*. Acta Crystallographica Section F-Structural Biology and Crystallization Communications 65: 357–360.

40. Yanagihara I, Nakahira K, Yamane T, Kaieda S, Mayanagi K, et al. (2010) Structure and functional characterization of *Vibrio parahaemolyticus* thermostable direct hemolysin. J Biol Chem 285: 16267–16274.

41. Srisucharitpanit K, Yao M, Promdonkoy B, Chimnaronk S, Tanaka I, et al. (2014) Crystal structure of BinB: A receptor binding component of the binary toxin from *Lysinibacillus sphaericus*. Proteins.

42. Alvarez C, Mancheno JM, Martinez D, Tejuca M, Pazos F, et al. (2009) Sticholysins, two pore-forming toxins produced by the Caribbean Sea anemone *Stichodactyla helianthus*: Their interaction with membranes. Toxicon 54: 1135–1147.

43. Kristan KC, Viero G, Dalla Serra M, Macek P, Anderluh G (2009) Molecular mechanism of pore formation by actinoporins. Toxicon 54: 1125–1134.

44. Rojko N, Kristan KC, Viero G, Zerovnik E, Macek P, et al. (2013) Membrane Damage by an alpha-Helical Pore-forming Protein, Equinatoxin II, Proceeds through a Succession of Ordered Steps. Journal of Biological Chemistry 288: 23704–23715.

45. Ottmann C, Luberacki B, Kufner I, Koch W, Brunner F, et al. (2009) A common toxin fold mediates microbial attack and plant defense. Proceedings of the National Academy of Sciences of the United States of America 106: 10359–10364.

46. Birck C, Damian L, Marty-Detraves C, Lougarre A, Schulze-Briese C, et al. (2004) A new lectin family with structure similarity to actinoporins revealed by the crystal structure of *Xerocomus chrysenteron* lectin XCL. Journal of Molecular Biology 344: 1409–1420.

47. Jones RT, Sanchez-Contreras M, Vlisidou I, Amos MR, Yang GW, et al. (2010) Photorhabdus adhesion modification protein (Pam) binds extracellular polysaccharide and alters bacterial attachment. Bmc Microbiology 10.

48. Hire RS, Hadapad AB, Dongre TK, Kumar V (2009) Purification and characterization of mosquitocidal *Bacillus sphaericus* BinA protein. J Invertebr Pathol 101: 106–111.

49. Kale A, Hire RS, Hadapad AB, D'Souza SF, Kumar V (2013) Interaction between mosquito-larvicidal *Lysinibacillus sphaericus* binary toxin components: Analysis of complex formation. Insect Biochemistry and Molecular Biology 43: 1045–1054.

50. Srisucharitpanit K, Inchana P, Rungrod A, Promdonkoy B, Boonserm P (2012) Expression and purification of the active soluble form of *Bacillus sphaericus* binary toxin for structural analysis. Protein Expression and Purification 82: 368–372.

51. Palma L, Muñoz D., Berry C., Murillo J., and Caballero P. (2014) Draft genome sequences of two *Bacillus thuringiensis* strains and characterization of a putative 41.9-kDa insecticidal toxin. Genome 6: 1490–1504.

52. Akiba T, Abe Y, Kitada S, Kusaka Y, Ito A, et al. (2009) Crystal Structure of the Parasporin-2 *Bacillus thuringiensis* Toxin That Recognizes Cancer Cells. Journal of Molecular Biology 386: 121–133.

53. Akiba T, Higuchi K, Mizuki E, Ekino K, Shin T, et al. (2006) Nontoxic crystal protein from *Bacillus thuringiensis* demonstrates a remarkable structural similarity to beta-pore-forming toxins. Proteins 63: 243–248.

54. Akiba T, Abe Y, Kitada S, Kusaka Y, Ito A, et al. (2009) Crystal structure of the parasporin-2 *Bacillus thuringiensis* toxin that recognizes cancer cells. J Mol Biol 386: 121–133.

55. Xu C, Wang BC, Yu Z, Sun M (2014) Structural Insights into *Bacillus thuringiensis* Cry, Cyt and Parasporin Toxins. Toxins (Basel) 6: 2732–2770.

56. de Maagd RA, Bravo A, Berry C, Crickmore N, Schnepf HE (2003) Structure, diversity, and evolution of protein toxins from spore-forming entomopathogenic bacteria. Annu Rev Genet 37: 409–433.

57. Donovan WP, J. C. Donovan, and A. C. Slaney (2000) *Bacillus thuringiensis* cryET33 and cryET34 compositions and uses thereof. USA: Monsanto Company.

58. Baum JA, Chu CR, Rupar M, Brown GR, Donovan WP, et al. (2004) Binary toxins from *Bacillus thuringiensis* active against the western corn rootworm, *Diabrotica virgifera virgifera* LeConte. Appl Environ Microbiol 70: 4889–4898.

59. Revina LP, Kostina LI, Dronina MA, Zalunin IA, Chestukhina GG, et al. (2005) Novel antibacterial proteins from entomocidal crystals of *Bacillus thuringiensis ssp. israelensis*. Can J Microbiol 51: 141–148.

60. Rossjohn J, Buckley JT, Hazes B, Murzin AG, Read RJ, et al. (1997) Aerolysin and pertussis toxin share a common receptor-binding domain. EMBO J 16: 3426–3434.

61. Broadwell AH, Baumann P (1987) Proteolysis in the gut of mosquito larvae results in further activation of the *Bacillus sphaericus* toxin. Appl Environ Microbiol 53: 1333–1337.

62. Nielsen-Leroux C, Charles JF (1992) Binding of *Bacillus sphaericus* binary toxin to a specific receptor on midgut brush-border membranes from mosquito larvae. Eur J Biochem 210: 585–590.

63. Kunthic T, Promdonkoy B, Srikhirin T, Boonserm P (2011) Essential role of tryptophan residues in toxicity of binary toxin from *Bacillus sphaericus*. Bmb Reports 44: 674–679.

64. Li HR, Olson M, Lin GF, Hey T, Tan SY, et al. (2013) *Bacillus thuringiensis* Cry34Ab1/Cry35Ab1 Interactions with Western Corn Rootworm Midgut Membrane Binding Sites. Plos One 8.

65. Herman RA, Scherer PN, Young DL, Mihaliak CA, Meade T, et al. (2002) Binary insecticidal crystal protein from *Bacillus thuringiensis*, strain PS149B1: effects of individual protein components and mixtures in laboratory bioassays. J Econ Entomol 95: 635–639.

66. DeLano WL (2006) PyMOL (v 0.99rc6). DeLano Scientific, LLC, San Carlos, CA.

Comparative Analysis of Human γD-Crystallin Aggregation under Physiological and Low pH Conditions

Josephine W. Wu[1]*, Mei-Er Chen[2], Wen-Sing Wen[3], Wei-An Chen[3], Chien-Ting Li[3], Chih-Kai Chang[3], Chun-Hsien Lo[3], Hwai-Shen Liu[3], Steven S.-S. Wang[3]*

[1] Department of Optometry, Central Taiwan University of Science and Technology, Taichung 40601, Taiwan, [2] Department of Entomology, National Chung Hsing University, Taichung 402, Taiwan, [3] Department of Chemical Engineering, National Taiwan University, Taipei 10617, Taiwan

Abstract

Cataract, a major cause of visual impairment worldwide, is the opacification of the eye's crystalline lens due to aggregation of the crystallin proteins. The research reported here is aimed at investigating the aggregating behavior of γ-crystallin proteins in various incubation conditions. Thioflavin T binding assay, circular dichroism spectroscopy, 1-anilinonaphthalene-8-sulfonic acid fluorescence spectroscopy, intrinsic (tryptophan) fluorescence spectroscopy, light scattering, and electron microscopy were used for structural characterization. Molecular dynamics simulations and bioinformatics prediction were performed to gain insights into the γD-crystallin mechanisms of fibrillogenesis. We first demonstrated that, except at pH 7.0 and 37°C, the aggregation of γD-crystallin was observed to be augmented upon incubation, as revealed by turbidity measurements. Next, the types of aggregates (fibrillar or non-fibrillar aggregates) formed under different incubation conditions were identified. We found that, while a variety of non-fibrillar, granular species were detected in the sample incubated under pH 7.0, the fibrillogenesis of human γD-crystallin could be induced by acidic pH (pH 2.0). In addition, circular dichroism spectroscopy, 1-anilinonaphthalene-8-sulfonic acid fluorescence spectroscopy, and intrinsic fluorescence spectroscopy were used to characterize the structural and conformational features in different incubation conditions. Our results suggested that incubation under acidic condition led to a considerable change in the secondary structure and an enhancement in solvent-exposure of the hydrophobic regions of human γD-crystallin. Finally, molecular dynamics simulations and bioinformatics prediction were performed to better explain the differences between the structures and/or conformations of the human γD-crystallin samples and to reveal potential key protein region involved in the varied aggregation behavior. Bioinformatics analyses revealed that the initiation of amyloid formation of human γD-crystallin may be associated with a region within the C-terminal domain. We believe the results from this research may contribute to a better understanding of the possible mechanisms underlying the pathogenesis of senile nuclear cataract.

Editor: Rizwan H. Khan, Aligarh Muslim University, India

Funding: This work was supported by the grants from the Ministry of Science and Technology, Taiwan (MOST 102-2221-E-002-161 and MOST 103-2221-E-002-208 to SSW, and NSC 101-2113-M-166-001-MY2 and MOST 103-2113-M-166-001-MY2 to JWW). The funders had no role in study design, data collection and analysis, decision to publish, or preparation of the manuscript.

Competing Interests: The authors have declared that no competing interests exist.

* Email: 107658@ctust.edu.tw (JWW); sswang@ntu.edu.tw (SSW)

Introduction

It is widely accepted that aggregation is a universal phenomenon that can occur to proteins of all types. Protein aggregation arises from a common mechanism whereby the normally folded proteins change conformation and results in partially unfolded intermediates that eventually aggregate by auto assembly to form either amorphous and/or fibril species [1,2]. Not only is protein aggregation a major problem in biotechnology products relating to protein expression, purification, and storage [3], it is also responsible for more than 40 human protein-deposition diseases that have been well documented to this day [4]. Among these so called protein conformational diseases is cataract, a major cause of visual impairment worldwide. Based on the World Health Organization (2011), cataract makes up 33% of global visual impairment (next to uncorrected refractive errors at 43%) and is

the leading cause of blindness in middle and low-income countries [5].

Cataract is the opacification of the eye's crystalline lens due to aggregation and precipitation of the crystallin proteins [6,7]. In the normal eye, the lens is a transparent refractive structure that serves to focus light onto the retina. It is capable of retaining transparency owing to a high concentration of crystallins that are arranged into short-range order. The absence of cellular organelles in the mature lens fiber cells also helps to minimize light scatter [8]. Additional contribution to lens transparency is provided by the unique structural and functional properties of the crystallins themselves. There are three types of crystallins in the mammalian lens: α-, β-, and γ- crystallins. α-Crystallin is a heat shock protein that function as a molecular chaperone to prevent other proteins from aggregating and insolubilizing under stressful conditions [9,10]. To stay true to its chaperone function, the protein has adopted high conformational flexibility and structural disorder to accommodate its interactions with target proteins, which includes the β- and γ- crystallins. Both β & γ-crystallins belong to the same superfamily and are considered structural proteins that, when maintained in their native globular state and arranged in densely-packed fashion, are responsible for preserving clarity of the crystalline lens. As the eye lens ages, structures of the crystallin proteins begin to change due to a variety of environmental factors, hence disrupting the orderly arrangements of protein packing that kept the lens in its transparent state.

Several theories on the mechanisms of cataract formation at the molecular level have been put forth. As the lens ages, crystallins are subjected to environmental insults that result in structural modifications or damages, leading to incorrect interactions, unfolding, oligomerization, and aggregation of proteins. Processes that can occur in the aging lens and have detrimental effects on the native structures of lens proteins include photooxidation (by UV radiation), deamidation, disulfide bond formation, and cleavage [11]. Oxidative damage is the process whereby reactive oxygen species are coupled with photooxidation and/or conversion of sulfhydryl groups to form half-cystine disulfide groups [12]. It has been found to be a major contributor to cataract formation in aged lens in which the level of glutathione is significantly reduced [13]. Another common process that causes damages to the crystallins is deamidation, where negative charge is introduced at the site of glutamine and asparagine causing the proteins to form cataractous aggregates [14]. Protease cleavage of crystallins that involves calpains has also been observed to be associated with senile nuclear cataract [15]. The consequence of the above-mentioned mechanisms of cataractogenesis is the disruption of the orderly arrangement within the crystalline fiber cells and the development of opacity in the once transparent lens structure, hence cataract.

Of the crystallin proteins, γ-crystallin has the simplest structure, existing as a monomer of four Greek key motifs rich in anti-parallel β-sheets. The molecular weight of the protein is approximately 20 kDa (173 amino acids), with a fold akin to many immunoglobulins [16,17]. Human gamma D-crystallin (HγD-crys) is the third most commonly expressed γ-crystallin in the human lens [9]. It is highly stable at neutral pH, but like many proteins, can form aggregation under certain conditions. Although the main form of aggregates found in the cataractous lens is of the amorphous type, HγD-crys has recently been observed to form amyloid fibrils as well. *In vitro* aggregation of HγD-crys has been noted in guanidine hydrochloride (GdnHCl) under physiological temperature and pH [17], as well as under acidic pH [18]. Low pH condition, without the presence of denaturant, leads to partially or fully unfolded species that form amyloid fibrils. These fibril aggregates have previously been characterized by various biophysical methods (e.g., Congo red, FTIR, X-ray diffraction, TEM, and 2D-IR) [10,18,19]. In any case, both full protein and (to a lesser extent) isolated C-terminal and N-terminal domains of HγD-crys are capable of forming amyloid fibrils under acidic pH condition [18]. The above-mentioned findings are proofs of concept demonstrating that despite high structural stability, HγD-crys (like many other proteins) also has the potential to reorganize and form amyloid structure in destabilizing conditions. Therefore, it has been suggested that this pathway may be an additional process contributing to the development of cataract with aging [10].

Of the various hypotheses that have been proposed on the mechanisms behind the development of age-related cataract, one that has not been fully explored is the possibility of low pH-induced cataract. Currently, two views exist in regards to acidic pH and cataractogenesis. One involves the possibility of partially degraded HγD-crys forming fibrillar aggregates in the low pH

environment of lysozomal compartment during lens fiber cell differentiation, which may be involved in the early stages of cataract formation [18]. Another hypothesis has to do with the decreased pH in the lens nucleus overtime leading to loss of α-crystallin chaperone capability to protect HγD-crys from aggregation, thus resulting in senile nuclear cataract formation [20,21]. Regardless of the mechanisms of cataract formation, the only accepted form of treatment currently available is the surgical removal of the opaque lens and replacement with an artificial lens. However, such procedure is not without risks of complications and is often inaccessible in low and middle-income countries. Therefore, in order to seek for other potential therapeutic strategies to treating cataract, a more thorough understanding of the crystallin aggregation process leading to cataractogenesis is imperative.

The current study is aimed at a more extensive investigation into HγD-crys aggregation in neutral and low pH. We first demonstrated that, except at physiological condition of pH 7.0 and 37°C, the aggregation of HγD-crys was observed to be augmented upon incubation, as revealed by turbidity measurements. Next, the types of aggregates (fibrillar or non-fibrillar aggregates) formed under different incubation conditions were identified using ThT fluorescence spectroscopy and transmission electron microscopy (TEM). We found that HγD-crys can be induced to form amyloid fibrillar species at acidic pH but not neutral pH. In addition, a number of spectroscopic techniques including far-UV circular dichroism (CD) spectroscopy, 1-anilinonaphthalene-8-sulfonic acid (ANS) fluorescence spectroscopy, and intrinsic or tryptophan fluorescence spectroscopy, were used to characterize the structural and conformational features in different incubation conditions. Finally, molecular dynamics (MD) simulations were performed to gain some insights into the molecular mechanism of the initial stage in the process of HγD-crys fibril formation. Not only does our work compare HγD-crys aggregation under different pH and temperature conditions, it is also the first to fully characterize its fibrillogenesis under low pH setting and brings all past studies of its kind into perspective.

Materials and Methods

Materials

Salts, tryptone, yeast extract, and chromatography columns were purchased from Sigma (USA). Kanamycin, imidazole, and isopropyl β-D-thiolgalactorpyranoside (IPTG) were obtained from Biobasic (Canada). EZ Ni-agarose 6 resin was obtained from Lamda Biotech (USA). All other chemicals were of reagent grade and obtained from Sigma (USA) unless otherwise specified.

Expression and purification of HγD-crys protein

Bacterial expression and purification of the recombinant proteins has been described previously [22]. The plasmid pQE1 containing the 6×His-tagged HγD-crys gene was provided by Dr. Jonathan King's laboratory at Massachusetts Institute of Technology [23,24] and was transformed into *E. coli* strain BL21 (DE3) using the heat shock method with high transformation efficiency. The 6×His-HγD-crys gene fragment from plasmid pQE1 was amplified by polymerase chain reaction (PCR) using primers that introduced *Nde*I and *Bam*HI restriction sites (forward primer, 5'-GAGGAGAAATTAA<u>CATATG</u>AAACATCACCATCA-3'; reverse primer, 5'-<u>GC</u>TTTGTTAGCAGCC<u>GGATCC</u>AAAT-TAAGAA-3'). The PCR procedure comprised a denaturation step at 94°C for 5 min, followed by 15 cycles of denaturation at 94°C for 1 min, annealing at 55°C for 40 sec, and extension at 72°C for 90 sec. The PCR products and pET30b(+) were digested with restriction enzymes (*Nde*I and *Bam*HI) and purified by electrophoresis through a 1.3% agarose gel. The resultant purified PCR products were ligated to the resultant digested plasmid pET30b(+) in a reaction containing 2 μL of T4 ligase, 2 μL of 10 mM ATP, and 10 μL of 2X T4 ligase reaction buffer, resulting in the generation of pEHisHγD-crys. The resultant pEHisHγD-crys, transformed into the *E. coli* strain BL21 (DE3) using the heat shock method, procured a high transformation efficiency and was used to express the 6×His-HγD-crys protein.

In a typical experiment, a single colony of *E. coli* strain BL21 (DE3) harboring plasmid pEHisHγD-crys was inoculated in 50 mL LB medium (10% tryptone, 5% yeast extract, 10% NaCl) containing the appropriate antibiotic (kanamycin 30 μg/mL) and grown with shaking at 200 rpm at 37°C. 1 mL overnight cultures were used to inoculate 100 mL of fresh LB medium, and these cultures were grown at 37°C. After reaching an optimal $OD_{600\ nm}$, bacterial cultures were induced at 30°C by the addition of IPTG. Cell lysis was performed by ultrasonication (1 s of plus-on and 1 s plus-off for 30 min) and the insoluble material was removed by centrifugation (13000 rpm for 30 min). The supernatant (the soluble part) was collected and passed through a Supelco liquid chromatography column (Sigma-Aldrich, USA) and EZ Ni-agarose 6 resin (Lamda Biotech, USA). The purified recombinant protein solution was dialyzed against salt solution (136.7 mM NaCl, 2.68 mM KCl, 0.01% (w/v) sodium azide, pH 7.0) and the resultant 6×His-HγD-crys stock solution was stored at 4°C.

Preparation of HγD-crys sample solutions and determination of protein concentration

For the sake of comparison with previous studies on HγD-crys aggregation associated with cataract formation done by our group and others [10,25], 1 mg/mL HγD-crys protein concentration was chosen for the aggregation experiments under various pH and temperature conditions. The sample solutions were prepared by diluting the stock solutions with salt solution (136.7 mM NaCl, 2.68 mM KCl, 0.01% (w/v) sodium azide, pH 7.0). The protein concentrations of HγD-crys sample solutions were determined by the bicinchoninic acid assay (BCA) using bovine serum albumin (BSA) as a standard [26].

Turbidity measurement

Turbidity measurements of samples were performed by monitoring the absorbance at 360 nm [27,28]. 1 mL of HγD-crys samples (1 mg/mL) taken at different times were added to a 1 cm light-path quartz cuvette. The analyses were carried out using a Cary Eclipse UV/VIS spectrophotometer (Varian, USA). Three measurements were performed, and the mean and standard deviation were obtained.

Thioflavin T (ThT) fluorescence measurement

The stock solution of ThT at a concentration of 20 mM was prepared in ethanol protected from light prior to use, and the concentration was determined spectrophotometrically using the molar extinction coefficient at 416 nm of 26600 M^{-1} cm^{-1} [29]. Phosphate buffered saline (PBS) with 0.01% (w/v) sodium azide was used to dissolve ThT to a final concentration of 20 μM. 40 μL of protein samples taken at different times were added to 960 μL of ThT solution (20 μM) and briefly mixed with vortex. The ThT fluorescence emission intensity at 485 nm of the resultant mixture was recorded for 60 sec using the excitation wavelength of 440 nm on a Cary Eclipse Fluorescence Spectrophotometer (Varian, USA).

Transmission electron microscopy (TEM) analysis

An aliquot of 5 μL of HγD-crys samples for TEM analysis were withdrawn from the working solutions and applied on a carbon-stabilized, formvar coated grid for 30 sec. Excess samples were removed by applying ashless filter papers at the edge of the grids and the grids were negatively stained with 1% uranyl acetate in distilled de-ionized water (Electron Microscopy Sciences, USA) for another 30 sec. After removing the excess stain, the grids were left to air-dry for at least 30 min and then examined and photographed on a Hitachi H-7650 transmission electron microscope with a Gantan model 782 CCD Camera (Tokyo, Japan) at an accelerating voltage of 100 kV.

Far-UV circular dichroism (CD) spectroscopy

The secondary structural changes of HγD-crys sample solutions were evaluated by far-UV CD spectroscopy. CD spectra of HγD-crys samples (0.1 mg/mL) were recorded after diluting 10-fold with de-ionized water over the wavelength range of 190–260 nm using a J-815 spectrometer (JASCO, Japan) with a 0.2 cm path length sample cell. All CD measurements were collected at room temperature using a bandwidth of 1.0 nm, a step interval of 0.1 nm, and a scanning speed of 50 nm/min. Each CD spectrum was the average of three scans. The secondary structure contents of HγD-crys samples were estimated using the CDSSTR algorithm with appropriate reference sets available from the DICROWEB website [30,31]. Control buffer scans were run in duplicate, averaged, and then subtracted from the sample spectra. All experiments were performed at room temperature. The results have been plotted as ellipticity (mdeg) versus wavelength (nm).

1-Anilinonaphthalene-8-sulfonic acid (ANS) fluorescence spectroscopy

100 μL HγD-crys sample solutions were mixed with 900 μL ANS working solution of 20 μM in PBS solution, and then the mixtures were incubated in the dark for 30 min at room temperature. ANS fluorescence intensities were recorded by exciting samples at 380 nm and emissions were recorded between 420 and 580 nm on a Cary Eclipse fluorescence spectrophotometer (Varian, USA). All measurements were repeated at least three times. The representative ANS fluorescence intensity was taken at the average emission wavelength (AEW), which accounts for both changes in intensity and spectrum envelop. The determination of AEW was carried out using the following equation:

$$AEW = \frac{\sum(F_i \times \lambda_i)}{\sum F_i}$$

where F_i is the ANS fluorescence emission intensity at wavelength λ_i.

Intrinsic or tryptophan fluorescence spectroscopy

Intrinsic or tryptophan fluorescence intensities of HγD-crys sample solutions at 0.1 mg/mL were recorded with a Cary Eclipse fluorescence spectrophotometer (Varian, USA) using a quartz cuvette with a path length of 1 cm. The spectra between 300 and 400 nm were recorded upon exciting the samples at 280 nm (for intrinsic fluorescence) or 295 nm (for tryptophan fluorescence). The excitation and emission slits were both set to 5 nm.

Thermally induced equilibrium denaturation

Evaluation of HγD-crys's thermal stability was accomplished by heating HγD-crys samples under different conditions. The thermally induced unfolding transition of the protein samples

(0.1 mg/mL) was determined by monitoring the changes in intrinsic fluorescence emission over the range of 15–100°C with a heating rate of 1°C/min and an equilibrium time of 1 min. Intrinsic fluorescence spectra were recorded every 2°C between 300 and 420 nm at the excitation wavelength of 280 nm on a Cary Eclipse fluorescence spectrophotometer (Varian, USA). Given that AEW is more sensitive than total fluorescence intensity in characterizing the structural change of local environment [32], the average emission wavelength (AEW) instead of the total fluorescence intensity of HγD-crys intrinsic fluorescence spectra was used as the key variable.

Dynamic light scattering (DLS)

DLS experiments were used to characterize the size distribution of HγD-crys samples. HγD-crys samples were poured into small-volume (4 mL) disposable cuvettes with a 1 cm light path. DLS measurements were carried out using a Zetasizer Nano-ZS (Malvern Instruments, U.K.) with the appropriate settings of viscosity and refractive index at 0.89 centipoises and 1.59, respectively. Samples were illuminated with a laser at the wavelength of 633 nm. The DLS intensities of samples at a 173° scattering angle in kilo counts per second were collected for 20 runs with 20-s duration each run and then averaged. The collected data were analyzed to obtain the size distributions using the Non-negative Least Squares (NNLS) method.

Bioinformatics prediction of potential protein-protein interaction sites

The Protein–Protein Interface Prediction (PPIPRED) server (http://bmbpcu36.leeds.ac.uk/ppi_pred/) [33] was used to predict the potential protein-protein interaction sites on the HγD-crys structure. The application uses support vector machine to train datasets of proteins with known binding sites, then cross reference the results with surface patch analysis to predict protein-protein binding sites base on criteria, such as surface topography, sequence conservation, electrostatic potential, hydrophobicity, residue interface propensity, solvent accessible surface area. PPIPRED scores potential interaction sites on three levels: the most probable interaction sites (highlighted in red), the next most likely sites (in yellow) and the third most likely sites (in green). With over 60 citations, its performance measure has been tested on large number of datasets throughout the years.

Molecular dynamics (MD) simulations and analyses

3D coordinate file of HγD-crys (PDB code: 1HK0) was used as a starting structure for the MD simulation runs performed with GROMACS v. 4.5.3 software [34,35]. Simulations were carried out under the isothermal-isobaric (NPT) ensemble with a set pressure of 1 bar. Particle Mesh Ewald (PME) method was used to account for long-range electrostatic interactions [36]. Radius cutoff of 1.4 nm was used for Lennard-Jones interactions. Bond lengths were constrained with the LINCS algorithm [37]. Each protein was solvated in a 7 nm×7 nm×7 nm cubic box with spc model under periodic boundary conditions. HγD-crys in pH 2 condition has a total charge of +1, while neutrally charged in pH 7; each system was neutralized by replacing water molecules with sodium counterions that simulated the experimental setting (see under **Preparation of HγD-crys sample solutions and determination of protein concentration** for detail). Simulations were performed using GROMOS energy function (GROMOS 96 45a3 force field) with 2 fs time steps [38]. At least eight separate simulation runs of 150 ns each were obtained through 75 million simulation steps.

HγD-crys in pH 7 and pH 2 were simulated under various temperature settings ranging from 310 K~425 K with a total simulation time of more than 1 μm. Simulations under various temperature settings were performed to get a general picture of the trend in the protein structural changes and to find the optimal temperature condition in which molecular insights can be gained. As the overall character and order of events in protein unfolding process are known to be conserved across temperatures [39], we mainly present the results obtained from the 343 K trajectory as it is the temperature that is closest to the experimental condition used (55°C) and allows for detailed observable conformational changes in the time scale used in our study.

Graphical visualization was performed with Discovery Studio 3.5 visualizer (Accelrys Inc., San Diego, CA). Analyses of backbone root-mean-square deviation (RMSD) and secondary structure based on DSSP were performed for all trajectories to examine the conformational changes that occur between the different pH settings. Structures averaged throughout simulation time was calculated in reference to the initial structure in RMSD, while secondary structure was monitored according to the criteria of Kabsch and Sander [40].

Statistical analysis

All data are expressed as means ± standard deviations (S.D.) of n independent determinations. If X_i refers to the individual data points, M is the mean, then standard deviation (SD) can be calculated by the following formula:

$$SD = \sqrt{\frac{\sum_{i=1}^{n}(X_i - M)^2}{n-1}}$$

The standard deviations, which describe the typical average difference between the data points and their mean, were used as the error bars shown in the figure. Specific n values (n≥5) are reported in the figure legends. The significance of the results was determined with one tailed Student's t-test assuming unequal variances given n independent measurements. Unless otherwise noted, significance was determined as p<0.01. The statistical analyses were conducted using Excel or KaleidaGraph software.

Results

Effects of incubation temperature and pH on the aggregation of HγD-crys as revealed by turbidity measurement

The extent of HγD-crys aggregation as a function of incubation time at various pH and temperatures was evaluated by measuring the turbidity of HγD-crys samples. We demonstrate in Figure 1 that HγD-crys sample showed no significant change in the absorbance at 360 nm under the physiological condition (37°C, neutral pH) spanning over 2 days, suggesting that aggregated species were not produced. However, when the incubation pH dropped (from 7.0 to 2.0) or temperature elevated (from 37°C to 55°C), the turbidity of HγD-crys samples dramatically increased with prolonging incubation time. For example, at the onset of incubation, the absorbance at 360 nm of 1 mg/mL HγD-crys was found to be ~0.047, ~0.048, or ~0.041 at pH 7.0 and 55°C, pH 2.0 and 37°C, or pH 2.0 and 55°C, respectively, whereas the absorbance was raised to ~0.075, ~0.084, or ~0.087 at pH 7.0 and 55°C, pH 2.0 and 37°C, or pH 2.0 and 55°C, respectively, after incubation for 48 hr (see Figure 1). Our findings suggest that,

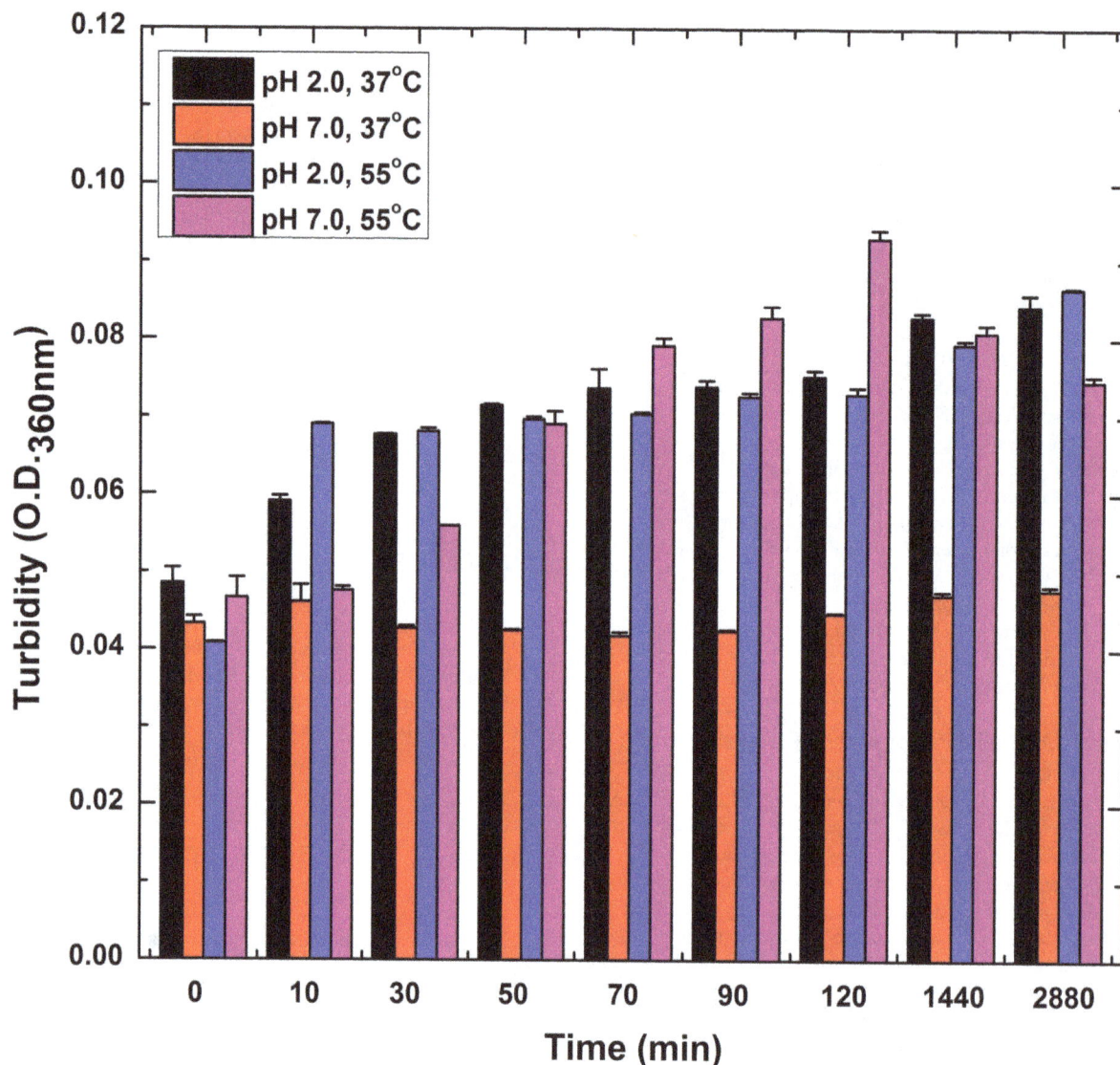

Figure 1. Turbidity measurement of the human γD-crystallin (HγD-crys) samples measured as a function of incubation time. Samples of HγD-crys at 1 mg/mL (45 μM) were incubated in different conditions (temperature = 37 or 55°C; pH = 2.0 or 7.0). The turbidity of the samples was evaluated by monitoring the absorbance of the sample solutions at the wavelength of 360 nm. After 2 days of incubation, aggregates were formed in all samples except for the ones under pH 7.0 and 37°C. The means ± standard deviations (S.D.) of at least 5 independent measurements (n≥5) are presented in the figure. The error bars used in the figure are the standard deviations of the data obtained from all independent measurements. The values of standard deviations were calculated by the formula listed in the Statistical Analysis of Materials and Methods section.

when the incubation temperature increases and/or pH drops from the physiological condition (pH 7.0 and 37°C), prolonged incubation leads to HγD-crys aggregation.

Effects of incubation temperature and pH on the formation of HγD-crys amyloid fibrils as revealed by ThT fluorescence spectroscopy

Our turbidity results clearly showed the formation of HγD-crys aggregates at acidic pH and/or high temperature. To further explore if the aggregated species are of the unordered amorphous aggregates or ordered amyloid fibril type, we monitored the changes of ThT fluorescence emissions of HγD-crys samples under different incubation conditions. ThT, a standard fluorescent dye that exhibits an increase in fluorescence intensity upon binding to amyloid structures, was introduced as a molecular probe to detect the formation of amyloid fibrils. While the exact

mode of binding is not completely clear, ThT is believed to bind to the grooves formed on the surface of amyloid fibrils by aligned rows of side chains [41]. Figure 2 depicts the time evolved ThT fluorescence intensity of HγD-crys samples under different incubation conditions. As shown in the figure, all HγD-crys samples exhibited comparable fluorescence intensity at the onset of incubation (t = 0). It is evident that the ThT fluorescence emission intensity of HγD-crys sample at pH 2.0 significantly increased in 10 min and reached a plateau thereafter, thus indicating the presence of amyloid structure when HγD-crys was incubated under the condition of pH 2.0 and 37°C. A similar trend in the profile of ThT fluorescence emission versus incubation period was observed in the samples upon incubation at a higher temperature 55°C, although with a faster growth rate during the initial incubation period. On the contrary, regardless of the incubation temperature used, almost no increase in ThT fluorescence signal was noted at pH 7.0 as well as for the dissolving solvent alone under the assay conditions (data not shown). Our results indicate that the observed elevation in emitted ThT fluorescence is associated with acidic conditions. In addition, HγD-crys sample incubated at pH 2.0 and 55°C displayed the highest ThT fluorescence emission intensity among all groups tested. Assuming that increased ThT fluorescence is correlated with the formation of amyloid fibril, our ThT fluorescence results would strongly suggest that low pH induces fibrillogenesis of HγD-crys.

Morphological characterization of HγD-crys samples under different incubation conditions as revealed by transmission electron microscopy (TEM)

Our preceding results suggest that the ThT fluorescence emission of HγD-crys sample under pH 2.0 was markedly

increased, which is a positive indication of amyloid fibrillogenesis. Given that a number of factors have been reported to interfere with ThT fluorescence [42], it would be reckless to conclude that HγD-crys has a *bona fide* amyloid fibril-forming ability in acidic condition soley based on the observed ThT fluorescence enhancement. Therefore, transmission electron microscopy (TEM) was performed to provide further supporting evidence for the acid-induced HγD-crys fibrillogenesis. Presented in Figure 3 are representative micrographs of HγD-crys samples at pH 2.0/37°C, pH 2.0/55°C, pH 7.0/37°C, and pH 7.0/55°C. It is evident that, after 1 hr-incubation at pH 2.0 and 37°C, HγD-crys sample showed the morphological features of typical un-branched amyloid fibrils with approximately 10 nm in diameter and several μm in length, as seen in Figure 3A. In addition, TEM visualization of the HγD-crys sample taken from the pH 2.0 and 55°C condition revealed that a greater amount of fibrillar species

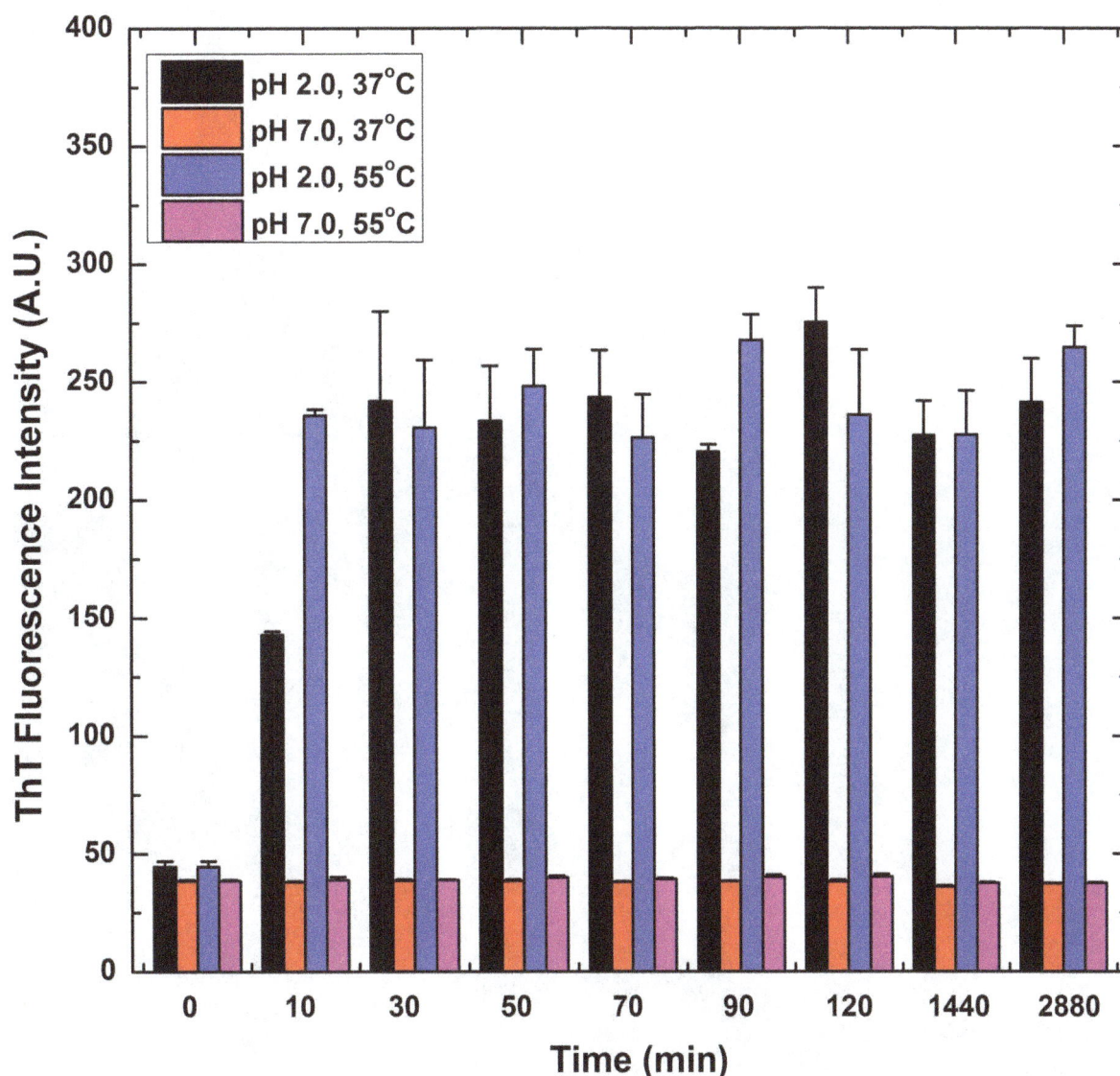

Figure 2. ThT fluorescence emission intensity measurement of HγD-crys sample measured as a function of incubation time. Samples of HγD-crys at 1 mg/mL (45 μM) were incubated in different conditions (temperature = 37 or 55°C; pH = 2.0 or 7.0). Regardless of temperature, the ThT fluorescence emission intensity was observed to increase for the samples incubated in pH 2.0, but remained the same for the ones in pH 7.0. Data points are presented as the means ± standard deviations (S.D.) of at least 5 independent measurements (n≥5) in the figure. The error bars shown in the figure are the standard deviations of the data obtained from all independent measurements. The values of standard deviations were calculated by the formula listed in the Statistical Analysis of Materials and Methods section.

was observed relative to that obtained at pH 2.0 and 37°C (see Figure 3B). In contrast, while a variety of granular species (~25–50 nm in diameter) appeared in the sample incubated at higher temperature of 55°C and pH 7.0, no fibrillar species were detected in the samples of HγD-crys incubated under neutral pH (shown in Figures 3C and 3D). Therefore, the TEM analysis reproducibly demonstrates a positive connection between the ThT fluorescence emission results and the amount of fibrils observed.

Effects of incubation temperature and pH on the tertiary structure of HγD-crys samples as revealed by ANS fluorescence spectroscopy and tryptophan/intrinsic fluorescence spectroscopy

To gain insights into the effects of incubation temperature and pH on the conformational changes of HγD-crys, ANS and tryptophan/intrinsic fluorescence spectra of HγD-crys were also recorded. We recorded the time evolution of ANS fluorescence emission at the average emission wavelength upon excitation at 380 nm. The hydrophobic fluorescent dye, ANS, has been commonly utilized to demonstrate the presence of partially folded conformations of globular proteins and probe for structural properties and solvent exposure of the hydrophobic surfaces [43–45]. The preferential binding of ANS to hydrophobic clusters

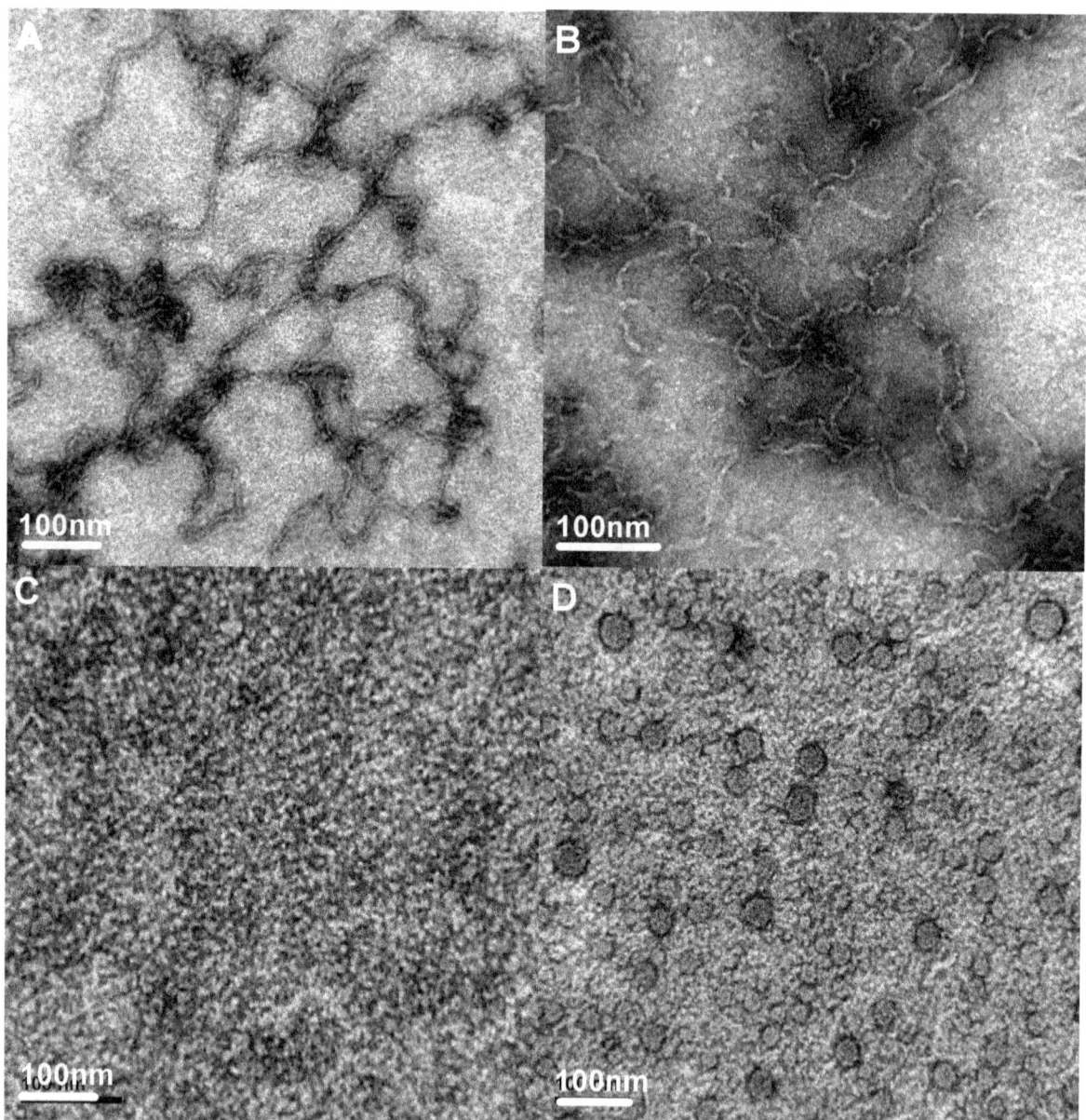

Figure 3. Representative negative staining transmission electron micrographs (TEM) of HγD-crys samples under different incubation conditions. (A) HγD-crys incubated at pH 2.0, 37°C for 1 hr; (B) HγD-crys incubated at pH 2.0, 55°C for 1 hr; (C) HγD-crys incubated at pH 7.0, 37°C for 1 hr; and (D) HγD-crys incubated at pH 7.0, 55°C for 1 hr. Aggregation formed under pH 2.0 conditions resembles fibril morphology, while retaining granular appearance under pH 7.0 conditions. The scale bar represents 100 nm.

gives rise to an enhancement in fluorescence emission accompanying a blue shift of the spectral maximum or average emission wavelength [43,44,46]. We show in Figures 4A and 4B that, under the condition of pH 7.0 (37 or 55°C), both average emission wavelength (AEW) and ANS fluorescence intensity remained almost unchanged during the 2 days of incubation. Furthermore, as seen in Figure 4B, the ANS fluorescence intensities at the average emission wavelength (or surface hydrophobicity) of HγD-crys samples were remarkably low, signifying that protein hydrophobic sites are hidden inside the compactly folded protein structure. However, incubating HγD-crys at pH 2.0 led to a drastic blue-shift in the average emission wavelength (AEW) and a pronounced enhancement in ANS fluorescence emission. These changes suggest that more hydrophobic regions were solvent-exposed in HγD-crys under the acidic conditions, probably due to conformational changes in the protein leading to a partial loss of tertiary structure.

Tryptophan fluorescence of protein, mainly due to high sensitivity of tryptophan to changes in its microenvironment, can be measured when samples are excited at 295 nm. This method has been widely employed in studies involving ligand binding, folding-unfolding, and protein conformational changes [47–49]. To further gain insights into the differences in conformation of HγD-crys incubated under various conditions, the intrinsic fluorescence spectra of samples (measured under excitation wavelength of 280 nm) were also recorded. We show in Figure 5A that there was almost no change in the tryptophan fluorescence spectra of the HγD-crys samples incubated under neutral pH and 37°C condition for 2 days. However, for samples under the same pH condition but incubation at 55°C (Figure 5C), a noticeable increase in the fluorescence emission and wavelengths of emission maximum were observed after 1 day of incubation time. As for the samples incubated in pH 2 conditions (Figures 5B and 5D), a much greater increase in the maximum tryptophan fluorescence intensity was observed in comparison to the samples incubated in neutral pH. This change occurred as early as 30 minutes into the incubation time regardless of temperature. It is interesting to note that under pH 2 and 55°C, the peak intensity begins to diminish slightly with longer incubation time past 30 minutes. We speculate that this phenomenon may be due to the rapid formation of larger aggregates in this high temperature that eventually precipitated out of the solution phase, thereby reducing the amount of samples that could be measured. In addition to a pronounced increase in fluorescence emission, prolonged incubation at pH 2.0 led to a considerable red-shift in the wavelengths of emission maximum (λ_{max}) (i.e., λ_{max} was found to increase from ~328 nm at 0 hr to ~349 nm at 2 day). This clearly indicates an extensive exposure of tryptophan residues to solvent upon incubation under the acidic condition [50]. Likewise, our HγD-crys samples excited at 280 nm revealed similar trends in the fluorescence spectra (shown in Figure S1A–D) as the ones obtained from tryptophan fluorescence spectroscopy with excitation wavelength at 295 nm (shown in Figure 5A–D).

From our ANS fluorescence and intrinsic/tryptophan fluorescence findings, we can conclude that tryptophans and, in general, hydrophobic regions are greatly solvent-exposed in the HγD-crys samples incubated at pH 2.0, clearly demonstrating that the conformation (or tertiary structure) of native HγD-crys was markedly affected by the decrease in pH [51].

Effects of incubation temperature and pH on the secondary structure of HγD-crys samples as revealed by circular dichroism (CD) spectroscopy

To further understand the role that incubation temperature and pH play in the secondary structural changes of HγD-crys, the far-UV circular dichroism spectra of HγD-crys samples under different conditions were monitored. At the beginning of incubation time, native HγD-crys sample shows a predominance of β-sheet secondary structure as manifested by the far-UV CD spectrum (shown in Figures 6A and 6B). The absorption minimum is at ~218 nm and the overall profile is in agreement with previous published results [52,53]. Regardless of incubation period (0–2 day) and temperature, a negligible difference in the CD spectra of HγD-crys samples was perceived in pH 7.0 (Figure 6A) indicating that changes in secondary structure were insignificant. However, we show in Figure 6B that, when the incubation pH dropped to 2.0, HγD-crys samples exhibited a structural transition, resulting in a prominent alteration in the relative secondary structural proportions. Regardless of the incubation temperature used, the far-UV CD spectra at pH 2.0 displayed a substantially different shape in which a shift in absorption minimum from ~218 nm to ~207 nm was observed along with a pronounced increase in signal. From 2 hr of incubation and on, the intensity of the absorption minimum remained consistent indicating that the changes in secondary structure were insignificant. To better quantify the structural transition, the far-UV CD spectra of all HγD-crys samples obtained at 0 and 2 hr were further de-convoluted using the software available from the DICROWEB website [30] and the results of secondary structure content of HγD-crys samples incubated at different pH values for 2 hr are listed as follows: (1) at pH 2.0: 7% α-helix, 32% β-sheet, 17% turns, and 44% unordered; (2) at pH 7.0: 4% α-helix, 43% β-sheet, 22% turns, and 31% unordered.

Effects of incubation temperature and pH on the size distribution of HγD-crys samples as revealed by dynamic light scattering (DLS)

We determined the size distribution of the HγD-crys samples under different incubation conditions using dynamic light scattering (DLS) and the results are depicted in Figures S1A–S1F. We demonstrate in Figure S2A and S2D that, regardless of the incubation pH used, the distribution of species with similar sizes were observed at the beginning of incubation. However, under the same incubation temperature (37 or 55°C), HγD-crys samples that were subjected to the conditions of pH 2.0 and pH 7.0 displayed considerably different size distributions after two days of incubation, as shown in Figures S2B–C and S2E–F. Evidently, aggregation of HγD-crys in acidic conditions at 55°C seemed to yield a single broader population of species existing in a range of sizes between ~10 and ~200 nm. In contrast, a bimodal distribution of population size (with peaks positioned at ~30 nm and ~200 nm) was detected in HγD-crys samples upon incubation at pH 7.0 and 55°C.

Thermal denaturation behaviors of HγD-crys samples under different incubation conditions

To explore the effects of incubation condition (e.g., pH and temperature) on the conformational stability of HγD-crys, thermally induced equilibrium unfolding of the HγD-crys samples under different incubation conditions was investigated. Given that the average emission wavelength (AEW) is a more sensitive probe for structural changes of local environment than the total intrinsic fluorescence, we monitored the AEW of recorded intrinsic

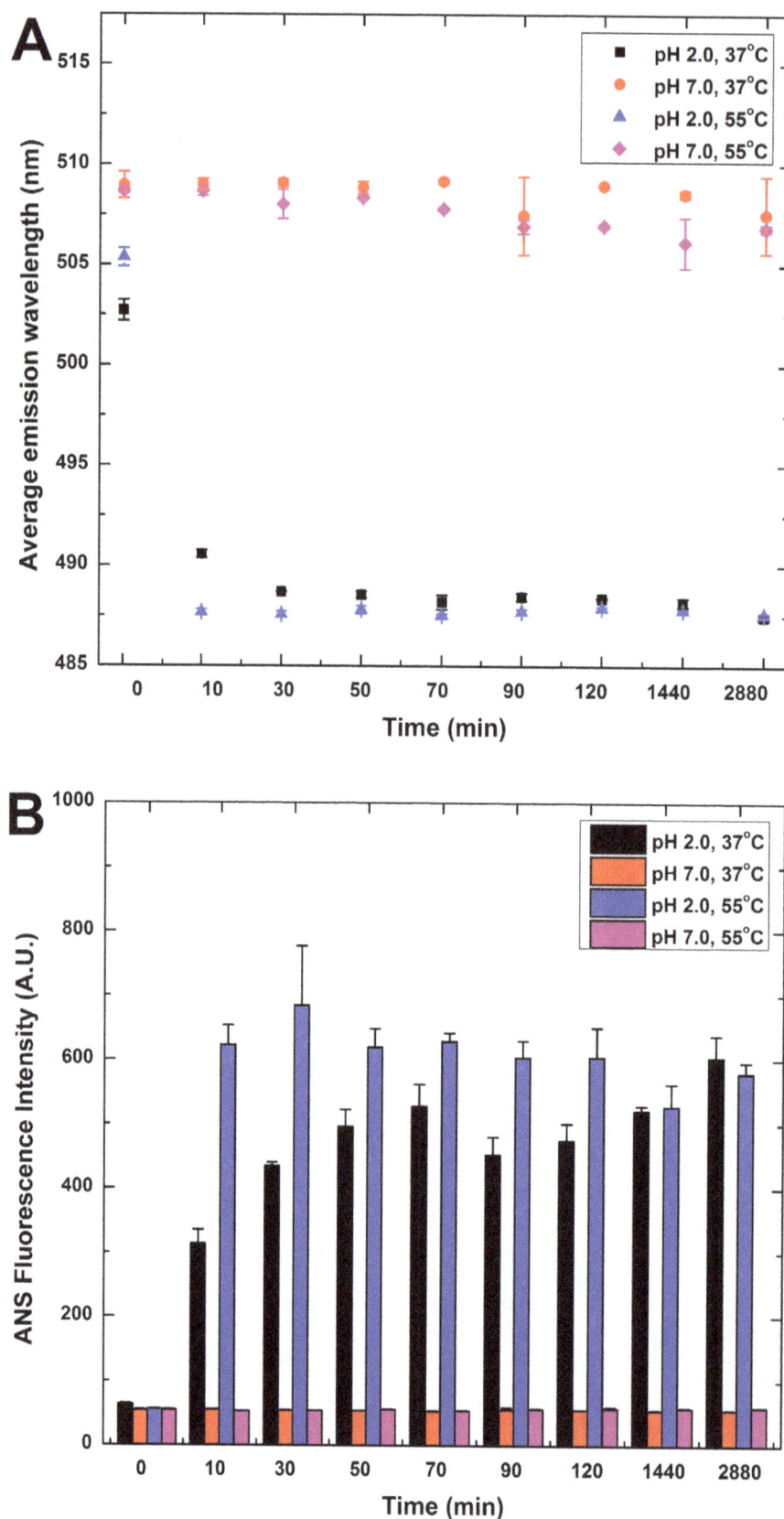

Figure 4. ANS fluorescence intensity measurement of HγD-crys samples. Surface hydrophobicity of HγD-crys samples were monitored by (A) Average emission wavelength (AEW) of HγD-crys sample as a function of incubation time and (B) ANS fluorescence emission intensity of HγD-crys sample as a function of incubation time. Samples were incubated under temperatures of 37 and 55°C; pH settings of 2.0 and 7.0. Incubation under pH 2.0 settings resulted in a blue-shift in AEW and an increase in ANS fluorescence emission not detected in samples under pH 7.0. Data points are

presented as the means ± standard deviations (S.D.) of at least 5 independent measurements (n≥5) in the figure. The error bars used in the figure are the standard deviations of the data obtained from all independent measurements. The values of standard deviations were calculated by the formula listed in the Statistical Analysis of Materials and Methods section.

fluorescence spectra as a function of temperature to obtain the thermal unfolding profiles/curves for the HγD-crys samples. Our spectral results revealed the temperature-induced structural unfolding of HγD-crys as demonstrated in Figure S3. The sigmoidal dependence of AEW with temperature was observed in all the HγD-crys samples. Our data evidently suggested that the thermally induced denaturation/unfolding from the folded to denatured/unfolded state of HγD-crys under the condition of pH 2.0 or 7.0 can be adequately described by a simple two-state model/process with cooperative characteristics. Comparison of the AEW-versus-temperature curves of HγD-crys samples indicated that the unfolding/denaturation curve shifted toward left when the incubation pH dropped from 7.0 to 2.0, suggesting that

the HγD-crys sample at pH 2.0 exhibits lower thermal stability than that at pH7.0 [54].

Potential mechanism of HγD-crys fibrillogenesis as revealed by molecular dynamics simulations

HγD-crys is highly stable at neutral pH due to its unique structural arrangement consisting of two domains (N-terminal and C-terminal) each harboring two Greek key anti-parallel β-sheet motifs (shown in Figure S4). The two structurally similar domain pairs with high internal symmetry of primary and tertiary structures are believed to have been form during the course of evolution by gene duplication and fusion [55,56]. Such structural

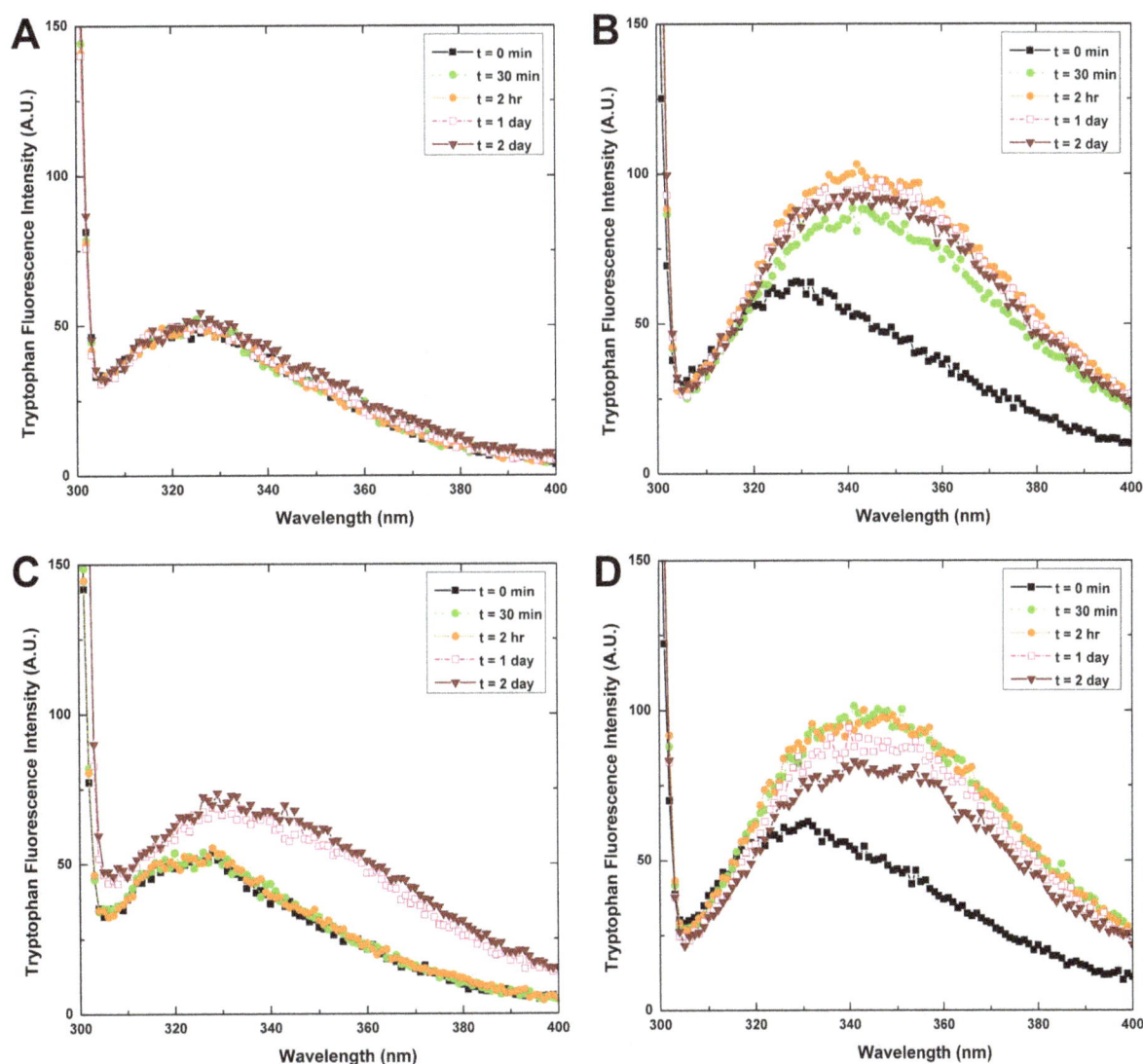

Figure 5. Tryptophan fluorescence intensity measurement of HγD-crys samples (0.1 mg/mL) under different incubation conditions. (A) HγD-crys incubated at pH 7.0, 37°C; (B) HγD-crys incubated at pH 2.0, 37°C; (C) HγD-crys incubated at pH 7.0, 55°C; and (D) HγD-crys incubated at pH 2.0, 55°C. The fluorescence spectra between 300 and 400 nm were recorded upon exciting the samples at 295 nm. Incubation under pH 2.0 settings resulted in a rapid and greater increase in fluorescence emission, as well as a more noticeable red-shift in the wavelengths of emission maximum than those of the samples under pH 7.0.

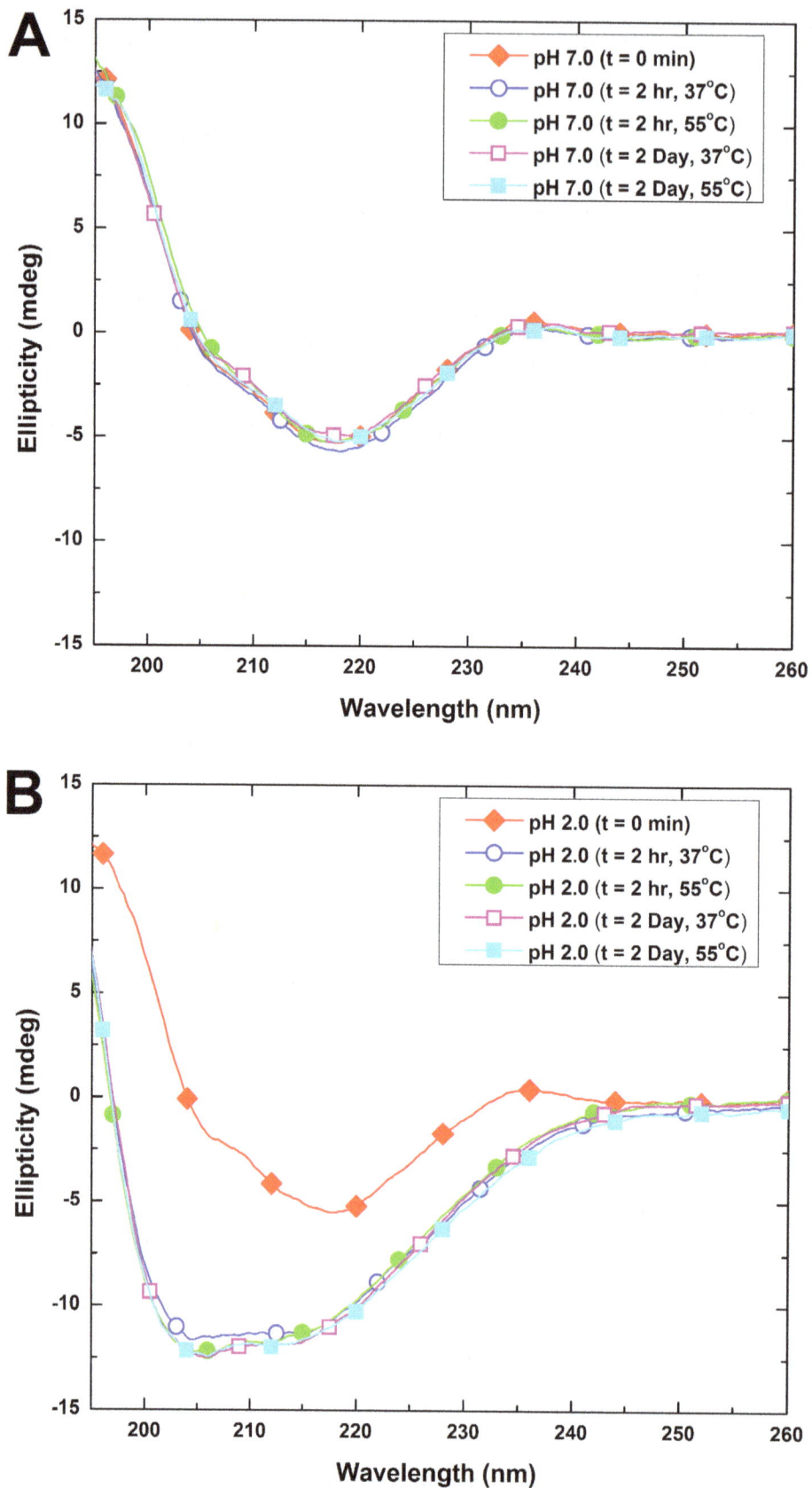

Figure 6. Far-UV CD spectra of HγD-crys samples under different incubation conditions. (A) HγD-crys incubated at pH 7.0 and (B) HγD-crys incubated at pH 2.0. The incubation temperatures used were 37 and 55°C. The incubation periods were 0, 2 hr, and 2 day. Incubation under pH 2.0 settings resulted in a large shift in absorption minimum and an increase in signal intensity not seen in samples under pH 7.

fold is highly stable and has been found repeatedly in a wide variety of β-sheet-rich proteins [57]. An important contribution to the stability of the HγD-crys monomer is the hydrophobic interface between the N-terminal domain (Ntd) motif 2 and C-terminal domain (Ctd) motif 4, which has been shown to play a crucial role in holding the two domains intact [58–61].

We first verify the accuracy of the PPIPRED tool by predicting the potential protein-protein interaction sites on the HγD-crys structure. In agreement with the previous studies [62–65], residues around and at the domain interface region of Ntd motif 2 and Ctd motif 4 (shown in Figure 7) were identified as the most probable sites for protein-protein interaction based on the six properties listed in the methods section. The second most probable sites were found, for the most part, in motifs 2 and 4 (regions surrounding the interdomain interface) as well as part of Ntd motif 1. The third most probable sites are located predominantly in Ctd motif 3.

To gain a better understanding of the structural changes that occur at low pH and high temperature, we performed MD simulations at both low and neutral pH (pH 2.0 and 7.0). Under physiological conditions (pH 7.0 and 310 K), the RMSD values remained stable at around 0.17 nm (shown in Figure 8A). As for the pH 2.0 condition, the RMSD began to rise around 70 ns, reaching a plateau of about 0.33 nm at 100 ns. Although the RMSD values increased for both pH conditions at the temperature of 343 K (shown in Figure 8B), they reached a plateau around 0.25 nm at pH 7.0 while peaking and remaining equilibrated at 0.37 nm starting from an earlier time frame

~55 ns at pH 2.0. These results show that our MD simulations follow a general trend found in our experimental results, whereby HγD-crys undergoes greater conformational change at pH 2.0 than at pH 7.0, and that this change occurs more rapidly at higher temperature.

We, then, examined the secondary structural changes in the two different pH and temperature settings. In 310 K condition, HγD-crys under both pH settings exhibited no significant structural changes, except for some transitions from helix to turn and vice versa in certain loop regions (Figures S5A–B). This is further evidenced when we superimposed the three dimensional structures of the HγD-crys every 10 ns to get a better picture of how the secondary structures change overall throughout the simulation time (Figures 9A and 9B). As can be seen, the conformations under pH 7.0 condition did not deviate greatly from the original structure at the beginning of the simulation time. Although at pH 2.0, slightly more deviation from the original structure was perceived throughout the duration of the simulation, the overall predominant secondary β-structure of the Greek key motif was still retained. The minute changes in secondary structures detected under physiological temperature for both pH settings reinforces the concept that HγD-crys has a highly stable conformation and the structure remained intact within the 150 ns of simulation time, even under low pH 2.0 setting. This is consistent with the results from past biophysical denaturation/unfolding study showing that γD-crys is resistant up to one week of incubation in acidic pH under physiological temperature [10]. However, heating acceler-

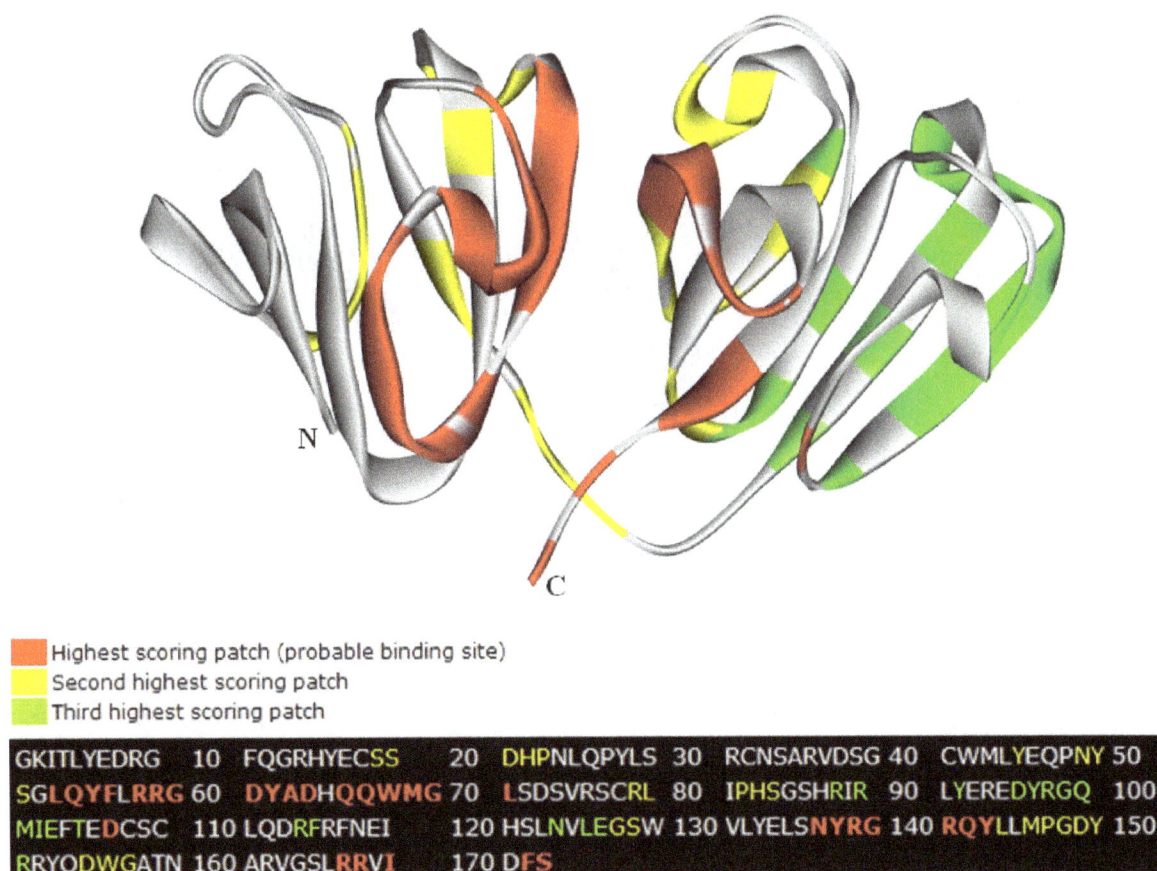

■ Highest scoring patch (probable binding site)
■ Second highest scoring patch
■ Third highest scoring patch

GKITLYEDRG	10	FQGRHYECSS	20	DHPNLQPYLS	30	RCNSARVDSG	40	CWMLYEQPNY	50
SGLQYFLRRG	60	DYADHQQWMG	70	LSDSVRSCRL	80	IPHSGSHRIR	90	LYEREDYRGQ	100
MIEFTEDCSC	110	LQDRFRFNEI	120	HSLNVLEGSW	130	VLYELSNYRG	140	RQYLLMPGDY	150
RRYQDWGATN	160	ARVGSLRRVI	170	DFS							

Figure 7. Potential protein-protein interaction sites predicted by PPIPRED based on surface topography, sequence conservation, electrostatical potential, hydrophobicity, residue interface propensity, and solvent accessible surface area. The three highest probable binding sites are color-coded red, yellow, and green. The regions are mapped onto the 3D-structure and residue sequence of HγD-crys.

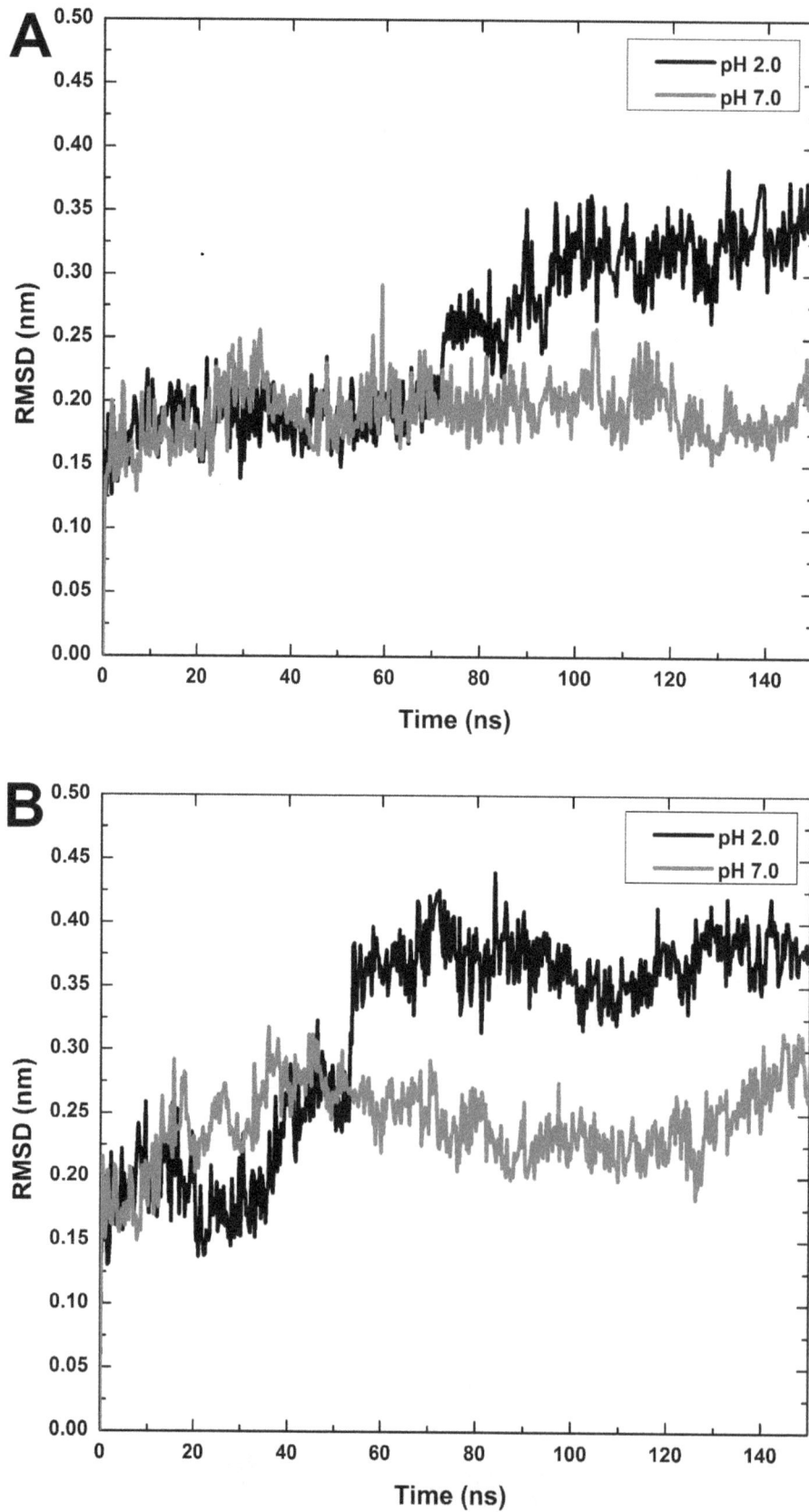

Figure 8. Root mean square deviation (RMSD) of HγD-crys in pH 7 and 2 under (A) 310 K and (B) 343 K as a function of 150 ns simulation time. RMSD values of HγD-crys increased up to the ranges of 0.35~0.40 nm for HγD-crys under pH 2 settings and remained within the ranges of 0.2~0.3 nm under pH 7 settings.

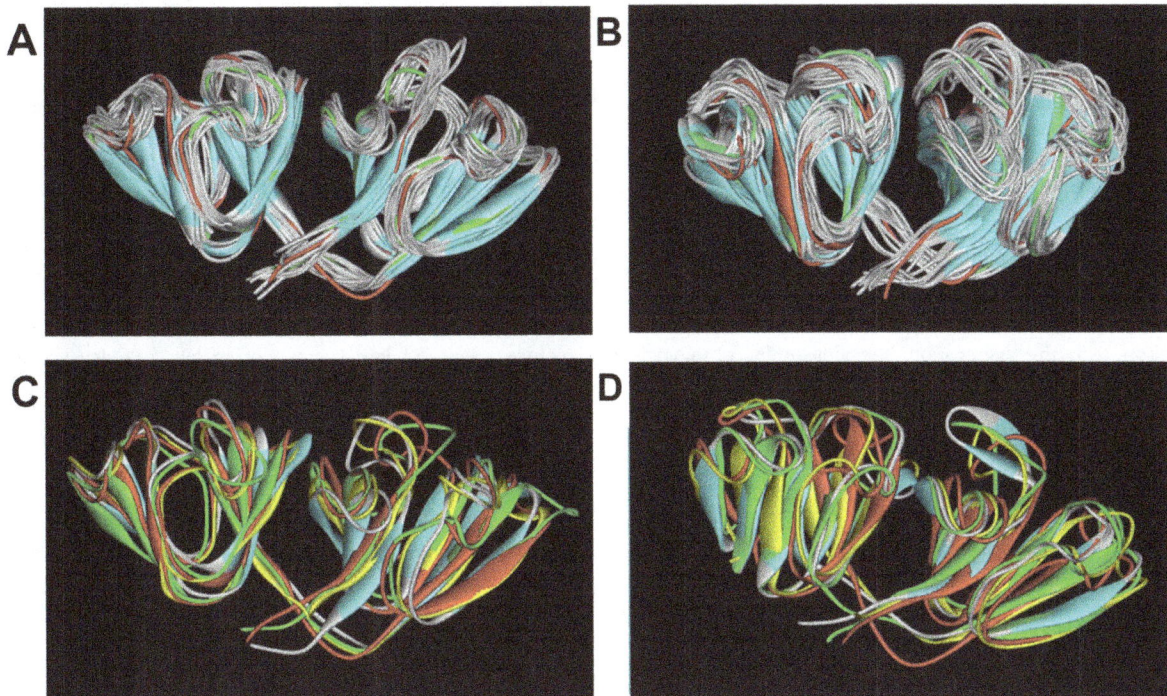

Figure 9. Superimposed HγD-crys structures throughout 150 ns simulation time. Snapshots of structures in 310 K, (A) pH 7 and (B) pH 2, respectively. Frames are taken every 10 ns with the starting structure displayed in red and the last structure in green. Snapshots of structures in 343 K, (C) pH 7 and (D) pH 2, respectively. Frames are taken every 50 ns with the starting structure displayed in red, 50 ns in yellow, 100 ns in green, and 150 ns in cyan. Structures in pH 7 and 2 under 310 K show no noticeable changes. Greater structural changes are observed in 343 K, with the appearance of antiparallel β-strands spanning residues 153–160 in the last 30 ns of simulation for HγD-crys in pH 2 condition.

ates the formation of amyloid fibril, as seen in other experiments as well as ours [10,61]. We, therefore, also raised the temperature of our simulation to speed up the process of protein conformational change under the two pH settings examined.

When the temperature was raised to 343 K, most of the β-structure was still retained under pH 7.0 condition (see Figures S5C and 9C), but underwent greater structural deviation in pH 2.0 (see Figures S4D and 9D). Specifically, β-structures spanning residues 43–53 began to undergo a change in conformation, in which parts of the β-strands and turn within the region became bend structure starting around 40 ns of simulation time. In addition, the region spanning residues 144–160 (that was predominantly helical) became turns/bends and developed a region of antiparallel β-strands corresponding to residues 153–160 in the last 30 ns seconds of simulation time (depicted in Figure S5D). This β-structure development from α-helix is clearly seen in the conformation acquired from the last nanosecond of simulation time (shown in Figure 9D, structure in cyan). Incidentally, residues spanning 155−157 within this segment were also previously identified by PPIPRED as one of the second most likely potential sites of protein-protein interaction next to the domain-domain interface that was predicted as the most probable site.

Discussion

Various investigations have indicated that amyloid fibrillogenesis is a process that can occur in crystallin proteins. For instance, the presence of amyloid fibrils has been noted in the interior fiber cells of normal murine lenses through the binding and/or staining of amyloidophilic dyes, Congo red and Thioflavin T [66]. These fibrils were identified to be of γ-crystallin origin. Sandilands *et al.*

(2002) have reported that an inherited γ-crystallin mutant caused early onset murine cataract in mice and formed inclusions containing filamentous material in the lens that displayed high Congo red binding affinity, which is indicative of the presence of amyloid fibrillar species [67]. The same group also found that the purified truncated protein (a mutant of murine γB-crystallin with its fourth Greek key motif absent) exhibited amyloid fibril-forming propensity *in vitro* [67]. Using the wild-type bovine α-, β-, and γ-crystallins as the models, Meehan and coworkers experimentally examined the in vitro fibrillogenic/aggregative properties of these proteins upon exposure to denaturing/destabilizing conditions. They observed that the full-length bovine α- and β-crystallins assembled to form amyloid fibrils, whereas fibrils derived from bovine γ-crystallin proteins were composed of both full-length and fragmented bovine γ-crystallin proteins [10]. In addition, it has been shown that mutant crystallin protein has a higher propensity for amyloid fibril formation than its wild-type counterpart [68]. Further experimental evidence originating from recent studies using individual γ-crystallins and/or their subunits has demonstrated that HγD-crys and human γC-crystallin (HγC-crys) are prone to fibrillate in the acidic environment [18]. The fact that HγD-crys proteins are capable of forming amyloid fibrils both *in vivo* and *in vitro* may imply a potential role in the pathogenesis of cataract.

Based on one theory, amyloid fibril formation of HγD-crys under acidic pH is highly relevant to the low pH lysosomal compartment of differentiating lens fiber cells where damaged HγD-crys are degraded [18]. Under such condition, the damaged proteins, which may have the potential of being amyloidogenic partially-unfolded intermediates, can serve as nucleating centers for fibril formation. This phenomenon has also been described for other proteins such as transtherytin [69]. Thus, the process initiates the early stages of cataract by disrupting the fiber cell organization causing light scatter and image degradation at the retina. It may take decades for this process to begin and develop into cataract on the macroscopic scale, thus in order to achieve a more thorough understanding of how HγD-crys behaves under the different aggregated states relevant to cataractogenesis, pH 2 was chosen as the setting to procure the fibril form of aggregation. Although the physiological pH within the cell (even within the lysosomal compartment) does not go as low as pH 2, this pH was chosen because we found it to be the optimal pH condition that will promote fibrillogenesis of HγD-crys, thus allowing for a full comparison with the aggregation behavior under physiological pH. After testing out various pH settings (data not shown), we found that pH 2 procures the most favorable condition for amyloidogenesis of HγD-crys; thus, providing the basis for our subsequent experiments.

HγD-crys with 6×His tag was purified and used in this study. Although there may be a concern that the fusion partner may interfere with the outcome of the experiments performed, it was previously reported that the presence of 6×His tag on recombinant HγD-crys neither affect the expression of the protein itself nor the ability of the protein to fold into the native state during the purification process. Moreover, analyses using fluorescence and circular dichroism spectroscopies revealed almost no difference in the structures between HγD-crys with and without 6×His tags [23]. We also conducted experiments, in which the N-terminal 6×His tag was first removed using dipeptidyl aminopeptidase I (DAPase-I), the enzyme that was reportedly utilized to cleave His-tag from recombinant proteins [70,71] The untagged HγD-crys was then purified. The structural features of the resulting untagged HγD-crys and its 6×His-tagged counterpart were further compared. Our results showed that both the His-tagged and untagged HγD-crys gave almost identical CD spectra (data not shown) and intrinsic fluorescence spectra (data not shown), suggesting that the inclusion of 6×His tag has negligible influence on the secondary and tertiary structures of HγD-crys [22], which further supports the previous report.

Due to the highly stable nature of the Greek key motifs in HγD-crys, it may take a long time for the protein to form aggregation if left alone under physiological condition (pH 7.0 and 37°C). Therefore, in order to better understand and compare the differences in aggregation behaviors of HγD-crys under varied pH in a reasonable time frame, we sped up the process by thermally inducing aggregation at the temperature of 55°C. Although we could have arbitrarily picked any temperature that is higher than the physiological 37°C for our comparison, we settled on 55°C because this was the temperature that allows the extent of aggregation in HγD-crys under both pH conditions to reach similar states within the 2-day time period tested (see Figure 1), thus enabling us to perform a thorough comparison of the effects of varied environment on the structural changes/aggregation behavior of HγD-crys protein. As seen in our TEM results, elevating the incubation temperature did not alter the aggregation pathways that the proteins were predestined for in the different pH settings. Under neutral pH, HγD-crys forms granular aggregates of various sizes ranging from ~25–50 nm with hydrophobic sites buried within the tightly packed protein structures. Based on our turbidity measurements, the aggregation process is slower under neutral environment than acidic environment, again signifying the inherent stability of the native HγD-crys conformation in physiological pH. Whereas HγD-crys aggregation is of the amorphous kind in neutral pH, low pH induced amyloidosis with partially unfolded structure of increased solvent-exposed hydro-

phobic regions. This is reminiscent of the HγD-crys fibrous aggregates induced by guanidinium hydrochloride [17].

It is known that the bulk of protein material removed from cataractus lens is in the form of amorphous aggregates [72,73]. This may not come as a surprise when one considers that the lowest pH of 6.5 detected in the nucleus [20,72] is not too different from the neutral pH of 7.0 explored by our study, which also induces amorphous aggregation in HγD-crys. While our experimental condition is an over simplified model of what's going on in the lens, it is undeniable that the inherent aggregation property of HγD-crys under neutral pH is of the amorphous kind. However, as previously mentioned, fibrils that may form in the lysozomal compartments of lens cortex during lens cell development and maturation may eventually move into the lens nucleus if they are not completely degraded before the lens cells mature and become devoid of their intracellular organelles. It is possible that remnants of these fibrils in the cells become precursors that provide the seeds to initiate the formation of amorphous aggregates under close to neutral pH of the lens nucleus. As the line dividing amorphous and fibrous aggregation is a thin one [74], the possibility is there that one form may trigger the mass production of another when the condition is right.

It was found previously that, in certain cases, global structural unfolding is unnecessary for the initiation of protein aggregation [75–77]. Simply altering the environment, thus changing the solvent exposure of certain aggregation-prone regions in the native protein, is enough to allow protein-protein interactions to commence leading to massive aggregation formation [4,78]. Thus, localized regions can serve as nucleation sites for β-amyloid formation. Although the Greek key motifs in HγD-crys are predominantly β-structures, we hypothesize that a possible reorganization of localized region(s) leads to the formation of exposed β-structures capable of initiating aggregation. This is consistent with the findings of a recent study that examined the structure of acid-induced HγD-crys amyloid fibrils by two-dimensional (2D) IR spectroscopy. The results from the study showed that nucleation and extension of fibrils were mainly contributed by part of the protein's Ctd [19]. According to the structural model suggested by Moran *et.al.* (2012), each protein molecule provides two β-strands from the Ctd to form β-sheets of the amyloid fibril core, while rest of the protein's secondary structure assumes a disordered state of loops and coil. As the main structural fold of HγD-crys is in the form of four homologous Greek-key motifs comprised of antiparallel β-sheets, the native conformation is already dominated by β-structures. Therefore, in the process of amyloid fibrillogenesis, localized structural reorganization may require majority of the native β-sheets to unfold and allow certain regions to reform into β-sheets characteristics of amyloid fibrils, as suggested by the IR spectroscopy study. This is also consistent with our far-UV CD results showing an overall decrease in β-structures and increase in unordered structure in the process of HγD-crys fibril formation.

Despite the fact that previous 2D IR spectroscopy study has revealed that Ctd forms β-sheets of the acid-induced HγD-crys fibril, specific region(s) within the domain involved in the fibril formation is still unknown. We, therefore, performed bioinformatic prediction and MD simulations to give us further insights into this matter. From the results obtained through our computational approaches, we speculate that a region in Ctd motif 4 corresponding to residues 153−160, which forms antiparallel β-strands during our simulation under low pH setting, may play an important role in the initiation of amyloid fibril formation. Further analysis is warranted to understand how it contributes to the mechanism of acid-induced HγD-crys fibrillogenesis and its implication in the formation of age-related nuclear cataract.

Supporting Information

Figure S1 Intrinsic fluorescence intensity measurement of HγD-crys samples (0.1 mg/mL) under different incubation conditions. (A) HγD-crys incubated at pH 7.0, 37°C; (B) HγD-crys incubated at pH 2.0, 37°C; (C) HγD-crys incubated at pH 7.0, 55°C; and (D) HγD-crys incubated at pH 2.0, 55°C. The fluorescence spectra between 300 and 400 nm were recorded upon exciting the samples at 280 nm. Incubation under pH 2.0 settings resulted in a rapid and greater increase in fluorescence emission, as well as a more noticeable red-shift in the wavelengths of emission maximum than those of the samples under pH 7.0.

Figure S2 The effect of incubation condition on size distribution of human γD-crystallin (HγD-crys) samples. Samples of HγD-crys at 1 mg/mL (45 μM) were incubated in different conditions (temperature = 37 or 55°C; pH = 2.0 or 7.0). The turbidity of HγD-crys sample was evaluated using dynamic light scattering (DLS). (A) pH 7.0, t = 0 hr; (B) pH 7.0, 37°C, t = 2 days; (C) pH 7.0, 55°C, t = 2 days; (D) pH 2.0, t = 0 hr; (E) pH 2.0, 37°C, t = 2 days; and (F) pH 2.0, 55°C, t = 2 days.

Figure S3 Thermal unfolding/denaturation profiles of HγD-crys as a function of temperature. The average

emission wavelengths for samples incubated at pH 2.0 or 7.0 were extracted from intrinsic fluorescence spectra to monitor for changes over the incubation period of 2 days. The intrinsic fluorescence spectra between 300 and 400 nm of all HγD-crys samples were recorded at the excitation wavelength of 280 nm. Fitting of the apparent thermodynamic values followed a two-state model with a shift toward lower temperatures from pH 7.0 to pH 2.0 conditions.

Figure S4 Structure of the HγD-crys monomer with the two domains specified in red dashed boxes and the four Greek key motifs labeled.

Figure S5 Time evolution of secondary structure during 150 ns of simulation time.

Acknowledgments

We are grateful to the staffs of Technology Commons, College of Life Science, National Taiwan University (NTU) for help with transmission electron microscopy (TEM).

Author Contributions

Conceived and designed the experiments: JWW SSW. Performed the experiments: MEC WSW WAC CTL CKC CHL. Analyzed the data: JWW WSW CKC. Contributed reagents/materials/analysis tools: HSL. Contributed to the writing of the manuscript: JWW SSW.

References

1. Dobson CM (1999) Protein misfolding, evolution and disease. Trends Biochem Sci 24: 329–332.
2. Fink AL (1998) Protein aggregation: folding aggregates, inclusion bodies and amyloid. Fold Des 3: R9–23.
3. Chennamsetty N, Voynov V, Kayser V, Helk B, Trout BL (2009) Design of therapeutic proteins with enhanced stability. Proc Natl Acad Sci U S A 106: 11937–11942.
4. Chiti F, Dobson CM (2006) Protein misfolding, functional amyloid, and human disease. Annu Rev Biochem 75: 333–366.
5. Organization WH (2011) Visual impairment and blindness report. Fact Sheet N 282: WHO.
6. Shearer TR, Ma H, Fukiage C, Azuma M (1997) Selenite nuclear cataract: review of the model. Mol Vis 3: 8.
7. Santhoshkumar P, Udupa P, Murugesan R, Sharma KK (2008) Significance of interactions of low molecular weight crystallin fragments in lens aging and cataract formation. J Biol Chem 283: 8477–8485.
8. Oyster CW (1999) The human eye structure and function: Sinauer Associates, Inc.
9. Horwitz J (1992) Alpha-crystallin can function as a molecular chaperone. Proc Natl Acad Sci U S A 89: 10449–10453.
10. Meehan S, Berry Y, Luisi B, Dobson CM, Carver JA, et al. (2004) Amyloid fibril formation by lens crystallin proteins and its implications for cataract formation. J Biol Chem 279: 3413–3419.
11. Hanson SR, Hasan A, Smith DL, Smith JB (2000) The major in vivo modifications of the human water-insoluble lens crystallins are disulfide bonds, deamidation, methionine oxidation and backbone cleavage. Exp Eye Res 71: 195–207.
12. Takemoto L (1996) Increase in the intramolecular disulfide bonding of alpha-A crystallin during aging of the human lens. Exp Eye Res 63: 585–590.
13. Truscott RJ (2005) Age-related nuclear cataract-oxidation is the key. Exp Eye Res 80: 709–725.
14. Hains PG, Truscott RJ (2010) Age-dependent deamidation of lifelong proteins in the human lens. Invest Ophthalmol Vis Sci 51: 3107–3114.
15. Biswas S, Harris F, Dennison S, Singh J, Phoenix DA (2004) Calpains: targets of cataract prevention? Trends Mol Med 10: 78–84.
16. Crabbe MJ, Goode D (1995) Protein folds and functional similarity; the Greek key/immunoglobulin fold. Comput Chem 19: 343–349.
17. Kosinski-Collins MS, King J (2003) In vitro unfolding, refolding, and polymerization of human gammaD crystallin, a protein involved in cataract formation. Protein Sci 12: 480–490.
18. Papanikolopoulou K, Mills-Henry I, Thol SL, Wang Y, Gross AA, et al. (2008) Formation of amyloid fibrils in vitro by human gammaD-crystallin and its isolated domains. Mol Vis 14: 81–89.
19. Moran SD, Woys AM, Buchanan LE, Bixby E, Decatur SM, et al. (2012) Two-dimensional IR spectroscopy and segmental 13C labeling reveals the domain structure of human gammaD-crystallin amyloid fibrils. Proc Natl Acad Sci U S A 109: 3329–3334.
20. Eckert R (2002) pH gating of lens fibre connexins. Pflugers Arch 443: 843–851.
21. Poon S, Rybchyn MS, Easterbrook-Smith SB, Carver JA, Pankhurst GJ, et al. (2002) Mildly acidic pH activates the extracellular molecular chaperone clusterin. J Biol Chem 277: 39532–39540.
22. Wen WS, Hsieh MC, Wang SSS (2011) High-level expression and purification of human gamma D-crystallin in Escherichia coli. Journal of the Taiwan Institute of Chemical Engineers 42: 547–555.
23. Kosinski-Collins MS, Flaugh SL, King J (2004) Probing folding and fluorescence quenching in human gammaD crystallin Greek key domains using triple tryptophan mutant proteins. Protein Sci 13: 2223–2235.
24. Pande A, Pande J, Asherie N, Lomakin A, Ogun O, et al. (2000) Molecular basis of a progressive juvenile-onset hereditary cataract. Proc Natl Acad Sci U S A 97: 1993–1998.
25. Wang Y, Petty S, Trojanowski A, Knee K, Goulet D, et al. (2010) Formation of amyloid fibrils in vitro from partially unfolded intermediates of human gammaC-crystallin. Invest Ophthalmol Vis Sci 51: 672–678.
26. Smith PK, Krohn RI, Hermanson GT, Mallia AK, Gartner FH, et al. (1985) Measurement of protein using bicinchoninic acid. Anal Biochem 150: 76–85.
27. Wang SS, Wen WS (2010) Examining the influence of ultraviolet C irradiation on recombinant human gammaD-crystallin. Mol Vis 16: 2777–2790.
28. Ellozy AR, Ceger P, Wang RH, Dillon J (1996) Effect of the UV modification of alpha-crystallin on its ability to suppress nonspecific aggregation. Photochemistry and Photobiology 64: 344–348.
29. Darghal N, Garnier-Suillerot A, Salerno M (2006) Mechanism of thioflavin T accumulation inside cells overexpressing P-glycoprotein or multidrug resistance-associated protein: Role of lipophilicity and positive charge. Biochemical and Biophysical Research Communications 343: 623–629.
30. Whitmore L, Wallace BA (2004) DICHROWEB, an online server for protein secondary structure analyses from circular dichroism spectroscopic data. Nucleic Acids Research 32: W668–W673.
31. Whitmore L, Wallace BA (2008) Protein secondary structure analyses from circular dichroism spectroscopy: Methods and reference databases. Biopolymers 89: 392–400.
32. Wang GZ, Dong XY, Sun Y (2009) The role of disulfide bond formation in the conformational folding kinetics of denatured/reduced lysozyme. Biochemical Engineering Journal 46: 7–11.
33. Bradford JR, Westhead DR (2005) Improved prediction of protein-protein binding sites using a support vector machines approach. Bioinformatics 21: 1487–1494.

34. Van Der Spoel D, Lindahl E, Hess B, Groenhof G, Mark AE, et al. (2005) GROMACS: fast, flexible, and free. J Comput Chem 26: 1701–1718.

35. Abraham MJ, Gready JE (2011) Optimization of parameters for molecular dynamics simulation using smooth particle-mesh Ewald in GROMACS 4.5. J Comput Chem 32: 2031–2040.

36. Essmann U, Perera L, Berkowitz ML, Darden T, Lee H, et al. (1995) A Smooth Particle Mesh Ewald Method. Journal of Chemical Physics 103: 8577–8593.

37. Hess B, Bekker H, Berendsen HJC, Fraaije JGEM (1997) LINCS: A linear constraint solver for molecular simulations. Journal of Computational Chemistry 18: 1463–1472.

38. Schuler LD, Daura X, Van Gunsteren WF (2001) An improved GROMOS96 force field for aliphatic hydrocarbons in the condensed phase. Journal of Computational Chemistry 22: 1205–1218.

39. Day R, Bennion BJ, Ham S, Daggett V (2002) Increasing temperature accelerates protein unfolding without changing the pathway of unfolding. J Mol Biol 322: 189–203.

40. Kabsch W, Sander C (1983) Dictionary of protein secondary structure: pattern recognition of hydrogen-bonded and geometrical features. Biopolymers 22: 2577–2637.

41. Bourhim M, Kruzel M, Srikrishnan T, Nicotera T (2007) Linear quantitation of A beta aggregation using Thioflavin T: Reduction in fibril formation by colostrinin. Journal of Neuroscience Methods 160: 264–268.

42. Hudson SA, Ecroyd H, Kee TW, Carver JA (2009) The thioflavin T fluorescence assay for amyloid fibril detection can be biased by the presence of exogenous compounds. Febs Journal 276: 5960–5972.

43. Liu CP, Li ZY, Huang GC, Perrett S, Zhou JM (2005) Two distinct intermediates of trigger factor are populated during guanidine denaturation. Biochimie 87: 1023–1031.

44. Smoot AL, Panda M, Brazil BT, Buckle AM, Fersht AR, et al. (2001) The binding of bis-ANS to the isolated GroEL apical domain fragment induces the formation of a folding intermediate with increased hydrophobic surface not observed in tetradecameric GroEL. Biochemistry 40: 4484–4492.

45. Sirangelo I, Bismuto E, Tavassi S, Irace G (1998) Apomyoglobin folding intermediates characterized by the hydrophobic fluorescent probe 8-anilino-1-naphthalene sulfonate. Biochim Biophys Acta 1385: 69–77.

46. Semisotnov GV, Rodionova NA, Razgulyaev OI, Uversky VN, Gripas AF, et al. (1991) Study of the "molten globule" intermediate state in protein folding by a hydrophobic fluorescent probe. Biopolymers 31: 119–128.

47. Eftink MR (1991) Fluorescence Techniques for Studying Protein-Structure. Methods of Biochemical Analysis 35: 127–205.

48. Eftink MR (1998) The use of fluorescence methods to monitor unfolding transitions in proteins. Biochemistry-Moscow 63: 276–284.

49. Di Stasio E, Bizzarri P, Misiti F, Pavoni E, Brancaccio A (2004) A fast and accurate procedure to collect and analyze unfolding fluorescence signal: the case of dystroglycan domains. Biophysical Chemistry 107: 197–211.

50. Varshney A, Ahmad B, Rabbani G, Kumar V, Yadav S, et al. (2010) Acid-induced unfolding of didecameric keyhole limpet hemocyanin: detection and characterizations of decameric and tetrameric intermediate states. Amino Acids 39: 899–910.

51. Rabbani G, Ahmad E, Zaidi N, Khan RH (2011) pH-dependent conformational transitions in conalbumin (ovotransferrin), a metalloproteinase from hen egg white. Cell Biochem Biophys 61: 551–560.

52. Kosinski-Collins MS, Flaugh SL, King J (2004) Probing folding and fluorescence quenching in human gammaD crystallin Greek key domains using triple tryptophan mutant proteins. Protein Sci 13: 2223–2235.

53. Pande A, Pande J, Asherie N, Lomakin A, Ogun O, et al. (2000) Molecular basis of a progressive juvenile-onset hereditary cataract. Proc Natl Acad Sci USA 97: 1993–1998.

54. Rabbani G, Kaur J, Ahmad E, Khan RH, Jain SK (2014) Structural characteristics of thermostable immunogenic outer membrane protein from Salmonella enterica serovar Typhi. Appl Microbiol Biotechnol 98: 2533–2543.

55. Wistow G (1990) Evolution of a protein superfamily: relationships between vertebrate lens crystallins and microorganism dormancy proteins. J Mol Evol 30: 140–145.

56. Rosinke B, Renner C, Mayr EM, Jaenicke R, Holak TA (1997) Ca^{2+}-loaded spherulin 3a from Physarum polycephalum adopts the prototype gamma-crystallin fold in aqueous solution. J Mol Biol 271: 645–655.

57. Ohno A, Tate S, Seeram SS, Hiraga K, Swindells MB, et al. (1998) NMR structure of the Streptomyces metalloproteinase inhibitor, SMPI, isolated from Streptomyces nigrescens TK-23: another example of an ancestral beta gamma-crystallin precursor structure. J Mol Biol 282: 421–433.

58. Mills IA, Flaugh SL, Kosinski-Collins MS, King JA (2007) Folding and stability of the isolated Greek key domains of the long-lived human lens proteins gammaD-crystallin and gammaS-crystallin. Protein Sci 16: 2427–2444.

59. Moreau KL, King J (2009) Hydrophobic core mutations associated with cataract development in mice destabilize human gammaD-crystallin. J Biol Chem 284: 33285–33295.

60. Das P, King JA, Zhou R (2011) Aggregation of gamma-crystallins associated with human cataracts via domain swapping at the C-terminal beta-strands. Proc Natl Acad Sci U S A 108: 10514–10519.

61. Das P, King JA, Zhou R (2010) beta-Strand interactions at the domain interface critical for the stability of human lens gammaD-crystallin. Protein Sci 19: 131–140.

62. Flaugh SL, Kosinski-Collins MS, King J (2005) Contributions of hydrophobic domain interface interactions to the folding and stability of human gammaD-crystallin. Protein Sci 14: 569–581.

63. Flaugh SL, Kosinski-Collins MS, King J (2005) Interdomain side-chain interactions in human gammaD crystallin influencing folding and stability. Protein Sci 14: 2030–2043.

64. Flaugh SL, Mills IA, King J (2006) Glutamine deamidation destabilizes human gammaD-crystallin and lowers the kinetic barrier to unfolding. J Biol Chem 281: 30782–30793.

65. Kong F, King J (2011) Contributions of aromatic pairs to the folding and stability of long-lived human gammaD-crystallin. Protein Sci 20: 513–528.

66. Frederikse PH (2000) Amyloid-like protein structure in mammalian ocular lenses. Current Eye Research 20: 462–468.

67. Sandilands A, Hutcheson AM, Long HA, Prescott AR, Vrensen G, et al. (2002) Altered aggregation properties of mutant gamma-crystallins cause inherited cataract. The EMBO journal 21: 6005–6014.

68. Zhang W, Cai HC, Li FF, Xi YB, Ma X, et al. (2011) The congenital cataract-linked G61C mutation destabilizes gamma D-crystallin and promotes non-native aggregation. Plos One 6.

69. Colon W, Kelly JW (1992) Partial denaturation of transthyretin is sufficient for amyloid fibril formation in vitro. Biochemistry 31: 8654–8660.

70. Block H, Kubicek J, Labahn J, Roth U, Schafer F (2008) Production and comprehensive quality control of recombinant human Interleukin-1beta: a case study for a process development strategy. Protein Expr Purif 57: 244–254.

71. Kingsbury JS, Klimtchuk ES, Theberge R, Costello CE, Connors LH (2007) Expression, purification, and in vitro cysteine-10 modification of native sequence recombinant human transthyretin. Protein Expr Purif 53: 370–377.

72. Al-Ghoul KJ, Lane CW, Taylor VL, Fowler WC, Costello MJ (1996) Distribution and type of morphological damage in human nuclear age-related cataracts. Exp Eye Res 62: 237–251.

73. Gilliland KO, Freel CD, Lane CW, Fowler WC, Costello MJ (2001) Multilamellar bodies as potential scattering particles in human age-related nuclear cataracts. Mol Vis 7: 120–130.

74. Rousseau F, Schymkowitz J, Serrano L (2006) Protein aggregation and amyloidosis: confusion of the kinds? Current Opinion in Structural Biology 16: 118–126.

75. Brubaker WD, Freites JA, Golchert KJ, Shapiro RA, Morikis V, et al. (2011) Separating instability from aggregation propensity in gammaS-crystallin variants. Biophys J 100: 498–506.

76. Chiti F, Dobson CM (2009) Amyloid formation by globular proteins under native conditions. Nat Chem Biol 5: 15–22.

77. Sahin E, Jordan JL, Spatara ML, Naranjo A, Costanzo JA, et al. (2011) Computational design and biophysical characterization of aggregation-resistant point mutations for gammaD crystallin illustrate a balance of conformational stability and intrinsic aggregation propensity. Biochemistry 50: 628–639.

78. Wu JW, Liu HL (2012) In silico Investigation of the Disease-Associated Retinoschisin C110Y and C219G Mutants. J Biomol Struct Dyn 29: 1–23.

Frequency and Fitness Consequences of Bacteriophage Φ6 Host Range Mutations

Brian E. Ford [1,2,¶], Bruce Sun [1¶¤a], James Carpino [1¤a], Elizabeth S. Chapler [1], Jane Ching [1¤b], Yoon Choi [1¤c], Kevin Jhun [1¤d], Jung D. Kim [1¤e], Gregory G. Lallos [1], Rachelle Morgenstern [1¤f], Shalini Singh [1], Sai Theja [1], John J. Dennehy [1*¤a]

1 Biology Department, Queens College of the City University of New York, New York, New York, United States of America, 2 The Graduate Center of the City University of New York, New York, New York, United States of America

Abstract

Viruses readily mutate and gain the ability to infect novel hosts, but few data are available regarding the number of possible host range-expanding mutations allowing infection of any given novel host, and the fitness consequences of these mutations on original and novel hosts. To gain insight into the process of host range expansion, we isolated and sequenced 69 independent mutants of the dsRNA bacteriophage Φ6 able to infect the novel host, *Pseudomonas pseudoalcaligenes*. In total, we found at least 17 unique suites of mutations among these 69 mutants. We assayed fitness for 13 of 17 mutant genotypes on *P. pseudoalcaligenes* and the standard laboratory host, *P. phaseolicola*. Mutants exhibited significantly lower fitnesses on *P. pseudoalcaligenes* compared to *P. phaseolicola*. Furthermore, 12 of the 13 assayed mutants showed reduced fitness on *P. phaseolicola* compared to wildtype Φ6, confirming the prevalence of antagonistic pleiotropy during host range expansion. Further experiments revealed that the mechanistic basis of these fitness differences was likely variation in host attachment ability. In addition, using computational protein modeling, we show that host-range expanding mutations occurred in hotspots on the surface of the phage's host attachment protein opposite a putative hydrophobic anchoring domain.

Editor: Mark J. van Raaij, Centro Nacional de Biotecnologia - CSIC, Spain

Funding: This work was supported by the National Science Foundation Faculty Early Career Award #1148879 (JJD), Professional Staff Congress of the City University of New York Award #62886-00-40 (JJD), and National Science Foundation Division of Environmental Biology Award #0804039 (JJD). The funders had no role in study design, data collection and analysis, decision to publish, or preparation of the manuscript.

Competing Interests: The authors have declared that no competing interests exist.

* Email: john.dennehy@qc.cuny.edu

¶ These authors are co-first authors on this work.

¤a Current address: The New York Stem Cell Foundation, New York, New York, United States of America
¤b Current address: University of Maryland School of Pharmacy, Baltimore, Maryland, United States of America
¤c Current address: Smilow Research Center, New York University Medical Center, New York, New York, United States of America
¤d Current address: Icahn School of Medicine at Mount Sinai, New York, New York, United States of America
¤e Current address: Epidemiology and Public Health, City University of New York School of Public Health, New York, New York, United States of America
¤f Current address: Mailman School of Public Health, Columbia University, New York, New York, United States of America

Introduction

After a long period of steady decline, mortality due to infectious disease increased over the past several decades, largely because of the emergence of new infectious diseases including HIV [1,2]. Of these new diseases, a disproportionate number have been viruses [3,4]. Because of their high mutation rates and vast population sizes, viruses have higher probabilities of acquiring the requisite mutation(s) allowing infection of novel hosts than do other types of pathogens [5]. A common fear is that a highly transmissible and virulent virus will spread pandemically among humans, causing widespread mortality and economic damage. Thus, there is a strong motivation to understand and predict virus emergence.

Virus emergence is a two-step process. A virus first mutates to gain the ability to infect a new host, and then fully emerges by achieving positive population growth on that host via adaptation [6]. Theoretical modeling has shown that emergence probabilities are highly sensitive towards the type of mutation(s) required to productively infect a novel host [7]. Emergence events requiring single nucleotide substitutions are far more likely to occur than those that require several simultaneous point mutations or recombination [8]. While mutations altering virus host specificity can involve large-scale genomic rearrangements, most virus host shifts likely entail the modification of a small number of virus receptor amino acid residues [9]. In fact, single nucleotide substitutions are often sufficient to expand a virus's host range [10]. If this mechanism of host range expansion were common, the number of host range expanding mutations and their frequency of appearance would be important parameters governing the probability of emergence of a potential human pathogen.

Few studies have systematically determined the type, number, frequency, and fitness consequences of host range expanding mutations for any particular virus-host combination [11]. Such data can aid the parameterization of evolutionary ecological model of virus emergence. Factoring in other parameters, such as transmission rates and population densities, may allow quantitative predictions of the likelihood a particular virus is able to emerge on a new host. This type of prioritization is critical before allocating resources to interdict potential pathogenic viruses before they emerge.

Here we use an experimental model system, the bacteriophage (phage) $\Phi 6$, to determine number, frequency, fitness and structural consequences of mutations allowing infection of a novel host. Phage $\Phi 6$ (family Cystoviridae) is a dsRNA virus with a tripartite genome divided into Small (2,948 bp), Medium (4,061 bp) and Large (6,374 bp) segments [12–14]. Mutations allowing $\Phi 6$ to infect novel hosts have been localized to the gene encoding the P3 protein on the Medium segment [11,15].

Two previous studies have systematically examined $\Phi 6$ host range expansion [11,15]. Duffy et al. isolated 10 $\Phi 6$ host range mutants on each of three different *Pseudomonas* host strains including *Pseudomonas pseudoalcaligenes* [15]. Genetic sequencing revealed that all mutations occurred in the P3 attachment protein. Moreover, the authors reported that 7 of the 9 host range mutations imposed a fitness cost on the canonical host, indicative of antagonistic pleiotropy. Ferris et al. isolated 40 $\Phi 6$ host range mutants on *P. glycinea*, of which 16 contained novel mutations [11]. The authors used a statistical approach to predict the existence of a further 39 mutations that were missed by their screen. In addition, they observed broad fitness costs on the canonical host in agreement with Duffy et al. [11,15].

Our study builds on each of these earlier studies in order to present a more complete picture of host range expansion in $\Phi 6$. We isolated 69 independent $\Phi 6$ mutants able to infect the novel host, *P. pseudoalcaligenes*, and sequenced the entire P3 gene for each of them in order to determine the number, location, and frequency of host range expanding genotypes. For a subset of unique mutant genotypes, we quantified plaque size on *P. pseudoalcaligenes*, and reproductive capacity (fitness) and host attachment rate on both the canonical host *P. phaseolicola* and the novel host, *P. pseudoalcaligenes*. Finally, we used protein structural modeling to predict the effects of host range mutations on P3 attachment protein structure. Our work comes to qualitatively different conclusions than earlier work, namely that: 1) the coupon collector's model, as currently construed, cannot successfully predict the number of potential host range mutations; 2) there are fewer than expected mutations allowing host range expansion; and 3) fine-grained host infectibility cannot be accurately predicted by phylogeny. Furthermore, we show that $\Phi 6$ host range mutations usually occur in hotspots on the face of the P3 attachment protein opposite a hydrophobic anchoring domain. We propose that mutations on this surface allow $\Phi 6$ to bind novel host receptors.

Methods and Materials

Study Organisms and Culture Conditions

Cystovirus $\Phi 6$ used in these experiments is a descendant of the strain originally isolated from bean straw in 1973 [16]. $\Phi 6$'s host of isolation was the Gram-negative bacterium, *P. syringae* pathovar *phaseolicola* (ATCC # 21781; hereafter PP) [16]. In our study, we used a nonpermissive host, *P. pseudoalcaligenes* East River isolate A (hereafter ERA), to isolate $\Phi 6$ host range mutants (HRMs). The ERA receptor to which $\Phi 6$ binds has not been determined, but is likely the ERA pili. Two other nonpermissive hosts, *P. syringae* pv. *tomato* (hereafter TOM) and *P. syringae* pv. *atrofaciens* (hereafter

ATRO), were used in some assays. All bacteria and virus stocks were obtained from Paul Turner, Yale University, New Haven, CT.

All phages and bacteria were propagated in lysogeny broth (LB: 10 g NaCl, 10 g Bacto tryptone, and 5 g Bacto yeast extract per liter of water) at pH 7. Bacterial cultures were initiated by transferring a single colony from a streak plate into 10 mL LB in a sterile 50 mL flask capped with a 20 mL beaker. Culture flasks were incubated with shaking (120 rpm) at 25°C for 18 hours, allowing bacteria to attain stationary-phase density ($\sim 6 \times 10^9$ cells mL^{-1}).

Virus Stock Preparation

High-titer phage lysates were prepared by adding 1 µL stock lysate and bacteria (200 µL of PP/ATRO/TOM or 20 µL ERA) to 3 mL top agar (LB with 0.7% Bacto agar; stored as liquid at 45°C, solidifies at 25°C), and pouring onto 35 mL bottom agar (LB with 1.5% Bacto agar) in a sterile Petri dish. After 24 hours at 25°C, the resulting plaques were harvested and resuspended in 4 mL of LB, followed by 10 min centrifugation at RCF = 1400×g to pellet agar and bacterial debris. Bacteria-free lysates were obtained by filtering the supernatant through a 0.22 µm filter (Durapore; Millipore, Bedford, MA). Phage particles per mL in the lysates were quantified via serial dilution and titering. Plaques were counted on plates where 30–500 plaques were visible. The number of plaque forming units per mL (pfu mL^{-1}) in the original lysate was obtained by multiplying the number of plaques times the dilution factor. Lysates were mixed 1:1 (v/v) with 80% glycerol and were stored at −20°C.

Host Range Mutant Frequency

The frequency of HRMs in a phage population was estimated by plating a known number of wildtype $\Phi 6$ on a lawn of a nonpermissive host and counting the resulting number of plaques. Each plaque represents the descendants of a single HRM in the parent population. To perform this assay, a single plaque was picked off a lawn of PP and placed in 1 mL LB. This mixture was serially diluted and plated on PP to estimate phage pfu mL^{-1}. Subsequently, 10^7 or (10^8 for TOM) phages were plated on lawns of the nonpermissive hosts, ERA, TOM and ATRO. Typically, a $\Phi 6$ plaque contains $\sim 5 \times 10^8$ pfu/mL so sufficient phage for plating were easily obtained [17]. Following 48 hrs growth, plaques were counted to estimate the number of spontaneous HRMs among the descendants of a single phage. This assay was repeated at least twenty times per nonpermissive host strain. Mutant frequency was calculated by dividing the number of plaques observed by the initial inocula. The resulting data were analyzed using a one-way analysis of variance (ANOVA) model with host as a factor. A Tukey-Kramer honest significant difference test was applied post-hoc to ascertain significant differences among mean mutant frequencies on each host type.

Host Range Mutant Isolation

Each HRM was isolated independently to minimize bias due to a "jackpot effect" where multiple descendants of the same mutational event appear in a population [18]. A $\Phi 6$ lysate was serially diluted and plated on a PP lawn such that only a few widely spaced plaques appeared on the bacterial lawn. A single plaque was picked at random and placed in 1 mL LB. After vortexing, 100 µL of the plaque suspension was added to 20 µL ERA and 3 mL top agar and plated. Following 48 hrs growth, a single HRM plaque was picked from the ERA lawn and suspended in 500 µL LB. 10 µL from this solution was plated on an ERA lawn to obtain phage lysate for RNA extraction.

500 μL 80% glycerol was added to the remainder which was then stored at −20°C. This protocol was repeated 69 times to obtain 69 independent HRM isolates.

RNA Extraction and Sequencing

To sequence the region of the Medium segment encoding the P3 protein, 3 mL phage lysate from each mutant was concentrated by centrifuging at RCF = 100,000×g for 3 hrs at 4°C using a Beckman TL-100 ultracentrifuge. The supernatant was discarded and the pellet was resuspended in 150 μL nuclease-free water. RNA was extracted using a QIAamp Viral RNA Mini Kit (QIAGEN, Valencia, CA). Phage RNA was reverse transcribed using random hexamer primers and Superscript III reverse transcriptase (Life Technologies, Grand Island, NY), and the resulting cDNA was used as template for PCR. Three sets of oligonucleotide primers corresponding to bases 1298–2142, 2042–3052, and 2877–3873 of Φ6's Medium segment were used for the PCR amplification of the region encoding the P3 host attachment protein. PCR product was purified for sequencing using ExoSAP-It (Affymetrix, Santa Clara, CA). PCR product was sequenced in both directions with a minimum of 3-fold replication (6-fold coverage). Sequencing was performed at the DNA Analysis Facility on Science Hill at Yale University. Sequence data were analyzed using Geneious Pro Ver. 5.4 [19] and MEGA Ver. 5.05 [20]. Chromatograms were verified via MacVector Ver. 12.5.1 bioinformatics software.

Mutant Characterization

We phenotypically characterized HRM genotypes by determining plaque size on ERA, and by assaying reproductive capacity (fitness) and attachment rates on ERA and PP. Of our unique mutant genotypes, we did not assay the three mutants whose mutations were not identified (see Table 1). Furthermore, we only assayed one mutant in situations where differences between genotypes were attributable to synonymous amino acid substitutions. Finally, for one mutant, stored frozen lysate degraded due to a freezer failure following sequencing, and viable phage could not be recovered for phenotypic characterization. In sum, we phenotypically characterized 13 of our 17 unique HRM genotypes.

Plaque Size Estimates

Plaque sizes for 13 HRM genotypes were estimated from digital photographs of plaques formed on ERA. All LB plates used for plaque assays were poured at the same time and were weighed to maintain consistency. Each mutant's lysate was diluted and plated such that between 20 to 100 plaques formed on the ERA lawn after 48 hrs growth. Digital photographs were taken using a Kodak Gel Logic 440 digital imaging system. ImageJ software (NIH, Bethesda, MD; http://rsb.info.nih.gov/ij/) was used to estimate the total area of the plaque. For each genotype, at least 35 plaque size estimates were made across 3 plates.

Mutant Fitness on Native and Novel Hosts

We assayed absolute fitness for 13 of our Φ6 mutants and the wildtype on native and novel hosts using traditional plating methods. Here 10^5 phages were added to 3×10^8 host cells in 10 mL LB and incubated at 25°C with rotary shaking (120 rpm) for 24 hrs. All assays were replicated 5x. Bacteria-free lysates were obtained by centrifuging 3 mL culture at RCF = 2.75×g for 10 min to pellet bacterial debris, then passing the supernatant through a 0.22 μm filter. Phage particles per mL in the lysates were quantified via serial dilution and titering on host lawns. For

each assayed mutant genotype, we estimated absolute fitness using the equation, $W = \ln\left(\frac{N_t}{N_i}\right)$, where N_i is the starting number of phage and N_t is the total number of progeny phage produced during the infection period.

Attachment Rate Assays

The rate of attachment to native and novel hosts of 13 mutant HRM genotypes was measured using a centrifugation method. This method relies on the fact that, following centrifugation, attached phages are pelleted with host cells, while unattached phages remain in the supernatant. The decline of unattached phage over time is quantified to give the rate of phage attachment to host cells. In this assay, 10^3 phages, which were titered on the same host as the assay host, were mixed with 5×10^9 exponentially growing host cells in 10 mL LB with 3-fold replication. The mixture was incubated with orbital shaking for 40 min. Immediately after mixing and every 10 min thereafter, a 1 mL sample from the mixture was centrifuged (RCF = 1,700×g) for 1 min to pellet the cells then 100 μL from the supernatant was plated on a PP lawn. The attachment rate constant (k) is calculated as

$$k = \frac{\ln\left(\frac{N_f}{N_i}\right)}{Ct}$$

where N_i is the total number of phage added, N_f is the number of unattached phage, C is the concentration of bacteria, and t is the incubation time in minutes.

Attachment Protein P3 3D Structure Prediction

Three-dimensional structures of canonical bacteriophage Φ6 ancestral strain P3 attachment protein were predicted by homologue modeling based on nucleotide sequences submitted to the online I-TASSER server (http://zhanglab.ccmb.med.umich.edu/I-TASSER) [21,22]. I-TASSER generates three-dimensional (3D) atomic models from multiple threading alignments and iterative structural assembly simulations. Default parameters were used for the I-TASSER submission. The P3 amino acid sequence was submitted to the transmembrane structure prediction Dense Alignment Surface software (DAS) website (http://www.sbc.su.se/~miklos/DAS/maindas.html). DAS uses low-stringency dot-plots of the query sequence against a collection of non-homologous membrane proteins using a previously derived, special scoring matrix to identify transmembrane helices of integral membrane proteins. Default parameters were used for the DAS submission.

Results

Mutation Frequency

Φ6 HRMs were readily isolated by plating wildtype phages on lawns of the nonpermissive hosts, ERA, TOM and ATRO. The frequency in which ERA-infective HRMs appeared in populations of Φ6 phages was 1.15×10^{-6} (n = 21, SD ±5.219×10^{-7}). This value is only slightly lower than Φ6's spontaneous mutation rate, 2.7×10^{-6} [23]. Rates on TOM and ATRO were 1.39×10^{-7} (n = 21, SD ±1.489×10^{-8}) and 4.45×10^{-7} (n = 21, SD ±1.001×10^{-7}) respectively. We conducted a one-way ANOVA on $\log_{10}$ mutant frequency with bacterial host strain as a factor, and found significant differences in mutant frequency across different host strains (F = 198.54, DF = 2, P<0.0001). A Tukey-Kramer post hoc test with α = 0.05 revealed that all compared means were significantly different from each other. Interestingly, HRMs

Table 1. Phage φ6 Mutations Allowing Infection of Novel Host ERA.

Nucleic Acid (Amino Acid)	N	G22A (E8K)	A23G (E8G)	A23C (E8A)	G24T (E8D)	G24C (E8D)	T137C (F46S)	T138A (F46L)	T177C (D59D)	A389G (Q130R)	T481C (S161P)	C896G (S299W)	C1588T (L530L)	A1661G (D554G)	A1661C (D554A)	C1663T (L555F)	Non P3
Single (n=56)	20	■															
	24		■														
	4			□													
	3				□												
	1					□											
	1									■							
	2													■			
	1														□		
Double (n=8)	1	■								■							
	1	■												■			
	1	■								■							
	1						■	□									
	3		■							■							
	1			□								□					
Triple (n=2)	1		■							■						■	
	1								○		■		○				
Unknown	3																▲
Total # Observed	69	23	28	5	3	1	1	1	1	7	1	1	1	3	1	1	3

In this table, the columns show the nucleotide and amino acid substitutions found in all 69 ERA host range mutants. Rows show unique genotypes. N is the number of mutants with a particular genotype. The last row shows the number of times a given mutation appears among all mutants. Transitions are indicated by closed squares, transversions with open squares, synonymous substitutions with open circles and unknown changes with a filled triangle.

appeared more readily on the phylogenetic outgroup *P. pseudoalcaligenes* ERA than they do on other conspecific *P. syringae* pathovars such as *P. syringae atrofaciens* and *P. syringae tomato* [24].

We found at least 17 unique genotypes among the 69 ERA HRMs isolated and partially sequenced (Table 1). Three HRMs had no mutations in the sequenced region of the genome, thus we count them as, at a minimum, one unique genotype. Out of a combined 78 identified mutations from three studies, the majority resulted in nonsynonymous substitutions in the P3 amino acid sequence. Only 2 synonymous substitutions were identified (Table 1). This result conforms to Duffy et al.'s report of only 1 synonymous substitution among 31 mutations [15]. Synonymous substitution frequencies were similar between the two studies (2.5% vs. 3.2%).

Mutation Substitution Frequency

Among all nucleotide substitutions identified by our screen, the estimated transition/transversion bias R was 1.90. At the 8th residue, at least 5 possible substitutions (G22A, A23G, A23C, G24T and G24C) allow infection of ERA. However, of all substitutions observed, the majority were transitions (51 vs. 9). These results suggest that host range mutations allowing infection of ERA are heavily biased towards transitional substitutions. The significance of this finding is not clear, and may simply be a consequence of spontaneous deamination.

87% (60 of 69) of all mutants possessed a mutation at the 8th amino acid residue in the P3 protein. Only 9 mutants (4 single, 1 double, 1 triple and 3 unknowns) did not show a mutation at the 8th residue. This imbalance is higher than was observed in Duffy et al.'s study, where only 14 of 30 mutants isolated on ERA possessed mutations at the 8th residue [15]. However, we note that Duffy et al.'s study did not control for the "jackpot effect" and 20 of 30 mutants were isolated from hosts other than ERA. The dominance of a single residue is not unprecedented. Ferris et al. reported that 12 of 40 mutants isolated on *P. glycinea* showed a mutation at the 554th residue [11]. Across all 3 studies and 5 different hosts, there seems to be 3 "hotspots" for host range mutations in Φ6. 85.4% of all amino acid substitutions occurred close to the 8th (54.7%), 138th (16.7%) and 544th (14%) residues (Fig. 1A; Table 2).

Phenotypic Change Analysis

Changes in mass, electrical charge and hydrophobicity presumably can alter host receptor binding by changing the protein's tertiary structure and altering protein-protein interactions. In Table 3, we compiled the phenotypic characteristics of all amino acid substitutions allowing infection of ERA observed in this study and in Duffy et al. [15]. Using this data, we performed a paired t-test on amino acid mass for each mutation with strain type (wildtype or mutant) as a factor. Both factors had significant effects on amino acid mass. Substituted amino acids in mutants had significantly less mass than the original amino acids in the wildtype strain (t = 6.73, DF = 77, P<0.0001). This effect was most pronounced in mutation hotspots (F = 7.25, DF = 71, P< 0.0001). Perhaps lower mass substitutions permit greater flexibility at the host binding site.

Electrostatic interactions between host and phage proteins are most likely the basis of phage attachment. If so, we expect that charge changes incurred by host range mutations should be consistently in the same direction. A X^2 test was used to determine whether chemical properties of substituted amino acids differed significantly from the random expectation based on the amino acid composition of the P3: 9.16% acidic, 8.69% basic, 24.53% hydrophilic, and 57.45% hydrophobic. We found that mutant amino acids were significantly more likely to be basic or hydrophilic than expected by chance ($X^2 = 110.008$, DF = 3, P<0.0001). Furthermore, the frequency of mutations occurring at acidic residues was disproportionately high (81/106 or 76%). Ferris et al. also observed a greater than expected number of loss of charge mutations [11]. We speculate that these chemical changes make the P3 protein's host-binding site more permissive for binding host receptors.

P3 3D Structure Prediction

Little is known regarding how host range expanding amino acid substitutions affect phage attachment protein structure. We used I-TASSER [21,22] and DAS modeling software [25] to predict structural features of the P3 protein. I-TASSER generates three-dimensional atomic models from multiple threading alignments and iterative structural assembly simulations based on homology to solved structures (Fig. 1B). The predicted model's confidence score (C-score) was −2.12, which is intermediate confidence where scores range from high (2) to low (−5) confidence. When predicting known structures, and using a C-score cutoff >−1.5 for the models of correct topology, both false positive and false negative rates are below 0.1 [21]. While our C-score did not meet this threshold, we are confident that the probability of an incorrect structure is still low. Our view is supported by the ability of the predicted structure to provide a biologically plausible interpretation of the mechanistic basis of host range expansion.

DAS modeling software predicts transmembrane protein segments based on low-stringency dot-plots of query sequences against a collection of non-homologous membrane proteins using a previously derived, scoring matrix. Although P3 is soluble [26], DAS predicted a 21 amino acid hydrophobic membrane-interactive domain at residues 271 to 291. Based on the fact that, on the predicted structure, this domain extends out from the P3 core (Fig. 1B), we venture that domain likely anchors the P3 protein to the integral membrane protein P6 [27], thus we will refer to it as the hydrophobic anchoring domain, or HAD. All host range mutations occurred on the face opposite the HAD, suggesting that the opposite surface binds the host receptor, and that mutations in this region allow infection of novel hosts. However, this hypothesis assumes that amino acid substitutions do not substantially alter the protein shape, and that residues on this face in the ancestor would remain on this face in the mutant. Figure 1C suggests our conjecture is valid as the E8G mutant's predicted structure does not show major structural rearrangements compared to the wildtype. Interestingly, the most common host range mutations found in our study alter the surface charge at this location from negative to neutral, hinting at a proximate mechanism for host range expansion (Fig. 1C).

Plaque Size

We isolated HRMs by visually identifying and picking plaques off lawns of the nonpermissive host, ERA. Our results showed that host range mutations were heavily biased towards the 8th residue. One possible criticism of our mutant isolation process is that it may have been biased towards certain mutations simply because these mutants formed larger plaques that were more likely to be spotted by the sampler. To test this hypothesis, we determined from digital photographs the average plaque size for 13 of 17 of our identified HRMs. Mean plaque size for our mutants ranged from 3.5 to 10.3 mm^2 (Table 4). We performed an ANOVA of mean plaque size with mutant frequency as a factor, and the results confirmed that plaque size did not predict mutant frequency. While we did find significant differences in plaque size among genotypes, the two most frequent genotypes found by our study (E8K and E8G)

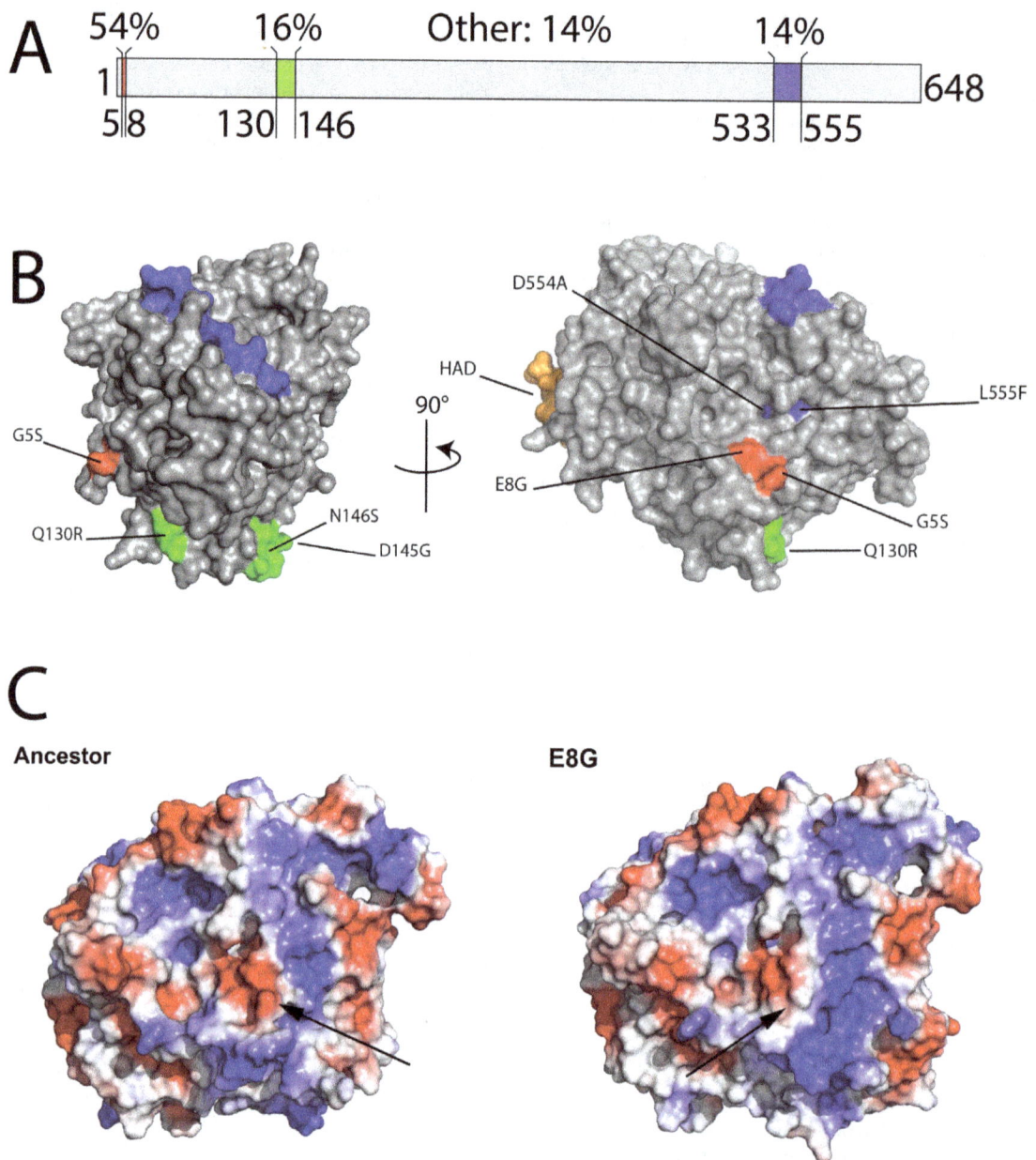

Figure 1. Spatial models of Φ6 P3 protein mutants. Panel A: Three host range mutation hotspots (accounting for 86% of all mutations) are highlighted in this linear representation of the 648 amino acid sequence of the Φ6 P3 gene. The remaining 14% of mutations are not shown. Panel B: Space-filling representations of the Φ6 P3 protein are shown as predicted by I-TASSER. Colored regions correspond to the mutation hotspots depicted in Panel A. A putative hydrophobic anchoring domain (HAD) is shown in orange. In our model, the hydrophobic anchoring domain penetrates Φ6's outer lipid membrane to bind inner membrane protein P6. Panel C: Surface electrical charges of E8G mutant contrasted with ancestor. Space-filling representations showing predicted surface electrical charges for the Φ6 E8G host range mutant and its ancestor were estimated using I-TASSER. Positively- and negatively-charged regions are depicted in blue and red respectively. Arrows indicate the predicted location of the mutated 8th residue. The most prominent difference between the mutant and the ancestor is the greater surface positive charge at the presumed host binding domain.

ranked 4th and 9th respectively in mean plaque size. These data imply that mutant sampling was not biased. Furthermore, we did not observe any correlation between fitness and plaque size.

Mutant Fitness on Original and Novel Hosts

The fitness consequences of host range expanding mutations will play a large role in the ability of these mutants to persist in host populations [28]. With this in mind, we estimated the absolute fitness of 13 of our mutant genotypes on the canonical host, PP, and the novel host, ERA (Table 4). A one-way ANOVA of absolute fitness with strain as a factor revealed significant differences among strain fitness on both ERA (Fig. 2A; $F = 40.64$, $DF = 12$, $P < 0.0001$) and PP (Fig. 2B; $F = 3.515$, $DF = 12$, $P = 0.0008$), but mean fitness on ERA was not correlated with mean fitness on PP nor was fitness on ERA correlated with the number of mutations a mutant possessed. In fact, genotypes containing multiple mutations tended to be less fit than those with single mutations, although this trend was not significant. Matching

Table 2. Host Range Substitution Hotspots in φ6 P3 Protein[a].

Substitution	This Study	Duffy	Ferris	N	Frequency[b]	Combined Frequency[b]
G5S	0	0	2	2	1.3%	54.7%
E8K	23	4	1	28	18.7%	
E8G	28	9	5	42	28%	
E8D	4	0	0	4	2.7%	
E8A	5	1	0	6	4%	
Q130R	7	0	0	7	4.7%	16.7%
A133V	0	9	0	9	6.0%	
D145G	0	0	3	3	2.0%	
N146S	0	0	6	6	3.9%	
D533A	0	0	1	1	0.7%	14.0%
D535N	0	0	1	1	0.7%	
D554G	3	1	8	12	8.0%	
D554A	1	0	1	2	1.3%	
D554V	0	0	1	1	0.7%	
D554N	0	0	2	2	1.3%	
L555F	1	0	1	2	1.3%	
Others	9	6	7	22	14.7%	14.7%
Total	81	30	39	150	100%	100%

Amino acid substitutions close together in the primary sequence are grouped together. We combine data from this study with two other studies of φ6 host range expansion. N is total number of times a substitution was observed across all studies. Frequency is percentage of total substitutions a particular substitution was observed. Combined frequency is percentage of total substitutions constituted by substitutions in a particular region of the primary sequence. Others category includes substitutions found outside substitution hotspots.
[a]Data compiled from this study, Duffy et al. 2006 and Ferris et al. 2007 [11,15].
[b]Some frequencies rounded off to nearest tenth percent.

previous results, fitness on PP was, in all but one case, less than that of the ancestor [11,15]. These results are indicative of antagonistic pleiotropy, implying a tradeoff in fitness between infection of PP and ERA. In addition, the coefficient of variation (i.e. standard deviation/mean; CV) in mutant fitness was considerably greater on ERA as opposed to PP (CV: 0.402 versus 0.015). This suggests that mutations expanding the host range have a much wider range of fitness effects on the novel host.

Attachment Rate

Bacteriophages initiate infections of host cells by binding to receptors on the surface of the bacterial outer membrane. As such, the host attachment rate is a critical factor in the ecological success of a phage. We measured the rate of phage attachment to the original and novel hosts for 13 mutant genotypes (Table 4). A one-way ANOVA of the rate of attachment to ERA with mutant genotype as a factor revealed significant differences among the strains (F = 10.17, DF = 12, P<0.0001). In addition, we regressed attachment rate against mutant fitness on ERA to determine if the two were correlated. Since our HRMs most likely differ only by mutations in the P3 host attachment protein, we expected that improved attachment would lead to increased fitness. Indeed, for 13 mutant strains whose fitnesses and attachment rates were estimated, fitnesses on ERA were correlated with ERA attachment rates (Fig. 3; F = 11.91, DF = 1, P = 0.0062). However, the linear regression model accounted for roughly half of the variance in attachment rate (R^2 = 0.54). These results are not surprising given the difficulty of precisely estimating the Φ6 attachment rate. Nevertheless, the results conform to our expectation of a positive correlation between fitness and host binding ability. By contrast,

attachment rates of the various mutants to PP were not significantly different, nor were they correlated with mutant fitnesses on this host. These results might be expected given the relatively narrow range of fitness differences on PP (Fig. 2; Table 4).

The rates of attachment to PP were significantly greater than attachment rates to ERA (One-way ANOVA: F = 216.7, DF = 1, P<0.0001). This latter result matches expectations since mutant fitness on PP is approximately an order of magnitude greater than that on ERA [17]. Presumably, the switching of receptor types and the lack of adaptation to an ERA receptor may account for the significant differences in mutant fitness on the different host types. However, attachment rates to PP and ERA were not correlated, implying that mutations that increase binding to ERA do not necessarily increase or decrease binding to PP.

Discussion

Φ6 Host Range Mutation Frequency

Understanding the genetic basis of virus host range expansion is critical to predicting the emergence of potentially dangerous viruses. The genetic distance a virus must cross to gain the ability to infect a novel host may be a dominant factor determining the probability of emergence. Not all viruses readily infect novel hosts [29]. For example, many mycobacteriophages isolated on *Mycobacterium smegmatis* are unable to infect *M. tuberculosis*, even when large numbers of phage are plated [30]. Presumably infection of *M. tuberculosis* requires several simultaneous mutations or even the recombination of whole genes or gene systems. By contrast, many viruses are able to infect novel hosts via single nucleotide substitutions [10,31–36]. This minimal genetic distance

Table 3. Amino Acid Substitutions Associated with φ6 Host Range Expansion on ERA.

Amino Acid Substitution	TS[a]	Du[b]	N	Average Mass		Electrochemical Properties		Hydrophobicity Index	
				Wildtype	Mutant	Wildtype	Mutant	Wildtype	Mutant
Glutamic Acid to Lysine (E8K)	23	4	27	129.1	128.2	Acidic (−) Polar	Basic (+) Polar	−3.5	−3.9
Glutamic Acid to Glycine (E8G)	28	9	37	129.1	57.1	Acidic (−) Polar	Neutral Nonpolar	−3.5	−0.4
Glutamic Acid to Aspartic Acid (E8D)	4	0	4	129.1	115.1	Acidic (−) Polar	Acidic (−) Polar	−3.5	−3.5
Glutamic Acid to Alanine (E8A)	5	1	6	129.1	71.1	Acidic (−) Polar	Neutral Nonpolar	−3.5	1.8
Aspartic Acid to Alanine (D35A)	0	2	2	115.1	71.1	Acidic (−) Polar	Neutral Nonpolar	−3.5	1.8
Phenylalanine to Serine (F46S)	1	0	1	147.2	87.1	Neutral Nonpolar	Neutral Polar	2.8	−0.8
Phenylalanine to Leucine (F46L)	1	0	1	147.2	113.2	Neutral Nonpolar	Neutral Nonpolar	2.8	3.8
Glutamine to Arginine (Q130R)	7	0	7	128.1	156.2	Neutral Polar	Basic (+) Polar	−3.5	−4.5
Alanine to Valine (A133V)	0	9	9	71.1	99.1	Neutral Nonpolar	Neutral Nonpolar	1.8	4.2
Serine to Proline (S161P)	1	0	1	87.1	97.1	Neutral Polar	Neutral Nonpolar	−0.8	−1.6
Serine to Threonine (S246T)	0	2	2	87.1	101.0	Neutral Polar	Neutral Polar	−0.8	−0.7
Serine to Tryptophan (S299W)	1	0	1	87.1	186.2	Neutral Polar	Neutral Slightly Polar	−0.8	−0.9
Lysine to Threonine(K311T)	0	1	1	128.2	101.0	Basic (+) Polar	Neutral Polar	−3.9	−0.7
Glycine to Serine (G515S)	1	0	1	57.1	87.1	Neutral Nonpolar	Neutral Polar	−0.4	−0.8
Aspartic Acid to Glycine (D554G)	3	1	4	115.1	57.1	Acidic (−) Polar	Neutral Nonpolar	−3.5	−0.4
Aspartic Acid to Alanine (D554A)	1	0	1	115.1	71.1	Acidic (−) Polar	Neutral Nonpolar	−3.5	1.8
Leucine to Phenylalanine (L555F)	1	0	1	113.1	147.2	Neutral Nonpolar	Neutral nonpolar	3.8	2.8

Amino acid substitutions found to allow φ6 infection of Pseudomonas pseudoalcaligenes ERA from data obtained by two separate studies. N is number of times substitution was observed across the two studies. Average mass is residue weight of the original and substituted amino acid [67]. Hydrophobicity index was obtained from Kyte and Doolittle [68]. Negative numbers are more hydrophilic; positive numbers are more hydrophobic.
[a] = This study; [b] = Duffy et al. 2006 [15].

Table 4. Phenotypic Characteristics of Bacteriophage φ6 Mutants.

Strain	Mutation	Frequency of Mutant	Fitness ERA	Fitness PP	Plaque Size ERA[a]	Attachment rate to ERA[b]	Attachment rate to PP
S8	D554G	2	3.31	14.73	6.89	2.58×10^{-11}	4.79×10^{-11}
S68	E8D	4	1.97	14.61	5.69	2.22×10^{-11}	4.72×10^{-11}
S53	E8A	4	3.97	14.57	7.8	4.80×10^{-11}	4.68×10^{-11}
S46	Q130R/S299W	1	0.98	14.80	9.26	1.68×10^{-11}	5.05×10^{-11}
S42	S161P/L555F	1	2.20	14.93	9.17	2.40×10^{-11}	5.31×10^{-11}
S4	Q130R	1	1.27	14.74	6.28	2.05×10^{-11}	5.41×10^{-11}
S30	E8K/Q130R	1	2.51	14.62	3.54	2.49×10^{-11}	5.71×10^{-11}
S28	E8A/F46L	1	2.47	15.21	5.94	2.53×10^{-11}	5.63×10^{-11}
S26	D554A	1	4.47	14.77	7.85	2.39×10^{-11}	5.63×10^{-11}
S154	E8G	24	4.12	14.92	6.29	2.59×10^{-11}	5.33×10^{-11}
S14	E8K/D554G	1	2.08	14.79	6.56	2.09×10^{-11}	4.50×10^{-11}
S13	E8K/F46S	1	3.16	14.85	10.25	2.24×10^{-11}	4.60×10^{-11}
S117	E8K	20	3.91	14.66	8.93	2.51×10^{-11}	4.39×10^{-11}
φ6 WT	n/a	n/a	n/a	15.07	n/a	n/a	4.91×10^{-11}

[a] = in mm^2; [b] = Attachment rate (k) units are per milliliter per cell (or per phage) per minute.

A

B

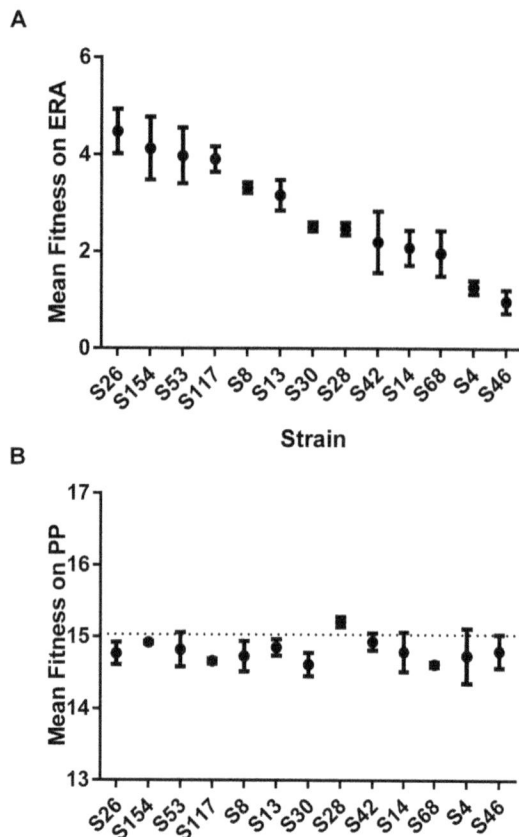

Figure 2. Mutant absolute fitness on canonical and novel hosts. Panel A: Absolute fitness of 13 Φ6 host range mutants on the novel host, ERA. Each point is the mean of 5 replicate measurements of fitness. Bars are ±1SE. Panel B: Absolute fitness of 13 Φ6 host range mutants on the canonical host, PP. Each point is the mean of 5 replicate measurements of fitness. Fitness of wildtype Φ6 is shown by the dotted line for comparison. Bars are ±1SE.

Figure 3. Mean ERA attachment rate (k) is plotted against phage Φ6 fitness on ERA. Attachment to ERA was correlated with fitness on ERA for Φ6 host range mutants. Each point is the mean of 3 replicate measurements. Dotted lines show 95% confidence intervals.

can easily be traversed because viral population sizes and mutation rates allow them to search available sequence space rapidly. The phage Φ6 is an excellent model to study virus emergence via single nucleotide substitutions because such HRMs are easily isolated, sequenced, and characterized in the laboratory [11,15].

In this study, we found that Φ6 HRMs appear on ERA at a rate (1.17×10^{-6}) slightly lower than the estimated Φ6 mutation rate of 2.7×10^{-6} per nucleotide per generation. Thus our figure seems somewhat low given that there are multiple possible mutations allowing host range expansion in the Φ6 genome (Table 1). However, Chao et al.'s estimate was derived from the frequency of revertants from an amber mutation (sus297), and it was assumed that there was only one way to revert [23]. If there are multiple ways to revert from Chao's et al.'s amber mutation, then theirs is an overestimate of the mutation rate. Moreover, Chao et al. estimated the mutation rate at a single locus, but the mutation rate may vary across the genome [37,38]. At any rate, it is clear that, given their potentially enormous population sizes, Φ6 HRMs can be isolated relatively easily.

Our results indicate that there is considerable variation in the ability of Φ6 to mutate to infect nonpermissive host strains. While there are certainly strong coarse-grained trends in infectivity, e.g., Φ6 seems mainly restricted to the pseudomonads [39], infectivity within this group is currently unpredictable. Phage Φ6 is better able to mutate to infect *P. pseudoalcaligenes* ERA, a distant

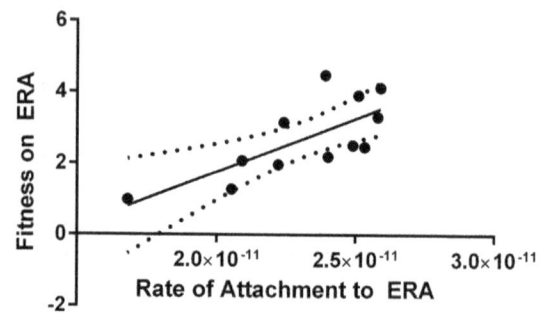

relative of *P. syringae* pv *phaseolicola* [24], than two pathovars from the same species, *P. syringae* pv *tomato* and *P. syringae* pv *atrofaciens* [40]. Duffy et al. and Cuppels et al. found many examples of other *P.syringae* pathovars nonpermissive for Φ6 even at high plating densities [15,39]. For example, Duffy et al. were unable to isolate HRMs on at least 8 *P. syringae* pathovars despite plating over 10^{10} Φ6 phages on each pathovar [15]. Similar results were obtained by Cuppels et al. [39]. It would appear that phylogeny is a poor predictor of infectivity, at least at the fine scale level within the pseudomonads. Φ6's ability to expand its host range appears to be somewhat idiosyncratic, which is to be expected given myriad possible outcomes for parasite-host coevolution [41]. It may be that the *P. syringae* strains have experienced recent coevolution with Φ6 or its close relatives, and thus have acquired resistance to infection to these phages. By contrast, more distantly Pseudomonads may not have recently experienced consistent Φ6 infection, therefore remain relatively sensitive to this virus.

The frequency of mutants lacking mutations in the P3 (4.3%) was similar to that found in Ferris et al.'s study (2.5%) [11]. These results provide strong evidence that the P3 sequence is the primary, but not exclusive, determinant of host range among phage Φ6 [42]. While it is tempting to speculate that additional host range mutations might be found in membrane fusion protein P6, Duffy et al. sequenced the P6 for 30 Φ6 HRMs and found no mutations [15]. As of publication, no other candidate genes for host range expansion on ERA have been explicitly identified in Φ6; however one study has reported that a mutation allowing infection of ERA was localized to the large segment [43]. This segment contains a gene encoding an RNA-dependent RNA polymerase and genes associated with RNA packaging and procapsid assembly [12].

The number of ways a virus can mutate to infect a novel host is an important parameter in predicting its potential for emergence [28]. Using a method based on the coupon collector's problem of statistical theory, Ferris et al. estimated the total number of possible mutations that allow Φ6 to infect a novel host, *P. glycinea* [11]. The coupon collector's problem can be informally stated as: Given *n* coupons, how many coupons will need to be sampled before each coupon is observed at least once [44]? One assumption of the coupon collector's problem is that all coupons are equally likely. This assumption does not hold for genetic mutations as some types are more likely than others are. Ferris et al. accommodate this simplification by adjusting the equation to account for differences in the probabilities of transitions and transversions. Since they found 19 distinct genotypes among their 40 independent samples, they estimated that further sampling would uncover an additional 36 mutations [11]. If Ferris et al.'s estimates are correct, it would mean that 1.3% of all possible nonsynonymous substitutions in P3 confer the ability to infect ERA (i.e., 55 of 4,380 potential nonsynonymous changes expand host range).

Although their HRMs were isolated on a different host, *P. glycinea*, both their study and ours found similar frequencies of transitions among all mutations (90% in Ferris et al., 84% in our study). However, out of 69 HRMs, we found only 17 distinct genotypes. Ferris et al. isolated almost the same number of distinct genotypes in half as many samples [11], which may be a consequence of the different hosts of isolation. Since Duffy et al. observed 10 unique genotypes out of 30 isolates (33% unique) [15] and we observed at least 17 unique genotypes out of 69 (26% unique), the implication is that more unique genotypes would be found with further sampling. However, a closer inspection of our data suggests otherwise. 8 of 17 of our unique genotypes were only unique because of second- or third-site mutations. If we consider only those mutations that are *sensu stricto* necessary for infection of ERA, we only find a combined 13/99 (13%) unique genotypes among our and Duffy et al.'s study [15]. In fact, we only found 3 unique *sensu stricto* substitutions not found by Duffy et al. study and they found 6 not identified in ours.

If 1.3% of all possible nonsynonymous substitutions allowed Φ6 to infect ERA, we would expect to see more unique genotypes among our isolates. Our results also indicate that some mutations occur far more frequently than expected by chance even if differences in transitions and transversions are accounted for. One possibility is that low fitness HRMs are eliminated by within plaque selection and consequently are not represented in the mutant collection sampled. We have no means to ascertain the validity of this hypothesis at this point, but it could be an interesting question to approach by deep sequencing of single HRM plaques. However, at the same time, it seems likely that additional factors that are not currently well understood, such as RNA structure, codon bias and variation in the mutation rate across the genome, influence the probability of mutation at any particular locus. Nonetheless, Ferris et al.'s method is a valuable step forward towards the estimation of an important parameter relevant to virus emergence.

Mutation Hotspots

We found that mutations expanding the host range of phage Φ6 were more likely to appear in certain regions of the P3 gene than others. Such mutation hotspots have been observed among virus drug resistance [45–47], host range [48,49], hemagglutinin [50], capsid [51], and core antigen genes [52] among others. Mutation hotspots are evidence of strong positive selection for substitutions that provide an adaptive advantage in a particular environment [53,54]. Growth on a novel host should impose strong positive selection for nonsynonymous substitutions at loci associated with host range expansion. Thus, we can use the frequency of mutations found in our survey to identify regions of the P3 protein that are important in attachment to a host receptor. 85.4% of all mutations identified by our study and by Duffy et al. [15] were found in just three regions (near 8th, 133rd and 554th residues) of the P3 gene (Fig. 1A; Table 2). We venture that these hotspots on the P3 protein are important in host range determination among Φ6 phages.

Structural Speculations

We used the structural modeling software I-TASSER [21,22] to predict the structure of the P3 protein from its amino acid sequence. The resulting structure showed homology to bacterial alcohol dehydrogenase quinoproteins [55–57]. Interestingly, in the best-fit model, our putative mutation hotspots were located close together on one face of the ~60 Å diameter P3 protein (Fig. 1B). Residues 8 and 130 were located at the surface 18 Å from each other, and residue 554 was located subsurface about 15 Å from residue 8 and 23 Å from residue 130. Other less frequently observed mutations also occur near this region (Fig. 1B). We propose that this region of the P3 protein is a host-binding domain and directly interacts with host receptors. This supposition is supported by the fact that the host binding domain is diametrically opposite the hydrophobic anchoring domain (residues 271–291) predicted by DAS (Fig. 1B). The most parsimonious explanation is that this domain serves to anchor the P3 to the integral membrane protein P6 [27], which leaves the putative host binding domain exposed to the environment.

Mutations allowing infection of ERA may not significantly alter the tertiary structure of the P3 protein. I-TASSER structural modeling did not show any major structural rearrangements in predicted structures for mutant strains. Rather mutations may alter the host-binding domain's electrical charge from negative to positive or neutral (Fig. 1C). This difference in electrical charge may allow mutant Φ6 to bind the ERA host receptor. The presumptive ERA receptor is its pilus, but this has not been definitively determined. If the ERA receptor were indeed the pilus, it would be interesting to know if its electrical properties are appreciably different from those of the pilus of PP. Moreover, it is plausible that neutral or positive electric charges and smaller mass amino acids confer more flexibility to the binding region, allowing a greater variety of structures to be bound [58]. It would be interesting to determine if host range expanding mutations more frequently result in the substitution of small for large amino acids or alter the charge of the binding site.

Fitness on Native and Novel Hosts

Fitness on native and novel hosts was assessed using standard flask productivity assays. Phage Φ6 HRMs showed a broad range of fitness values on ERA, some of which were significantly different from the others (Fig. 2A). Mutant fitnesses on the native host, PP, were much greater than those on ERA (Fig. 2B). Since Φ6 is presumably well adapted to native but not novel hosts, these results meet our expectations. Supporting these results, we found that the coefficient of variation (CV) of mutant fitnesses on PP was much lower than CV of mutant fitnesses on ERA. These results conform to theoretical expectations that there should be less variation in fitness values close to a fitness peak on an adaptive landscape [59]. Directional selection should erode the variation in fitness as a population increases in fitness in a particular environment. Thus, a virus that is adapted to a particular host should have lower variation in fitness on that host as opposed to a host to which it is not well adapted.

We found that, in concert with previous studies [11,15], mutations expanding the Φ6 host range usually reduced fitness on the original host, PP. On average, HRM fitness on PP was reduced about 2.5% compared to the wildtype. Negative genetic associations between host types is an example of antagonistic pleiotropy [60,61]. The adage that "a jack of all trades is a master of none" is well supported, at least among Φ6 host infections. However, the ultimate cause of host specialism or generalism remains opaque. Intuitively one would imagine that a broader host range would produce greater returns than a narrow one as long as the reduction in productivity on a single host was offset by an increase in overall productivity [62]. With regard to the present system, it seems unlikely that the relatively minor cost in fitness on the original host imposed by host range expansion should outweigh the benefits of an expanded host range. Moreover, we isolated one mutant (S28) whose fitness on the canonical host actually increased following the acquisition of a mutation permitting infection of ERA. Why then are broad host range phages relatively rare? The rarity of generalism may be a result of the interaction of widespread habitat patchiness, reduced dispersal and the ubiquity of local adaptation [63]. If these general trends hold, competition within a patch should favor the evolution of specialism. This hypothesis should be amenable to testing via experimental evolution studies.

As a rule, we might expect that novel hosts will present a greater challenge to virus reproduction than native hosts, a conclusion that is supported by many examples in the literature [64–66]. Novel hosts may represent ecological sinks, defined as habitats where the basic reproductive rate is <1. Our fitness results support this conjecture, and suggest that Φ6 probably experiences a broader range of sink conditions on ERA than it does on PP. Consequently, Φ6 population extinction is more likely in a habitat populated by ERA than one populated by PP [17]. Given the many HRM genotypes over a broad range of fitness values, Φ6 should be a valuable system to test hypotheses regarding virus emergence [28].

Attachment to Native and Novel Hosts

With the exception of the three non-P3 mutants, the mutant strains are most likely isogenic outside the host attachment protein region. The differences in fitness are expected to result mainly from differences in binding efficiency to the host receptor. Our results indicate that different suites of mutations had highly divergent attachment rates and fitnesses on the novel host (Fig. 2 and 3). Nonetheless, a regression of phage fitness on ERA against attachment rate to ERA revealed a significant positive correlation. Ferris et al. reported a similar result for Φ6 infecting *P. glycinea* [11]. These results make intuitive sense as mutants that are better able to bind to the host are expected to reproduce at a higher rate. Moreover, attachment to ERA was significantly lower than to PP, which is also reflected in the large differences in fitness.

Implications for Disease Emergence

This study and other recent studies of Φ6 host range expansion suggest several generalizations. First, phylogeny may only allow relatively coarse-grained predictions of virus host range. Phage Φ6's ability to mutate to infect close relatives was frequently worse than its ability to infect distant relatives. Second, nonsynonymous substitutions allowing host range expansion may occur at hotspots in the host attachment protein. This prediction makes intuitive sense as host attachment relies on binding affinity between host and virus proteins. In addition, many host range-expanding mutations may not result in large structural rearrangements in host attachment proteins. Rather, amino acid substitutions may result in more subtle changes in protein surface charges, allowing binding to different host proteins. Furthermore, the number of nonsynonymous substitutions allowing host range expansion is probably relatively small considering the number of possible substitutions. Nonetheless, the relatively high virus mutation rate allows viruses to rapidly acquire host range expanding mutations despite their relative rarity. Finally, initial fitness on a novel host is usually much less than that on the original host, and antagonistic pleiotropy among host range mutations is common. This generalization conforms to our expectations since evolutionary tradeoffs in different habitats are anticipated to be ubiquitous.

Acknowledgments

We thank Paul Turner for providing phage and bacterial strains and Tim Short for technical advice. Constructive criticism from Paul Gottlieb and three anonymous reviewers was much appreciated. This work was completed in part using equipment in the Core Facility for Imaging, Cellular and Molecular Biology at Queens College.

Author Contributions

Conceived and designed the experiments: BEF JJD. Performed the experiments: BEF BS J. Carpino ESC J. Ching YC KJ JDK GGL RM SS ST JJD. Analyzed the data: BEF BS J. Ching KJ GGL RM JJD. Wrote the paper: BEF BS J. Carpino GL JJD.

References

1. Christensen KLY, Holman RC, Steiner CA, Sejvar JJ, Stoll BJ, et al. (2009) Infectious disease hospitalizations in the United States. Clinical Infectious Diseases 49: 1025–1035.
2. Armstrong GL, Conn LA, Pinner RW (1999) Trends in infectious disease mortality in the United States during the 20th century. JAMA-Journal of the American Medical Association 281: 61–66.
3. Woolhouse M, Gaunt E (2007) Ecological origins of novel human pathogens. Critical Reviews in Microbiology 33: 231–242.
4. Cleaveland S, Laurenson MK, Taylor LH (2001) Diseases of humans and their domestic mammals: pathogen characteristics, host range and the risk of emergence. Philosophical Transactions of the Royal Society of London Series B-Biological Sciences 356: 991–999.
5. Woolhouse MEJ, Haydon DT, Antia R (2005) Emerging pathogens: the epidemiology and evolution of species jumps. Trends in Ecology and Evolution 20: 238–244.
6. Antia R, Regoes RR, Koella JC, Bergstrom CT (2003) The role of evolution in the emergence of infectious diseases. Nature 426: 658–661.
7. Gandon S, Hochberg ME, Holt RD, Day T (2012) What limits the evolutionary emergence of pathogens? Philosophical Transactions of the Royal Society of London Series B-Biological Sciences 368.
8. Alexander HK, Day T (2010) Risk factors for the evolutionary emergence of pathogens. Journal of the Royal Society Interface 7: 1455–1474.
9. Holmes EC, Drummond AJ (2007) The evolutionary genetics of viral emergence. Wildlife and Emerging Zoonotic Diseases: the Biology, Circumstances and Consequences of Cross-Species Transmission 315: 51–66.
10. Baranowski E, Ruiz-Jarabo CM, Pariente N, Verdaguer N, Domingo E (2003) Evolution of cell recognition by viruses: A source of biological novelty with medical implications. Advances in Virus Research, Vol 62 62: 19–111.
11. Ferris MT, Joyce P, Burch CL (2007) High frequency of mutations that expand the host range of an RNA virus. Genetics 176: 1013–1022.
12. Mindich L, Nemhauser I, Gottlieb P, Romantschuk M, Carton J, et al. (1988) Nucleotide sequence of the large double stranded RNA segment of bacteriophage Φ6 - genes specifying the viral replicase and transcriptase. Journal of Virology 62: 1180–1185.
13. Mindich L, Nemhauser I, Gottlieb P, Romantschuk M, Carton J, et al. (1988) Nucleotide sequence of the large dsRNA segment of bacteriophage Φ6 - genes specifying the viral replicase and transcriptase. Journal of Virology 62: 1180–1185.
14. McGraw T, Mindich L, Frangione B (1986) Nucleotide sequence of the small double stranded RNA segment of bacteriophage Φ6 - novel mechanism of natural translational control. Journal of Virology 58: 142–151.
15. Duffy S, Turner PE, Burch CL (2006) Pleiotropic costs of niche expansion in the RNA bacteriophage Φ6. Genetics 172: 751–757.
16. Vidaver AK, Koski RK, Vanetten JL (1973) Bacteriophage Φ6 - lipid containing virus of *Pseudomonas phaseolicola*. Journal of Virology 11: 799–805.
17. Dennehy JJ, Friedenberg NA, Holt RD, Turner PE (2006) Viral ecology and the maintenance of novel host use. The American Naturalist 167: 429–439.
18. Luria SE, Delbruck M (1943) Mutations of bacteria from virus sensitivity to virus resistance. Genetics 28: 491–511.
19. Kearse M, Moir R, Wilson A, Stones-Havas S, Cheung M, et al. (2012) Geneious Basic: an integrated and extendable desktop software platform for the organization and analysis of sequence data. Bioinformatics 28: 1647–1649.
20. Tamura K, Peterson D, Peterson N, Stecher G, Nei M, et al. (2011) MEGA5: molecular evolutionary genetics analysis using maximum likelihood, evolutionary distance, and maximum parsimony methods. Molecular Biology and Evolution 28: 2731–2739.
21. Zhang Y (2008) I-TASSER server for protein 3D structure prediction. BMC Bioinformatics 9: 40.
22. Roy A, Kucukural A, Zhang Y (2010) I-TASSER: a unified platform for automated protein structure and function prediction. Nat Protoc 5: 725–738.
23. Chao L, Rang CU, Wong LE (2002) Distribution of spontaneous mutants and inferences about the replication mode of the RNA bacteriophage Φ6. Journal of Virology 76: 3276–3281.

24. Mulet M, Lalucat J, Garcia-Valdes E (2010) DNA sequence-based analysis of the *Pseudomonas* species. Environmental Microbiology 12: 1513–1530.

25. Cserzo M, Wallin E, Simon I, von Heijne G, Elofsson A (1997) Prediction of transmembrane alpha-helices in prokaryotic membrane proteins: the dense alignment surface method. Protein Engineering 10: 673–676.

26. Stitt BL, Mindich L (1983) Morphogenesis of bacteriophage Φ6 - a presumptive viral membrane precursor. Virology 127: 446–458.

27. Kenney JM, Hantula J, Fuller SD, Mindich L, Ojala PM, et al. (1992) Bacteriophage Φ6 envelope elucidated by chemical cross-linking, immunodetection, and cryoelectron microscopy. Virology 190: 635–644.

28. Dennehy J (2009) Bacteriophages as model organisms for virus emergence research. Trends in Microbiology 17: 450–457.

29. Ayora-Talavera G, Shelton H, Scull MA, Ren JY, Jones IM, et al. (2009) Mutations in H5N1 Influenza virus hemagglutinin that confer binding to human tracheal airway epithelium. PLoS One 4.

30. Sampson T, Broussard GW, Marinelli LJ, Jacobs-Sera D, Ray M, et al. (2009) Mycobacteriophages BPs, Angel and Halo: comparative genomics reveals a novel class of ultra-small mobile genetic elements. Microbiology-Sgm 155: 2962–2977.

31. Cox J, Putonti C (2010) Mechanisms responsible for a ΦX174 mutant's ability to infect *Escherichia coli* by phosphorylation. Journal of Virology 84: 4860–4863.

32. Baranowski E, Ruiz-Jarabo CM, Domingo E (2001) Evolution of cell recognition by viruses. Science 292: 1102–1105.

33. Li ZJ, Chen HL, Jiao PR, Deng GH, Tian GB, et al. (2005) Molecular basis of replication of duck H5N1 influenza viruses in a mammalian mouse model. Journal of Virology 79: 12058–12064.

34. Subbarao EK, London W, Murphy BR (1993) A single amino acid in the PB2 gene of influenza A virus is a determinant of host range. Journal of Virology 67: 1761–1764.

35. Jonah G, Rainey A, Natonson A, Maxfield LF, Coffin JM (2003) Mechanisms of avian retroviral host range extension. Journal of Virology 77: 6709–6719.

36. Yamada S, Suzuki Y, Suzuki T, Le MQ, Nidom CA, et al. (2006) Haemagglutinin mutations responsible for the binding of H5N1 influenza A viruses to human-type receptors. Nature 444: 378–382.

37. Baer CF, Miyamoto MM, Denver DR (2007) Mutation rate variation in multicellular eukaryotes: causes and consequences. Nature Reviews Genetics 8: 619–631.

38. Ellegren H, Smith NGC, Webster MT (2003) Mutation rate variation in the mammalian genome. Current Opinion in Genetics & Development 13: 562–568.

39. Cuppels DA, Vanetten JL, Lambrecht P, Vidaver AK (1981) Survey of phytopathogenic pseudomonads for a restriction and modification system active on the double-stranded ribonucleic acid phage Φ6. Current Microbiology 5: 247–249.

40. Bono LM, Gensel CL, Pfennig DW, Burch CL (2013) Competition and the origins of novelty: experimental evolution of niche-width expansion in a virus. Biology Letters 9.

41. Dennehy JJ (2012) What can phages tell us about host-pathogen coevolution? International Journal of Evolutionary Biology 2012: 396165.

42. Gottlieb P, Metzger S, Romantschuk M, Carton J, Strassman J, et al. (1988) Nucleotide sequence of the middle dsRNA segment of bacteriophage Φ6 - placement of the genes of membrane associated proteins. Virology 163: 183–190.

43. Mindich L, Mackenzie G, Strassman J, McGraw T, Metzger S, et al. (1985) cDNA cloning of portions of the bacteriophage Φ6 genome Journal of Bacteriology 162: 992–999.

44. Dawkins B (1991) Siobhan's problem: the coupon collector revisited. The American Statistician 45: 76–82.

45. Yamada S, Matsumoto Y, Takashima Y, Otsuka H (2005) Mutation hot spots in the canine herpesvirus thymidine kinase gene. Virus Genes 31: 107–111.

46. Sasadeusz JJ, Tufaro F, Safrin S, Schubert K, Hubinette MM, et al. (1997) Homopolymer mutational hot spots mediate herpes simplex virus resistance to acyclovir. Journal of Virology 71: 3872–3878.

47. Ng TI, Shi Y, Huffaker HJ, Kati W, Liu Y, et al. (2001) Selection and characterization of varicella-zoster virus variants resistant to (R)-9-[4-Hydroxy-2-(hydroxymethy)butyl] guanine. Antimicrobial Agents and Chemotherapy 45: 1629–1636.

48. Wu K, Peng G, Wilken M, Geraghty RJ, Li F (2012) Mechanisms of host receptor adaptation by severe acute respiratory syndrome coronavirus. Journal of Biological Chemistry 287: 8904–8911.

49. Hall AR, Scanlan PD, Buckling A (2011) Bacteria-phage coevolution and the emergence of generalist pathogens. The American Naturalist 177: 44–53.

50. Hoeper D, Kalthoff D, Hoffmann B, Beer M (2012) Highly pathogenic avian influenza virus subtype H5N1 escaping neutralization: more than HA variation. Journal of Virology 86: 1394–1404.

51. Tapparel C, Cordey S, Junier T, Farinelli L, Van Belle S, et al. (2011) Rhinovirus genome variation during chronic upper and lower respiratory tract infections. PLoS One 6.

52. Yuang TTT, Shih C (2000) A frequent, naturally occurring mutation (P130T) of human hepatitis B virus core antigen is compensatory for immature secretion phenotype of another frequent variant (I97L). Journal of Virology 74: 4929–4932.

53. Chattopadhyay S, Weissman SJ, Minin VN, Russo TA, Dykhuizen DE, et al. (2009) High frequency of hotspot mutations in core genes of *Escherichia coli* due to short-term positive selection. Proceedings of the National Academy of Sciences of the United States of America 106: 12412–12417.

54. Hughes AL, Nei M (1988) Pattern of nucleotide substitution at major histocompatibility complex class I loci reveals overdominant selection. Nature 335: 167–170.

55. Keitel T, Diehl A, Knaute T, Stezowski JJ, Höhne W, et al. (2000) X-ray structure of the quinoprotein ethanol dehydrogenase from *Pseudomonas aeruginosa*: basis of substrate specificity. J Mol Biol 297: 961–974.

56. Ghosh M, Anthony C, Harlos K, Goodwin MG, Blake C (1995) The refined structure of the quinoprotein methanol dehydrogenase from *Methylobacterium extorquens* at 1.94 A. Structure 3: 177–187.

57. Oubrie A, Rozeboom HJ, Kalk KH, Huizinga EG, Dijkstra BW (2002) Crystal structure of quinohemoprotein alcohol dehydrogenase from *Comamonas testosteroni*: structural basis for substrate oxidation and electron transfer. J Biol Chem 277: 3727–3732.

58. Petsko GA, Ringe D (2004) Protein Structure and Function. Sunderland, MA: New Science Press.

59. Fisher RA (1930) The Genetical Theory of Natural Selection. Oxford, UK: Clarendon Press.

60. Rose MR (1982) Antagonistic pleiotropy, dominance and genetic variation. Heredity 48: 63–78.

61. Cooper VS, Lenski RE (2000) The population genetics of ecological specialization in evolving *Escherichia coli* populations. Nature 407: 736–739.

62. Kassen R (2002) The experimental evolution of specialists, generalists, and the maintenance of diversity. Journal of Evolutionary Biology 15: 173–190.

63. Dennehy JJ (2014) What ecologists can tell virologists. Annu Rev Microbiol.

64. Sokurenko EV, Gomulkiewicz R, Dykhuizen DE (2006) Opinion - Source-sink dynamics of virulence evolution. Nature Reviews Microbiology 4: 548–555.

65. Chattopadhyay S, Feldgarden M, Weissman SJ, Dykhuizen DE, van Belle G, et al. (2007) Haplotype diversity in "source-sink" dynamics of *Escherichia coli* urovirulence. Journal of Molecular Evolution 64: 204–214.

66. Williams PD (2010) Darwinian interventions: taming pathogens through evolutionary ecology. Trends in Parasitology 26: 83–92.

67. Lide DR (1991) Handbook of Chemistry and Physics. Boca Raton, FL: CRC Press.

68. Kyte J, Doolittle RF (1982) A simple method for displaying the hydropathic character of a protein. Journal of Molecular Biology 157: 105–132.

16

Combining Physicochemical and Evolutionary Information for Protein Contact Prediction

Michael Schneider, Oliver Brock*

Robotics and Biology Laboratory, Department of Electrical Engineering and Computer Science, Technische Universität Berlin, Berlin, Germany

Abstract

We introduce a novel contact prediction method that achieves high prediction accuracy by combining evolutionary and physicochemical information about native contacts. We obtain evolutionary information from multiple-sequence alignments and physicochemical information from predicted *ab initio* protein structures. These structures represent low-energy states in an energy landscape and thus capture the physicochemical information encoded in the energy function. Such low-energy structures are likely to contain native contacts, even if their overall fold is not native. To differentiate native from non-native contacts in those structures, we develop a graph-based representation of the structural context of contacts. We then use this representation to train an support vector machine classifier to identify most likely native contacts in otherwise non-native structures. The resulting contact predictions are highly accurate. As a result of combining two sources of information—evolutionary and physicochemical—we maintain prediction accuracy even when only few sequence homologs are present. We show that the predicted contacts help to improve *ab initio* structure prediction. A web service is available at http://compbio.robotics.tu-berlin.de/epc-map/.

Editor: Yang Zhang, University of Michigan, United States of America

Funding: This work was supported by the Alexander-von-Humboldt Foundation through an Alexander-von-Humboldt professorship, http://www.humboldt-foundation.de/web/start.html, and a NIH Grant (1 R01 GM076706), http://www.nih.gov/. The funders had no role in study design, data collection and analysis, decision to publish, or preparation of the manuscript.

Competing Interests: The authors have declared that no competing interests exist.

* Email: oliver.brock@tu-berlin.de

Introduction

Protein contact prediction identifies potential residue pairs in spatial proximity in the native protein—without knowledge of the native structure itself.

Accurate contact prediction is of great interest and value, as even partial knowledge of residue-residue contacts for a target protein enables the computation of that protein's native structure [1,2]. Information about native contacts can also be used to guide conformational space search in *ab initio* protein structure prediction [3,4]. Contact prediction therefore represents an important intermediate step towards the long-standing goal of tertiary structure prediction [5–7].

There are five broad categories of contact prediction methods: contact prediction from evolutionary information, from sequence-based machine-learning algorithms, from template structures, from structure prediction decoys and by integrating sequence and structural restraints. They differ in the type of information they use to make predictions.

- *Contact prediction from evolutionary information* leverages the fact that two contacting residues are likely to co-evolve to maintain structural integrity of the protein. Thus, co-evolution signals in multiple-sequence alignments (MSAs) can reveal contacting residues in the protein structure.
- *Machine-learning-based methods* exploit evolutionary sequence information in a slightly different way. They identify common sequence patterns occurring around contacting amino acids. These patterns can be learned and recognized to make contact predictions.
- *Template-based methods* leverage the information contained in structure databases, such as the PDB [8]. They search these databases for appropriate structural templates, using sequence matching or threading and then extract contact information from the retrieved templates.
- *Ab initio protein structure prediction methods* use conformational space search and the *physicochemical information* captured in the energy function to make predictions about contacting residues. These methods generate many low-energy candidate structures and use simple occurrence statistics to identify native contacts.
- *Methods that integrate sequence and structural restraints* use sequence-based predictions and additionally take structural restraints into account. Structural restraints are derived from prior knowledge about protein structures or from templates.

Most of the aforementioned categories of contact prediction methods rely on a single source of information. When no valuable information is available from that source, prediction accuracy deteriorates. This effect is drastic if the number of sequences in the alignment is insufficient or if the correct template cannot be retrieved [9–11]. In contact prediction from *ab initio* predicted structures, the quality of information depends on the ability of a search procedure to identify low-energy regions in the energy landscape. If no appropriate regions can be discovered, contact prediction performs poorly.

Methods that use evolutionary information and methods that use physicochemical information are on opposite sides of the spectrum of approaches to contact prediction. Evolutionary information methods are accurate if many sequences are available, but not effective if this information is absent. On the other hand, physicochemical information methods perform well even when only few sequences are available, but do not benefit as much from sequence data as evolutionary methods. Evolutionary and physicochemical information are largely orthogonal information sources. Thus, the combination of those information sources should unlock the synergistic potential of both approaches to perform highly accurate contact prediction.

In this article, we introduce a novel contact prediction method, *EPC-map*, that predicts contacts using two sources of information: evolutionary information from multiple sequence alignments and information from physicochemical energy potentials (EPC-map stands for using Evolutionary and Physicochemical information to predict Contact maps). EPC-map relies on GREMLIN [10], an established method for sequence-based contact prediction, to leverage evolutionary information. To identify and leverage physicochemical information, we present a novel, machine-learning based classifier that uses a graph-based encoding of the structural context of contacts. This classifier distinguishes native from non-native contacts in *ab initio* decoys with unprecedented accuracy. A graphical outline of our method is shown in Figure 1.

In our experiments with 528 proteins, EPC-map reaches 53.2% accuracy on the top scoring $L/5$ predicted long-range contacts (where L is the length of the protein), increasing the accuracy by 7.8% relative to the state of the art. In our analysis, EPC-map is also the best performing method on proteins from CASP10. Furthermore, we show that EPC-map performs better than contemporary methods, regardless how many sequences are available. We further show that physicochemical information improves prediction in cases where deep alignments are not available, effectively alleviating the main weakness of evolution-based contact prediction. To achieve this, EPC-map does not use any structural information from homologous sequences and is therefore effective when templates are not available. Finally, we show that EPC-map predicted contacts improve *ab initio* tertiary structure prediction.

Related Work

We review the different approaches to contact prediction following the categorization established above. Of particular interest in our review is the source of information that is leveraged and under which circumstances the methods are applicable.

Contact prediction from evolutionary information

The earliest methods for contact prediction were based on evolutionary information from multiple sequence-alignments (MSAs). They exploit knowledge of correlated mutations and phylogeny to predict contacts. Over the course of evolution, destabilizing mutations of a specific residue are frequently accompanied by matching mutations of spatially close residues. This is appealing from a theoretical perspective, but initially yielded only poor prediction accuracy [12]. This low accuracy was caused by the presence of transitive correlations obfuscating the information about direct correlations. More recent approaches therefore separate direct and transitive correlations by estimating the inverse covariance matrix of the MSA [13,14]. This achieves

high accuracy but requires on the order of $5L$ sequences for alignment, where L is the length of the protein measured in amino acids [9,10]. Thus, evolutionary methods are of limited use when only few related sequences are available. For many of these proteins with few sequences, the PDB does not contain any structural homologs [10]; these proteins would therefore benefit most from accurate contact predictions.

Contact prediction from sequence-based machine learning

Methods in this category employ machine learning to identify sequence patterns indicative of contacts in the protein structure. They vary in the machine learning algorithm they employ and in their training procedure. Researchers have used neural networks [15–17], support vector machines [11,18], hidden-Markov models [19], or random forests [20] to devise prediction algorithms. More recent approaches employ deep learning architectures [21] and deep learning combined with boosting [7]. Improvements in prediction accuracy stem from the application of novel machine learning algorithms, larger training sets, data preprocessing, and/ or better training procedures.

Methods in this category are robust when only few sequences are available; they consistently perform well for *ab initio* predictions in the CASP experiments [22,23]. Nevertheless, these predictors are not routinely used in structure prediction.

Contact prediction from template structures

A decisively different approach to contact prediction uses information from template structures, making explicit use of structural information available in databases. Methods from this category identify template structures by sequence matching or threading and derive contact predictions from the obtained template [11,24]. If a good template is found, predictions are highly accurate. This accuracy is further improved through the use of multiple templates [11]. Even though template structures are used to predict contacts, the template retrieval step is essentially based on sequence information, rendering these methods unsuitable for novel folds. The failure to identify a good template then leads to significant loss in prediction accuracy [11].

Contact prediction from *ab initio* protein structure prediction

Methods based on *ab initio* protein structure prediction sample an energy function to generate many candidate protein structures, called decoys. As native contacts are energetically favorable, they should occur more frequently in these decoys than non-native contacts. It is possible to obtain accurate contact predictions by analyzing the distribution of contacts in those decoys [25]. This approach effectively leverages another source of information than all sequence-dependent methods: The physicochemistry captured in energy functions and encoded in the decoy structures. This approach has been applied successfully to derive consensus distances [26] and energy-weighted occurrences of residue-residue contacts [27]. Related work uses sampling statistics and machine learning to predict native β-strand contacts [28]. Recently, highly accurate predictions were obtained based on simple occurrence frequencies [25]. Strikingly, this simple heuristic achieved the highest prediction accuracy in an *ab initio* setting at the CASP9 experiment [22] and was among the best in CASP10 [23].

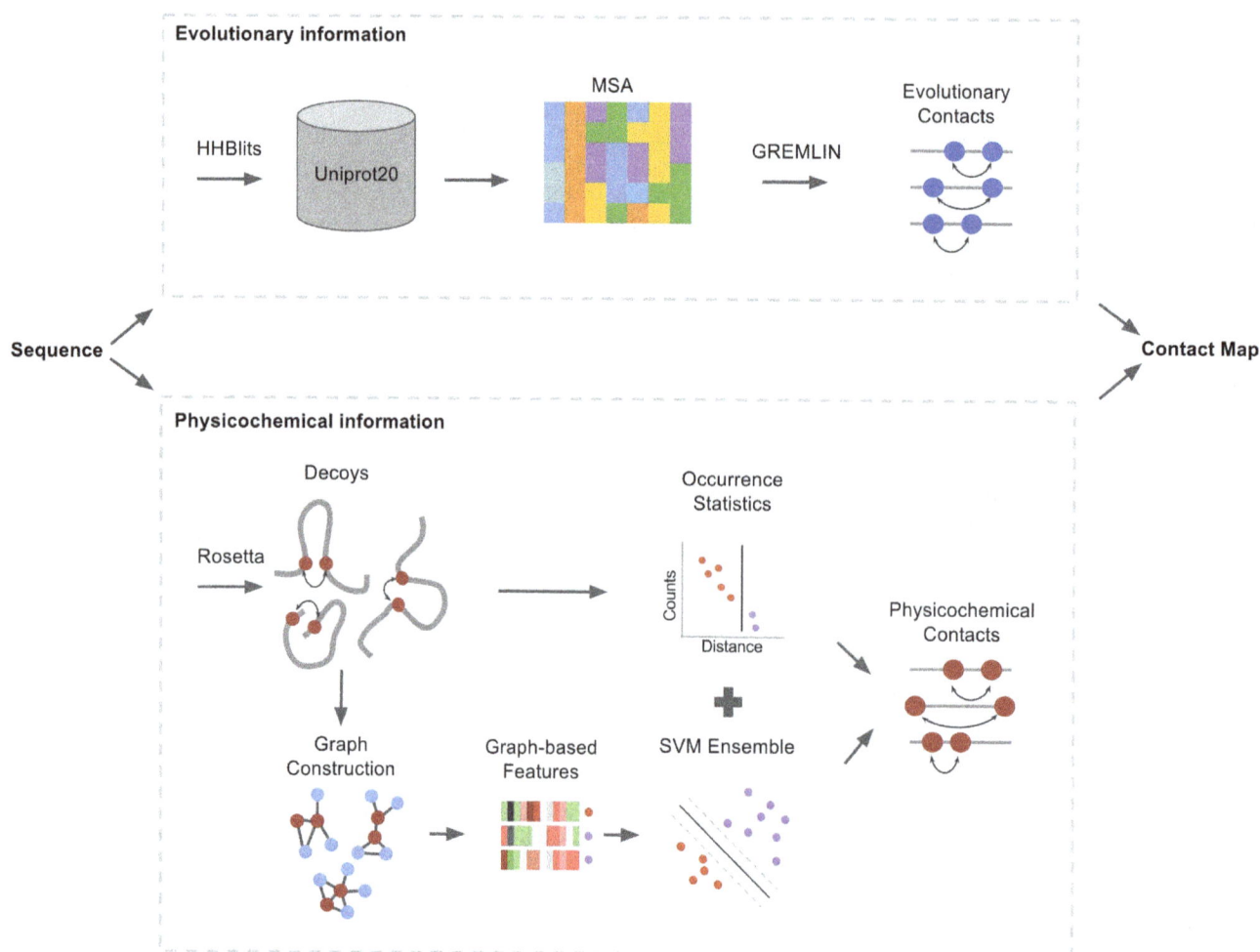

Figure 1. Flowchart overview of EPC-map, combining evolutionary information (upper box) and physicochemical information (lower box). For evolutionary contact prediction, multiple-sequence alignments are constructed by searching the Uniprot20 database with HHblits. GREMLIN is then used to predict contacts from the alignments. For physicochemical contact prediction, decoys are generated with Rosetta. From each decoy, contact graphs are constructed and feature input vectors computed. An SVM ensemble predicts the contact probability from each feature vector. The SVM probability and occurrence statistics predict physicochemical contacts. Lastly, evolutionary and physicochemical contact prediction are combined to form the output of EPC-map.

Contact prediction by integrating sequence and structural restraints

Recent sequence-based methods improve contact prediction by integrating structural restraints into the prediction procedure. Structural restraints have been used as prior probabilities in a pseudo-likelihood approach [10] and by enforcing structural restraints with integer linear programming [29]. These approaches show that the use of structural restraints allows for more accurate contact prediction. Karakas et al. integrate sequence and structure information by coupling sequence-based neural networks with structural templates from fold-recognition [24].

The combination of structure and sequence information is an emerging route towards improved contact prediction. However, combining explicit structure and physicochemical information from structure prediction decoys with evolutionary information has not been attempted yet. As we will show, this approach results in significant performance improvement.

Methods

Contact definition and evaluation

Two residues are defined to be in contact if their C_β atoms (C_α for glycine) are within 8 Å in the native structure of a protein. We investigate medium-range and long-range contact predictions. In medium-range (long-range) contacts, the contacting residues are separated by $12-23$ (>23) residues in the sequence. For evaluation, we consider the top scoring fraction of $L/10$, $L/5$ and $L/2$ predicted contacts, where L is the length of the protein. Our performance metrics are accuracy ($\text{Acc} = \text{TP}/(\text{TP+FP})$) and coverage ($\text{Cov} = \text{TP}_{frac}/\text{TP}_{total}$). Here, TP are true positives: contacts that are predicted and also in contact in the native structure. FP are false positives: contacts that are predicted but not in contact in the native structure. TP_{frac} are the true positives in the top scoring fraction of the predicted contacts, while TP_{total} are the total true positives for a protein. Our main analysis focuses on the top scoring long-range contacts as they are most valuable in structure modeling.

Generation of multiple-sequence alignments

Our method relies on multiple-sequence alignments in two ways. First, the contact graph (defined below) contains information from these alignments, e.g. local sequence conservation of a particular residue. Second, our method incorporates information from sequence-based contact prediction obtained from GREMLIN [10]. These multiple-sequence alignments are generated by searching the query sequence with HHblits [30] (version 2.0.11) against a clustered Uniprot [31] database with maximum pairwise sequence similarity of 20% (dated March 2013).

Evolutionary contact information

We use GREMLIN [10] to obtain evolutionary information contact scores. We obtained a version of GREMLIN from the authors and run it with default parameters.

Decoy generation

The generation of protein decoys is the first step towards leveraging physicochemical information for contact prediction. We use the standard *ab initio* protocol in Rosetta version 3.2 to generate decoys [32]. Rosetta performs fragment assembly with a reduced representation of the protein chain, using a knowledge-based force field. Decoys are refined in an all-atom phase, adding side chains and minimizing the decoys' energy in a hybrid physical/knowledge-based all-atom potential.

Decoys for training. To generate decoys for training purposes, we use three independent Rosetta runs, each with a different strength of native bias [33]. The goal is to obtain decoys that are far away, relatively close, and very close to the native structure. This will provide us with a good decoy training set, containing a diverse set of positive and negative examples.

The native bias is introduced by three different fragment libraries. We quantify the impact of these biases by using the GDT_TS as a measure of quality of the five lowest-energy decoys for each protein. The GDT_TS measure ranges from 0 if two structures are completely dissimilar to 100 for a perfect structural match. In the first fragment library, we exclude proteins homologous to the target sequence. Decoys obtained using this fragment library are usually far away from the native structure (min/max/mean/median GDT_TS of 10.7/73.5/26.5/24.3). To obtain the second fragment library, we allow fragments from homologous proteins to be included, leading to decoys closer to the native structure (min/max/mean/median GDT_TS of 11.2/99.3/28.1/25.4). We enrich the third fragment library with fragments from the native structure itself, enabling decoys that are even closer to the native structure of the target (min/max/mean/median GDT_TS of 10.4/99.8/37.0/31.2). Note that this use of the native bias is acceptable, as we are only generating decoys for training purposes. For each type of fragment library, we generate 200 decoys, resulting in 600 decoys per protein. From each of these sets of 200 decoys, we retain the 3% with lowest energy.

Decoys for prediction. To perform contact prediction, we generate 1000 decoys without homologous fragments, i.e. using the first fragment library, and retain the top 2% based on energy. Thus, our approach does not use any structural information from homologous sequences such as templates. Therefore, we set EPC-map apart from methods that use threading to find templates and extract contacts from them.

Contact graphs for feature generation

Past research has demonstrated the effectiveness of decoy-based contact predictions using simple occurrence statistics [25].

However, these statistics were gathered on entire decoys, selected by their energy. The energy criterion favors all occurring contacts equally, even the non-native ones, making it difficult to differentiate between native and non-native contacts.

Our approach is based on the insight that the discrimination of native and non-native contacts in decoys must improve significantly if information from the decoy's energy is complemented with information specific to the individual contacts. Our main assumption is that this information is captured by the immediate structural environment of a contact. Thus, we would like to characterize this local environment and learn how to differentiate native from non-native environments.

To characterize the properties of a contact's neighborhood, we use undirected graphs (refer to Figure 2 for the remainder of this section). In these graphs, nodes correspond to residues and edges connect contacting residues. Nodes and edges are labeled with physicochemical, structural and evolutionary characteristics; these labels are described in the supporting information (Text S1 and Tables S1–S2 in File S1). First, we consider the neighborhood of individual residues. The neighborhood of residue i is defined as all residues up to two positions away in sequence, i.e. residues $i-2, i-1, i, i+1, i+2$, as well as all residues in contact with those, according to the definition of a contact given in Methods. For α-helices, the $i-4$, i, $i+4$ residues are used instead to include the residues with the same facing towards the contact on subsequent helix turns. We capture this notion of neighborhood of residue i in a neighborhood graph N_i (Figures 2A and B).

To capture the local context of a contact C_{ij} between residue i and j, we use two different kind of graphs. The shared neighborhood graph (SN_{ij}) captures the shared neighborhood of the residues i and j in the context of their sequential neighbors $\{i-2, i-1, i+1, i+2\}$ and $\{j-2, j-1, j+1, j+2\}$, thus being the intersection of N_i and N_j: $SN_{ij} = N_i \cap N_j$ (Figure 2C). Additionally, we capture the local context that directly influences the residues i and j by the immediate neighborhood graph (IN_{ij}) that is formed by all residues in immediate contact to residues i and j (Figure 2D).

These graphs are the fundamental data structure of our method. We then apply machine learning to learn to discriminate native from non-native graphs (i.e. contacts). The success of learning critically depends on the expressiveness of the employed features, which we will describe next.

Overview of used features

To capture contact characteristics in decoys, we define $n = 48$ features, each representing a measurable property of residues in contact. Each feature consists of one or several binary or real-valued inputs. All of these inputs are joined into a single vector, which serves as input to a support vector machine (SVM) during training and testing. Note that features defined on graphs are evaluated for the shared neighborhood graphs and immediate neighborhood graphs separately. Thus, each graph feature is present two times in the final input vector which has a length of $m = 228$.

Features are categorized into seven groups: pairwise, graph topology, graph spectrum, single node, node label statistics, edge label statistics, and whole protein features. We will briefly motivate each of these groups (see also Table 1).

Pairwise features capture properties of the amino acids i and j, such as the chemical type, secondary structure, and solvent accessibility.

We use topological and spectral graph features to characterize the underlying contact network. For example, nodes in well-packed regions of a decoy will tend to have a higher degree than

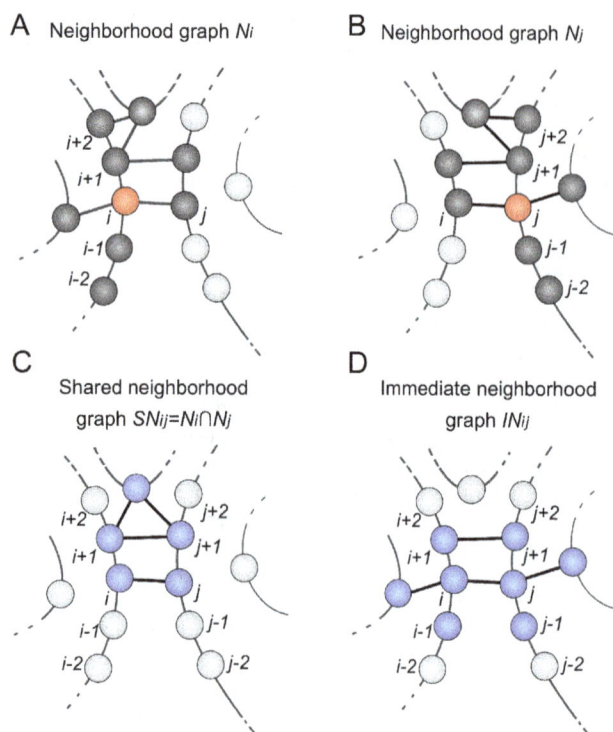

Figure 2. Definition of graphs used to model the neighborhood of the contacting residues i and j: Nodes represent residues (circles), edges represent contacts (solid black lines). A: The neighborhood graph N_i for residue i contains all residues in contact with residues $i-2, i-1, i, i+1$, and $i+2$ (dark grey). **B:** The neighborhood graph N_j. **C:** The shared neighborhood graph SN_{ij} for the contact between residues i and j is defined by the intersection of N_i and N_j. Residues that belong to SN_{ij} are shown in blue. Shared neighborhood graphs capture the local context of the shared neighborhood of the contacting residues. **D:** The immediate neighborhood graph IN_{ij} is defined by all residues that are in contact to i or j. Residues that belong to IN_{ij} are shown in blue. Immediate neighborhood graphs capture the direct neighborhood of the contacting residues.

those in poorly-packed regions. Consequently, contacts in well-packed regions have a higher likelihood of being native. This can be measured by the average degree centrality of the graph.

Node and edge label statistics extract additional information from contact networks to complement topological considerations. For example, native contacts in the protein core should be embedded in a network of hydrophobic residues. This property is captured by the distribution of the chemical nature of neighboring nodes.

Whole protein features specify information about the protein at hand, such as amino acid composition, secondary structure composition and protein chain length.

The individual features from each group are listed in Tables S3-S9 in File S1. A detailed description of each feature and its implementation is also provided in the supporting information (Text S2 in File S1).

Software used for feature generation. Several of our features are based on external software. Solvent accessibility and free solvation energies are computed with POPS [34]. We use STRIDE [35] to obtain secondary structure and hydrogen bonding assignments from decoys. Sequence conservation features are computed as described by Fischer et al. [36]. Some of our pairwise features are inspired by Cheng et al. [18] and we use a contact potential introduced in [20]. We construct graphs by using the NetworkX Python package [37] and use the SVM library of scikit-learn [38]. Finally, many of our topological and spectrum features have been shown to be effective for graph classification [39].

Next, we describe how we use these features to train a support vector machine with physicochemical contact information.

SVM training with physicochemical contact information

A challenging aspect in using support vector machines for contact prediction is that the contact prediction learning problem is inherently imbalanced i.e. there are many more non-native than native contacts in the decoys of our training set (see respective section for details on training set construction). Random under-sampling is a common technique to cope with the unbalanced learning problem [40]. However, performing random under-sampling leads to information loss because many training instances are not used for learning. Furthermore, the resulting learner might be biased towards the specific training sample, leading to high

Table 1. Overview of the features used for contact prediction. A detailed description of the features is given in the supporting information.

Group	Feature examples	Number of inputs
Pairwise	Chemical type, secondary structure, solvent accessibility, sequence separation, hydrogen bonding, sequence separation from N/C-terminus, contact potential, distance, average distance in ensemble, mutual information	49
Graph topology[a]	Number of nodes, number of edges, average degree centrality, average closeness centrality, average betweenness centrality, graph radius, graph diameter, average eccentricity, number of end points, average clustering coefficient	10
Graph spectrum[a]	Largest two eigenvalues, number of different eigenvalues, sum of eigenvalues, energy of adjacency matrix	5
Single node[a]	Degree, closeness centrality, betweenness centrality, sequence conservation and sequence neighborhood conservation for i and j	10
Node label statistics[a]	Chemical type of residues, secondary structure descriptors, solvent accessibility, hydrogen bonding, average free solvation energy, 4-bin solvation energy distribution, entropy of labels, neighborhood impurity degree, average distance from centroid, sequence conservation, sequence neighborhood conservation	43
Edge label statistics[a]	Link impurity, 5-bin mutual information distribution, cumulative mutual information, 3-bin contact potential distribution	12
Whole protein	Amino acid composition, secondary structure composition, length class	29

[a]Graph-based features.

variance. We reduced this effect by training an SVM ensemble, with each SVM performing its own random undersampling. First, proteins in the training set are randomly split into five non-overlapping subsets. Second, a SVM classifier is trained for each subset with random undersampling by selecting 50 native and 150 non-native contacts. This procedure handles the imbalanced training problem by random undersampling and reduces the effect from information loss and variance by using multiple SVM instances. This yields an SVM ensemble of five different SVM classifiers, one for each subset. Each subset uses approximately 30.000 training instances. Each input in the input vector is normalized by subtracting the mean and dividing by standard deviation.

We use a binning procedure [41] to obtain calibrated probability estimates. The raw SVM output values between the 5 and 95 percentile are grouped into ten bins. Then, the probability of a native contact is computed separately for each bin. We find that this procedure improves prediction accuracy compared to Platt's method [42].

We use the Gaussian kernel for training and determine the cost and the kernel parameter γ by 10-fold cross-validation on the EPC-map_train data set, optimizing the long-range $L/5$ accuracy. Furthermore, probability estimates by binning are obtained by 5-fold cross-validation on the training set. We find that $c = 10$ for the soft margin parameter and the Gaussian kernel parameter $\gamma = 0.001$ yield the best performance.

SVM prediction of contacts from physicochemical information

To perform contact prediction for a protein, we consider the 2% lowest-energy decoys generated by Rosetta, using a homology-free fragment library. Each contact present in each decoy is scored by the SVM ensemble. The probability $p(C_{ij})$ for an individual contact C_{ij} in one decoy is given by:

$$p(C_{ij}) = \frac{1}{l} \sum_{k=1}^{l} p_{\text{SVM}}^{k}(C_{ij}),$$

where $p_{\text{SVM}}^{k}(C_{ij})$ is the probability output value of the k-th SVM.

Note that the same contact may appear in multiple decoys. The final score of a contact is the average score over all decoys in the *decoy ensemble* containing that contact:

$$S_{\text{ENS},ij} = \frac{1}{m} \sum_{n=1}^{m} p(C_{ij}^{n}),$$

where $S_{\text{ENS},ij}$ is the score of the contact between residues i and j, C_{ij}^{n} is the contact in the n-th decoy, $p(C_{ij}^{n})$ is the output of the SVM ensemble for the contact C_{ij} in the n-th decoy, and m is the number of decoys containing the contact.

Combination of evolutionary and physicochemical information for contact prediction

Finally, we combine evolutionary and physicochemical information to predict contacts. The output of the SVM system is combined with the frequency f_{ij} of contact C_{ij} occurring in the ensemble and the score output value of GREMLIN $S_{\text{GREMLIN},ij}$:

$$S_{\text{EPC-map}}(C_{ij}) = \beta(\alpha f_{ij} + (1-\alpha)S_{\text{ENS},ij}) + (1-\beta)S_{\text{GREMLIN},ij}.$$

The α and β parameters are found by optimizing the $L/5$ accuracy of long-range contacts by five-fold cross-validation on the training set. Output values from GREMLIN scale differently, depending on how many sequences are available. Furthermore, the performance of GREMLIN is highly dependent on the number of available sequences. Predicted contacts from GREM-LIN perform well in template discrimination tasks if $5L$ or more sequences are available [10]. This indicates that GREMLIN is accurate for proteins with more than $5L$ sequences, but does not consistently perform well if less sequences are found. Thus, separate β parameters are tuned for proteins with $<5L$ sequences and for proteins with $\geq 5L$ sequences. With this procedure, we supplement GREMLIN's predictions that are already accurate (when many sequences are available) and compensate for GREMLIN's loss in accuracy when only few sequences are available. Final parameters are: $\alpha = 0.425$ and $\beta = 0.275$ for proteins with $<5L$ and $\beta = 0.35$ for proteins with $\geq 5L$ sequences, respectively. The output of our algorithm is the list of contacts in rank order based on their score.

Data sets

We compiled a non-redundant training set, EPC-map_train, to provide patterns of native and non-native contacts for learning. In addition, we used six test sets (EPC-map_test, D329, SVMCON_-test, CASP9-10_hard, CASP10 and CASP10_hard) to evaluate the performance of our method.

EPC-map_train. The training set consists of protein chains culled from the PDB using PISCES [43]. The resulting set contains chains of 50–150 amino acids with at most 25% sequence identity and 0–2 Å resolution for X-ray structures. We limited ourselves to smaller chains for training to facilitate rapid method development and testing. From this set we removed: *a)* chains containing chain breaks (a chain break is defined as a distance larger than 4.2 Å between Cα atoms of two residues adjacent in sequence [20]) and *b)* chains with extended structures and chains whose structure is significantly determined through packing to other chains in their PDB structure or interior bound ligands. To avoid structural redundancy, we performed pairwise structural alignment with Deepalign [44] and removed chains that had a GDT_TS of 60 or more to any other chain in the training set (if the aligned region comprised more than 60% of the smaller protein).

Finally, we removed all chains from the training set that had more than 25% sequence identity or a GDT_TS of 60 or more to any of the chains in the EPC-map_test, D329, SVMCON_test and CASP9-10_hard test sets. From the remaining chains, we removed 15% randomly to form the test set EPC-map_test. The final training set consists of 742 chains. All of our predictions on the CASP10/CASP10_hard data set are performed with a version of EPC-map that uses the 727 training proteins dated before CASP10 (May 2012).

EPC-map_test. This test set contains 132 chains randomly selected from the training set as described above. The proteins in this set were not used for training.

D329. The D329 data set [20] consists of 329 chains of varying sizes (55–458 amino acids).

SVMCON_test. The SVMCON_test data set is comprised of 48 medium-sized protein chains (46–198 amino acids) [18]. We excluded one protein (1aaoA), because it is listed as a theoretical model in the PDB.

CASP9-10_hard. We used 16 protein chains from the CASP9 experiment and four protein chains from the CASP10 experiment (20 total). Chains in this set contain *only* free modeling domains (FM category in CASP) or difficult template-based modeling

(TBM/FM category) domains. Note that proteins containing at least one FM or TBM/FM domain are also excluded from this set. These proteins are among the most difficult modeling targets, because they do not have many sequence homologs, templates or have unusual folds. Since there is no template information available for these proteins, they represent cases for which contact prediction might be most useful.

CASP10. We used 104 proteins for which crystal structures are available from the CASP website at the time of this study.

CASP10_hard. We also evaluated our approach on a subset of the CASP10 data set by taking difficult proteins from CASP10. Unfortunately, only four proteins are available from CASP10 that are exclusively comprised by FM or TBM/FM domains. Thus, we selected all CASP10 proteins that contain at least one FM or TBM/FM domain for this evaluation. This results in 14 protein chains.

Importantly, the CASP10 and CASP10_hard data sets allow us to compare our results to all groups that participated in CASP10. The results from the CASP10 methods are available from the CASP website. All of our predictions on the CASP10/CASP10_hard data set are performed with a version of EPC-map that only uses databases and proteins that are dated before CASP10 (May 2012). This allows us to make a fair comparison of EPC-map with all other methods that only had information available that is dated May 2012 or earlier.

Modeling of contact restraints in Rosetta

In addition to the accuracy of contact prediction, we also quantify the benefits gained from the predicted contacts in *ab initio* protein structure prediction. We use contacts as distance restraints in *ab initio* Rosetta calculations. In other words, we include in Rosetta's energy function the degree to which predicted contacts are present in a decoy. However, contact predictions are likely to contain false positives. Therefore, we do not penalize the violation of a particular predicted contact. Instead, we devise an energy term to maximize the number of satisfied contacts for a given conformation. This is accomplished by incorporating a modified Lorentz function L into the energy function of Rosetta:

$$L(d_{ij}) = \begin{cases} -\dfrac{1}{\pi} \dfrac{\frac{1}{2}w}{(d_{ij}-l)^2 + (\frac{1}{2}w)^2} & if \quad d_{ij} < l \\[3ex] -\dfrac{1}{\pi} \dfrac{\frac{1}{2}w}{(\frac{1}{2}w)^2} & if \quad l < d_{ij} \le u \\[3ex] -\dfrac{1}{\pi} \dfrac{\frac{1}{2}w}{(d_{ij}-u)^2 + (\frac{1}{2}w)^2} & if \quad u < d_{ij} \end{cases}$$

where d_{ij} is the distance between residues i and j in the decoy. Further parameters of the function are the lower bound l, the upper bound u and the half width w. We use $l = 1.5$ Å, $u = 8$ Å and $w = 1.0$. The w parameter regulates how quickly the energy bonus decreases when d_{ij} is not within the lower/upper bounds, with w being the half-width i.e., the violation in where the $E_{max}/2$ is still rewarded with E_{max} being the maximum energy bonus. If the restraint is satisfied, the full energy bonus is rewarded. Restraints that are only mildly violated ($d_{ij} - l \le 2w$ or $u - d_{ij} \le 2w$) result in a decreased energy bonus. The contribution of the restraint falls back to zero in case of significant violation ($l - d_{ij} > 2w$ or $d_{ij} - u > 2w$). Note that the former case ($l - d_{ij} > 2w$) actually does not apply when modeling contact restraints with $l = 1.5$ Å and $w = 1.0$. However, the potential is designed for general use and for other parameters (for example $l = 10$ and $u = 20$) the $l - d_{ij} > 2w$ case is meaningful.

Source files that implement our scoring function are available on request.

Results and Discussion

We evaluate the performance of EPC-map on the six test data sets described in Methods. We first evaluate our method on proteins of CASP10 and compare our results with the top methods from the CASP10 experiment. We then analyze the prediction performance of EPC-map on the remaining data sets. Then, we discuss how the performance of EPC-map varies with sequence alignment depth and protein chain length. Furthermore, we discuss the limitations of EPC-map and show that predicted contacts from EPC-map improve *ab initio* structure prediction.

We measure performance by evaluating the accuracy and coverage of the top scoring $L/10$, $L/5$ and $L/2$ contacts from each prediction method, where L is the length of the protein. Because long-range contacts are of most value in structure modeling [45], our main discussion focuses on long-range contacts.

Performance on test data sets

We first evaluate the contact prediction performance on proteins from the CASP10 experiment. This allows us to compare our approach with several methods that participated in CASP10. We downloaded the results of the CASP10 methods from the CASP10 website. EPC-map does not use any information from structural homologs. Therefore, it is appropriate to compare the performance of EPC-map with that of several other sequence-based methods.

We have selected the six methods that showed top performance in the CASP10 experiment [23], submitted predictions for all targets, and –to the best of our knowledge– did not use templates and/or server models for contact prediction.

We have chosen the following six methods: Group 305 (server name: IGB-Team, program name: CMAPpro) [21], Group 222 (server name: MULTICOM-construct, program name: DNCON) [7], Group 358 (server name RaptorX-Roll), Group 113 (server name: SAM-T08 Server, program name: SAM-T08) [47], Group 314 (server name: Proc_S4), Group 424 (MULTICOM-Novel, program name: NNcon) [16]. IGB-Team and MULTICOM-construct use deep networks to predict contacts. RaptorX-Roll uses a context-specific distance-based statistical potential [46], Proc_S4 uses random forests to predict contacts and is based on the original method that also participated in CASP9 [20]. SAM-T08 and MULTICOM-Novel are based on neural networks.

Additionally, we include PhyCMAP, a recent method based on random forests and physical constraints that has been shown to outperform current methods on the CASP10 dataset [44]. Furthermore, we include PSICOV [14] and GREMLIN [10] which predict contacts from evolutionary information. We run locally installed versions of PSICOV and GREMLIN; we use PhyCMAP by its web service. Finally, we evaluate contact prediction by occurrence frequencies on the decoys we generate, which we will refer to as Counting. This is similar to some of the most accurate contact predictors from the recent CASP experiments [22,23]. The main difference is that decoy-based methods in the CASP setup use a consensus approach with decoys from several tertiary prediction servers that might use different energy functions, sampling methods and/or templates. The decoys from our Counting approach all stem from Rosetta *ab initio* generated decoys.

Unless stated otherwise in this section, we refer to the accuracy of the top scoring $L/5$ long-range contacts.

Figure 3 summarizes the long-range $L/5$ contact prediction performance on the CASP10 data set. Detailed information about the medium- and long-range performance on different L/n cutoffs is given in Tables S10 and S11 in File S1. EPC-map reaches a mean accuracy of 0.492, the second-best method (GREMLIN) reaches a mean accuracy of 0.448, followed by PhyCMAP with 0.325 mean accuracy. MULTICOM-construct(DNCON), the best performing method of the CASP10 experiment [23], has a mean accuracy of 0.285 on the CASP10 dataset. Thus, EPC-map is 4.4% more accurate than GREMLIN and 20.7% more accurate than MULTICOM-construct(DNCON) on the entire CASP10 dataset.

Of the 104 proteins in CASP10, 14 proteins contain domains that are classified by having free modeling or difficult free-modeling/template-based domains. For this kind of proteins, contact prediction is most useful because the structure cannot modeled by only using templates. For these difficult proteins (CASP10_hard) the top performing methods of the CASP10 experiment predict contacts with 0.165–0.192 accuracy. GREM-LIN and PhyCMAP are competitive on this dataset with 0.203 and 0.200 mean accuracy, respectively. EPC-map reaches 0.246 accuracy, improving on GREMLIN by 4.3% and being 5.4% more accurate than the best CASP10 method on this difficult data set.

Ideally, we would compare our method with the best methods of the CASP10 experiment on all test sets. Unfortunately, standalone

versions of many of the best-performing CASP10 methods were not available to us at the time of this study and their server implementations are not designed for the evaluation of hundreds of proteins. Thus, for the remainder of this study, we only evaluate methods that are available as a standalone version or server that allows for high-throughput contact prediction. This includes NNcon, PhyCMAP, Counting, PSICOV and GREMLIN. PhyCMAP and GREMLIN perform on par or better than the top methods of the CASP10 experiment and can therefore be considered to be state of the art (see Figure 3). Therefore, comparing EPC-map with these methods provides a fair estimate of state-of-the-art performance. Note that for the remaining data sets, we use sequences and training proteins dated after CASP10 in EPC-map to evaluate the current capabilities of our method.

Figure 4 summarizes the long-range $L/5$ contact prediction accuracies, grouped by the remaining data sets (CASP9-10_hard, EPC-map_test, D329, SVMCON_test). For detailed analysis, refer to Tables S12–S15 in File S1.

We structure our further discussion of prediction performance based on data set difficulty, as judged by the distribution of available sequences in the MSA, i.e. alignment depths (Figure 5). For the most difficult data set, CASP9-10_hard, EPC-map (mean accuracy 0.322) improves the mean prediction accuracy by 9.7% over the next best method (see Figure 4). Interestingly, neither the best structure-based method (Counting) nor the best method that uses evolutionary information (GREMLIN) delivers good results

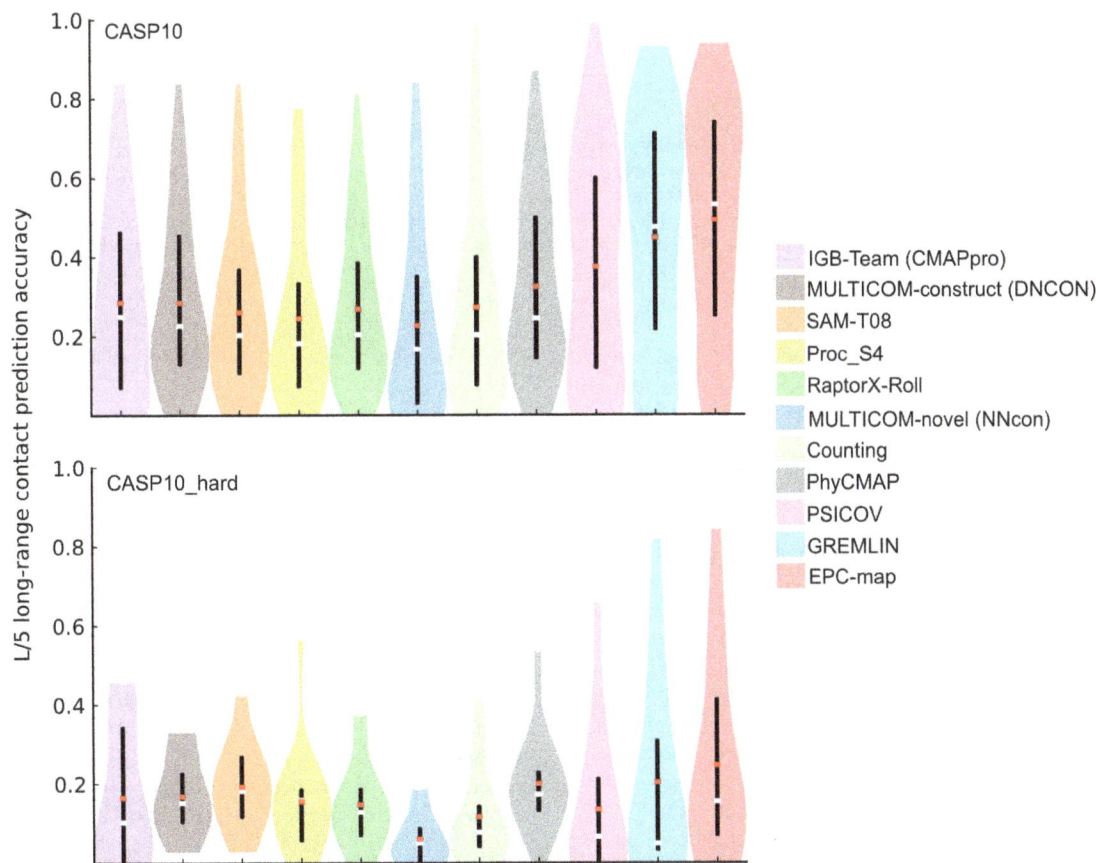

Figure 3. Prediction performance overview for the CASP10 and CASP10_hard data sets. The figure shows the long-range contact prediction performance of the top scoring L/5 contacts. Different methods are shown as color coded violin plots. The lower and upper end of the black vertical bars in each violin denote the accuracy at the 25 and 75 percentile, respectively. White horizontal bars indicate the median, red horizontal bars the mean accuracy. The distribution of the prediction accuracies for individual proteins is indicated by the shape of the violin.

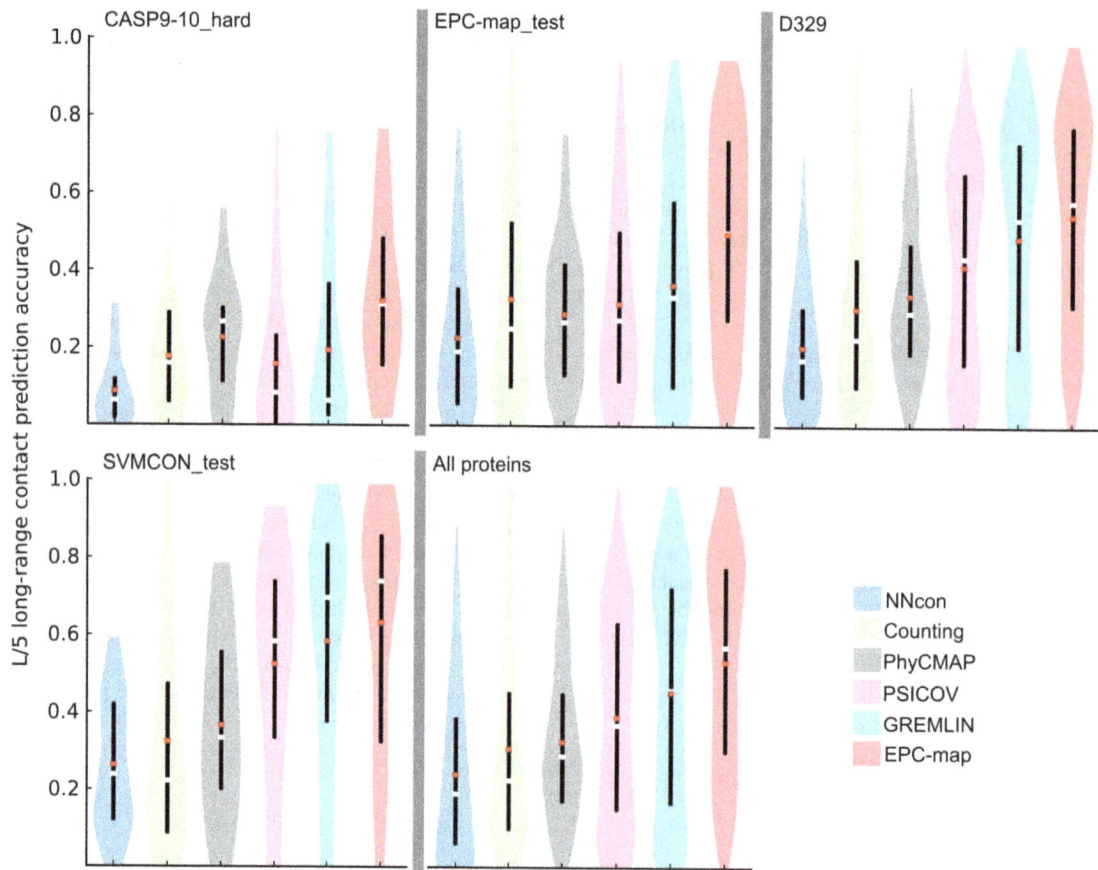

Figure 4. Prediction performance overview for the CASP9-10_hard, EPC-map_test, D329 and SVMCON_test data sets. The figure shows the long-range contact prediction performance of the top scoring L/5 contacts. Different methods are shown as color coded violin plots. The lower and upper end of the black vertical bars in each violin denote the accuracy at the 25 and 75 percentile, respectively. White horizontal bars indicate the median, red horizontal bars the mean accuracy. The distribution of the prediction accuracies for individual proteins is indicated by the shape of the violin. Data sets are sorted from difficult (CASP9-10_hard) to easy (SVMCON_test). The last panel shows the pooled results for all proteins from these data sets.

for this data set (mean accuracies of 0.173 and 0.193, respectively). However, the combination approach taken by EPC-map unlocks the potential of both, evolutionary and physicochemical information methods.

On the EPC-map_test data set, EPC-map (mean: 0.496, median: 0.5) performs on average 13.3% better than GREMLIN (mean: 0.363, median: 0.333), the second-best performing method. In this data set, 45% of the proteins have alignments with fewer than $3.2L$ sequences. These proteins are difficult to predict with evolutionary methods [9,10], showing that the approach taken by EPC-map is well suited for proteins with a low number of sequences.

For the easier data sets, D329 and SVMCON_test, the improvements are less pronounced but EPC-map still outperforms the second-best method by 5.9% and 4.8%, respectively. These two data sets contain many proteins with deep alignments, leading to the robust performance of methods relying on evolutionary sequence information. However, the additional physicochemical information leveraged by EPC-map leads to further performance improvements.

Averaged over 528 proteins from the CASP9-10_hard, EPC-map_test, D329 and SVMCON_test data sets, EPC-map reaches 53.2% mean accuracy and 57.1% accuracy at the median for top $L/5$ predicted long-range contacts. The second best is GREMLIN

with 45.4% mean accuracy and 46% median accuracy. Thus, EPC-map improves the mean accuracy by 7.8% and the median accuracy by 11.1%. Additionally, predictions with $L/5$ accuracy higher than 0.3 are more frequent for EPC-map (394 cases, 74%) then for GREMLIN (338 cases, 64%). We also find that EPC-map significantly improves the medium-range contact prediction accuracy (see Tables S10–S15 in File S1) in most cases.

EPC-map achieves superior performance by integrating the physicochemical information of the energy function of structure prediction with the evolutionary sequence information from multiple sequence alignments.

Dependence of contact prediction accuracy on alignment depth and sequence length

In addition to the performance analysis on various data sets, we further analyzed the prediction performance as a function of other factors, such as alignment depth and sequence length. For this analysis, we used the proteins from the CASP9-10_hard, EPC-map_test, D329 and SVMCON_test data sets.

Figure 6 shows the prediction performance with increasing alignment depth. The performance of all methods increases with the amount of available sequences. Evolutionary methods (PSICOV, GREMLIN), perform poorly in cases with less than $1L$ sequences, while being clearly superior to decoy-based

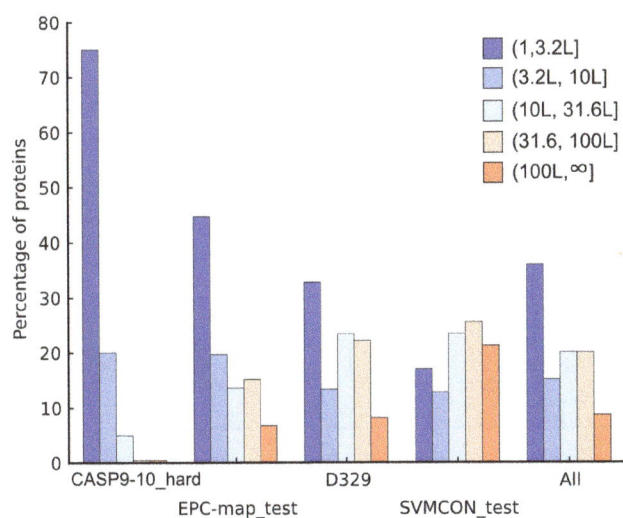

Figure 5. Alignment depth composition of the CASP9-10_hard, EPC-map_test, D329 and SVMCON_test data sets. Proteins are grouped into bins based on their number of sequences in the alignment. Colors correspond to a particular bin, from dark blue (few sequences) to red (many sequences). Data sets are sorted from difficult (CASP9-10_hard) to easy (SVMCON_test). The last panel shows the pooled results.

(Counting) and machine-learning based methods (NNcon, PhyC-MAP) in cases with more than $5L$ sequences. On the other hand, decoy-based and machine-learning based methods perform robustly in the $(1-1L]$ and $(1L-5L]$ intervals, but do not benefit as much from more than $5L$ sequences as evolutionary methods. EPC-map improves prediction accuracy over the second best method, regardless how many sequences are available. This makes EPC-map a versatile approach to contact prediction that performs robustly for proteins with low and high numbers of homologous sequences.

Furthermore, we analyzed the performance of the SVM component and its contribution to the overall performance of the system. First, we analyze the performance of the single SVMs that are part of the SVM ensemble. The individual SVM classifiers reach accuracies between 0.267–0.287. Using an SVM ensemble improves the accuracy to 0.332 (see Table S16 in File S1). Thus the SVM ensemble improves prediction accuracy over the single SVMs. Furthermore, training five SVMs with small data subsamples facilitates faster training compared to a single SVM that uses all training data.

Second, we omit the SVM component and only combine Counting and GREMLIN scores. The β value is re-tuned in the same fashion as described in Methods. We find that the improvement by the SVM component is most pronounced when few sequences are available (see Table 2). In case of fewer than $1L$ alignments, the SVM component improves the mean long-range $L/5$ prediction accuracy by 3.4%. Our experiments show that physicochemical information is most helpful if insufficient sequences are available for sequence-based methods, effectively compensating their major shortcoming. We believe that combination of physicochemical and evolutionary information is an attractive route to advance the currently rapid evolving field of contact prediction.

Naturally, any decoy based-method (such as EPC-map) depends on the quality of the generated decoys. Generating decoys for larger proteins is more difficult and they are likely poorer in

quality. This might affect the prediction quality of decoy-dependent methods more than sequence-based methods. Figure 7 shows the prediction performance of EPC-map, GREMLIN, Counting and Counting +SVM (which is the SVM component from EPC-map) versus the sequence length of the proteins. The improvement in prediction accuracy by EPC-map is most pronounced for proteins smaller than 250 amino acids. For smaller proteins, Counting is performing better due to higher quality decoys. In part, this accounts for the good performance of EPC-map on shorter targets. However, the SVM component of EPC-map consistently improves pure decoy-based prediction over Counting by leveraging physicochemical information (see Figure 7). EPC-map is still more accurate than GREMLIN for longer proteins, but the performance improvement over GREMLIN is less pronounced, probably due to lower-quality decoys. Nevertheless, EPC-map is still ahead or on par with GREMLIN for larger proteins.

Limitations of EPC-map

The computational most intense step of EPC-map is the generation of decoys for contact prediction. In the construction of our training and test sets, we limited the maximum length of the proteins to 150 amino acids to allow for faster training and testing. For proteins with 250 residues, Rosetta needs approximately ten minutes per decoy which results in 7 CPU days for 1000 decoys. We run Rosetta on a compute cluster with 100 nodes, thus we need about 100 minutes for decoy generation for a protein of this size. In contrast, feature generation and prediction with by the SVM ensemble is quite fast and takes only a couple of minutes on a single CPU. However, EPC-map is computationally much more intense than sequence-based methods.

On the one hand, this might render EPC-map unsuitable for some applications, such as proteome-wide analysis of protein contacts. On the other hand, decoy generation can be easily parallelized and run on low-cost commodity clusters with sufficient speed for many practical applications.

However, the main purpose of contact prediction is to aid *ab initio* tertiary structure prediction, which naturally requires substantial computational resources. Thus, the required computation power might already be available to many laboratories that work on *ab initio* structure prediction. For this application, EPC-map contacts might even save computational time needed in *ab initio* structure prediction by guiding conformational space search towards the native state. In any case, if the computational requirements exceed available resources, EPC-map predicted contacts can be obtained from our web service at http://compbio.robotics.tu-berlin.de/epc-map/.

Improvement of *ab initio* structure prediction by using predicted contacts

The main purpose of contact prediction is to aid tertiary structure prediction. We tested the impact of including information from EPC-map predictions into *ab initio* Rosetta calculations for the 132 proteins from EPC-map_test. We model contacts as distance restraints using a bounded Lorentz function (see Methods for details). This function assigns an energy bonus to satisfied restraints. If a restraint is not satisfied, the energy bonus falls back to zero. This implies that restraints violated by a large margin are simply neglected, compensating the detrimental effect of false positive contact predictions. Example configuration files and commands of our contact-guided structure prediction setup are provided in the supporting information.

For each target, we generate 1000 decoys with contact restraints and 2000 without restraints. We use 2000 decoys in the second

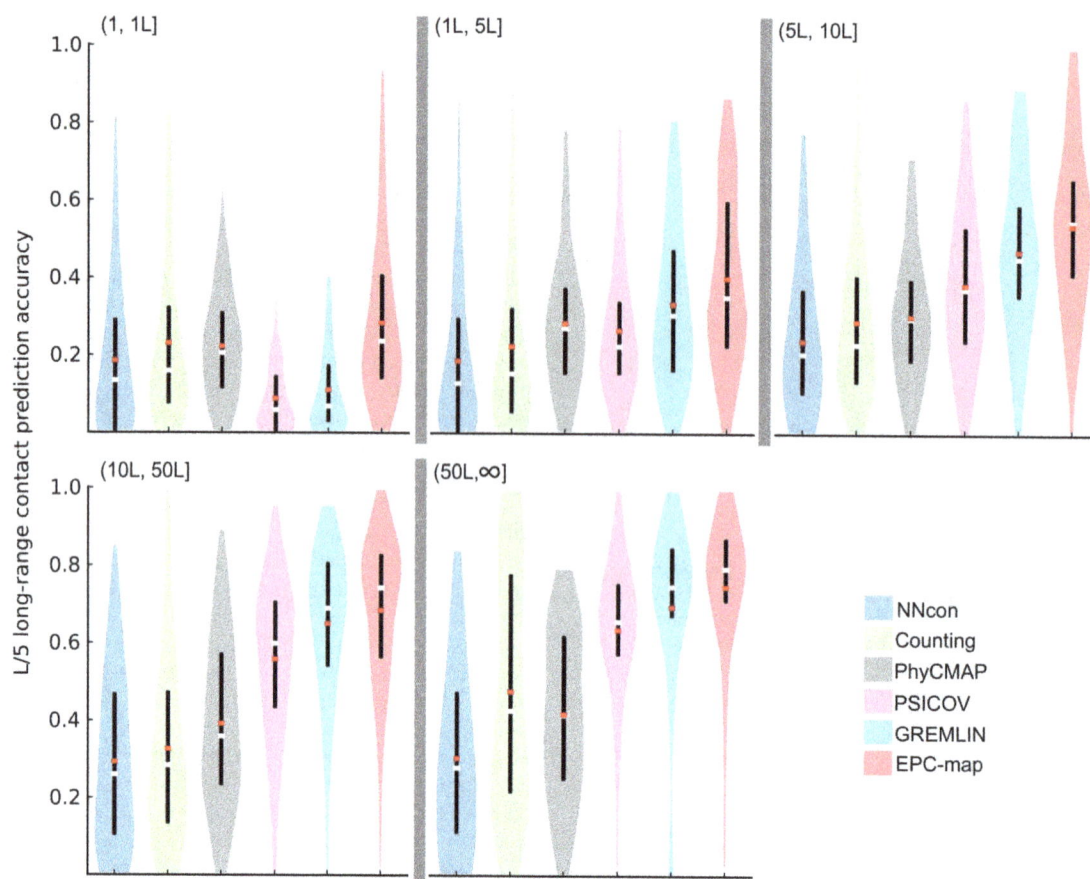

Figure 6. Prediction performance for proteins with increasing sequence alignment depth. Results are shown for all proteins pooled from the CASP9-10_hard, EPC-map_test, D329 and SVMCON_test data sets. Different methods are shown as color coded violin plots. The lower and upper end of the black vertical bars in each violin denote the accuracy at the 25 and 75 percentile, respectively. White horizontal bars indicate the median, red horizontal bars the mean accuracy. The distribution of the prediction accuracies for individual proteins is indicated by the shape of the violin. EPC-map is consistently more accurate than the other tested methods, regardless how many sequences are available.

case to make a fair comparison in terms of sampling, since 1000 decoys were already used to predict contacts with EPC-map. We varied the number of predicted contacts to study the influence of the accuracy/coverage trade-off on structure prediction accuracy. We found that using $1.5L$ contacts gives the best performance.

The prediction improvements are depicted in Figure 8. We used the GDT_TS measure to quantify the quality of the predicted structures. The GDT_TS measure ranges from 0 if two structures are completely dissimilar to 100 for a perfect structural match. At a GDT_TS of 50 or more, a prediction is considered to capture the native topology. The average GDT_TS of contact-guided

Table 2. Contribution of the SVM component to contact prediction.

Method	Range	Acc(SE)/Cov[L/10]	Acc(SE)/Cov[L/5]	Acc(SE)/Cov[L/2]
120 proteins with (1, 1L) sequences				
with SVM	Long	0.335(0.023)/0.038	0.278(0.019)/0.062	0.205(0.014)/0.110
w/o SVM	Long	0.305(0.024)/0.035	0.244(0.019)/0.055	0.188(0.015)/0.102
102 proteins with (1L, 5L) sequences				
with SVM	Long	0.471(0.025)/0.045	0.395(0.022)/0.076	0.279(0.015)/0.134
w/o SVM	Long	0.475(0.026)/0.045	0.388(0.022)/0.073	0.280(0.016)/0.133
306 proteins with >5L sequences				
with SVM	Long	0.741(0.012)/0.071	0.678(0.012)/0.131	0.530(0.011)/0.253
w/o SVM	Long	0.739(0.012)/0.070	0.679(0.012)/0.131	0.528(0.011)/0.250

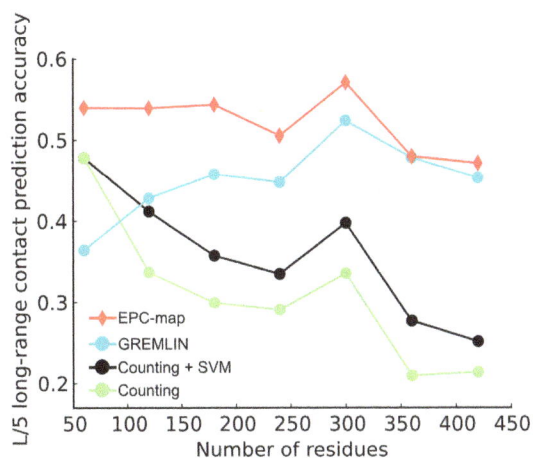

Figure 7. Dependence of prediction accuracy on sequence length. EPC-map is more accurate or on par with GREMLIN, irrespective of sequence length. The performance increase over GREMLIN is most pronounced for proteins smaller than 250 residues. Counting performs better on smaller proteins. The SVM component of EPC-map consistently improves the contact prediction from decoys over Counting by leveraging physicochemical information.

Rosetta increases to 40.9 compared to 33.1 using standard Rosetta (paired Student t-test p-value $< 10^{-10}$), an absolute improvement of 7.8%. The GDT_TS increases by more than 10 for 41 of the 132 proteins. In 24 cases, the GDT_TS increase is higher than 20. In addition, for 21 proteins the GDT_TS transitions from well below 50 to 50 or higher. In these cases, the combination of EPC-map predicted contacts and Rosetta allows for the folding of proteins that could not be modeled with Rosetta alone. Thus, our results show that contact information from EPC-map readily enhances structure prediction performance.

We also notice that the prediction of one structure deteriorates by 10 GDT_TS when predicted contacts are used. In this case,

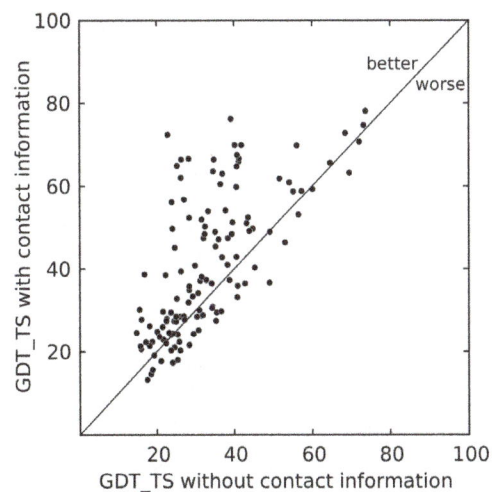

Figure 8. Comparison of *ab initio* structure prediction of 132 proteins from EPC-map_test with and without predicted contacts: each data point corresponds to the GDT_TS of the lowest-energy structure generated with and without the use of EPC-map predicted contacts. EPC-map increases the average prediction accuracy by 7.8% from 33.1 to 40.9 GDT_TS (paired Student's t-test p-value $< 10^{-10}$).

most of the predicted contacts are wrong and mislead tertiary structure prediction.

Figure 9 shows three example proteins for which the combination of EPC-map and Rosetta yielded significant improvements in prediction accuracy. For these examples, only few homologous sequences are available (less then $1.5L$). For each protein, we show the contact map obtained by EPC-map with true and false positives, the prediction of Rosetta without the inclusion of contact information, and the prediction based on contact information.

Without contacts, Rosetta fails to find the correct topology for the dissimilatory sulfite reductase D (PDB|1ucrA, Figure 9A). The structure modeled with contacts from EPC-map resembles the native topology and has minor deviations in loop regions. However, the most C-terminal helix is still incorrectly modeled.

Rosetta predictions without contacts capture the general topology for the *E. coli* SSB-DNA polymerase III (PDB|3sxuB, Figure 9B), but fail to arrange the β-sheet topology. In contrast, with the help of the predicted contacts from EPC-map, more native-like β-sheet topologies can be sampled. The most C-terminal part of the structure is wrongly oriented by an incorrectly formed anti-parallel β-sheet. Nevertheless, the GDT_TS of the modeled structure increases from 33.0 to 53.9.

The most prominent feature of the GIT1 paxillin-binding domain(PDB|2jx0A, Figure 9C) is the four-helix bundle. In this case, Rosetta cannot model the topology correctly, especially of the second helix and fails to find the fine-tuned packing (GDT_TS 36.5). Predicted contacts from EPC-map guide Rosetta to the native topology (GDT_TS 58.0). However, the EPC-map guided prediction shows deviations from the native structure in the loop regions and N-terminus.

These experiments with contact-guided structure prediction demonstrate the potential of coupling EPC-map's contact prediction with structure prediction. The strategy of interleaving structure and contact prediction might be a promising future route to improve *ab initio* structure modeling.

Conclusion

We presented EPC-map, a contact prediction method that achieves unprecedented prediction accuracy by combining evolutionary information from multiple-sequence alignments with physicochemical information from structure prediction methods. By combining two sources of information, our method improves prediction accuracy when compared to state-of-the-art algorithms. At the same time, we show that one source of information is able to compensate for the performance degradation induced by poor quality of the other source. This alleviates the main short-coming of popular evolution-based contact predictors, whose performance depends on the availability of many homologous sequences. We therefore believe that combining evolutionary and physicochemical information is an attractive route to improve contact prediction and reducing the need for deep alignments.

Key to the performance improvements achieved by our method is a graph-based representation of the characteristics of the local contact neighborhood to leverage physicochemical information. We use the graph-based representation to distill information about graph topology and label statistics into vector-based features. An SVM model is trained with these features to distinguish native from non-native contacts in *ab initio* decoys with unprecedented accuracy. We then combine this physicochemistry-based system with an evolutionary-based method to an approach that leads to substantial performance improvements over methods that only use a single source of information.

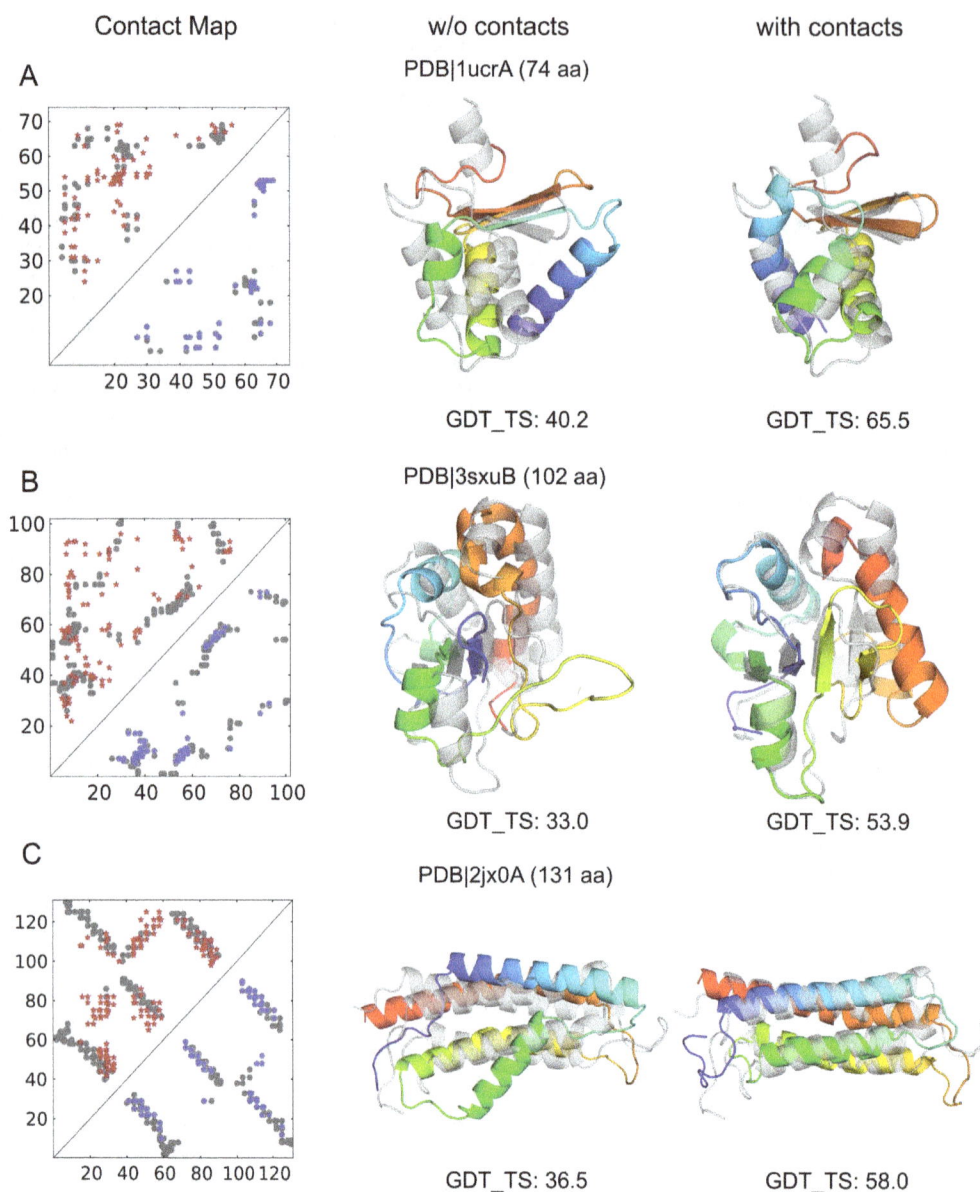

Figure 9. Tertiary structure prediction improvement of the dissimilatory sulfite reductase D (PDB|1ucrA), of the *E. coli* SSB-DNA polymerase III (PDB|3sxuB) and of the GIT1 paxillin-binding domain (PDB|2jx0A). Contact maps show false positive predictions in the upper triangle (red), true positive predictions in the lower triangle (blue) and native contacts in grey. For the shown predictions, native structures are shown in grey and predicted structures are colored from N-terminus (blue) to C-terminus (red). The predictions correspond to the lowest-energy structure generated without use of contacts (middle column) and with EPC-map predicted contacts (right column).

Using this strategy, EPC-map achieves 53.2% mean accuracy for top $L/5$ predicted long-range contacts over 528 proteins, 7.8% higher than the second-best method. Furthermore, EPC-map outperforms other top methods on proteins from CASP10. We showed that EPC-map displays improved performance regardless of the available alignment size, but is particulary effective if less than $5L$ or even $1L$ sequences are available. The predicted contacts improve *ab initio* structure prediction by guiding search in the conformational space towards the native state.

Our method is build to extract physicochemical contact information from structure decoys. One can expect that the quality of contact prediction increases as the quality of the generated decoys increases. Thus, we suggest that alternating between tertiary structure and contact prediction might be a

promising route to incrementally increase the quality of contact information and of the resulting structural models.

Supporting Information

File S1   Text S1, Graphs for modeling physicochemical context. **Text S2,** Features used and their generation. **Text S3,** Setup and example files for contact-guided Rosetta predictions. **Table S1,** Summary of node labels. **Table S2,** Summary of edge labels. **Table S3,** Pairwise features between contacting residues. **Table S4,** Graph topology features. **Table S5,** Graph spectrum features. **Table S6,** Single node features. **Table S7,** Node label statistics. **Table S8,** Edge label statistics. **Table S9,** Whole protein features. **Table S10,** Contact prediction performance of several methods on the CASP10 data set (104 proteins). **Table S11,** Contact prediction performance of several methods on the CASP10_hard data set (14 proteins). **Table S12,** Contact prediction performance of EPC-map, Counting, GREMLIN, PSICOV, PhyCMAP and NNcon on the CASP9-10_hard data set (20 proteins). **Table S13,** Contact prediction performance of EPC-map, Counting, GREMLIN, PSICOV, PhyCMAP and NNcon on the EPC-map_test data set (132 proteins). **Table S14,** Contact prediction performance of EPC-map, Counting, GREMLIN, PSICOV, PhyCMAP and NNcon on the D329 data set (329 proteins). **Table S15,** Contact prediction performance of EPC-map, Counting, GREMLIN, PSICOV, PhyCMAP and NNcon on the SVMCON_test data set (47 proteins). **Table S16,** Accuracies of the single SVM classifiers and the Ensemble SVM on 528 proteins from the CASP9-10_hard, EPC-map_test, D329 and SVMCON_test data sets.

File S2 Dataset S1, Proteins used for training of EPC-map. **Dataset S2,** Proteins used for validation of EPC-map. **Dataset S3,** Proteins from CASP10 used for validation of EPC-map.

Dataset S4, Proteins from CASP10 containing at least one FM or FM/TBM domain. **Dataset S5,** Proteins from CASP9 and CASP10 containing only FM or FM/TBM domains. **Dataset S6,** The D329 data set contains proteins from literature that are used to access contact prediction performance. **Dataset S7,** The SVMCON_test data set contains proteins from literature that are used to access contact prediction performance.

Acknowledgments

We thank Sergey Ovchinnikov for providing GREMLIN. We also thank Ines Putz, Mahmoud Mabrouk, Tim Werner, Lila Gierasch and Anne Gershenson for many useful comments on the manuscript.

Author Contributions

Conceived and designed the experiments: MS OB. Performed the experiments: MS. Analyzed the data: MS. Contributed reagents/materials/analysis tools: MS. Wrote the paper: MS OB.

References

1. Vassura M, Margara L, Di Lena P, Medri F, Fariselli P, et al. (2008) Reconstruction of 3D structures from protein contact maps. IEEE/ACM Trans Comput Biol Bioinform 5: 357–367.
2. Li W, Zhang Y, Skolnick J (2004) Application of sparse NMR restraints to large-scale protein structure prediction. Biophys J 87: 1241–1248.
3. Wu S, Szilagyi A, Zhang Y (2011) Improving protein structure prediction using multiple sequence-based contact predictions. Structure 19: 1182–1191.
4. Kosciolek T, Jones DT (2014) De novo structure prediction of globular proteins aided by sequence variation-derived contacts. PLoS ONE 9: e92197.
5. Hamilton N, Huber T (2008) An introduction to protein contact prediction. Methods Mol Biol 453: 87–104.
6. Vassura M, Margara L, Lena PD, Medri F, Fariselli P, et al. (2008) FT-COMAR: fault tolerant three-dimensional structure reconstruction from protein contact maps. Bioinformatics 24: 1313–1315.
7. Eickholt J, Cheng J (2012) Predicting protein residue-residue contacts using deep networks and boosting. Bioinformatics 28: 3066–3072.
8. Bernstein FC, Koetzle TF, Williams GJ, Meyer E F, Brice MD, et al. (1977) The protein data bank: a computer-based archival file for macromolecular structures. J Mol Biol 112: 535–542.
9. Marks DS, Hopf TA, Sander C (2012) Protein structure prediction from sequence variation. Nat Biotechnol 30: 1072–1080.
10. Kamisetty H, Ovchinnikov S, Baker D (2013) Assessing the utility of coevolution-based residue– residue contact predictions in a sequence- and structure-rich era. PNAS 110: 15674–15679.
11. Wu S, Zhang Y (2008) A comprehensive assessment of sequence-based and template-based methods for protein contact prediction. Bioinformatics 24: 924–931.
12. Goebel U, Sander C, Schneider R, Valencia A (1994) Correlated mutations and residue contacts in proteins. Proteins 18: 309–317.
13. Marks DS, Colwell LJ, Sheridan R, Hopf TA, Pagnani A, et al. (2011) Protein 3D structure computed from evolutionary sequence variation. PLoS ONE 6: e28766.
14. Jones DT, Buchan DWA, Cozzetto D, Pontil M (2012) PSICOV: precise structural contact prediction using sparse inverse covariance estimation on large multiple sequence alignments. Bioinformatics 28: 184–190.
15. Punta M, Rost B (2005) PROFcon: novel prediction of long-range contacts. Bioinformatics 21: 2960–2968.
16. Tegge AN, Wang Z, Eickholt J, Cheng J (2009) NNcon: improved protein contact map prediction using 2D-recursive neural networks. Nucleic Acids Res 37: W515–W518.
17. Vullo A, Walsh I, Pollastri G (2006) A two-stage approach for improved prediction of residue contact maps. BMC Bioinformatics 7: 180.
18. Cheng J, Baldi P (2007) Improved residue contact prediction using support vector machines and a large feature set. BMC Bioinformatics 8: 113.
19. Björkholm P, Daniluk P, Kryshtafovych A, Fidelis K, Andersson R, et al. (2009) Using multi-data hidden markov models trained on local neighborhoods of protein structure to predict residue-residue contacts. Bioinformatics 25: 1264–1270.
20. Li Y, Fang Y, Fang J (2011) Predicting residue-residue contacts using random forest models. Bioinformatics 27: 3379–3384.
21. Di Lena P, Nagata K, Baldi P (2012) Deep architectures for protein contact map prediction. Bioinformatics 28: 2449–2457.
22. Monastyrskyy B, Fidelis K, Tramontano A, Kryshtafovych A (2011) Evaluation of residue–residue contact predictions in CASP9. Proteins 79: 119–125.
23. Monastyrskyy B, D'Andrea D, Fidelis K, Tramontano A, Kryshtafovych A (2013) Evaluation of residue–residue contact prediction in CASP10. Proteins 82: 138–153.
24. Karakas M, Woetzel N, Meiler J (2010) BCL::ContactLow confidence fold recognition hits boost protein contact prediction and de novo structure determination. J Comp Biol 17: 153–168.
25. Eickholt J, Wang Z, Cheng J (2011) A conformation ensemble approach to protein residue-residue contact. BMC Bioinformatics 11: 38.
26. Samudrala R, Xia Y, Huang E, Levitt M (1999) Ab initio protein structure prediction using a combined hierarchical approach. Proteins 3: 194–198.
27. Zhu J, Zhu Q, Shi Y, Liu H (2003) How well can we predict native contacts in proteins based on decoy structures and their energies? Proteins 52: 598–608.
28. Blum B, Jordan MI, Baker D (2010) Feature space resampling for protein conformational search. Proteins 78: 1583–1593.
29. Wang Z, Xu J (2013) Predicting protein contact map using evolutionary and physical constraints by integer programming. Bioinformatics 29: 266–273.
30. Remmert M, Biegert A, Hauser A, Soeding J (2012) HHblits: lightning-fast iterative protein sequence searching by HMM-HMM alignment. Nat Meth 9: 173–175.
31. Apweiler R, Bairoch A, Wu CH, Barker WC, Boeckmann B, et al. (2004) UniProt: the universal protein knowledgebase. Nucleic Acids Res 32: 115–119.
32. Rohl CA, Strauss CEM, Misura KMS, Baker D (2004) Protein structure prediction using Rosetta. Meth Enzymol 383: 66–93.
33. Tyka MD, Jung K, Baker D (2012) Efficient sampling of protein conformational space using fast loop building and batch minimization on highly parallel computers. J Comput Chem 79: 2483–2491.
34. Cavallo L, Kleinjung J, Fraternali F (2003) POPS: a fast algorithm for solvent accessible surface areas at atomic and residue level. Nucleic Acids Res 31: 3364–3366.
35. Frishman D, Argos P (1995) Knowledge-based protein secondary structure assignment. Proteins 23: 566–579.
36. Fischer JD, Mayer CE, Soeding J (2008) Prediction of protein functional residues from sequence by probability density estimation. Bioinformatics 24: 613–620.
37. Hagberg AA, Schult DA, Swart PJ (2008) Exploring network structure, dynamics, and function using networkX. Proceedings of the 7th Python in Science Conference. p. 11–15.
38. Pedregosa F, Varoquaux G, Gramfort A, Michel V, Thirion B, et al. (2011) Scikit-learn: Machine Learning in Python. JMLR 12: 2825–2830.
39. Li G, Semerci M, Yener B, Zaki MJ (2012) Effective graph classification based on topological and label attributes. Stat Anal Data Min 5: 265–283.
40. He H, Garcia E (2009) Learning from imbalanced data. IEEE Trans Knowl Data Eng 21: 1263–1284.
41. Zadrozny B, Elkan C (2001) Obtaining calibrated probability estimates from decision trees and naive bayesian classifiers. In: Proceedings of the Eighteenth International Conference on Machine Learning. p. 609–616.
42. Platt JC (1999) Probabilistic outputs for support vector machines and comparisons to regularized likelihood methods. In: Advances in large margin classifiers. MIT Press. p. 61–74.
43. Wang G, Dunbrack J Roland L (2003) PISCES: a protein sequence culling server. Bioinformatics 19: 1589–1591.
44. Wang S, Ma J, Peng J, Xu J (2013) Protein structure alignment beyond spatial proximity. Sci Rep 3.

45. Sathyapriya R, Duarte JM, Stehr H, Filippis I, Lappe M (2009) Defining an essence of structure determining residue contacts in proteins. PLoS Comput Biol 5: e1000584.

46. Zhao F, Xu J (2012) A position-specific distance-dependent statistical potential for protein structure and functional study. Structure 20: 1118–1126.

47. Karplus K (2009) SAM-T08, HMM-based protein structure prediction. Nucleic Acids Res 37: W492–W497.

17

TALEs from a Spring – Superelasticity of Tal Effector Protein Structures

Holger Flechsig*

Department of Mathematical and Life Sciences, Graduate School of Science, Hiroshima University, Higashi-Hiroshima, Japan

Abstract

Transcription activator-like effectors (TALEs) are DNA-related proteins that recognise and bind specific target sequences to manipulate gene expression. Recently determined crystal structures show that their common architecture reveals a superhelical overall structure that may undergo drastic conformational changes. To establish a link between structure and dynamics in TALE proteins we have employed coarse-grained elastic-network modelling of currently available structural data and implemented a force-probe setup that allowed us to investigate their mechanical behaviour in computer experiments. Based on the measured force-extension curves we conclude that TALEs exhibit superelastic dynamical properties allowing for large-scale global conformational changes along their helical axis, which represents the *soft* direction in such proteins. For moderate external forcing the TALE models behave like linear springs, obeying Hooke's law, and the investigated structures can be characterised and compared by a corresponding spring constant. We show that conformational flexibility underlying the large-scale motions is not homogeneously distributed over the TALE structure, but instead soft spot residues around which strain is accumulated and which turn out to represent key agents in the transmission of conformational motions are identified. They correspond to the RVD loop residues that have been experimentally determined to play an eminent role in the binding process of target DNA.

Received June 17, 2014; **Accepted** September 6, 2014; **Published** October 14, 2014

Funding: This work was supported by a Grant-in-Aid for Scientific Research on Innovative Areas "Spying minority in biological phenomena" (23115007) and Platform for Dynamic Approaches to Living System of the Ministry of Education, Culture, Sports, Science and Technology, Japan. http://www.mext.go.jp/english/. The funders had no role in study design, data collection and analysis, decision to publish, or preparation of the manuscript.

Competing Interests: The author has declared that no competing interests exist.

* Email: holgerflechsig@hiroshima-u.ac.jp

Introduction

TAL (transcription activator-like) effectors are proteins that are secreted in plants by bacteria of the *Xanthomonas* genus. Upon injection into cells they are able to activate transcription of specific target plant genes which may be beneficial for bacterial infection [1,2]. Hence, a considerable amount of scientific attraction to this protein is owing to its role in the disease of various plant types including important crops as well [3,4]. Furthermore, artificial TALEs engineered to target prescribed sequences offer interesting applications in genome editing [5–7].

The common molecular architecture of TALEs consists of canonical two-helix repeats, each of them involved in the recognition of one specific DNA base, that are arranged around a central axis to form an overall superhelical protein structure that wraps around a central groove in which duplex DNA can be bound [8,9].

Apparently, adequate understanding of the mechanisms underlying the operation in these important proteins requires a combination of various experimental approaches, including structure determination and biochemical manipulation plus analyses, with detailed methods of molecular modelling. On the other side, the structure-function relationship of proteins, i.e. the principle of how the three dimensional folded protein conformation defines its functional activity, may reveal surprisingly simple patterns. In this regard molecular machines and motors represent

a prime example. Their modular architecture, consisting of rigid domains connected by more flexible joints, gives rise to well-organised relative internal motions through which the particular function is implemented [10,11].

In view of this conception one may ask the seemingly simple question which is probably most appealing from the perspective of a physicist: Given the fact that TALEs look like a spring do they also exhibit spring-like dynamics and what are the benefits of such a structure in terms of its elastic properties? Those questions which refer to the mechanical aspects of the operation of TALE proteins are addressed in this paper.

As we remember from high school physics we can probe the properties of a deformable object by holding it at one end and put a weight on the other end. The force generated by the weight would then induce an extension of that object, i.e. a deviation from its natural length, which can be easily measured. Using a variety of different weights allows to trace the dependence of the extension from the applied force, a relation which is typically used to discuss the objects' elastic properties.

Here, we have performed such experiments in modelling simulations based on the currently available crystal structures of TALEs. For the protein dynamics we have employed the coarse-grained elastic-network description in which the protein is represented as a network of beads connected via deformable strings [12–14]. Despite gross simplifications present in such models, they have been proven to perform remarkably well in the prediction of functional chemo-mechanical motions in proteins [15–19]. It should be stressed that proteins modelled as elastic networks still present highly complex systems which generally can only be treated numerically and, as shown previously, may exhibit strong nonlinearities in their conformational dynamics [20].

Our analysis was performed for four different TALE structures. We have considered the artificially engineered dHax3 TALE in its free form and in the conformation which was co-crystallized with DNA [8]. Both structures contain 11 TAL repeats which complete one helical turn, but the corresponding pitch is found to be substantially different. Furthermore we have taken into account the structure of the PthXo1 TALE from the rice pathogen *Xanthomonas oryzae*, which was determined in the presence of duplex DNA and reveals an overall two-turn helical shape formed by 23.5 repeats [9]. To allow comparison with dHax3 we have also constructed a shortened one-turn version of PthXo1 (named PthXo1* throughout the paper, see Methods section). All investigated TALE structures are listed in Table 1.

Methods

We considered the structures of two TALE proteins, that of the artificially engineered dHax3 in its DNA-free form (PDB ID 3V6P) and in a conformation determined in the presence of DNA (3V6T), and that of PthXo1 from the rice pathogen *Xanthomonas oryzae* which was co-crystallized with its DNA target (3UGM). In Table 1 all investigated TALE structures are listed. All figures that display protein conformations in this paper were prepared with the VMD software [21].

Network construction and dynamics

The elastic network of a TALE protein was obtained by replacing each amino acid of the corresponding structure by a single bead that was placed at the position of the respective alpha-carbon atom (denoted by $\vec{R}_i^{(0)}$ for bead i). Then each two beads were connected by a deformable string if their spatial distance was below a prescribed interaction radius r_{int}. The constructed network of dHax3 consisted of $N = 373$ beads (corresponding to Gly^{303}-Gly^{675}) and the PthXo1 network had $N = 789$ beads (Gly^{234}-Asp^{1032}). In the PthXo1* network the shortened structure (Gly^{234}-Cys^{623}) was considered and $N = 388$ (a summary is given in Table 1). All networks were constructed using an interaction radius of 8 Å.

The total elastic energy of the network $U = \sum_{i<j}^{N} \frac{A_{ij}}{2}(d_{ij} - d_{ij}^{(0)})^2$ is the sum over all string contributions, where N is the number of network beads and $A_{ij} = 1$, if beads i and j are connected by a string, and $A_{ij} = 0$ else. Here, $d_{ij}^{(0)} = |\vec{R}_i^{(0)} - \vec{R}_j^{(0)}|$ is the natural length of a string connecting beads i and j (as extracted from the

respective PDB file) and $d_{ij} = |\vec{R}_i - \vec{R}_j|$ is its corresponding deformed length, with $\vec{R}_i = (x_i, y_i, z_i)^T$ being the actual position vector of bead i. The energy given above represents a rescaled energy in which the dependency from the stiffness constant (which was the same for all strings) was removed. Neglecting thermal fluctuations and hydrodynamical interactions, the dynamics of the network can be described by a set of Newton equations considered in the over-damped limit [17]. For bead i the equation of motion is $\gamma \frac{d}{dt}\vec{R}_i = \vec{F}_i + \vec{f}_i^{ext}$, where $\vec{F}_i = -\frac{\partial}{\partial \vec{R}_i} U = -\sum_j^N A_{ij} \frac{d_{ij} - d_{ij}^{(0)}}{d_{ij}}(\vec{R}_i - \vec{R}_j)$ are the internal forces generated by deformed strings which are connected to bead i and $\vec{f}_i^{ext}$ is an external force which can be applied to that bead. Using a rescaled time we can remove the friction coefficient γ (the same for all beads) on the left hand side of the equations of motion. Due to the energy rescaling the forces have the dimension of lengths in our description. Note that the network dynamics is generally nonlinear since distances are nonlinear functions of the spatial coordinates, i.e. $d_{ij} = \sqrt{(x_i - x_j)^2 + (y_i - y_j)^2 + (z_i - z_j)^2}$. In order to follow the dynamical evolution of the network under external forcing, i.e. to obtain the position of all network beads at every time moment, the equations of motion were numerically integrated. In the simulations we have implemented a first-order integration scheme with a time-step of 0.1.

Force-probe setup

To implement our force-probe experiment we have immobilised the network at the bead with index 0 (at one side of the protein) and applied an external force to a single bead with index f (at the other protein side); see Fig. 1. The force had magnitude F and the direction was chosen to coincide with a vector that connects the two selected beads, i.e. $\vec{f}_f^{ext} = F\vec{u}$ with the unit vector $\vec{u} = (\vec{R}_f - \vec{R}_0)/d_{f0}$. In the simulations we have varied the force magnitude F and, each time starting from the initial network, integrated the equations of motions until a steady conformation of the network with the applied force was reached. To detect the steady network state the following condition was implied: The root mean square displacement of the actual network beads with respect to that of the initial network was calculated every 1000th integration step. When the absolute change of two such subsequently determined values was below 0.0001 we have stopped the integration procedure, assuming the network conformation to be sufficiently close to the steady state. In the steady network conformation we have measured the protein extension as $\Delta X = d_{f0} - d_{f0}^{(0)}$. Relative extensions (given e.g. in Fig. 2) were calculated as $\Delta X_{rel} = (d_{f0} - d_{f0}^{(0)})/d_{f0}^{(0)}$.

In the simulations we wanted to stretch the TALE proteins along the 'spring axis', i.e. along the superhelical axis which

Table 1. Investigated Tal effector structures.

TALE	PDB ID	DNA	residues	TALE repeats	immobilized/forced residue	pitch (Å)
dHax3	3V6P	no	Gly^{303}-Gly^{675}	11 (1b to 12a)	Gly^{303}/Gly^{675}	60.5
dHax3-DNA	3V6T	yes	Gly^{303}-Gly^{675}	11 (1b to 12a)	Gly^{303}/Gly^{675}	35.7
PthXo1	3UGM	yes	Gly^{234}-Asp^{1032}	23.5 (-1b to 22b)	Gly^{234}/Lys^{1016}	74.7
PthXo1*	3UGM	yes	Gly^{234}-Cys^{623}	11.5 (-1b to 10b)	Gly^{234}/Gly^{608}	36.5

Summary of the TALE structures considered in the force-probe simulations. The pitch was measured as the distance between the alpha-carbon atoms of the immobilised and force residue.

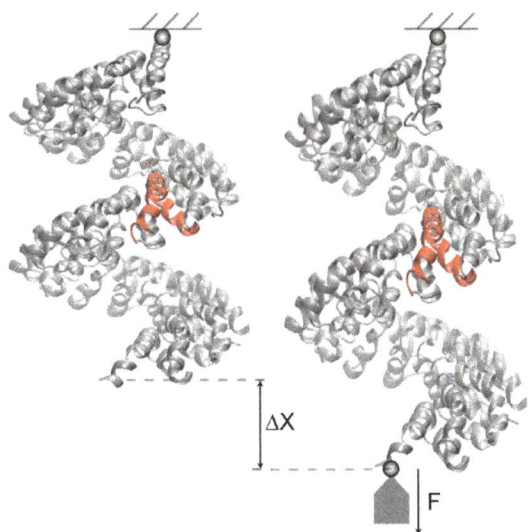

Figure 1. Schematic illustration of the *in silico* force-probe experiments. Exemplarily, the PthXo1 TAL effector is shown in ribbon representation with one end being immobilised and the other end exerted to a force caused by a fictitious weight applied. One selected TAL repeat is shown in red colour.

coincides with the orientation of the DNA groove. Therefore, the immobilised network bead (index 0) and the bead to which the force is applied (index f) were carefully chosen. For the dHax3 TALE structures the immobilised bead corresponded to residue Gly^{303} at the beginning of the 1b TALE repeat and the force was applied to the bead which corresponded to residue Gly^{675}, which is located at the end of TALE repeat 12a. For PthXo1 and PthXo1* the immobilised bead was Gly^{234} from the long helix of repeat −1. In the case of PthXo1* the forced bead was Gly^{608} from the long helix of repeat 10 and for the entire PthXo1 TALE the forced bead corresponded to residue Lys^{1016} from the long helix of repeat 22 (see summary in Table 1).

To investigate the distribution of deformations in the final deformed TALE networks (see Fig. 3 and Movie S1), we have assigned each bead i the value $\chi_i = (\sum_j^N A_{ij} \cdot |d_{ij} - d_{ij}^{(0)}|) / \sum_j^N A_{ij}$, which is the average absolute deformation of strings connected to bead i.

Random forces

A comparison between the dHax3 crystal structures in free and DNA-bound form suggests that the conformational dynamics underlying the functional activity in TALE proteins is dominated by motions along the superhelical axis, and indeed, the setup of our computer experiments was designed to probe the mechanical properties of TALE structures by generating motions in this particular direction. However, in a set of independent additional simulations we aimed to parse the space of generally possible conformational changes of the considered TALE proteins by exerting random external forces to the corresponding elastic networks. The resulting structural changes were then analysed in terms of deformations induced along the superhelical axis and in the lateral direction. Details of these particular simulations can be found in the Text S1.

Results

A schematic representation of the implemented force-probe setup and the molecular architecture of PtXo1, as an example of the considered TALE structures, is shown in Fig. 1. In the Methods section a detailed description of the performed computer experiments is provided.

The force-extension curves obtained from our simulations are displayed in Fig. 2. They show that through the application of the prescribed forces all four TALE structures were able to undergo significant stretching, revealing that they exhibit an intrinsic flexibility along their superhelical axis. To give an illustration of the performed computer experiments we provide the Movie S1 which shows large-scale global structural changes from a single simulation of force-induced stretching of the PthXo1 structure. Overall we find that three of the TALE proteins could be stretched to far more than twice their initial lengths. The DNA-free dHax3 protein presents an exception; here a force of the same magnitude could induce relative length changes of 'only' 52%, i.e. the extended structure is 'only' roughly half times longer than the initial one. We have checked whether the identified force-induced strong deformations were reversible and found that in all cases, after the release of the applied force and the immobilisation, the TALE protein returned to its particular initial structure. This finding indicates that the overall structural arrangement of TALEs must possess remarkable elastic properties which may give rise to large-amplitude conformational motions along the superhelical axis.

We observe that the dependence of the extension of TALE structures from the applied force is separated into two regimes. For moderate extensions, i.e. when deviations from the initial length were not too large, a linear relation to the force is found, which represents Hooke's law with the proportionality factor corresponding to the stiffness constant which can be assigned to the respective TALE structure. The validity of this linear behaviour differs for the four TALEs (see Fig. 2); for the dHax3-DNA and PthXo1* the range (as estimated from the force-extension curves) is up to relative extensions of 25%. Across that region and for deformations far beyond the initial protein length we identified a nonlinear regime in which the force grows super-proportional with the extension, i.e. the structure gets stiffer the more it is stretched.

The stiffness constants derived from a linear regression in the linear regime show that the full two-turn PthXo1 TALE is the softest structure as compared to the other TALEs. The single-turn PthXo1* is twice as stiff as PthXo1 but softer than dHax3-DNA by a factor of 1.5. The DNA-free dHax3 protein represents the stiffest of the investigated TALE structures. It should be noted that due to the coarse-grained protein description the absolute scale of the dimensionless stiffness constants is arbitrary.

Apparently, the global large-scale elongations of TALE structures in response to forcing represent a collective effect resulting from the accumulation of local structural deformations which are still small. In the elongated states, as we find, such deformations are inhomogeneously spread over the TALE conformation and multiple residues around which strain is accumulated and which define soft spots of the structure can be identified (results for dHax3 are shown in Fig. 3). They are located along the central DNA binding groove and belong to the short RVD-loops which connect the two alpha-helices of each TAL repeat and are critical for establishing contacts to DNA. Similar observations are made for PthXo1* and full PthXo1 (not shown in Fig. 3 but in the Movie S1 the PthXo1 structure was coloured according to the deformation pattern of its elastic network).

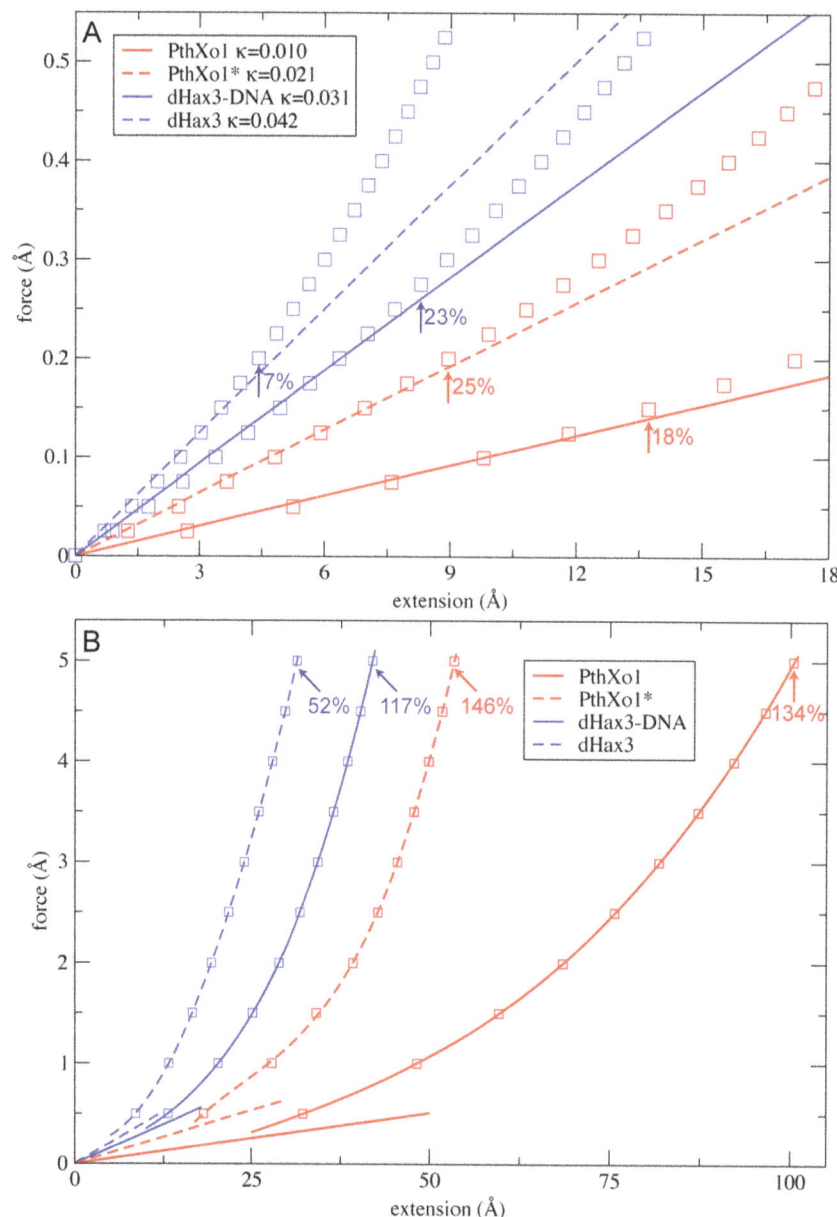

Figure 2. Force-extension relation for the four studied TALEs. A: Linear regression of the data and the derived stiffness constants are shown; the estimated validity range in terms of relative extensions (see Methods) is indicated for each TAL structure. B: Non-linear super-proportional force extension curves for large TALE deformations. For the force of magnitude $F = 5.0$ Å, the relative protein extensions are given.

We were curious whether conformational motions induced in our simple force-probe setup may cover aspects of the transition between the two dHax3 structures, that determined in the presence of DNA and the DNA-free form. The DNA-bound structure has a length of 35.7 Å (measured as the distance between immobilised and forced residue, which roughly corresponds to the structural pitch, see Methods), whereas the same length of the DNA-free conformation is 60.5 Å. As we see from Fig. 2, the force that would be needed to generate the corresponding extension of 24.8 Å in the dHax3-DNA structure cannot be determined from Hooke's law but instead has to be deduced from the nonlinear relation. From a cubic regression of the dHax3-DNA force-extension data in that regime we computed the corresponding magnitude of the force as 1.45 Å. In a single simulation we have applied this force to the dHax3-DNA elastic network and

compared the resulting extended structure with that of the DNA-free dHax3 crystal structure (see Fig. 4). We find that after superposition of their Cα-atoms they compare with a RMSD-value of 3 Å.

In the main force-probe setup always a single external force was applied to induce extensions of the TALE structures along the superhelical axis. However, to further probe the structural flexibility and test generally possible conformational changes we have performed additional independent simulations employing random external forcing (see Methods). Simulation details are summarised in Text S1 and results are shown in Figure S1. For all considered TALE structures we find that deformations generated along the superhelical direction are generally larger by one order of magnitude as compared to structural changes in the lateral direction (induced by the same random forces). Therefore, our

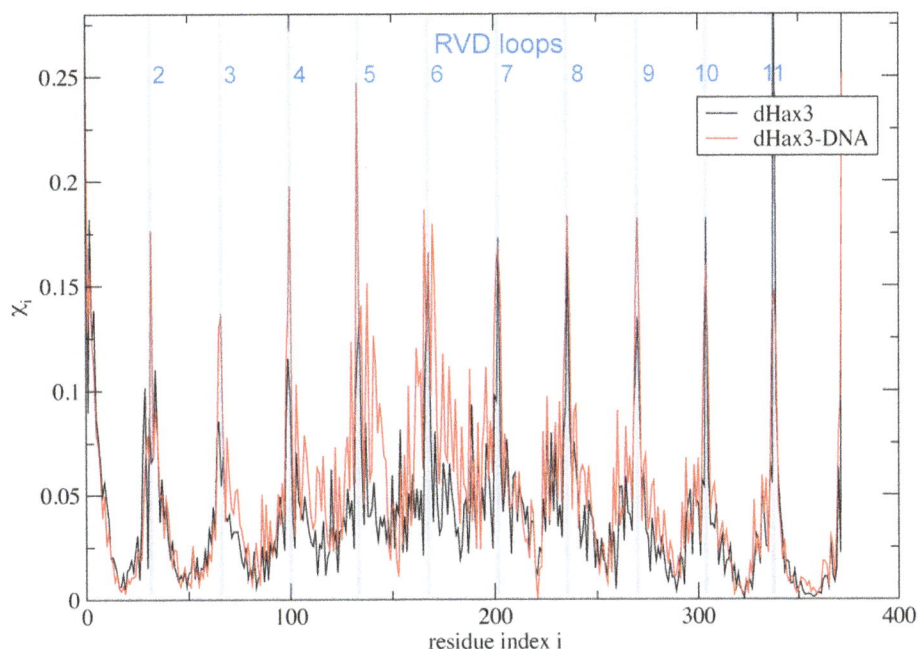

Figure 3. Distribution of local conformational changes in the elongated dHax3 structures obtained after application of an external force with magnitude F = 1.0 Å. The deformation value for the residues is plotted (see Methods, lines are used for better visibility). Positions of the RVD loops are indicated by blue lines.

findings evidence that conformational dynamics of the considered TALEs is dominated by motions along the superhelical axis which apparently represent the *soft* direction in such proteins.

Discussion

In this paper we report results of specific force-probe computer experiments performed for four Tal-effector (TALE) protein structures. Their molecular architecture shows a superhelical arrangements of basic structural repeats, and, crystal structures obtained for a particular TALE have indicated pronounced conformational flexibility along the helical axis [8]. Inspired by their spring-like shape we aimed to investigate whether TALEs also exhibit spring-like dynamical properties. For that purpose we have implemented a very basic setup of computer experiments in which a TALE protein was immobilised at one end and exerted to

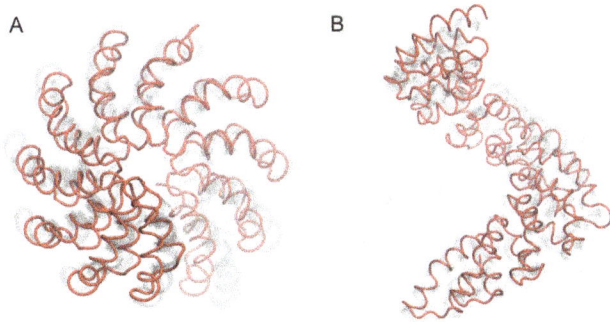

Figure 4. Comparison of dHax3 TALE structures. An *in silico* structure (shown as red tube) obtained from force-induced stretching of the dHax3-DNA conformation is superimposed with the DNA-free crystal structure of dHax3 (shown as transparent grey ribbons). Top view (A) and side view (B) perspective is shown.

a force at the opposite site. For the TALE dynamics we have employed the coarse-grained elastic-network description in which the protein is viewed as a meshwork of beads connected via deformable strings. This approach emphasises the mechanical aspects of protein function and allows for an efficient numerical implementation in computer experiments.

Our analysis based on the evaluation of force-extension curves shows that the investigated TALE structures indeed exhibit an enormous flexibility along the superhelical axis and even very large deformations are found to be elastic, i.e. the structural changes are reversible. The dynamical properties of TALEs resemble that of mechanical springs; in particular Hooke's law, i.e. the linear dependence of the extension from the applied force is found to be valid for the considered protein structures. However, there are also differences to ordinary springs. The elastic deformability of a macroscopic mechanical spring, typically designed as a periodically wound metal wire, results solely from its particular helical shape and the spring material is usually stiff in itself. Proteins, however, represent *soft material* and the force-induced global deformations in TALE structures must effectively result from a collective amplification phenomena manifested by the accumulation of local conformational changes. To generate such internal motions requires forces acting on all protein residues and which each have to become larger to induce an overall larger global deformation. While the linear regime is valid for moderate TALE extensions, the external force applied in our simulations, which has to compensate all the internal forces, therefore grows in a super-proportional fashion for larger deformations, meaning that the structures become stiffer the more they are stretched. Similar response curves are well-known for mechanical progressive springs.

The stiffness constants for the four TALE structures obtained from the linear regime resemble what we know from macroscopic springs: A linear spring that is cut in halves can be stretched by only half the fraction by the same force, i.e. it is twice as stiff.

When comparing the full PthXo1 TALE, which represents a long spring with two helical turns, with the artificially shortened PthXo1*, which has only one helical turn, we find this behaviour to be reproduced also in the case of TALE protein springs. Comparing the properties of the PthXo1 and dHax3 TALEs is certainly complicated due to the lack of structural data. Although our attempt revealed that PthXo1* has the softer structure, the underlying reasons can only be speculated about. Since both structures comprise 11 TAL repeats per helical turn and the pitch differs only by 0.8 Å, the different elasticity may be owed to the fact that dHax3 is an engineered protein whereas PthXo1 represents a structure evolved under real biological conditions. The dHax3 TALE crystallised in the absence of DNA is found to be the stiffest spring and that for which the validity range of Hooke's law is most narrow. This property can be explained by the fact that this structure, having a pitch of 60.5 Å, already represents an extended spring as compared to its DNA-bound version, which has a much lower pitch of 35.7 Å. Hence, it responds to forcing by a larger stiffness.

Despite its approximate nature, our modelling could establish a link between the structural arrangement of TALE proteins and their dynamical properties and demonstrate that the spring-like shape benefits superelasticity. In our description, amongst other simplifications, we neglected the effect of DNA in the simulations (but we did so for both TALE proteins). It is therefore highly remarkable that our simple setup – i.e. the TALE is hold at one end and a single external force is applied at the opposite end – can reproduce to a decent degree the global structural changes of the DNA-associated transition in the dHax3 TALE. Furthermore our prediction that flexibility is not homogeneously distributed over the TALE structures, but instead soft regions are found along the central DNA binding groove, is consistent with the crystallographic B-factors of the free dHax3 conformation, which are systematically larger (i.e. residues are generally more mobile) for residues from the RVD loops. Our findings suggest that these soft spots represent key agents in the transmission of conformational motions in TALE structures. The enhanced flexibility of RVD loops may be critical for the recognition and binding of bases from the target DNA. Maybe such predictions can be checked in single molecule experiments. Generally it should be possible to probe the flexibility and elastic properties of TALE structures in experiments using e.g. appropriate atomic force microscopy setups.

The super-elastic properties of TALE structures revealed by this study are apparently relevant for the functional activity of these proteins, since they involve the ability to undergo enormous conformational changes along their superhelical axis. Moreover, we have shown that in this direction the TALE structures are *soft* whereas deformations in the lateral orientation, i.e. perpendicular to the superhelical axis, are found to be generally much less pronounced. Indeed, the crystal structures of the artificially engineered dHax3, representing the only TALE for which the DNA-bound and free protein could be determined to date, show that the free conformation is stretched by 70% as compared to the DNA-complexed form, thus providing evidence that the large-scale motions are linked to the interactions between the TALE and DNA. Nuclear magnetic resonance studies together with small-angle X-ray scattering analysis performed for the effector protein

PthA are in support of these observations [22] and a recently proposed model based on the analysis of AvrBs3 TALE mutants suggests relative motions of the TALE repeats upon DNA target scanning and in the process of DNA recognition and binding to the RVDs [23]. Furthermore, detailed molecular dynamics simulations of the dHax3 TALE have shown that open-close motions between the two ends of the superhelical structure constitute the dominant conformational dynamics [24]. Hence, from the current experimental and modelling studies the significance of conformational flexibility for binding DNA has emerged as a central aspect of TALE function.

Nonetheless, it should be stressed that in the present situation the knowledge about interactions between TALE proteins and DNA is still confined and functionally relevant mechanism such as recognition and binding of their target sequences are not yet sufficiently explored. In fact, the number of studies devoted to investigate such processes is yet limited and further investigations are needed to elucidate the operation principles of TAL effectors. Modelling studies providing a combination of detailed molecular dynamics descriptions of TALE proteins and their target DNA may indeed play an important role in future studies.

Supporting Information

Figure S1 Conformational changes of TALE structures in response to random external forcing. Each data point corresponds to a single deformed TALE network (50 realisations have been generated for each TALE structure). In such conformations the structural changes as compared to the corresponding original TALE were characterised in terms of the extension along the superhelical axis (shown on the horizontal axis) and that in the lateral direction (shown on the vertical axis). See Text S1 for details.

Text S1 Details of independent additional simulations performed to probe conformational changes of TALE structures in response to random external forcing.

Movie S1 Force-induced stretching of PthXo1. Large-amplitude deformations of the PthXo1 TALE in response to external forcing ($F = 1.0$ Å). The colour code represents the distribution of local deformations according to the computed deformation value of each residue in the final steady conformation (see Methods, red colour corresponds to a low value and blue colour to large values).

Acknowledgments

The author thanks Naoya Tochio and Yuichi Togashi for helpful discussions.

Author Contributions

Conceived and designed the experiments: HF. Performed the experiments: HF. Analyzed the data: HF. Contributed reagents/materials/analysis tools: HF. Wrote the paper: HF.

References

1. Kay S, Hahn S, Marois E, Hause G, Bonas U (2007) A bacterial effector acts as a plant transcription factor and induces a cell size regulator. Science 318: 648–651.
2. Boch J, Bonas U (2010) *Xanthomonas* AvrBs3 family-type III effectors: discovery and function. Annu Rev Phytopathol 48: 419–436.
3. Bogdanove AJ, Schornack S, Lahaye T (2010) TAL effectors: finding plant genes for disease and defense. Curr Opin Plant Biol 13: 394–401.
4. Dodds PN, Rathjen JP (2010) Plant immunity: towards an integrated view of plant-pathogen interactions. Nat Rev Genet 11: 539–548.
5. Scholze H, Boch J (2011) TAL effectors are remote controls for gene activation. Curr Opin Microbiol 14: 47–53.

6. Boch J (2011) TALEs of genome targeting. Nature Biotechnol 29: 135–136.
7. Bogdanove AJ, Voytas DF (2011) TAL effectors: customizable proteins for DNA targeting. Science 333: 1843–1846.
8. Deng D, Yan C, Pan X, Mahfouz M, Wang J, et al. (2012). Structural basis for sequence-specific recognition of DNA by TAL effectors. Science 335: 720–723.
9. Mak AN, Bradley P, Cernadas PA, Bogdanove AJ, Stoddard BL (2012) The crystal structure of TAL effector PthXo1 bound to its DNA target. Science 335: 716–719.
10. Alberts B (1998) The cell as a collection of protein machines: preparing the next generation of molecular biologists. Cell 92: 291–294.
11. Vale RD, Milligan RA (2000) The way things move: looking under the hood of molecular motor proteins. Science 288: 88–95.
12. Tirion MM (1996) Large amplitude elastic motions in proteins from a single-parameter, atomic analysis. Phys Rev Lett 77: 1905–1908.
13. Bahar I, Atilgan AR, Erman B (1997) Direct evaluation of thermal fluctuations in proteins using a single-parameter harmonic potential. Fold Des 2: 173–181.
14. Haliloglu T, Bahar I, Erman B (1997) Gaussian dynamics of folded proteins. Phys Rev Lett 79: 3090–3093.
15. Tama F, Sanejouand YH (2001) Conformational change of proteins arising from normal mode calculations. Protein Engineering 14: 1–6.
16. Zheng W, Doniach S (2003) A comparative study of motor-protein motions by using a simple elastic-network model. Proc Natl Acad Sci USA 100: 13253–13258.
17. Togashi Y, Mikhailov AS (2007) Nonlinear relaxation dynamics in elastic networks and design principles of molecular machines. Proc Natl Acad Sci USA 104: 8697–8702.
18. Flechsig H, Mikhailov AS (2010) Tracing entire operation cycles of molecular motor hepatitis C virus helicase in structurally resolved dynamical simulations. Proc Natl Acad Sci USA 107: 20875–20880.
19. Flechsig H, Popp D, Mikhailov AS (2011) *In silico* investigation of conformational motions in superfamily 2 helicase proteins. PLoS ONE 6(7): e21809.
20. Togashi Y, Yanagida T, Mikhailov AS (2010) Nonlinearity of mechanochemical motions in motor proteins. PLoS Comput Biol 6: e1000814.
21. Humphrey W, Dalke A, Schulten K (1996) VMD - Visual Molecular Dynamics. J Molec Graphics 14: 33–38.
22. Murakami MT, Sforça ML, Neves JL, Paiva JH, Domingues MN, et al. (2010) The repeat domain of the type III effector protein PthA shows a TPR-like structure and undergoes conformational changes upon DNA interaction. Proteins 78: 3386–3395.
23. Schreiber T, Bonas U (2014) Repeat 1 of TAL effectors affects target specificity for the base at position zero. Nucl Acids Res 42: 7160–7169.
24. Wan H, Hu J-p, Li K-s, Tian X-h, Chang S (2013) Molecular dynamics simulations of DNA-free and DNA-bound TAL effectors. PLoS ONE 8(10): e76045.

18

Solution NMR of MPS-1 Reveals a Random Coil Cytosolic Domain Structure

Pan Li[1⦾], Pan Shi[2⦾], Chaohua Lai[1], Juan Li[1], Yuanyuan Zheng[1], Ying Xiong[1], Longhua Zhang[1]*, Changlin Tian[1,2]*

1 Hefei National Laboratory of Microscale Physical Sciences, School of Life Sciences, University of Science and Technology of China, Hefei, Anhui, P. R. China, 2 High Magnetic Field Laboratory, Hefei institutes of Physical Science, Chinese Academy of Sciences, Hefei, Anhui, P. R. China

Abstract

Caenorhabditis elegans MPS1 is a single transmembrane helical auxiliary subunit that co-localizes with the voltage-gated potassium channel KVS1 in the nematode nervous system. MPS-1 shares high homology with KCNE (potassium voltage-gated channel subfamily E member) auxiliary subunits, and its cytosolic domain was reported to have a serine/threonine kinase activity that modulates KVS1 channel function via phosphorylation. In this study, NMR spectroscopy indicated that the full length and truncated MPS-1 cytosolic domain (134–256) in the presence or absence of n-dodecylphosphocholine detergent micelles adopted a highly flexible random coil secondary structure. In contrast, protein kinases usually adopt a stable folded conformation in order to implement substrate recognition and phosphoryl transfer. The highly flexible random coil secondary structure suggests that MPS-1 in the free state is unstructured but may require a substrate or binding partner to adopt stable structure required for serine/threonine kinase activity.

Editor: Michael Massiah, George Washington University, United States of America

Funding: This work was supported by the National Natural Science Foundation of China grant U1332138 to CT and National Natural Science Foundation of China young investigator grant 31300685 to PS. The funders had no role in study design, data collection and analysis, decision to publish, or preparation of the manuscript.

Competing Interests: The authors have declared that no competing interests exist.

* Email: cltian@ustc.edu.cn (CT); zlhustc@ustc.edu.cn (LZ)

⦾ These authors contributed equally to this work.

Introduction

Voltage-gated potassium (Kv) channels are widely expressed in nervous, cardiovascular and other tissues, and are crucial mediators of membrane excitability in virtually all mammals [1,2]. The highly diverse functions of Kv channels originate partly from the pore-forming α subunits. Additional regulatory proteins or β-subunits also modulate Kv channel properties including expression and distribution, sensitivity to stimulation, gating and pharmacological responses [3,4]. MinK-related peptides (MiRPs) are well-characterized KCNE family membrane proteins associated with modulation of voltage-gated potassium channels in heterologous systems such as the HERG, KCNQ1, HCN, and Kv4.2 subunits [5–8]. MPS-1 belongs to the conserved KCNE family of β-subunits and is the first MiRP-related β-subunit to be investigated in the *nematode Caenorhabditis elegans*. Four MPS-1, MPS-2, MPS-3 and MPS-4 are expressed exclusively in the *C. elegans* nervous system [9] and are essential for neuronal excitability [10].

MPS-1 is a single transmembrane domain β-subunit that modulates the voltage-gated pore-forming potassium channel KVS1 in *C. elegans* [11]. Secondary structure prediction indicates a typical mammalian MiRP-like topology consisting of an extracellular N-terminus, a single transmembrane domain, and an intracellular C-terminus [10]. MPS-1 colocalizes with KVS-1,

and this complex was shown to conduct less current, exhibited faster inactivation, and recovered slower from inactivation when compared to the KVS-1 alone [9]. Defects in either MPS-1 or KVS-1 can result in defective chemotaxis, disrupted mechano-transduction, and impaired locomotion [10].

As with other KCNE β-subunit family, the transmembrane domain of MPS-1 is both necessary and sufficient for MPS-1-KVS-1 complex formation [12]. Unlike other KCNE β-subunits, the cytoplasmic domain of MPS-1 was reported to possess serine/threonine kinase activity, and phosphorylation of KVS-1 by MPS-1 lowered the ability of the KVS-1 channel to be opened [13]. Sequence homology with AGC protein kinases (a family of protein kinase containing PKA, PKC and PKG)[14] shows two characteristic protein kinase motifs in the cytoplasmic domain: (1) the DFG (Asp-Phe-Gly) triplet, which contributes to the Mg^{2+} binding site of ATP-binding molecules, and (2) the HSD (His-Ser-Asp) triplet, which contributes to the catalytic function [14,15]. Mutating D178 to N in the Δ132-256 truncated version of MPS-1 abolished KVS-1 channel modulation [13], suggesting the MPS-1 cytosolic domain is essential for KVS-1 phosphorylation and consequent channel modulation. Auto-phosphorylation has also been reported for MPS-1, similar to observation with other protein kinases [16]. Consequently, two independent KVS-1 channel modulation mechanisms have been proposed to explain how MPS-1 may reduce the channel current. The first mechanism involves formation of a complex between KVS-1 and the MPS-1 transmembrane helices, as observed with other KCNE β-subunits. The second mechanism involves phosphorylation of KVS-1 by the serine/threonine kinase activity of the MPS-1 cytosolic domain [13]. Results of the current studies have provided insights in the latter mechanism.

In addition, MPS-1 was also reported to assemble with the voltage-gated K^+ channel KHT-1 (K^+ channel for habituation to tap) in the tactile neurons of nematodes. MPS-1 was reported to phosphorylate KHT-1 and modulation of KHT-1 current channeling properties was observed [17].

Despite numerous electrophysiological and biochemical studies on the interactions between MPS-1 and KVS-1 or other Kv channels, no structural studies have yet been performed that can probe the MPS-1 channel modulation mechanism at the atomic level. To this end, the structural properties of full length MPS-1(1–256) and two truncated versions (74–256; 134–256) were studied by NMR spectroscopy (Fig. 1A, B). The MPS-1(74–256) variant included an almost intact cytosolic domain, whereas the MPS-1(134–256) variant, prepared based on previous observations, abolished the KVS-1 channel modulation activity exhibited by the Δ132–256 MPS-1 variant [13]. Our studies included the use of n-Dodecylphosphocholine (DPC) detergent micelles to mimic the lipid bilayer environment of the cell membrane, since the choline head group of DPC is similar to those of native lipids. Surprisingly, secondary structure and backbone relaxation analysis indicated that the cytosolic domain adopted a highly flexible random coil secondary structure that is inconsistent with the stable fold required of a functional serine/threonine kinase. These results question the phosphorylation-based channel modulation mechanism, and suggest that further *in vivo* biophysical and protein–protein interaction experiments are required.

Materials and Methods

Cloning, Over-expression and Purification of Full-length and Truncated MPS-1

A DNA fragment encoding full-length MPS-1 was amplified by PCR from a *C. elegans* cDNA library. Truncation fragments MPS-1(74-256) and MPS-1(134–256) were amplified using pairs of primers with *Nde* I and *Xho* I at the 5′- and 3′-ends, respectively. The amplified full-length and truncated fragments were ligated into the pET-21b vector (Novagen, Co.). All constructs were confirmed by DNA sequencing. Expression constructs were transformed into BL21(DE3) Rosetta (Novagen Co.) and cells were incubated overnight at 37°C on Luria-Bertani (LB) agar containing ampicillin and chloramphenicol. Colonies were transferred to 4 mL LB medium containing ampicillin and chloramphenicol and incubated at 37°C overnight with shaking at 225 rpm. This culture was used to inoculate M9 minimal medium containing 100 μg/ml ampicillin and 100 μg/ml chloramphenicol, and cultures were grown at 37°C, 225 rpm until the OD_{600} reached 0.6. Flasks were then incubated at 25°C until the OD_{600} reached 0.8, and IPTG was added to a final concentration of 0.8 mM to induce protein expression, and the culture was incubated at 25°C for 12 h.

To achieve uniform isotope labeling during protein expression, 1 g/L [^{15}N]-NH₄Cl and 3 g/L [^{13}C] labeled glucose (Cambridge Isotope Laboratory, Andover, MA, USA) were added to M9 medium for ^{13}C/^{15}N labeling. For ^{15}N labeling only, unlabeled glucose was used.

Full length MPS-1(1–256) was expressed as inclusion bodies and bacterial cells were harvested by centrifugation at 4000 rpm for 20 min at 18°C. Cell pellets were resuspended in 40 mL lysis

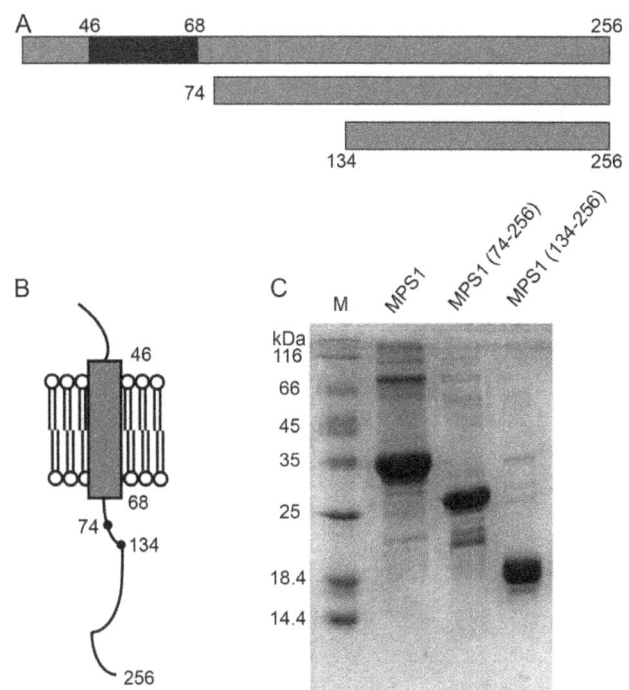

Figure 1. Schematic diagram of full length and truncated MPS-1. (A) Various MPS-1 constructs with MPS-1(1–256), MPS-1(74–256) and MPS-1(134–256) depicted from top to bottom. The black segment (residues 46 to 68) represents the predicted transmembrane helix; (B) Model of the MPS1 domains showing the two phosphorylation sites within the cytoplasmic C-terminal domain. (C) SDS-PAGE of the purified MPS-1 proteins. Lane 1: molecular weight marker; lanes 2-4: purified MPS-1(1–256), MPS-1(74–256) and MPS-1(134–256).

buffer (70 mM Tris–HCl, 300 mM NaCl, pH 8.0) and sonicated on ice using a VC500 probe sonicator (Sonics and Materials, Danbury, CT, USA) for a total of 10 min at a power level of 30% using 2 s pulses with 4 s between pulses. Five milligrams lysozyme, 1 mg DNase and 0.5 mg RNase were added, and the lysate was mixed at 4°C for 1 h before centrifugation at 16,000 rpm for 20 min at 4°C. The Pellet was washed three times by resuspending in 40 mL lysis buffer and centrifuged. The pellet was then resuspended in binding buffer (20 mM Tris–HCl, 100 mM NaCl, pH 8.0) containing 8 M urea and 0.2% (w/v) SDS and incubated at room temperature (25°C) for 2 h followed by centrifugation at 16,000 rpm for 20 min at 18°C to remove any insoluble debris. The supernatant was mixed with 5 mL Ni^{2+}-NTA resin (QIAgen, Valencia, CA, USA) and loaded onto a gravity-flow column (BIO-RAD, Hercules, CA, USA) equilibrated in 20 mM Tris–HCl, 200 mM NaCl, pH 8.0. Weakly bound *E. coli* proteins were eluted using 50 mL 20 mM Tris–HCl, 200 mM NaCl, 8 M urea, 0.2% (w/v) SDS, pH 8.0, followed by 40 mL wash buffer (20 mM Tris–HCl, 200 mM NaCl, pH 8.0) containing 0.2% (v/v) DPC (Anatrace, Maumee, OH, USA). On-column refolding of the protein was achieved during this step while the detergent was exchanged from SDS to DPC. Full-length MPS-1 was eluted using elution buffer (20 mM Tris–HCl, 200 mM NaCl, 250 mM imidazole, pH 8.0, 0.5% DPC).

The MPS-1(74–256) and MPS-1(134–256) truncations were over-expressed as soluble proteins in *E. coli*. Cells were harvested by centrifugation, suspended in lysis buffer, and lysed by sonication as described above. After centrifugation, supernatants were mixed with 5 mL Ni^{2+}-NTA resin (QIAgen) at 4°C for

30 min before loading onto a gravity-flow column (BIO-RAD) equilibrated with 20 mM Tris–HCl pH 8.0, 200 mM NaCl. Weakly bound *E. coli* proteins were eluted with 50 mL 20 mM Tris–HCl, 200 mM NaCl, pH 8.0, then with 40 mL wash buffer (20 mM Tris–HCl, 200 mM NaCl, 30mM imidazole, pH 8.0). The truncated proteins were eluted with 20 mM Tris–HCl, 200 mM NaCl, 250 mM imidazole, pH 8.0.

Protein concentration was determined by measuring the absorbance at 280 nm, and the purity was checked using standard SDS-PAGE. Purified proteins were concentrated using an Amicon Ultra-15 device with a 3000 MWCO (Millipore), and this was also used for buffer exchange into NMR buffer (50 mM NaH_2PO_4/Na_2HPO_4, pH 6.5). The final sample volume was 450 μL, to which 50 μL D_2O was added to yield a final concentration of 10% (v/v). A protein concentration of 1.0 mM was used for the multi-dimensional NMR experiments. 100 mM DPC was added to an equivalent [^{13}C, ^{15}N] labeled MPS-1(134–256) sample for multi-dimensional NMR experiments.

Solution NMR Data Acquisition and Backbone Resonance Assignment

Two-dimensional ^{1}H-^{15}N hetero-nuclear single quantum correlation spectroscopy (HSQC) experiments on ^{15}N-labeled full-length and truncated protein samples was performed at 25°C in a 600 MHz Bruker spectrometer equipped with a TXI probe. Standard Bruker HSQC pulse sequences with careful 90° pulse calibration were applied for data acquisition with 1024×256 complex points in the ^{1}H and ^{15}N dimensions, respectively. The acquired HSQC data were processed using nmrPipe [18] with a Gaussian window function in both dimensions and spectra were analyzed using nmrView [19].

A set of triple resonance multi-dimensional NMR experiments were conducted at 25°C for the [^{13}C, ^{15}N] labeled MPS-1(134–256) sample in 50 mM NaH_2PO_4/Na_2HPO_4, pH 6.5, in the presence or absence of 100 mM DPC. The NMR experiments included HNCO, HNCA, HN(CA)CO, HNCACB, CBCA(CO)NH, CC(CO)NH, HCC(CO)NH and HBHA(CO)NH. The data were processed using nmrPipe with both forward and backward linear prediction for resolution improvement in each indirect dimension. Multi-dimensional NMR spectra were processed using nmrPipe and analyzed using nmrView for backbone resonance assignment.

Secondary Structure Determination and Backbone Relaxation Analysis

The secondary structure of MPS-1(134–256) in the presence or absence of DPC micelles was determined using the TALOS+ program by inputting the chemical shift values of the backbone amide ^{1}H, ^{15}N, ^{13}CO, ^{13}C$_\alpha$, and ^{13}C$_\beta$ [20]. The relaxation parameter, ^{15}N longitudinal relaxation T_1, ^{15}N transverse relaxation T_2 and ^{1}H-^{15}N NOE, for backbone amide ^{15}N nuclei, were acquired using a 600 MHz Bruker spectrometer at 25°C. Backbone ^{15}N longitudinal relaxation T_1 values were calculated from a series of ^{1}H-^{15}N correlation spectra with 11.2, 61.6, 142, 243, 364, 525, 757, and 1150 ms relaxation evolution delays. Backbone ^{15}N transverse relaxation T_2 values were obtained from the spectra with 0, 17.6, 35.2, 52.8, 70.4, 105.6, and 140.8 ms delays. Relaxation constants and the associated experimental errors were derived from single exponential curve-fitting of the peak heights using nmrView. Steady-state ^{1}H-^{15}N NOE intensities were obtained from the ratio I_{NOE}/I_{NO-NOE}, where I_{NOE} and I_{NO-NOE} are the peak heights in the NOE spectra with and without proton saturation, respectively.

Results and Discussion

Protein Expression and Purification

Previously, MPS-1 was predicted to contain a single transmembrane helix (46–68) [10], and further secondary structure prediction using PSIPRED verified the presence of a long stretch of hydrophobic residues that likely form a helix (see file S1). The truncated MPS-1(74–256) variant includes the entire cytosolic domain of MPS-1. Physiological studies of KVS-1 channel modulation by MPS-1 demonstrated the essential role of the MPS-1 cytosolic domain involves residues after residue Pro132 [13], suggesting our structural studies of the MPS-1(134–256) variant may provide insights of how the function of KVS-1 may be modulated by the serine/threonine kinase activity of MPS-1. As control, we also studied the full-length protein, which was purified from inclusion bodies in the presence of DPC micelles [21,22]. In contrast, the truncated variants were purified in the soluble form in the presence or absence of DPC micelles. SDS-PAGE confirmed the successful purification of the proteins (Fig. 1C).

Comparison of Solution NMR Spectra of MPS-1(134–256) and MPS-1(1–256) in DPC Micelles

In the 2D ^{1}H-^{15}N HSQC spectrum, each (cross peak) corresponds to a pair of ^{1}H and ^{15}N nuclei in the protein backbone amide groups. Since both ^{1}H and ^{15}N chemical shift values are very sensitive to local variations in protein conformation, the distribution of cross peaks in ^{1}H-^{15}N HSQC spectrum can be used to evaluate protein conformation under different conditions. In this study, 2D ^{1}H-^{15}N HSQC spectra were acquired for MPS-1(1–256), MPS-1(74–256) and MPS-1(134–256) in both the presence and absence of detergent micelles (Fig. 2).

Due to the presence of the hydrophobic transmembrane helix, detergent micelles were required to maintain the stable conformation of full-length MPS-1(1–256) in aqueous buffer. Only 132 peaks were discernable (Fig. 2 A) instead of ~256. The unobserved peaks were possibly due to signal overlap or line-width broadening in the large micelles. The ^{1}H-^{15}N HSQC spectrum of MPS-1(134–256) showed 120 peaks (Fig. 2B), and most of the resonances overlapped with those of the full-length protein (Fig. S1A in file S1). The additional 12 cross speaks observed with the full-length MPS-1 presumably belonged to residues in other regions of the protein. The MPS-1(134–256) variant therefore shared a very similar tertiary structure in the cytosolic domain with the full-length protein, suggesting that the detergent micelles preserved the similar conformation of the full-length form.

HSQC Spectra of MPS-1(74–256) and MPS-1(134–256) Exhibited High Similarity

The 2D ^{1}H-^{15}N HSQC spectra of MPS-1(74–256) and MPS-1(134–256) (Fig. 2C, D) showed similarly dispersed resonances suggesting that both variants adopted a comparable structure in aqueous solution in the absence of detergent. Overlaying the spectra revealed the presence of only a few extra peaks in the MPS-1(74–256) spectra (Fig. S1B in file S1), presumably belonging to residues between position 74–134.

Backbone NMR Resonance Assignment of MPS-1 (134–256) in Presence or Absence of DPC Micelles

The HSQC spectra of MPS-1(134–256) in the presence (Fig. 2B) or absence (Fig. 2D) of DPC micelles showed clear differences in the distribution of cross peaks (Fig. S2 in file S1), possibly due to interactions between MPS-1(134–256) and the choline head groups of DPC micelles. In order to assign resonances for MPS-1(134–256)

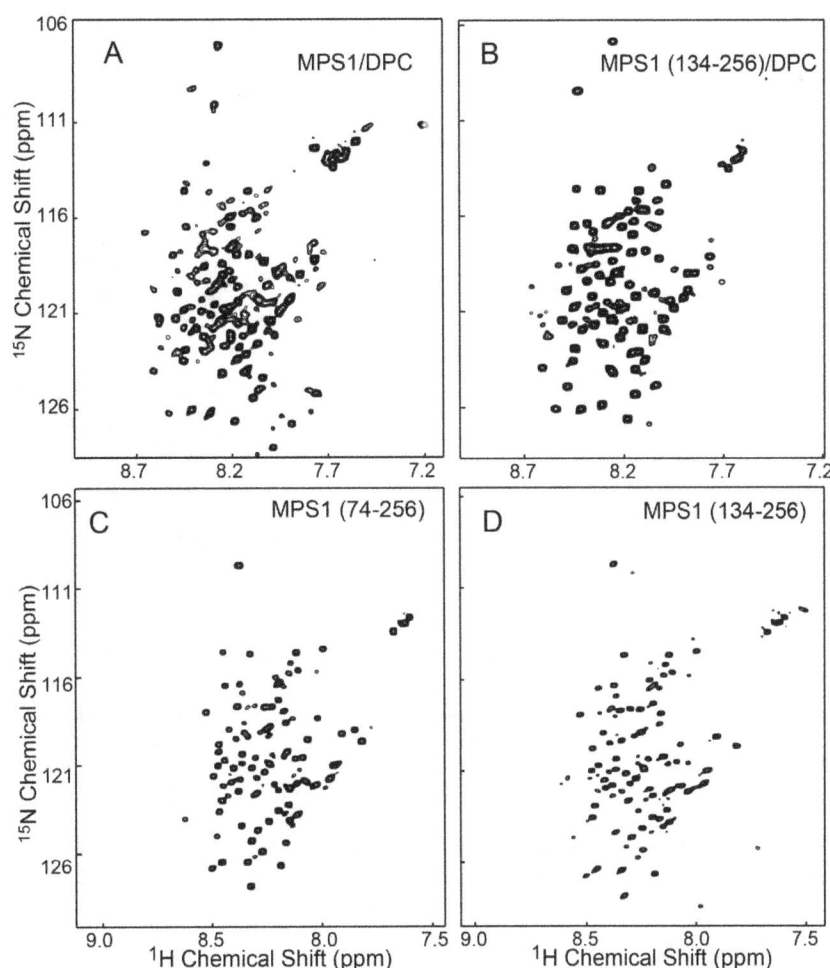

Figure 2. Two-dimensional ^{1}H-^{15}N HSQC spectra of ^{15}N labeled full-length and truncated MPS-1 under different buffer conditions at 25°C. (A) Full-length MPS-1(1–256) spectrum in DPC micelles; (B) MPS-1(134–256) spectrum in DPC micelles; (C) MPS-1(74–256) spectrum in aqueous buffer; (D) MPS-1(134–256) spectrum in aqueous buffer.

in the presence or absence of DPC micelles, two sets of triple resonance 3D NMR experiments were conducted with uniformly ^{13}C- and ^{15}N-enriched protein. A total of 90 backbone resonance assignments (NH, N, C$_\alpha$, C$_\beta$ and CO) were achieved for MPS-1(134–256) in aqueous buffer for 111 non-proline residues (Fig. 3A), while 66 backbone resonance assignments were achieved in the presence of DPC micelles (Fig. 3B). The first ten residues could not be assigned in either case, indicating a high degree of flexibility in this region. The lower number of resonances assigned in the presence of micelles was most likely due to extensive line-width broadening of hydrophobic residues interacting with the micelles. Nevertheless, a similar resonance distribution profile was observed for MPS-1(134–256) in either aqueous or detergent-associated conditions.

Secondary Structure and Backbone Relaxation Analysis of MPS-1(134–256)

Using the assigned backbone C$_\alpha$, C$_\beta$ and CO chemical shift values, the program TALOS+ [20] was used to estimate the dihedral angles φ and φ of the residues within MPS-1(134–256) in the presence or absence of DPC micelles [20]. Normally, a positive TALOS+ index value indicates a β-sheet secondary structure, while a negative value indicates a α-helical secondary structure.

Since the amplitude of the TALOS+ index is a measure of the validity of the predicted secondary structure, a value of zero indicates a random coil secondary structure. Surprisingly, most of the residues in MPS-1(134–256) had TALOS+ index close to zero (Fig. 4A), indicating a random coil secondary structure in both the presence and absence of DPC micelles. Three positive TALOS+ index values were observed for residues 154, 155 and 156 (the residue 21, 22 and 23 starting from residue 134) in the absence of micelles, indicating a β-sheet secondary structure in this region, which was consistent with the predicted β-extended structures around residues 154, 155 and 156 predicted using PRIPRED (see file S1).

To verify the secondary structure results, backbone relaxation analysis of MPS-1(134–256) was conducted in the presence or absence of DPC micelles. Backbone ^{15}N T$_1$ longitudinal relaxation time, T$_2$ transverse relaxation time, and steady-state ^{1}H-^{15}N NOE values were measured for uniformly ^{15}N-labeled protein (Fig. 4B). T$_1$ longitudinal relaxation values were clustered around 300 or 400 ms for MPS-1(134–256) in both the presence or absence of DPC micelles (Fig. 4B, upper row), whereas T$_2$ transverse relaxation values were approximately 150 or 250 ms (Fig. 4B, middle row). Steady-state ^{1}H-^{15}N NOE values were below zero for MPS-1(134–256), again in both conditions (Fig. 4B, lower row).

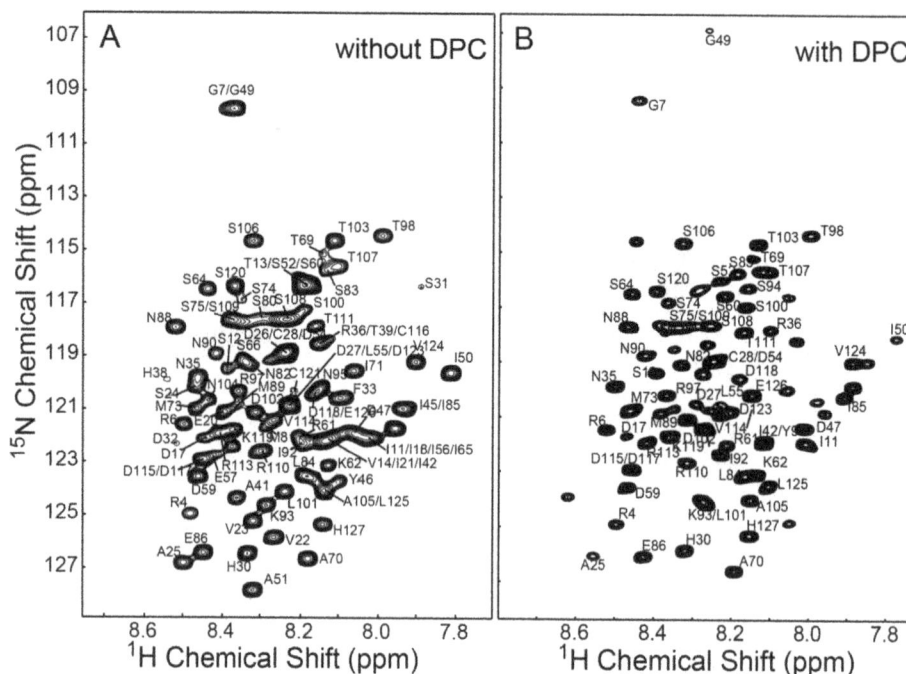

Figure 3. ^{1}H-^{15}N HSQC spectra showing the crosspeak assignments of MPS-1(134–256) in the absence or presence of DPC micelles. (A) Identifes of a total of 90 of 111 non-proline amino acids were based on (N)H, N, C$_\alpha$, C$_\beta$ and CO correlations were made for MPS-1(134–256); (B) In contrast, 66 amino acids were assigned for for MPS-1(134–256) in DPC micelles.

For a stably-folded protein of a similar size (T4 lysozyme, 130 residues), T_1 and T_2 values were approximately 500 ms or 150 ms, respectively [23]. Normally, the rotational correlation time of a globular protein (τ_c) shares a linear relationship with the T_1/T_2 ratio. The calculated T_1/T_2 ratio for MPS-1(134–256) in aqueous buffer ($\sim$1.6) was around half that of T4 lysozyme ($\sim$3.3), indicating that MPS-1(134–256) was not in a stably-folded conformation, but rather in a high flexibility unfolded state in aqueous buffer. The relatively large T_1/T_2 ratio of MPS-1(134–256) ($\sim$ 2.0) may be due to the high viscosity in the presence of detergent micelles. Another very strong indication of protein flexibility was the backbone steady-state ^{1}H-^{15}N NOE values. A positive value (greater than 0.7) indicates a rigid conformation, while a negative value indicates high flexibility (mostly around − 0.4). In contrast, many of the ^{1}H-^{15}N NOE values for T4 lysozyme were at least 0.8 [23].

The Highly Flexible Random Coil Conformation suggests MPS-1 is not a Ser/Thr Kinase

The backbone T_1 and T_2 relaxation and heteronuclear NOE values measured in this study indicated a highly flexible random coil secondary structure for the truncated MPS-1(134–256). Furthermore, TALOS+ predicted a disordered structure along the entire sequence in both the presence and absence of DPC micelles. This raises the important question of how an intrinsically unstructured C-terminal region could possibly possess the previously reported serine/threonine kinase activity. Several conserved residues were considered to be indicative of kinase activity. Residues D179, F180 and G181 are conserved among kinases and are involved in chelating the Mg^{2+} ions associated with ATP. Residues H162, S163, and D164 are believed to be catalytic residues that may act as general acids and bases important for phosphoryl transfer. However, for all serine/threonine kinases characterized to date, the conserved residues around the ATP

binding pocket reside in stable conformations and defined secondary structures [24]. The highly flexible random coil structure of the MPS-1(134–256) therefore is not consistent with this function. The random coil secondary structure prevented any further efforts to determine the 3D structure of MPS-1(134–256) in any more detail using conformational restraints.

All active serine/threonine kinase enzymes characterized to date have defined α-helical and β-sheet secondary structures that assemble into the tertiary and quaternary structures characteristic of the enzyme family. Protein kinase A, B, C, D, and G subfamilies all belong to the wider AGC family that are characterized by a C-terminal extension of the kinase domain that includes one or more regulatory phosphorylation sites that are important for kinase activity [14]. The AGC kinase was originally characterized by Steven Hanks and Tony Hunter in 1995, and defined the subgroup of Ser/Thr protein kinases most closely related to PKA, PKG and PKC based on sequence alignment of the catalytic kinase domain [14]. All AGC kinase domain structures determined to date exhibit the typical bilobal kinase fold first described for PKA [25]. Taken together, the random coil structure of the cytoplasmic domain of MPS-1 determined in this study strongly indicates that this protein is not an active serine/threonine kinase.

Despite the flexible random coil structure of MPS-1(134–256), there remains the possibility that MPS-1 adopts a functional secondary and tertiary structure upon forming a complex with the KVS-1 channel or other binding partner. Unfortunately, the large size of the tetrameric KVS-1 channel prevented us from investigating this potential interaction by solution NMR. MPS-1 shares high sequence similarity with other β-auxiliary proteins such as MiRP1 and MiRP3 that have been reported to associate with and modulate potassium channels [10]. Solution NMR and backbone relaxation studies of MiRP1 (KCNE2) illustrated a flexible C-terminal domain, and truncations of KCNE2 showed that this domain modulated the activity of KCNQ1 [26].

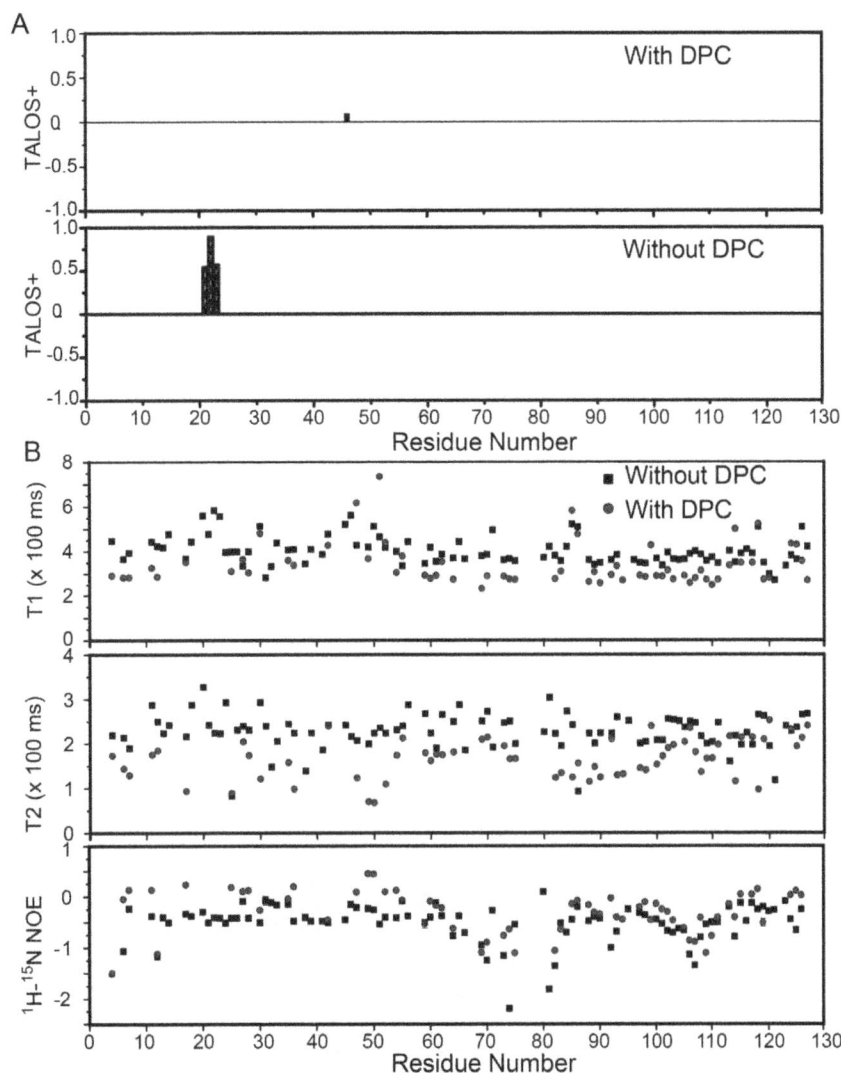

Figure 4. Secondary structure and backbone relaxation analysis of MPS-1(134–256) in the absence or presence of DPC micelles. (A) TALOS + secondary structure prediction were based on the assigned backbone ^{13}CO, $^{13}C_\alpha$, $^{13}C_\beta$ chemical shifts of MPS-1(134–256) in the presence or absence of DPC. Only random coil secondary structure was observed for MPS-1(134–256) in both aqueous buffer and DPC micelles. (See Fig. S3 in file S1 for the Y-axis scale from 0 to 0.3) (B) Distribution of the measured ^{15}N T_1 longitudinal and T_2 transverse relaxation times and steady-state ^1H-^{15}N NOE values along the primary sequence in the absence (black) or presence (red) of DPC micelles.

Protein domain flexibility can be important for protein-protein interactions, and intrinsically unstructured proteins can act as hubs upon which other proteins can bind. However, enzyme–substrate recognition and catalysis involve formation of a stable complex that requires some degree of rigidity [27]. Even so, the RS domain of the serine/arginine-rich splicing factor 1 was recently reported to perform a dramatic conformational switch from a fully disordered state to a partially rigidified arch-like structure upon phosphorylation [28]. Such a mechanism has yet to be reported for kinases, and would require the spontaneous formation of extensive α-helical regions. Further *in vivo* biophysical and functional studies of MPS-1 in *C. elegans* are necessary to confirm whether MPS-1 modulates KVS-1 channel current via serine/threonine kinase activity.

In summary, full-length and truncated MPS-1 were over-expressed in *E. coli* and purified in the presence or absence of DPC micelles. The similarity of the ^1H-^{15}N HSQC spectra of the full-length and MPS-1(134–256) truncated variant in DPC

micelles indicated that the C-terminal cytosolic domain shares a similar structure in both proteins. Backbone resonance assignment of MPS-1(134–256) in the presence or absence of DPC micelles provided the basis for site-specific secondary structure and relaxation analysis. TALOS+ and backbone relaxation experiments clearly showed that the C-terminal cytosolic domain of MPS-1 (residues 134–256) adopted a highly flexible random coil structure. Such a structure would not be expected to support the reported serine/threonine kinase activity, which indicates that MPS-1 does not modulate KVS-1 channel function via a phosphorylation mechanism.

Supporting Information

File S1 Supporting information. Figure S1, NMR spectra overlay of full-length and truncated MPS-1. Figure S2, NMR spectra overlay of MPS-1(134–256) in the absence and presence of DPC micelles. Figure S3, TALOS + secondary structure calculation prediction were based on the assigned backbone 13CO, 13Cα, 13Cβ chemical shifts of MPS-1(134–256) in the presence of DPC.

Acknowledgments

We thank Mr. Jiahai Zhang for his well maintenance of the solution NMR spectrometers in School of life sciences of University of Science and Technology of China. We appreciate the original plasmids of MPS1 from Prof. Federico Sesti in Robert Wood Johnson Medical School, University of Medicine and Dentistry of New Jersey, and Prof. Kewei Wang in department of neurobiology, health science center, Peking University.

Author Contributions

Conceived and designed the experiments: CT PS. Performed the experiments: CL PS. Analyzed the data: PL JL PS. Contributed reagents/materials/analysis tools: PL JL YZ YX. Wrote the paper: LZ PS CT.

References

1. Shah NH, Aizenman E (2014) Voltage-gated potassium channels at the crossroads of neuronal function, ischemic tolerance, and neurodegeneration. Transl Stroke Res 5: 38–58.
2. Blunck R, Batulan Z (2012) Mechanism of electromechanical coupling in voltage-gated potassium channels. Front Pharmacol 3: 166.
3. Li Y, Um SY, McDonald TV (2006) Voltage-gated potassium channels: regulation by accessory subunits. Neuroscientist 12: 199–210.
4. Torres YP, Morera FJ, Carvacho I, Latorre R (2007) A marriage of convenience: beta-subunits and voltage-dependent K+ channels. J Biol Chem 282: 24485–24489.
5. Abbott GW, Sesti F, Splawski I, Buck ME, Lehmann MH, et al. (1999) MiRP1 forms IKr potassium channels with HERG and is associated with cardiac arrhythmia. Cell 97: 175–187.
6. Tinel N, Diochot S, Borsotto M, Lazdunski M, Barhanin J (2000) KCNE2 confers background current characteristics to the cardiac KCNQ1 potassium channel. EMBO J 19: 6326–6330.
7. Yu H, Wu J, Potapova I, Wymore RT, Holmes B, et al. (2001) MinK-related peptide 1: A beta subunit for the HCN ion channel subunit family enhances expression and speeds activation. Circ Res 88: E84–87.
8. Zhang M, Jiang M, Tseng GN (2001) minK-related peptide 1 associates with Kv4.2 and modulates its gating function: potential role as beta subunit of cardiac transient outward channel? Circ Res 88: 1012–1019.
9. Cai SQ, Park KH, Sesti F (2006) An evolutionarily conserved family of accessory subunits of K+ channels. Cell Biochem Biophys 46: 91–99.
10. Bianchi L, Kwok SM, Driscoll M, Sesti F (2003) A potassium channel-MiRP complex controls neurosensory function in Caenorhabditis elegans. Journal of Biological Chemistry 278: 12415–12424.
11. Park KH, Hernandez L, Cai SQ, Wang Y, Sesti F (2005) A family of K+ channel ancillary subunits regulate taste sensitivity in Caenorhabditis elegans. Journal of Biological Chemistry 280: 21893–21899.
12. Wang Y, Sesti F (2007) Molecular mechanisms underlying KVS-1-MPS-1 complex assembly. Biophysical Journal 93: 3083–3091.
13. Cai SQ, Hernandez L, Wang Y, Park KH, Sesti F (2005) MPS-1 is a K+ channel beta-subunit and a serine/threonine kinase. Nat Neurosci 8: 1503–1509.
14. Pearce LR, Komander D, Alessi DR (2010) The nuts and bolts of AGC protein kinases. Nat Rev Mol Cell Biol 11: 9–22.
15. Johnson LN, Noble ME, Owen DJ (1996) Active and inactive protein kinases: structural basis for regulation. Cell 85: 149–158.
16. Zheng B, Larkin DW, Albrecht U, Sun ZS, Sage M, et al. (1999) The mPer2 gene encodes a functional component of the mammalian circadian clock. Nature 400: 169–173.
17. Cai SQ, Wang Y, Park KH, Tong X, Pan Z, et al. (2009) Auto-phosphorylation of a voltage-gated K+ channel controls non-associative learning. EMBO J 28: 1601–1611.
18. Delaglio F, Grzesiek S, Vuister GW, Zhu G, Pfeifer J, et al. (1995) NMRPipe: a multidimensional spectral processing system based on UNIX pipes. J Biomol NMR 6: 277–293.
19. Johnson BA (2004) Using NMRView to visualize and analyze the NMR spectra of macromolecules. Methods Mol Biol 278: 313–352.
20. Shen Y, Delaglio F, Cornilescu G, Bax A (2009) TALOS+: a hybrid method for predicting protein backbone torsion angles from NMR chemical shifts. J Biomol NMR 44: 213–223.
21. Tian C, Karra MD, Ellis CD, Jacob J, Oxenoid K, et al. (2005) Membrane protein preparation for TROSY NMR screening. Methods Enzymol 394: 321–334.
22. Tian C, Vanoye CG, Kang C, Welch RC, Kim HJ, et al. (2007) Preparation, functional characterization, and NMR studies of human KCNE1, a voltage-gated potassium channel accessory subunit associated with deafness and long QT syndrome. Biochemistry 46: 11459–11472.
23. Mine S, Ueda T, Hashimoto Y, Imoto T (2000) Analysis of the internal motion of free and ligand-bound human lysozyme by use of 15N NMR relaxation measurement: a comparison with those of hen lysozyme. Protein Sci 9: 1669–1684.
24. Adams JA (2001) Kinetic and catalytic mechanism of protein kinases. Chem Rev 101: 2271–2290.
25. Knighton DR, Zheng JH, Ten Eyck LF, Ashford VA, Xuong NH, et al. (1991) Crystal structure of the catalytic subunit of cyclic adenosine monophosphate-dependent protein kinase. Science 253: 407–414.
26. Li P, Liu H, Lai C, Sun P, Zeng W, et al. (2014) Differential modulations of KCNQ1 by auxiliary proteins KCNE1 and KCNE2. Sci Rep 4: 4973.
27. Kosinska U, Carnrot C, Eriksson S, Wang L, Eklund H (2005) Structure of the substrate complex of thymidine kinase from Ureaplasma urealyticum and investigations of possible drug targets for the enzyme. FEBS J 272: 6365–6372.
28. Xiang S, Gapsys V, Kim HY, Bessonov S, Hsiao HH, et al. (2013) Phosphorylation drives a dynamic switch in serine/arginine-rich proteins. Structure 21: 2162–2174.

Permissions

List of Contributors

Emanuel Maldonado
CIIMAR/CIMAR – Interdisciplinary Centre of Marine and Environmental Research, University of Porto, Porto, Portugal

Kartik Sunagar, Daniela Almeida, Vitor Vasconcelos and Agostinho Antunes
CIIMAR/CIMAR – Interdisciplinary Centre of Marine and Environmental Research, University of Porto, Porto, Portugal
Department of Biology, Faculty of Sciences, University of Porto, Porto, Portugal

Giacomo Bastianelli and Michael Nilges
Institut Pasteur, Unitéde Bioinformatique Structurale, Département de Biologie Structurale et Chimie, Paris, France
CNRS UMR 3528, Paris, France

Anthony Bouillon and Jean-Christophe Barale
Institut Pasteur, Unitéd'Immunologie Moléculaires des Parasites, Département de Parasitologie et de Mycologie & CNRS URA 2581, Paris, France
CNRS, URA2581, Paris, France

Christophe Nguyen and Dung Le-Nguyen
SYSDIAG, CNRS UMR3145 CNRS-BioRad, Montpellier, France

Mohd Rehan, Mohd A. Beg and Ghazi A. Damanhouri
King Fahd Medical Research Center, King Abdulaziz University, Jeddah, Kingdom of Saudi Arabia

Shadma Parveen
Bareilly College, M. J. P. Rohilkhand University, Bareilly, Uttar Pradesh, India

Galila F. Zaher
Department of Haematology, Faculty of Medicine, King Abdulaziz University, Jeddah, Kingdom of Saudi Arabia

Hugo G. Schmidt
Department of Biochemistry & Cambridge Systems Biology Centre, University of Cambridge, Cambridge, United Kingdom

Sven Sewitz and Karen Lipkow
Department of Biochemistry & Cambridge Systems Biology Centre, University of Cambridge, Cambridge, United Kingdom

Nuclear Dynamics Programme, The Babraham Institute, Cambridge, United Kingdom

Steven S. Andrews
Fred Hutchinson Cancer Research Center, Seattle, Washington, United States of America

Hyunjun Park, Brian M. Kevany, David H. Dyer, Michael G. Thomas and Katrina T. Forest
Department of Bacteriology, University of Wisconsin-Madison, Madison, Wisconsin, United States of America

Bashir A. Akhoon and Krishna P. Singh
Department of Bioinformatics, Systems Toxicology Group, CSIR-Indian Institute of Toxicology Research, Lucknow, India

Megha Varshney
Interdisciplinary Biotechnology Unit, Aligarh Muslim University, Aligarh, India

Shishir K. Gupta
Department of Bioinformatics, Biocenter, Am Hubland, University of Wu¨ rzburg, Wu¨ rzburg, Germany

Yogeshwar Shukla
Department of Proteomics, CSIRIndian Institute of Toxicology Research, Lucknow, India
Academy of Scientific and Innovative Research (AcSIR), New Delhi, India

Shailendra K. Gupta
Department of Bioinformatics, Systems Toxicology Group, CSIR-Indian Institute of Toxicology Research, Lucknow, India
Academy of Scientific and Innovative Research (AcSIR), New Delhi, India

Humberto J. Debat and Daniel A. Ducasse
Instituto de Patología Vegetal, Centro de Investigaciones Agropecuarias, Instituto Nacional de Tecnología Agropecuaria (IPAVE-CIAP-INTA), Córdoba, Argentina,

Mauro Grabiele, Patricia M. Aguilera and Dardo A. Marti
Instituto de Biología Subtropical, Universidad Nacional de Misiones (IBS-UNaM-CONICET), Posadas, Misiones, Argentina

Instituto de Biotecnología de Misiones, Facultad de Ciencias Exactas Químicas y Naturales, Universidad Nacional de Misiones (INBIOMIS-FCEQyN-UNaM), Misiones, Argentina

Rosana E. Bubillo
Estación Experimental Cerro Azul, Instituto Nacional de Tecnología Agropecuaria (EEA Cerro Azul-INTA), Misiones, Argentina

Mónica B. Otegui and Pedro D. Zapata
Instituto de Biotecnología de Misiones, Facultad de Ciencias Exactas Químicas y Naturales, Universidad Nacional de Misiones (INBIOMIS-FCEQyN-UNaM), Misiones, Argentina

Qiwei Li and Marina Vannucci
Department of Statistics, Rice University, Houston, Texas, United States of America

David B. Dahl
Department of Statistics, Brigham Young University, Provo, Utah, United States of America

Hyun Joo and Jerry W. Tsai
Department of Chemistry, University of the Pacific, Stockton, California, United States of America

Komal Kalani and Santosh Kumar Srivastava
Medicinal Chemistry Department, CSIR-Central Institute of Medicinal and Aromatic Plants, Lucknow, 226015 (U.P.) India
Academy of Scientific and Innovative Research (AcSIR), Anusandhan Bhawan, New Delhi, 110 001, India

Vikas Kushwaha, Richa Verma and P. K. Murthy
Division of Parasitology, CSIR-Central Drug Research Institute, Lucknow, 226001, UP, India

Feroz Khan
Metabolic & Structural Biology Department, CSIR-Central Institute of Medicinal and Aromatic Plants, Lucknow, 226015 (U.P.) India
Academy of Scientific and Innovative Research (AcSIR), Anusandhan Bhawan, New Delhi, 110 001, India

Mukesh Srivastava
Clinical and Experimental Medicine, Biometry section, CSIR-Central Drug Research Institute, Lucknow, 226001, UP, India

Pooja Sharma
Metabolic & Structural Biology Department, CSIR-Central Institute of Medicinal and Aromatic Plants, Lucknow, 226015 (U.P.) India

Ariel Erijman, Eran Rosenthal and Julia M. Shifman
Department of Biological Chemistry, The Alexander Silberman Institute of Life Sciences, The Hebrew University of Jerusalem, Jerusalem, Israel

Martin Carlsen and Peter Røgen
Department of Applied Mathematics and Computer Science, Technical University of Denmark, Kongens Lyngby, Denmark

Patrice Koehl
Department of Computer Science and Genome Center, University of California Davis, Davis, CA, United States of America

Xu Yan, Jun-Jie Huang, Zheng Zhou, Jie Chen and Yi Liang
State Key Laboratory of Virology, College of Life Sciences, Wuhan University, Wuhan, China

Matthew S. Kelker, Kenneth E. Narva, Steven L. Evans, Reetal Pai, David G. McCaskill, Nick X. Wang and Joshua C. Russell
Dow AgroSciences, LLC, Indianapolis, Indiana, United States of America

Colin Berry and Matthew D. Baker
Cardiff School of Biosciences, Cardiff University, Cardiff, Wales, United Kingdom

Cheng Yang and J. W. Pflugrath
Rigaku Americas Corporation, The Woodlands, Texas, United States of America

Matthew Wade and Tim J. Wess
School of Optometry & Vision Sciences, Cardiff University, Cardiff, Wales, United Kingdom

Mei-Er Chen
Department of Entomology, National Chung Hsing University, Taichung 402, Taiwan

Josephine W. Wu
Department of Optometry, Central Taiwan University of Science and Technology, Taichung 40601, Taiwan

Wen-Sing Wen, Wei-An Chen, Chien-Ting Li, Chih-Kai Chang, Chun-Hsien Lo3, Hwai-Shen Liu and Steven S.-S. Wang
Department of Chemical Engineering, National Taiwan University, Taipei 10617, Taiwan

Brian E. Ford
Biology Department, Queens College of the City University of New York, New York, New York, United States of America

The Graduate Center of the City University of New York, New York, New York, United States of America

Bruce Sun, James Carpino, Elizabeth S. Chapler, Jane Ching, Yoon Choi, Kevin Jhun, Jung D. Kim, Gregory G. Lallos, Rachelle Morgenstern, Shalini Singh, Sai Theja and John J. Dennehy
Biology Department, Queens College of the City University of New York, New York, New York, United States of America

Michael Schneider and Oliver Brock
Robotics and Biology Laboratory, Department of Electrical Engineering and Computer Science, Technische Universita˙t Berlin, Berlin, Germany

Holger Flechsig
Department of Mathematical and Life Sciences, Graduate School of Science, Hiroshima University, Higashi-Hiroshima, Japan

Pan Li, Chaohua Lai, Juan Li, Yuanyuan Zheng, Ying Xiong and Longhua Zhang
Hefei National Laboratory of Microscale Physical Sciences, School of Life Sciences, University of Science and Technology of China, Hefei, Anhui, P. R. China

Changlin Tian
Hefei National Laboratory of Microscale Physical Sciences, School of Life Sciences, University of Science and Technology of China, Hefei, Anhui, P. R. China High Magnetic Field Laboratory, Hefei institutes of Physical Science, Chinese Academy of Sciences, Hefei, Anhui, P. R. China

Pan Shi
HighMagnetic Field Laboratory, Hefei institutes of Physical Science, Chinese Academy of Sciences, Hefei, Anhui, P. R. China

Index

A

Acyltransferase Domain Structure, 47
Akt Signaling Pathway, 24
Amino Acid Sequence, 23, 28, 85, 91-92, 94, 188
Anti-filarial Agent, 97, 103
Atovaquone Drug Resistance, 57

B

Bacillus Thuringiensis, 151, 157, 163-165
Bacteriophage, 183-185, 191, 194-195
Bayesian Model, 85-86, 92-94
Binding Affinities, 110, 112-113, 118-119
Binding Modes, 66, 110
Biophysical Characterization, 151, 182

C

C-terminal Domains, 138
Caffeine Synthase, 69, 75, 77, 80, 82, 84
Cancer Therapy, 24-25
Cell-free Conversion System, 138
Circular Dichroism Spectroscopy, 166, 181
Coil Cytosolic Domain Structure, 218
Combine Tests, 1
Comparative Analysis, 166
Composite Modeling Method, 57
Computational Algorithms, 110
Computational Design, 10-11, 13, 23, 182
Computational Insights, 24
Cytochrome B (cyt B) Protein, 57

D

Dioecious Crop Tree, 69
Distance Measures, 120-121, 123-130, 135
Domain Replacement, 138-139, 149
Drug Discovery, 35, 66, 95, 97, 104
Drug-protein Interaction, 24

E

Evolutionary Biology, 1, 8, 195
Evolutionary Information, 196-199, 202-203, 205, 207
Extender Unit Binding, 47, 56

F

Fatal Prion Diseases, 138-139
Fibril Formation, 138-139, 142-143, 145-150, 167, 179-180, 182

Fitness Consequences, 183, 188
Folded Models, 120, 135

G

Gene Regulation, 36, 83
Graphical User Interface (gui), 1

H

Homology Modeling, 2, 4, 8, 10, 13, 17, 25
Host Attachment Protein, 183, 189, 194
Host Range Mutations, 183-184, 187, 192, 194
Human Cd-crystallin Aggregation, 166
Human Prion Proteins, 138
Hydrophobic Anchoring Domain, 183, 187-188, 193
Hydroxymalonyl-acyl Carrier Protein (acp), 47, 56

I

In Vivo Studies, 24, 97
Inhibitory Mechanism, 24-25, 32, 34
Insecticidal Proteins, 151, 163
Integrated Model, 36
Intersegmental Transfer, 36-38, 41-44

K

Knowledge-based Potentials, 120, 127, 135

L

Low Ph Conditions, 166

M

Malarial Parasites, 57
Multiprogram Platform, 1, 8-9
Mutation Scoring, 10

N

Natural Selection, 1-3, 7, 195

O

Opacification, 166
Orally Active Inhibitor, 24

P

Parasporal Crystalline Inclusions, 151
Peptidic Trypsin Inhibitor, 10
Peripheral Blood Microfilaremia, 97
Physicochemical Information, 196-199, 204-205, 207
Plasmodium Vivax, 10-11, 17, 20, 23

Point Mutations, 57-59, 113, 182-183

Polyketide Synthase, 47, 56

Protein Contact Prediction, 196, 209

Protein Primary Sequence, 85

Protein Structures, 2, 35, 42-43, 58, 67, 119-120, 124-125, 129, 135-136, 152, 160, 163-164, 180, 196-197, 211, 215

Protein-based Inhibitors, 10-11

Protein-protein Interactions, 48, 67, 110, 118-119, 180

Q

Quaternary Protein Structure, 36

R

Recognition Mechanism, 47

Refinement Approaches, 57

S

Selection Pressures, 1-2

Spatial Proximity, 196

Stand-alone Implementation, 85

Structure Prediction, 11, 59, 67, 75, 78, 80, 82, 85-86, 92, 95-96, 119-121, 135-137, 185, 187, 194, 196-198, 202, 205-210, 218, 223

Subtropical Tree Crop, 69

Superelasticity, 211, 216

Synergistic Potential, 197

T

Tal Effector, 211, 213, 217

Target Site Finding, 36

Thermodynamics Hypothesis, 120

Total Yield Production, 69

Transcription Activator-like Effectors (tales), 211

Transcription Factor Diffusion, 36, 39

Transcriptome Assembly, 69-70, 83

Transmissible Spongiform Encephalopathies, 138

U

Ursolic Acid, 97-98, 100, 102-103, 106, 108-109

V

Van Der Waals (vdw) Interactions, 111

Visual Impairment, 166, 181

W

Web Application, 85-86

X

X-ray Crystal Structure, 47

Y

Yerba Mate, 69-84

www.ingramcontent.com/pod-product-compliance
Lightning Source LLC
Chambersburg PA
CBHW080528200326
41458CB00012B/4371

9781632398130